Methods in Molecular Biology

Series Editor
John M. Walker
School of Life Sciences
University of Hertfordshire
Hatfield, Hertfordshire, AL10 9AB, UK

For further volumes:
http://www.springer.com/series/7651

RNA Sequence, Structure, and Function: Computational and Bioinformatic Methods

Edited by

Jan Gorodkin

University of Copenhagen, Center for non-coding RNA in Technology and Health, IKVH, Frederiksberg, Denmark

Walter L. Ruzzo

University of Washington, Dept. Computer Science & Engineering, Seattle, Washington, USA

Humana Press

Editors
Jan Gorodkin
Center for non-coding RNA in Technology
and Health, IKVH
University of Copenhagen
Frederiksberg, Denmark

Walter L. Ruzzo
Department of Computer Science
and Engineering
University of Washington
Seattle, WA, USA

ISSN 1064-3745 ISSN 1940-6029 (electronic)
ISBN 978-1-4939-6165-8 ISBN 978-1-62703-709-9 (eBook)
DOI 10.1007/978-1-62703-709-9
Springer New York Heidelberg Dordrecht London

Softcover reprint of the hardcover 1st edition 2014

Printed on acid-free paper

Humana Press is a brand of Springer
Springer is part of Springer Science+Business Media (www.springer.com)

Preface

The existence of noncoding RNA genes (ncRNAs) was proposed simultaneously with protein coding genes in 1961 by Jacob and Monod. Since then, a substantial focus has been on protein coding genes, while the area of ncRNA evolved more slowly and received less attention even though major breakthroughs were made, such as the discovery of RNA's ability to carry out catalytic function, which also gave rise to the hypothesis that life originated from an RNA world, given that RNA also can store genetic information. It was also revealed early on that RNA might often be more conserved in structure rather than sequence. It quickly became apparent that adding consideration of structure to RNA sequence analysis programs was far more computationally demanding than, for example, comparing DNA by the primary sequence as in the case of sequence alignment methods. Whereas a pairwise sequence alignment scales with the square of the length being aligned, folding a single sequence scales with the cube of the length. Thus, doubling the length makes the raw version of the alignment methods run four times longer, but RNA folding algorithms run eight times longer. Thus, a likely contribution to the relative neglect of RNA bioinformatics in the early days can probably be attributed to the fact that this is a harder problem than many other bioinformatics problems. Today, much faster computers plus meaningful heuristics have made it possible to engineer practical RNA bioinformatics tools. Though RNA bioinformatics is still in its early phase with respect to practicality of genome-scale analyses, computational tools might help in uncovering the extent of RNAs in genomes. Given that, for example, protein coding sequence makes up about 1.2% of the human genome and that most of the genome is transcribed, this leaves an enormous potential for noncoding transcripts that might carry out a function and thus qualify as an ncRNA.

The ncRNAs have now been recognized as an abundant class of genes which often function through their structure. Protein coding genes have also been recognized to contain RNA structural motifs or RNA structures involved in, for example, regulation. Since even before the word "bioinformatics" was coined, researchers have been developing tools and computational methodologies for the analysis of RNA sequences, for aiding RNA (secondary) structure determination, for functional studies, and for a range of subsequent disciplines rooted in the principles for RNA structure prediction. Recognizing that RNA structure is a characteristic feature of ncRNAs, these tools have enabled genome-scale, *in silico* screens for ncRNAs. Furthermore, the same basic principles underlying RNA folding algorithms have been extended to a range of related problems such as homology search, design of interfering RNAs, and prediction of RNA–RNA interactions, to mention some examples. This book addresses a range of these methodologies from both a practical point of view as well from a computational and algorithmic perspective. Traditionally, the computational methods were referred to as computational RNA biology. However, with the recent applications on genomic and transcriptomic data, the more applied side of computational RNA biology, focused on processing experimental data (especially high-throughput data) is more commonly covered by the term Bioinformatics. This book covers a substantial and relevant fraction of both these directions and addresses both the biologist interested in knowing more about RNA bioinformatics as well as the bioinformaticist interested in aspects of the "engine room."

Recent technological development pushes high-throughput data generation and motivates further improvement of the generally computational resource demanding programs in RNA bioinformatics, a cost "inherited" from the generic RNA folding algorithms. Whereas these issues are addressed and the concepts of many methods shown, it is beyond this book to enter the area of assembly and read mapping. Here, we walk through the key methods and principles of RNA bioinformatics. Whereas a substantial part of the methodologies originate in the principles employed for prediction of RNA secondary structure, they employ further layers for specific applications as well as restrictions to reduce computation time and memory requirements, for example. In particular, developments in this respect have pushed for making methods in computational RNA biology applicable within RNA bioinformatics. Here, we range from the methodologies to their actual applications.

The content of this book is organized as follows. Initially an introduction to RNA bioinformatics is given (Chapter 1). This is followed by a description of RNA 3D structure (Chapter 2) and the origin of RNA folding parameters (Chapter 3) constitutes a background. Chapters 4 and 5 describe folding of single sequences by energy and by probabilistic modeling using stochastic context-free grammars. Chapter 6 describes RNA databases based on structural alignments of RNAs. Following this, folding of multiple aligned sequences by two strategies (Chapters 7 and 8) show that foldings can be made more reliably in such instances. This is followed by approaches describing genomic annotation of structured RNAs by homology search (Chapter 9) and the search for class-specific ncRNAs (Chapter 10). Chapter 11 describes how to extract (RNA 2D) motifs from RNA structures and Chapter 12 introduces the concept of comparing RNA secondary structures. Chapters 13–15 introduce structural alignments ranging from the so-called Sankoff-based approaches to alternatives and finally to exploitation of this to search for RNA structures in genomic sequence with low signal of conservation at the sequence level. In Chapter 16 an in-depth introduction to the evolution of RNA structure is given. The following Chapter 17 introduces RNA editors for careful curation of RNA structural alignments. In Chapter 18, RNA 3D modeling is introduced. Strategies and principles for RNA–RNA interactions are introduced in Chapter 19 and the special case of microRNA target prediction in Chapter 21. Before that, in Chapter 20, microRNA gene finding is presented. In Chapter 22, design of siRNAs are introduced and the final Chapter 23 provides an overview for RNA–protein interactions.

Finally, we would like to acknowledge those who helped make this book possible: all of the chapter authors for their hard work and thoughtful contributions, the Series Editor, John Walker, for his continued encouragement, and our families for their constant support.

Frederiksberg, Denmark
Seattle, WA, USA

Jan Gorodkin
Walter L. Ruzzo

Contents

Contributors

EBBE S. ANDERSEN • *Department of Molecular Biology, Aarhus University, Aarhus, Denmark; Department of Molecular Biology and Genetics, Aarhus University, Aarhus, Denmark*
MIRELA ANDRONESCU • *Department of Genome Sciences, University of Washington, Seattle, WA, USA*
KIYOSHI ASAI • *Computational Biology Research Center (CBRC), National Institute of Advanced Industrial Science and Technology (AIST), Koto-ku, Tokyo, Japan*
Department of Computational Biology, Graduate School of Frontier Sciences, University of Tokyo, Kashiwa, Chiba, Japan
ROLF BACKOFEN • *Lehrstuhl fur Bioinformatik, Albert-Ludwigs-Universitat, Freiburg, Germany;*
Center for non-coding RNA in Technology and Health, University of Copenhagen, Frederiksberg C, Denmark
LARS E. BARQUIST • *Wellcome Trust Genome Campus, Wellcome Trust Sanger Institute, Hinxton, UK*
STEPHAN H. BERNHART • *Institute for Theoretical Chemistry, University of Vienna, Vienna, Austria;*
JANUSZ M. BUJNICKI • *Laboratory of Bioinformatics and Protein Engineering, International Institute of Molecular and Cell Biology, Warsaw, Poland;*
Laboratory of Bioinformatics, Faculty of Biology, Institute of Molecular Biology and Biotechnology, Adam Mickiewicz University, Poznan, Poland
CHRISTOPHER B. BURGE • *Massachusetts Institute of Technology, Cambridge, MA, USA*
ANNE CONDON • *Department of Computer Science, University of British Columbia, Vancouver, BC, Canada*
ROBIN C. FRIEDMAN • *Systems Biology Laboratory, Department of Genomes and Genetics, Institut Pasteur, Paris, France*
PAUL P. GARDNER • *School of Biological Sciences, University of Canterbury, Christchurch, New Zealand*
TANJA GESELL • *Max F. Perutz Laboratories, Department of Structural and Computational Biology, University of Vienna, Vienna, Austria;*
School of Engineering and Applied Sciences, Harvard University, Cambridge, MA, USA
ROBERT GIEGERICH • *Faculty of Technology and Center of Biotechnology, Bielefeld University, Bielefeld, Germany*
JAN GORODKIN • *Center for non-coding RNA in Technology and Health, IKVH, University of Copenhagen, Frederiksberg C, Denmark*

MICHIAKI HAMADA • *Department of Computational Biology, Graduate School of Frontier Sciences, University of Tokyo, Kashiwa, Chiba, Japan*
JAKOB HULL HAVGAARD • *Center for non-coding RNA in Technology and Health, IKVH, University of Copenhagen, Frederiksberg C, Denmark*
JANA HERTEL • *Bioinformatics Group, Department of Computer Science, University of Leipzig, Leipzig, Germany*
MARC P. HOEPPNER • *Department of Medical Biochemistry and Microbiology, Uppsala University, Uppsala, Sweden*
IVO L. HOFACKER • *Department of Theoretical Chemistry, University of Vienna, Vienna, Austria;*
Center for non-coding RNA in Technology and Health, University of Copenhagen, Frederiksberg C, Denmark
STEFAN JANSSEN • *Faculty of Technology and Center of Biotechnology, Bielefeld University, Bielefeld, Germany*
TEJAL JOSHI • *Technical University of Denmark, Lyngby, Denmark*
MANOLIS KELLIS • *Computer Science and Artificial Intelligence Lab, Massachusetts Institute of Technology, Cambridge, MA, USA*
ELEONORA KULBERKYTE • *Technical University of Denmark, Lyngby, Denmark*
DAVID LANGENBERGER • *Bioinformatics Group, Department of Computer Science, University of Leipzig, Leipzig, Germany*
RUNE LYNGSØ • *Department of Statistics, University of Oxford, Oxford, UK*
DAVID H. MATHEWS • *Department of Biochemistry and Biophysics, University of Rochester Medical Center, Rochester, NY, USA;*
Department of Biostatistics and Computational Biology, University of Rochester Medical Center, Rochester, NY, USA
QUAID MORRIS • *The Donnelly Centre, University of Toronto, Toronto, Canada*
ERIC P. NAWROCKI • *Janelia Farm Research Campus, Howard Hughes Medical Institute, Ashburn, VA, USA*
YANN PONTY • *CNRS, Laboratoire d'Informatique de l'X (LIX) UMR 7161, INRIA AMIB, Ecole Polytechnique, Palaiseau, France*
ANGELA RE • *University of Trento, Mattarello, Italy*
MAGDALENA ROTHER • *Laboratory of Bioinformatics, Faculty of Biology, Institute of Molecular Biology and Biotechnology, Adam Mickiewicz University, Poznan, Poland;*
Laboratory of Bioinformatics and Protein Engineering, International Institute of Molecular and Cell Biology, Warsaw, Poland
KRISTIAN ROTHER • *Laboratory of Bioinformatics and Protein Engineering, International Institute of Molecular and Cell Biology, Warsaw, Poland;*
Laboratory of Bioinformatics, Faculty of Biology, Institute of Molecular Biology and Biotechnology, Adam Mickiewicz University, Poznan, Poland
WALTER L. RUZZO • *Department of Computer Science and Engineering, University of Washington, Seattle, WA, USA;*
Department of Genome Sciences, University of Washington, Seattle, WA, USA;
Fred Hutchinson Cancer Research Center, Seattle, WA, USA;

Center for non-coding RNA in Technology and Health, University of Copenhagen, Frederiksberg C, Denmark

STEFANIE SCHIRMER • *Institute for Research in Immunology and Cancer (IRIC), Department of Computer Science and Operations Research, Universite de Montre al, Montre al, QC, Canada*

PETER SCHUSTER • *Institute fur Theoretische Chemie der Universitat Wien, Vienna, Austria;*
The Santa Fe Institute, Santa Fe, NM, USA

PAWEL SKIBA • *Laboratory of Bioinformatics, Faculty of Biology, Institute of Molecular Biology and Biotechnology, Adam Mickiewicz University, Poznan, Poland*

PETER F. STADLER • *Bioinformatics Group, Department of Computer Science, and Interdisciplinary Center for Bioinformatics, University of Leipzig, Leipzig, Germany;*
Max Planck Institute for Mathematics in the Sciences, Leipzig, Germany;
Fraunhofer Institute for Cell Therapy and Immunology, Leipzig, Germany;
Institute for Theoretical Chemistry, University of Vienna, Vienna, Austria;
Center for non-coding RNA in Technology and Health, University of Copenhagen, Frederiksberg C, Denmark;
Santa Fe Institute, Santa Fe, NM, USA

ZSUZSANNA SUKOSD • *Bioinformatics Research Centre, Aarhus University, Aarhus, Denmark*

HAKIM TAFER • *Institut fur Informatik, Universitat Leipzig, Leipzig, Germany*

DOUGLAS H. TURNER • *Department of Chemistry, The College of Arts and Sciences, Rochester, NY, USA*

STEFAN WASHIETL • *Computer Science and Artificial Intelligence Lab, Massachusetts Institute of Technology, Cambridge, MA, USA*

CHRISTOPHER T. WORKMAN • *Center for Biological Sequence Analysis, Technical University of Denmark, Lyngby, Denmark;*
Center for non-coding RNA in Technology and Health, University of Copenhagen, Frederiksberg C, Denmark

CHRISTIAN ZWIEB • *Department of Biochemistry, University of Texas Health Science Center San Antonio, San Antonio, TX, USA*

Chapter 1

Concepts and Introduction to RNA Bioinformatics

Jan Gorodkin, Ivo L. Hofacker, and Walter L. Ruzzo

Abstract

RNA bioinformatics and computational RNA biology have emerged from implementing methods for predicting the secondary structure of single sequences. The field has evolved to exploit multiple sequences to take evolutionary information into account, such as compensating (and structure preserving) base changes. These methods have been developed further and applied for computational screens of genomic sequence. Furthermore, a number of additional directions have emerged. These include methods to search for RNA 3D structure, RNA–RNA interactions, and design of interfering RNAs (RNAi) as well as methods for interactions between RNA and proteins.

Here, we introduce the basic concepts of predicting RNA secondary structure relevant to the further analyses of RNA sequences. We also provide pointers to methods addressing various aspects of RNA bioinformatics and computational RNA biology.

Key words Mutual information, RNA folding, RNA prediction evaluation, RNA secondary structure, RNA structure prediction

1 Introduction and Challenges

Non-coding RNA (ncRNA) or non-protein-coding genes were until a decade ago given much less attention than mRNAs. However, within the last decade this perception has changed and ncRNAs have proven to be an abundant class of genes carrying out several different types of functions ranging from being housekeeping genes to regulators, e.g., [1–4]. Interestingly, the ncRNAs might not necessarily be highly stable [5]. The size of ncRNAs also varies extremely from small RNAs of size ~20 nt, to sizes of ~100,000 nt [6]. Examples are microRNAs (miRNAs) which are ~22 nt long and involved in regulation of protein-coding genes (mRNA), small nucleolar RNAs (snoRNAs) which are of size 70–200 nt and involved in guiding base modifications of the 1,500–3,000 nt size ribosomal RNA (rRNA), small signal recognition particle RNA (SRP RNA) of size ~300 nt and part of the

Jan Gorodkin and Walter L. Ruzzo (eds.), *RNA Sequence, Structure, and Function: Computational and Bioinformatic Methods*, Methods in Molecular Biology, vol. 1097, DOI 10.1007/978-1-62703-709-9_1,

complex involved in recognizing the signal peptide of processed proteins, the long ncRNAs (lncRNAs) such as Xist RNA which is ~19,000 nt long, involved in the inactivation of the X chromosome, and *Air* RNA of size ~108,000 nt, involved in imprinting of protein-coding genes by an epigenetic mechanism [7].

A large part of the early work in RNA bioinformatics has been focused on providing aid in folding of single RNA molecules. However, within the past two decades, the repertoire of (sub)areas has expanded tremendously. RNA folding approaches have adapted to available genomic sequences and methods that can fold multiple sequences have been developed. Whereas these methods mainly are based on folding secondary structures (2D), methods for three-dimensional folding have emerged, e.g., [8–10], exploiting the increasing number of RNA structures available in PDB (Protein Data Bank) [11]. The area of identifying RNA–RNA interactions as well as RNA–protein interactions has taken off with great speed, with methods working on both single and multiple sequences. A challenge is the apparent lack of good, readily accessible data, especially for RNA–RNA interactions. Evolutionary information is also being incorporated into the methods and helps increasing the interpretation of data.

Much of the data applied for testing RNA bioinformatics methods have been structured RNAs, covering either entire ncRNA genes or structural elements harbored in the UTRs of mRNAs, for example riboswitches and iron response elements. New challenges are the emerging classes of long ncRNAs (lncRNAs) for which it is unclear how structured they are. In the case of HOTAIR, it appears to contain two highly structured domains involved in function, joined by a long linker region whose secondary structure is uncertain [12]. Interestingly, a recent study of the steroid receptor RNA revealed that it is a highly structured lncRNA [13]. There is emerging activity in the area of lncRNAs, which includes re-annotating EST sequences first considered as junk in the absence of protein-coding potential. The RIKEN mouse EST project did, however, systematically describe these as ncRNAs [14]. Later work revealed many lncRNAs to be long transcripts that are capped and poly-adenylated, like mRNAs, but however lack coding ability [15]. Recently, compilations of these have been made as well [16, 17].

In general, lncRNAs are not well annotated, often lacking well-defined start and stop positions, even though recent work has made some progress. However, recently more than 8,000 lncRNAs were identified [18]. The lncRNAs fall into three classes: introns; intergenic, that is in between protein-coding genes as long intervening non-coding RNAs (lincRNAs); and natural antisense transcripts (NATs) [19]. They seem to carry out very different functions. Furthermore, they are often not well conserved (at least

Fig. 1 An example of the 3D structure of the tetrahydrofolate (THF) riboswitch [27]. A riboswitch is an RNA structure, typically located in the 5′UTR of an mRNA. This riboswitch senses tetrahydrofolate and helps regulate the adjacent gene, in this case folT, involved in folate transport, as well as mRNAs encoding folC and folE, involved in its biosynthesis [25]. The structure was accessed via PDB [11] and the figure created using `jmol` [28]

not in sequence), which therefore makes it challenging to search for them. Recently, however, functional conservation was shown on some lncRNAs, using a combination of identification of stretches of sequence similarity, RNAseq data, and morpholino knockdown assays [20].

The recent explosion of ncRNA discovery is in strong contrast to the history where ncRNAs have been much ignored, right after they were proposed together with a model for regulation of protein-coding genes by Jacob and Monod in 1961 [21]. One may speculate about the reasons which in part might be due to technical limitations, experimental as well as computational. When the sequence of the human genome was published in 2001 [22] it was revealed that only about 1.2% encodes for proteins. Even this enormous potential for ncRNAs to be novel molecular players did not prompt a huge effort in the area. However, some effort, in part driven by bioinformatics, was the search for miRNAs, e.g., [23], as mentioned which is just one class of ncRNAs. Later, computational approaches revealed snoRNAs in Archea using their Eukaryotic homologs [24]. Bioinformatics has played a central role in the discovery of many riboswitches, e.g., the tetrahydrofolate (THF) riboswitch [25]. Its 3D structure, which has been recently resolved, is shown in Fig. 1, and the corresponding secondary structure of a sequence in the same RNA family is shown in Fig. 2 (in several alternative representations). Furthermore, the second largest bacterial ncRNA discovered to date was found using bioinformatic approaches [26].

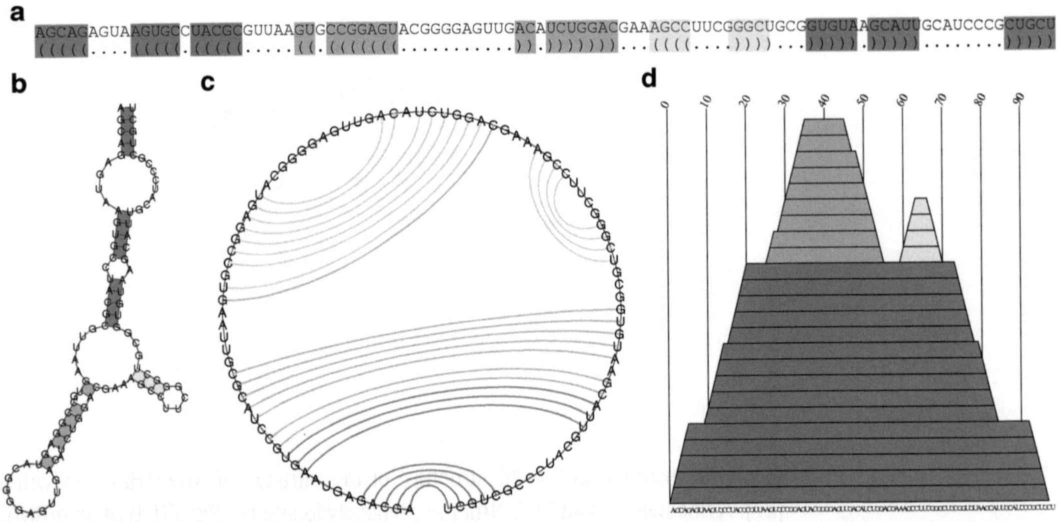

Fig. 2 Representations of the tetrahydrofolate riboswitch (Rfam accession RF01831 [29]). The matching *grayscale* indicates corresponding stems. The same *grayscale* represents the same base pairs and stems on the respective plots. (**a**) Dot bracket notation, (**b**) a secondary structure diagram, (**c**) a planar circle plot, (**d**) a mountain plot. The plots were made by tools from the RNA Vienna package [30] followed by manual editing of the generated postscript files

As indicated computational methods have been developed for both *homology search* and *de novo* search for ncRNAs. Searching for ncRNAs both by homology and by *de novo* hold their respective challenges. In both instances RNA secondary structure is an objective. In homology search, a model over the RNA secondary structure is typically created and exploited to find matches on genomic sequence. This can be migrated into search for class-specific RNA families, e.g., [31–33] where a combination of conserved sequence and structure motifs are used in a joint scoring scheme. In *de novo* search, this can only be meaningfully carried out in a comparative fashion, since in general it is hard (often impossible) to distinguish the folding of a single, functional RNA sequence from the folding of sequence with randomized nucleotide order. However, the use of comparative sequence information carries conceptual issues of its own.

Comparative sequence analysis typically starts from sequence-based alignments. However, when comparing structured RNAs, alignments taking the RNA secondary structure into account are often more suitable. This is also reflected in the recent computational strategies which make a trade-off between faster search in pre-generated sequence-based alignments versus computationally expensive searches conducting structural alignments e.g., [34, 35]. However, in both cases the main limitation is that "only" sequence information extracted from a global sequence-based comparison, such as what ends up in the UCSC browser [36], is searched in. This gives a number of challenges as the depth of the phylogeny might fragment the search space and where too few sequences might contain too little variation to detect traces of RNA secondary structure.

Methods that search for structured RNAs in genomic sequence, in one way or another, try to exploit the pattern of compensatory base changes in multiple organisms (e.g., an A-U in one sequence can have evolved to a G-C pair in another) [34]. Whereas this provides a signal to search for, this also, together with the already computationally expensive folding algorithms, adds to the consumption of computational resources [37]. Extracting patterns of compensating base changes is through existing databases, in particular Rfam [29]. Somewhat ironically, this structure variation is also being exploited in the search for RNA structures in genomic sequence, although the high computational time and memory resources (complexity) make this somewhat expensive. Still, conducting sequence similarity-based search approaches such as `BLAST` [38] is in general not sufficient for homology search and certainly useless for *de novo* search in genomic sequence. It was also shown that the energy of a single ncRNA (with miRNAs as a class of ncRNAs being an exception) in general is not sufficient to distinguish it from its background (shuffling of the nucleotides in the sequence) [39, 40]. Still, when employing multiple sequences it has been shown that real structure can to some extent be distinguished from a background [41] and methods to shuffle multiple alignments have been developed [42, 43].

Recent work promises to add more information by combining high-throughput (transcription) data with secondary structure methods [44]. This provides one approach towards combining experimental data with RNA bioinformatics methods, and other recently published methods promise to incorporate high-throughput RNA structure probing data [45, 46] directly into the folding algorithms [47, 48].

Hence, ncRNA is an active field and the "accompanying" RNA bioinformatics and computational RNA biology (terms we will not rigorously distinguish between) push in various directions to gain novel biological insight, e.g., [34]. Computational methods (before the term "bioinformatics" appeared) have evolved early on from aiding in folding prediction to whole genomic screens for RNA structure [37] and most recently to search for lncRNAs. Here, the focus will be on methods involving predicting of RNA structure, and how these approaches are applied to search for RNA structures in genomic sequence as a hallmark of either (a part of) an ncRNA or a regulatory structure in an mRNA. The same concepts are also applied to search for RNA–RNA interactions. RNA and protein interactions will also be covered. Future advances of the "traditional" methods which (for computational reasons) take only the *RNA secondary structure* into account will take whole 3D structure into account. In this chapter we will introduce the basic nomenclature for RNA secondary structure, the basic folding algorithm, and mutual information as a first step to identify base pairs in an RNA structure. We outline the type of methods employed in RNA bioinformatics and provide an overview of how the methods can be benchmarked.

2 RNA Secondary Structure Representations

An RNA *secondary* structure is defined as the set of base pairs which can be mapped in the plane. This means creating a list of all interacting pairs (ignoring base triples and quadruples). As an example, for the tetrahydrofolate riboswitch [25], this list is graphically illustrated in different ways in Fig. 2. A simple linear representation of the secondary structure is shown in Fig. 2a, where paired bases are marked by matching parentheses. For this class of structures a single base pair can uniquely be extracted from the parenthesis assignment line. A more common type of illustration is the diagram in Fig. 2b which more visually shows which base pairs *stack* on one another and provides some more explicit information about longer range interactions. This type of illustration is often used in publications. It is important not to be misled to believe that bases close or far apart are close or far a part in three-dimensional space. For example, there are known interactions between the top "internal loop" (unpaired bases) and the bottom "hairpin loop." Such an interaction is also called a *pseudoknot*.

An illustration useful for understanding the computational principles is the planar *circle plot* shown in Fig. 2c, where all the bases can be mapped to the plane and a line (arc) can be drawn between paired bases without any crossing each other. If we draw an arc reflecting the interactions between the internal loop and hairpin loop, we would have crossing arcs. In standard folding algorithms these are ignored due to their computational cost (and pragmatically excused since they typically form only a small fraction of the base pairs). Yet another illustration in Fig. 2d shows a mountain plot [49], which is a base pair "profile" over the sequence and is useful to see the depth of pairing as well as to compare the two structure assignments, for example a prediction and its reference (curated based on experimental data).

RNA secondary structure can further be represented as a tree as shown in Fig. 3. Such a representation can be useful in many respects, for example when comparing RNA structures [50] or in probabilistic modeling of RNA structure [51].

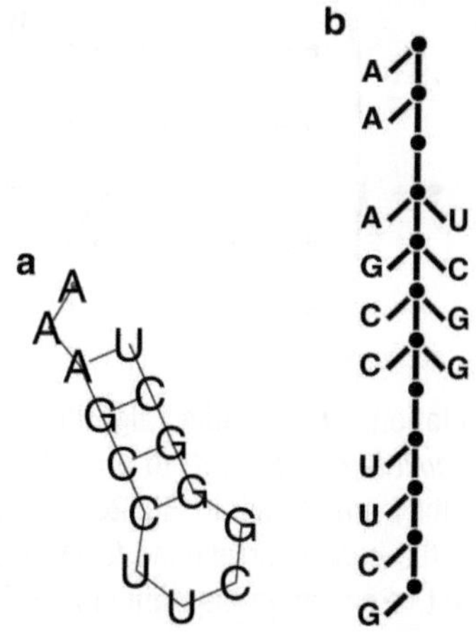

Fig. 3 Tree representation of RNA secondary structure. (**a**) The section of smallest stem in Fig. 2. (**b**) A tree representation of the same structure. More complicated RNA structures, for example branching stems, will also result in more complex trees, which can branch correspondingly

3 Basic Folding Algorithm

A *key* concept of RNA folding is to keep track of subsequences. That is, if we have a sequence N nucleotides (nt) long it can algorithmically be folded recursively, by first optimizing the structure on a subsequence. Strictly, the "folding" is first carried out for all subsequences of length 1 nt. These are then used to carry out the folding for all subsequences of length 2 nt and so on. Base pairing typically starts for subsequences of length 5, thus enforcing a minimum loop size of 3 bases. In the simplest form, base pairs for a sequence of length N are found from the subsequences one base shorter at either end, or from the subsequence one base shorter at both ends. In the two former cases no new base pairs can be introduced, but in the latter a single new base pair is allowed, if the two ends can form a base pair. The idea in this procedure is to keep track of the "best collection" of base pairs for each subsequence, where the best collection in the basic form is the most base pairs, and in the real folding algorithm, the subsequence with the lowest free energy.

3.1 The Nussinov Algorithm

Let us first consider a "toy version" of RNA folding, which simply tries to maximize the number of base pairs formed. Let $x_1, x_2, \ldots, x_N$ be the length N nucleotide sequence to be folded. We want to compute the maximum number of pairs that can be formed on the subsequence $x_{[i;j]}$ (i.e., nucleotides i through j, inclusive), assuming we have already computed it for all shorter subsequences $x_{[m;l]}$ with $i < m < l < j$. Clearly, we can obtain a

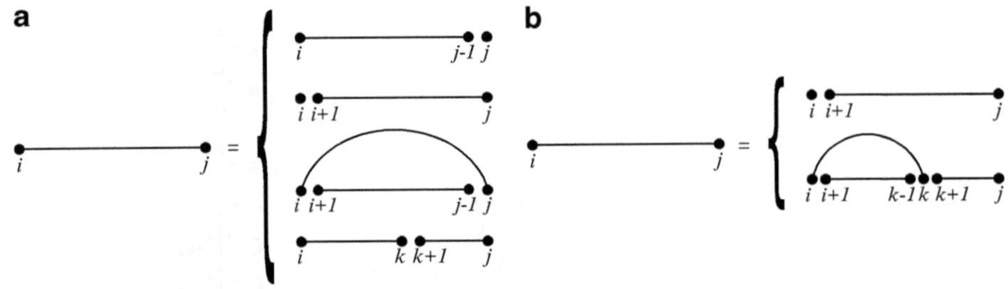

Fig. 4 Graphical representations of the folding algorithms. An *isolated bullet* refers to an unpaired nucleotide, and a *straight line* (*with bullets on the end*) to a subsequence with arbitrary structure. *Arcs* stand for base pairs. In contrast to how they are shown in Fig. 2c the sequence here is indicated as a *straight line*. (**a**) The plain Nussinov algorithm: the cases correspond to the four cases on the right of Eq. 1 and (**b**) unambiguous case corresponding to the two cases on the right side of Eq. 2

structure on $x_{[i;j]}$ by considering a structure on $x_{[i+1;j]}$ (or $x_{[i;j-1]}$) and adding an unpaired base at position i (or j). This of course does not change the number of pairs in the structure. Alternatively, if the nucleotides at positions i and j, $x[i]$ and $x[j]$, can base pair with each other, then a structure on $x_{[i;j]}$ can be obtained by enclosing a structure on $x_{[i+1;j-1]}$ with a new base pair (i, j), increasing the number of pairs by 1. There is only one further alternative: If i and/or j are paired, but not with each, then our structure must consist of two substructures, i.e., there must exist a position k, $i < k < j$, such that our structure can be split into a substructure on $x_{[i;k]}$ and another substructure $x_{[k+1;j]}$.

This recursive decomposition of structures is illustrated in Fig. 4a. It can readily be converted into a recursion, where we fill a matrix $E(i, j)$ containing the maximum number of pairs on the subsequence $x_{[i;j]}$. Figure 5 illustrates the resulting dependencies. Each cell above the diagonal represents a subsequence from position i to j and its value is the maximum number of base pairs. Thus, to obtain the maximum number of base pairs on the whole sequence, we have to fill the entire upper right half of the matrix. Figure 5a corresponds to the first three cases of Fig. 4a (adding an unpaired nucleotide or an enclosing pair), while Fig. 5b corresponds to the case where a structure is split into two components.

As further indicated by the black cells in the diagonals of Fig. 5 the number of base pairs on some cells is zero. In the figure all subsequences of length three or less have been initialized to zero, that is, for all $d, -1 \leq d \leq 2$ and all $i, 1 \leq i \leq N - d$, $E(i, i + d) = 0$. The technique of recursively computing and tabulating is commonly referred to as *dynamic programming*.

The dynamic programming solution to maximizing the number of base pairs, as described above, was first introduced by Nussinov [52] and is therefore commonly referred to as the *Nussinov* algorithm. It can be summarized in a recursion as follows.

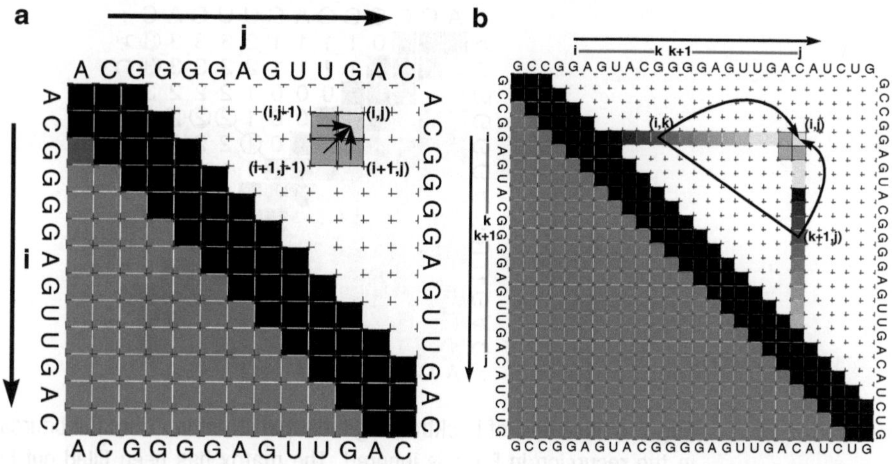

Fig. 5 Concepts of the Nussinov dynamic programming algorithm shown on respective subsections of the tetrahydrofolate (THF) riboswitch with Rfam accession RF01831. In both panels, the lower triangle is not used and the black diagonals illustrate prohibited cells where in this case a minimum loop size of $d_l = 2$ bases is an imposed constraint. **(a)** Illustrates how the resulting structure on the subsequence $x_{[i;j]}$ can be obtained from the structures on neighboring subsequences $x_{[i+1;j]}$, $x_{[i;j-1]}$, $x_{[i+1;j-1]}$. **(b)** Illustrates how the resulting structure on the subsequence $x_{[i;j]}$ may be obtained from combining two other structures on two subsequences $x_{[i;k]}$ and $x_{[k+1;j]}$. The respective cells of (i,k), $(k+1,j)$ and (i,j) are indicated and connected by a solid line and *joining arrows*. Here, the two subsequences of $x_{[5;11]}$ and $x_{[12;21]}$ are combined, as $k = 11$. In general, the values of cells in the corresponding greyscale color are to be added for successive k's. Note that some cells (the two light grey one's) in row i / column j pair with black cells near the diagonal, which are cells with value zero due to the initialization of loop size at least 2. Notice that k in this case spans $6, \ldots, 19$

Let $E(i,j)$ be the maximum number of base pairs for the subsequence $x_{[i;j]}$, then with $j - i > d_l$, the minimum loop length

$$E(i,j) = \max \begin{cases} E(i,j-1) \\ E(i+1,j) \\ E(i+1,j-1) + s(i,j) \\ \max_{i<k<j-1} \{E(i,k) + E(k+1,j)\}, \end{cases} \quad (1)$$

where $s(i,j) = 1$ when $x[i]$ and $x[j]$ potentially base pair and $s(i,j) = 0$ otherwise. The last term, combining two subsequences, is often referred to as a *bifurcation* or a branching structure. The different cases are illustrated graphically in Fig. 4a.

The recursion in Eq. 1 counts the number of base pairs for any subsequence until the whole (final) sequence from 1 to N has been reached. Thus, at this point the structure itself has not been computed. As for a pairwise sequence alignment, filling out the matrix gives the score. The alignment is obtained from backtracking. The same is the case here. Starting with the cell $(1, N)$ one can (e.g., while filling the matrix) keep track of which cells resulted in the value of $E(1, N)$. This could be either the "simple cases" $E(2, N)$, $E(1, N-1)$, $E(2, N-1)$, or simply a bifurcation at position k' and thus be the number of base pairs from

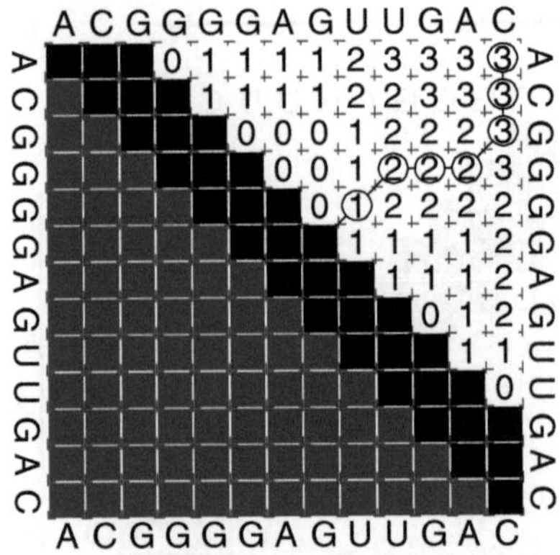

Fig. 6 An example of backtracking a hairpin structure. Here the bifurcating term in the recursion in Eq. 1 is ignored. The matrix has been filled out by starting at the first available diagonal of cells (assigning a value of zero to the *black diagonal*). Filling out the matrix results in a final score of 3, corresponding to 3 base pairs on the sequence. A backtrack is indicated by the *connected circles*. In general, the path of the backtrack is dependent on details of the program (specifically, on how ties are resolved)

the two subsequences $E(1, k')$ and $E(k' + 1, N)$. In the case where we came from the subsequence $x_{[2;N-1]}$ (which holds $E(2, N - 1)$ base pairs), we could only have arrived at $x_{[1;N]}$ (with $E(1, N)$ base pairs) if position 1 and N base paired. Thus we can keep a note on that and continue to work down the list. In case we came from either of the subsequences $x_{[1;N-1]}$ or $x_{[2;N]}$ (that is either $E(1, N) = E(1, N - 1)$, or $E(1, N) = E(2, N)$) we would not have registered any base pairs yet. If we came from the two subsequences $x_{[1;k']}$ and $x_{[k'+1;N]}$ (for which $E(1, N) = E(1, k') + E(k' + 1, N)$), we could keep track of a bifurcation point and then continue to backtrack the subsequences $x_{[1;k']}$ and $x_{[k'+1;N]}$ separately, having them each "emitting" base pairs and new branch points. An example for an unbranched structure (hairpin) is shown in Fig. 6 using the same sequence from Fig. 5a. The score (number of base pairs) is 3, since that's the number in the upper right corner (cell $(1, N)$). When backtracking one inspects where one could come from and works one's way to the main diagonal of the matrix and each anti-diagonal (down-left) step corresponds to a base pair. Following the circles connected by lines and starting from the upper right corner this structure `..(.((..))..)` on ACGGGGAGUUGAC is obtained. Note that it is possible to choose another path which would lead to another structure assignment, but still having three base pairs. The backtracking procedure has been described several places in the literature. A good overview is given in [53].

3.2 Towards a Folding Algorithm

A central issue in computational RNA biology is the computational resources (complexity) the algorithms need. Obviously, we are filling out a matrix of roughly $N^2/2$ cells, so for a sequences of length N the memory needed scales with N^2, that is, memory complexity is (big O) $O(N^2)$. Similarly a complexity can be found for the computation time, that is the time it takes to fill out the (upper part of) the matrix. For any of the three first terms on the right side of Eq. 1 we need to do only a constant amount of work (independent of N), so their total contribution to the cost scales as $O(N^2)$. However, due to the bifurcation term, we, for each cell need to consider an additional number cells which in number scale with the size of a given subsequence from position i to j. This adds another factor which scales with N, so the computation time reaches $O(N^3)$. This is the basic property in RNA folding algorithms based on dynamic programming and is a major challenge to overcome in a large-scale application. Note that the Nussinov algorithm does not take pseudoknots (crossing arcs) into account; this is also the case for most energy-based approaches. If pseudoknots are to be included, depending on the type of pseudoknots, the time/memory requirements will increase, for example up to $O(N^6)$ for certain classes of pseudoknots [54].

By closer inspection of Eq. 1 and the underlying decomposition (Fig. 4a), one will notice that it contains some redundancy, in the sense that the same structure can be produced in several ways. In other words our decomposition is *ambiguous*. For example, the subsequence with the structure assignment

```
.(((...)))....((((...)))).. 
```

can in the bifurcating term be computed from the two subsequences

```
.(((...)))   +   ....((((...))))..
```

or

```
.(((...)))..   +   ..((((...))))..
```

and several other combinations. Thus, each value in the matrix will be computed in many different ways. In our case this does not change results, nor does it affect computation time. Nevertheless, it can have undesired properties as described below.

Making sure that each cell is computed unambiguously is not essential for determining the structure, neither in the maximum base pair nor in the minimum free-energy (MFE) version, simply because the maximum (or minimum) of a list of values is unaffected if some values are duplicated. It becomes problematic, however, when one considers suboptimal structures, or even if we want to obtain *all* optimal structures. In this case backtracking will produce the same structure many times through different backtracking paths. The situation becomes even worse when one wants to compute quantities that are obtained by *summing* over all structures, such as the partition function (see below).

Fortunately, ambiguity can be easily avoided in most cases. Figure 4b shows a way to decompose structures uniquely. Using this decomposition, the recursion in Eq. 1 can be rewritten as

$$E(i,j) = \max \begin{cases} E(i+1,j) \\ \max_{i+3\leq k\leq j \text{ s.t. } s(i,k)=1} \{1 + E(i+1,k-1) \\ \quad +E(k+1,j)\}\,. \end{cases} \quad (2)$$

The reasoning supporting this version is simply that the ith nucleotide x_i is either base paired or it is not. If it is unpaired, then the upper term applies—count the maximal number of pairs in the shorter subsequence omitting x_i. On the other hand, if x_i is paired, then the lower term applies—x_i pairs with the nucleotide at some position k in the indicated range (e.g., $i + 3 \leq k$ enforces the minimum loop length restriction), with which it is allowed to pair ($s(i,k) = 1$), and the maximal number of base pairs formed is one more (for pair (i,k)) than the maximal number possible in the substrings $x_{[i+1,k-1]}$ plus $x_{[k+1,j]}$. The lower term simultaneously accounts for base pairs and bifurcating structures. Note that the terms at the boundaries, like $E(i+1,i)$, are set to zero by definition. It is also worth noting that the lower term in Eq. 2 is where the no-pseudoknot assumption is exploited by the algorithm. Under this assumption, the presence of the (i,k) base pair blocks interactions between nucleotides within $x_{[i+1,k-1]}$ and those outside of that subsequence, hence allowing $E(i+1,k-1)$ to be calculated in isolation.

The decompositions shown in Fig. 4 can also be recognized as a graphical representation of a grammar (more precisely a context free grammar) that generates structured RNA sequences. *See* Chapters 5 and 8 for further details.

Suboptimal structures can be much different from the optimal computed structure and sometimes only vary a bit. For a given sequence consider all possible structures (all possible dot-bracket strings) that can be assigned to a sequence. Each structure comes with its own probability correlated with its free energy. The *partition function* which can be interpreted as the ratio between the number of MFE structures and the number of all other structures, is used for this type of calculation. Equation 2 will yield identical results as the original version, but allows us to compute suboptimal structures without duplicates, and can be easily turned into a recursion for computing partition functions. A range of properties can be derived from the partition function including the probability of two nucleotides forming a base pair (roughly, considering how often they base pair over all possible structures). Like RNA folding, the partition function can also be computed

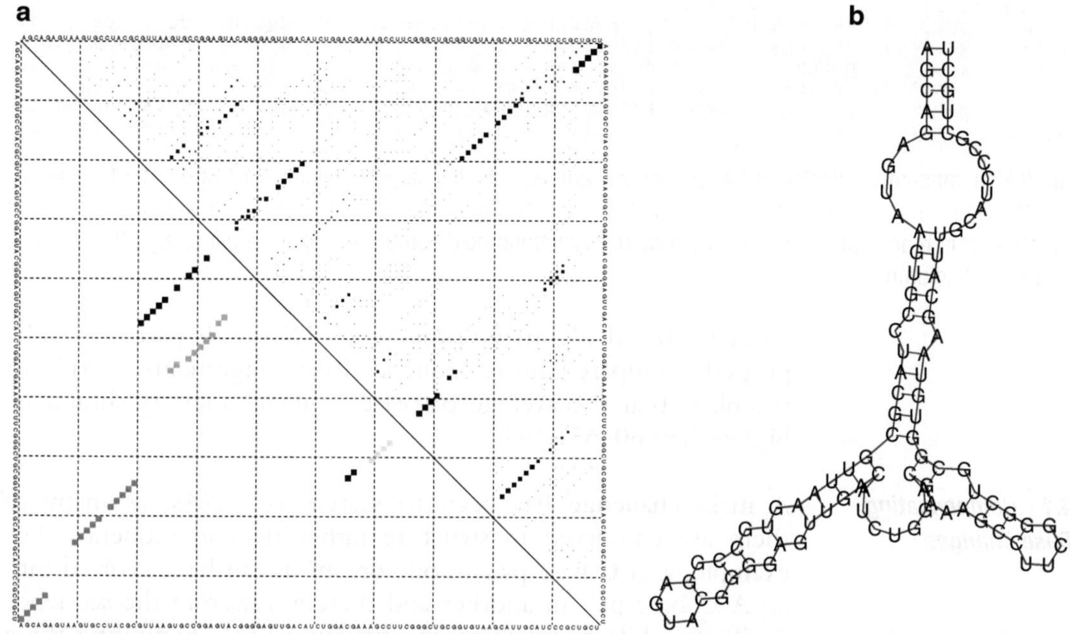

Fig. 7 Folding analysis of the riboswitch (Rfam accession RF01831). (**a**) A dotplot on the *upper triangle* showing the highest base pair probabilities by their size and on the *lower triangle* the base pairs from the MFE structure (*black*) and those (in *grayscale*) from the comparative-based analysis (in Fig. 2). (**b**) A diagram of the secondary structure predicted by MFE folding

by dynamic programming [55]. The derived probabilities are very useful when studying a sequence for its most preferred structures. An example is given in Fig. 7, where a *dotplot* (Fig. 7a) is compared to the predicted MFE structure (Fig. 7b) using the Vienna RNA folding package [30]. In the dotplot (upper triangle) all base pair probabilities exceeding a threshold are shown revealing a number of potential likely foldings. The lower triangle compares the MFE structure and the structure derived from comparative analysis (shown in Fig. 2). Note that these structures do not agree completely, reflecting that the comparative analysis can take more information into account than can be extracted from a single sequence.

4 Exploiting Comparative Information

Within the last decade comparative genomics has provided and driven knowledge generation by searching for conserved regions in the genome, as these indicate functionality. Genome browsers, e.g., from UCSC [36] and ENSEMBL [56] readily hold a lot of comparative information, both annotation of individual genomes as well as experimental and *in silico* data. Whereas the comparative information in genome browsers currently is built on

```
Position  1........10........20........30........40........50........60........70......
X00916.1  CCCGCGAUAGUUCAGAUUGGU-AGAACGGCGGACUGUAGAUCCGCAUGUCGCUGGUUCAAA--UCCGGCUCGCGGGA
D17540.1  GCGGGAGUAGCUCAGUU-GGU-AGAGCAUCGGCUUCCCAAGCCGAGGGUCGCGGGUUCGAG--UCCCGUCUCCCGCU
M10217.1  AAGCCUGCGGUGUUU-------GACAUGCCAGAUUGCAAAUCUCGAGA-AGCAA--ACGAAGGU-UUGCCGGGCUUC
X12631.1  CAAGGCAUAGCUAGUU--AG--UAAGUUUCUCCUUUACACGGAGACCA-CACUCGUGCAAU--UCGAGUUGUUUUGA
M26978.1  GCCGCCGUAGCUCAGUA-GGUAGGAGCGCUCGGCUGUUAACCGAGUGGUCGCAGGUUCGAG--UCCUGCCGGCGGCG
SS_cons   (((((((..((((.........)))).(((((.......))))).....(((((.......))))))))))))).
```

Fig. 8 Five randomly selected tRNA sequences among the 967 seed sequences in Rfam (10.1). The *top line* "Position" indicates the position in the alignment. The *bottom line* ("SS_cons") indicates the consensus structure. It is readily seen that several matching columns hold compensating base pairs, e.g., the first and next-to-last columns

sequence-based alignments, structured RNAs only partially can be properly compared using sequence-based alignments. A rule of thumb is that the average pairwise sequence identity should be higher than 60–65% [41].

4.1 Compensating Base Changes

A main challenge has been to search for RNAs, as many of them are conserved in structure rather than in sequence. For example, a G-C base pair in one organism can have evolved into an A-U base pair in another and thereby lowering the sequence similarity while maintaining the structure [34]. Exploiting these compensating base changes is a key feature used in determining the RNA structure. This was also employed in probably the first RNA bioinformatics work ever, prior to a Cold Spring Harbor meeting in 1966, where the *comparative* analysis of four tRNA sequences was presented with the aim of determining its structure [57]. The scientists looked not only at the sequence similarity of the sequences (to align them), but they also searched for corresponding base pairs, for there were compensating changes, such as depicted in Fig. 8.

4.2 Measuring and Visualizing Mutual Information

Given a "correct" alignment of RNA sequences, observed compensating changes are an important clue to the structure. A key problem, however, is exactly to get the core of the alignment correct. This has and still does depend on intensive manual intervention, and it remains as one the major problems in bioinformatics, in spite of the substantial improvement over the past decade in constructing methods for structural RNA alignments.

For a given alignment of RNA sequences, mutual information is an often employed measure. The idea is to compare the nucleotide content of two columns that (potentially) base pair, for example columns 11 and 26 of the tRNAs in Fig. 8. Given the compositions of each of the columns, the expectation to observe a particular base pair by chance can be computed and compared to the real observation. Considering the fraction (over the number of sequences) of base a in column i, $p_{a,i}$ and fraction of base b in column j, $p_{b,j}$, assuming independence of the column pair we would expect to observe the fraction of pairs between a and b to be $p_{a,i}p_{b,j}$. This can be compared to our observed fraction called

$p_{ab,ij}$ and doing this for all bases on positions i and j, the mutual information can be written as

$$M_{ij} = \sum_{a,b \in B} p_{ab,ij} \log \frac{p_{ab,ij}}{p_{a,i}\, p_{b,j}}, \tag{3}$$

where $B = \{A, C, G, U\}$ is the set of bases. Since $\sum_{a,b \in B} p_{ab,ij} = 1$ and $\sum_{a \in B} p_{a,i} = \sum_{b \in B} p_{b,j} = 1$, M_{ij} is an information content. The logarithm is often computed in base 2, to obtain the information content in bits. In practise, when there are gaps in the alignment these are either ignored or a gap compensating term can be introduced, e.g., [58, 59]. However, this implies that M_{ij}, strictly speaking, is no longer an information content, but nonetheless can be a useful measure of the compensating changes in columns i and j. An example (using the measure in Eq. 5) of how this can be useful to extract patterns of base pairs is shown in a mutual information plot in Fig. 9 on a set of the Riboswitch structures, there the upper part holds diagonal lines corresponding to the base pairs. (Note that the y-axis has been reversed relative to the dotplot to consistently capture the sequence conservation depicted along the axes; see the figure caption for details.)

Note that the mutual information content does capture *conserved base pairs*. In fact, as can be seen in Fig. 9 sequence conservation and compensatory changes are mutually exclusive. Sequence conservation can be measured in a way similar to mutual information: letting $p_{b,i}$ be the fraction of observed bases b in column i of the alignment, then the information content of column i is

$$I_i = \sum_{b \in B} p_{b,i} \log \frac{p_{b,i}}{q_b}, \tag{4}$$

where q_b is "background frequency" of b, for example extracted from the nucleotide content of a corresponding genome. If the background is equiprobable, that is $q_b = 0.25$ for all bases b, I_i reduces to 2 minus the *Shannon entropy* (when working in base 2 of the logarithm). The form in Eq. 4 is referred to as the *Kullback–Leibler divergence* or the relative entropy. The sequence conservation combined with the mutual information content can also be combined in an extended *sequence logo* [61], called a *structure logo* [62]. The version of the mutual information, referred to as a *covariation measure* content in a structure logo is the Kullback–Leibler divergence between two measures (expected and observed) of covariance given by

$$M_{ij} = \tilde{p}_{ij} \log \frac{\tilde{p}_{ij}}{E[\tilde{p}_{ij}]} + (1 - \tilde{p}_{ij}) \log \frac{(1 - \tilde{p}_{ij})}{(1 - E[\tilde{p}_{ij}])}, \tag{5}$$

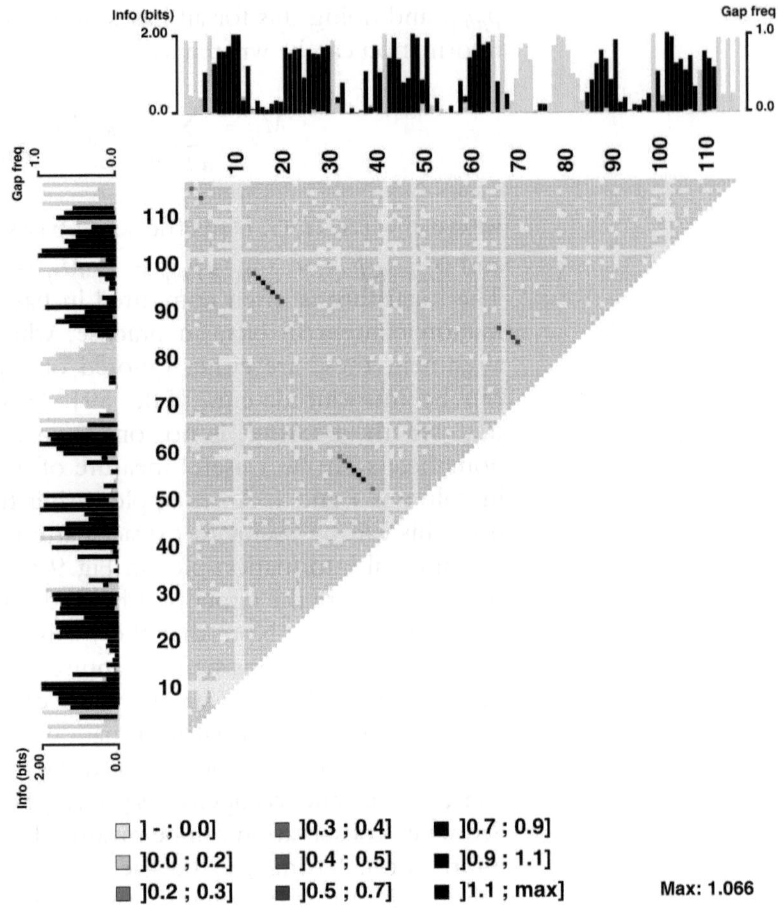

Fig. 9 Mutual information plot of sequences from the THF riboswitch. The alignment was for this purpose cleaned for sequences with very long inserts. The mutual information measure is shown in the variant shown in Eq. 5. The *grayscale* indicates the degree of mutual information. Along the axis is plotted a sequence conservation profile in terms of sequence information from Eq. 4 (in *black*) as well as the amount of gap content (*gray*). The plot was generated using `MatrixPlot` [60]

where $\tilde{p}_{ij} = \sum_{a,b\in B} C_{ab} p_{ab,ij}$ is the "accumulated" base pair content shared by positions i and j and $E[\tilde{p}_{ij}] = \sum_{a,b\in B} C_{ab} p_{a,i}\, p_{b,j}$ the expected base pair content. The term C_{ab} is one when a and b and one the base pairs A-U, C-G, or G-U (or the reverse of these) and zero otherwise. Note that gap content is implicitly included in the second term. The structure logo combines this measure with sequence conservation, by adding the shared mutual information among the position base pairings as shown in Fig. 10.

An alternative logo, `RNAlogo` merges the mutual information and sequence conservation directly in the RNA secondary structure diagram, thereby making it possible to directly assess conservation

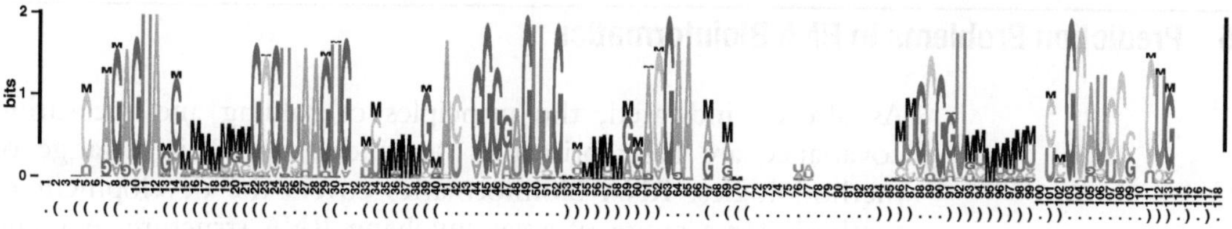

Fig. 10 Structure logo of sequences used in Fig. 9. The total height of the content in each position is the sum of I and $\frac{1}{2}M$. The height is therefore the sum of two measures and itself not an information content. The height of the symbols for bases is shown in proportion to the fraction they occupy at that position. Note that there is a higher proportion of "M" in the positions which are less conserved and involved in base pairing. The structure shown is the same as in Fig. 2

and consensus structure [63]. For example, long range interactions can readily be assessed for which positions are due to strict sequence conservation and which are due to compensating changes.

A variety of covariation measures have over time been proposed. These include a normalization of the mutual information from Eq. 3 by the sequence information contribution for a position, e.g., i (by $-\sum_{a \in B} p_{a,i} \log p_{a,i}$) to allow for individually considering the contributions for positions i and j which in general are asymmetric [64]. Using this measure higher order structure correlations have been identified.

A later published alternative was incorporated directly into the `RNAalifold` program [65] with the other computed energies from folding and is therefore negative. The measure uses a 16×16 matrix $D_{ab,a'b'}$, to weight the different types of base pairs, where fully compensatory pairs obtain a value of −2 kcal/mol (i.e., the Hamming distance between ab, $a'b'$ is 2), −1 kcal/mol if one base changes (e.g., U-A to U-G, and the Hamming distance between ab, $a'b'$ is 1), and zero otherwise. The final covariation measure is

$$M_{ij} = \sum_{ab,a'b' \in B} p_{ab,ij} D_{ab,a'b'} p_{a'b',ij}. \qquad (6)$$

In [59] these and other measures were compared and in particular a "correction" of gap content was introduced and subtracted from the measures. While these measures in a strict sense are not information measures these ***approximate mutual information measures*** were shown to have increased ability to identify base pairs in a set of alignments when compared to the known (curated) structure assignments. In a recent study it has further been shown that by extending C_{ab} in Eq. 5 to represent branch length of the evolutionary tree, while keeping the gap correction from [59], that the overall ability to identify base pairs is further improved [66]. In [66] the logos are further extended to cover RNA–RNA interactions by `RIlogo`.

5 Prediction Problems in RNA Bioinformatics

As already indicated, the principles of folding and measuring covariance are basic principles "extended upon" in a range of methodologies. RNA bioinformatics covers the development of methods for a range of areas involving RNA structure. A range of methods exist for various types of problems.

The strategy to search for 2D structure or 2D folding has in particular been through energy folding of which popular implementations such as `RNAfold` [30] and `mfold` [67] exist. The goal with 2D structure prediction is try to predict the base pairs. In the basic problem pseudoknots are ignored. They can be included within a dynamic programming framework extending on the energy folding algorithms, e.g., [54, 68] as well as with a variety of other types of approaches, e.g., [69–71]. Depending on the type of pseudoknot the corresponding algorithms come with corresponding complexity, for example the Rivas and Eddy algorithm is $O(N^6)$ in time and $O(N^4)$ in memory for a sequence of length N. Recent work takes advantage of *sparsification* in which some combinations in computing the optimal score can be neglected [72]. Recent implementations of sparsification in basic RNA folding algorithm have also been proposed [73].

Methods to predict the structure from multiple aligned RNA sequences attempt to exploit the pattern of compensating base pairs, as already mentioned with `RNAalifold` above, which combines this with energy calculations. A range of methods for structural alignments, that is, building an alignment of RNA sequences while simultaneously including RNA secondary structure have also been proposed and are reviewed in Chapters 13–15. Some of these methods can also deal with pseudoknots [74].

Another emerging area is the recognition of 3D motifs in which people are trying to recognize small structural motifs, e.g., [75] typically consisting of non-Watson–Crick and G-U wobble base pairs [76] and therefore appear as large internal loop on secondary structure predictions. Recently a prediction program `RMdetect` has been made available [75]. Emerging work is trying to take these types of base pairs into account in the 2D folding algorithms leading to more complex versions of the "standard" folding algorithms, e.g., [77].

In RNA 3D structure prediction the goal is to predict the correct positions of the atoms. Approaches include (but are not limited to) building the RNA structure from comparative modeling of 3D structures [9] to direct prediction of tertiary structure from a single sequence [78]. Predicting 3D structure via the secondary structure predictions including non-canonical base pairs has also been proposed [8]. Strategies inspired from protein 3D structure prediction were also exploited from RNA 3D structure through a combination of sampling conformations along with the full-atom energy function from the Rosetta framework [79]. In line with blind tests for protein structure predictions, a CASP for RNA has been carried out, where RNA 3D structure prediction is evaluated by blind tests [80].

The work towards genome annotation for ncRNA covers annotation of ncRNA genes as well as structural elements in UTRs of mRNAs. The annotation of ncRNA genes "traditionally" refers to structural ncRNA genes, but has broadened to include genes containing structural elements, or in short searching for structured RNAs and thereby also covering elements in UTR regions [34]. Approaches to search for long ncRNAs without using features of RNA structures have been carried out using other signals such as splice sites [81]. Thus genome annotation for structured RNAs ranges from (1) (structural 2D) homology search of structured ncRNA genes, (2) structured RNA elements in UTR regions, (3) class-specific searches, and (4) *de novo* search for novel RNA structures. For (1) and (2) a careful structural probabilistic model can be extracted from an existing multiple alignment of related RNA sequences and used to search in the genomic sequences for more or completely new ncRNAs in case of unannotated genomes [82]. The modeling is typically based on *Stochastic Context-Free Grammars* (SCFGs) which can be considered as a model framework extending hidden Markov models, where probabilities for long range interactions corresponding to base pairs are incorporated on top of those linear in the sequence. In a loose sense one can think of a tree representation akin to Fig. 3 where each bullet represents a probability distribution on nucleotides or pairs, typically extracted from a set of aligned sequences (e.g., by tedious and time-consuming work by the curator). Other strategies extending on SCFGs have also been implemented, such as the `CONTRAfold` program for 2D folding, using a class of probabilistic models called conditional log-linear models [83]. It was demonstrated that using "folding parameters" extracted from curated data gave better (secondary) structure predictions than methods based on standard thermodynamic parameters.

To reduce search time in homology search, a `BLAST` [84] or HMM [82, 85] filter can be imposed and the computationally more expensive structural search employed in regions around remote sequence similarity. Still, when searching for matches in phylogenies far away, the structure itself may vary; `BLAST` for the remote sequence similarity might in some cases be of almost the same quality as employing structural similarity [86]. For class-specific searches, more context can be taken into account. For example, when searching for tRNAs as a class, the specific tRNA is not predicted, but its promoter region is taken into account in the screen to lower the number of false positives [87], a problem which also is present in the homology search. MicroRNAs are examples where the specific version (with specific seed) is not considered, for example in `RNAmicro` [88]. For *de novo* search completely new RNA structures are searched for. This is typically done either by exploiting multiple sequences that are already aligned based on their sequence similarity, or by conducting structural alignments on corresponding, but unaligned sequences [37]. In both cases the goal is to obtain support for compensating changes.

Efforts in RNA–RNA interactions also come in several different flavors and make use of the same principles as in folding single and multiple sequences. Searching for RNA–RNA interactions holds some of the same challenges as for predicting pseudoknots. For single sequences a joining linker can be employed which is neutrally scored in the model and the two sequences can then be co-folded as if they were one sequence [89]. Other methods compute the interaction directly, e.g., [90, 91], but ignore intramolecular parings to save time, while others again employ more complicated models at the expensive of increased computational resources [92]. Also, this problem can be addressed by a partition function [93]. More recently, versions for RNA–RNA interactions on two sets of multiple alignments were introduced [94–98]. Like for class-specific ncRNA search, similar work for target predictions has been made, for example for microRNAs [99–101] and snoRNAs [102, 103].

Predicting RNA–protein interactions has been tried, both from an RNA as well and a protein perspective. Starting from RNA binding proteins, a number of *in silico* methods have predicted which RNAs might bind; conversely, given an RNA sequence others have tried to find which proteins might bind to it e.g., [104].

A summary of the main prediction problems is listed in Table 1. Many of the problems strongly depend on well-curated data, which is often tedious to obtain. For example, RNA–RNA interaction data are often only published as image files, e.g., [105], which makes downstream analysis difficult. A future challenge will be to provide means for researchers to overcome this task with a reasonable effort.

6 Benchmarking

In order to assess the quality of RNA structure or folding programs it is necessary to benchmark them. This typically involves comparing the predicted structure to a known structure on a test set. For 2D or secondary structure predictions this would reduce to comparing dot-bracket notations between the prediction and the one from a known sequence. The choice of comparison of predicted and known structures as well as choice of data each comes with their own set of issues. For the predicted and known

Table 1
The table provides an overview of central subareas of RNA bioinformatics (left column) with pointer to chapters and references (right column) for further details

Prediction problem	Pointer
2D structure prediction of single sequences	C1; C3; C4; C5
3D structure prediction and modeling of single sequences	C2; C18
2D structure prediction of multiple aligned sequences	C7; C8
2D structural alignments of unaligned sequences	C13; C14; C15; C17
Folding kinetics	C4
2D and 3D structure motif detection	C11
Detection of RNA regulatory motifs	C15
RNA genefinding	C15
RNA homology search	C9
Search for class-specific ncRNAs	C10; C20
RNA (homology) structure comparison	C12; C18
RNA structure evolution	C16
RNA–RNA interactions	C19; C21
siRNA design	C22
RNA–protein interactions	C23
RNA databases	C6
Prediction of pseudoknots	e.g., [69–72]
Beyond 2D, not yet 3D RNA structure prediction	e.g., [8, 77]
RNA 3D motif determination	e.g., [75, 106]
Folding with constraint of high-throughput probing data[a]	e.g., [47, 48, 107]
SNPs in RNA structure	e.g., [108–113]
RNA sequence design (beyond siRNA design)	e.g., [79]

[a]In the context of identifying ncRNAs

structures, one might be happy if a base is predicted to belong to the correct helix, or more conservatively be interested in whether a particular base pair has been correctly predicted or not. The latter seems to be what most people do in the literature (for 2D predictions). For the choice of benchmark data it seems to be an eternal struggle in finding RNA sequences with well-curated (and even better experimentally verified) base pair assignments. The best information today is the one which can be extracted from multiple alignments, where an expert in that particular RNA has been involved in curation and the base pair assignment accomplished through a combination of studying compensatory changes while including experimental knowledge.

A main issue is that it is hard and thankless work to curate structural alignments of RNA and once these have been made there is a need for maintaining them. Some data sets have been made, such as the Bralibase initiatives covering various aspects of RNA structure prediction and homology search methods [114–116]. Currently, most people seem to extract data from Rfam [29] and clean them one way or the other as well using already extracted data sets sometimes from the papers to which a comparison is made to the corresponding prediction method. Even the most well-maintained structural RNA databases, e.g., RNAse P RNA [117] and SRP RNA [118], sometimes need cleaning. The cleaning might not be because the curators did a poor job, but simply because some sequence is evolutionarily remote compared to the remaining sequences and therefore can have large insertions or deletions. Obviously a single sequence sticking out because of a large insert can make it problematic to obtain a good base pattern for and it might have to be discarded before training and testing a prediction method. The families might also vary structurally and will have to be subdivided accordingly to extract meaningful consensus structures, as is the case for RNAse P RNA and SRP RNA. Initiatives like the RNA structure alignment ontology have been made in the attempt to push forward developing representations exceeding the alignment-based paradigm as well as to facilitate 2D and 3D data integration [119] and the recent RNAstar is a useful direction [120].

6.1 Benchmarking RNA Structure Predictions

For benchmarking by comparing explicit base pairs between two assignments, the most commonly reported measures are based on the following numbers: *True positives* TP (number of predicted base pairs that are base pairs); *False positives* FP (number of predicted base pairs that are not base pairs); *True negatives* TN (number of pairs predicted not to base pairs that do not base pair); and *False negatives* FN (number of pairs predicted not to base pairs that do base pair). From these numbers a range of measures can then be constructed. The measures include the sensitivity (SEN) and positive predictive values (PPV):

$$\mathrm{SEN} = \frac{\mathrm{TP}}{\mathrm{TP} + \mathrm{FN}} \quad ; \quad \mathrm{PPV} = \frac{\mathrm{TP}}{\mathrm{TP} + \mathrm{FP}} \,, \tag{7}$$

which both takes values in the interval between zero and one.

The *F*-measure (or F1-score), defined as

$$F = \frac{2 \cdot \mathrm{SEN} \cdot \mathrm{PPV}}{\mathrm{SEN} + \mathrm{PPV}} = \frac{2}{\frac{1}{\mathrm{SEN}} + \frac{1}{\mathrm{PPV}}}, \tag{8}$$

is recognized as the harmonic mean of SEN and PPV as is also used in the literature, e.g., [121].

Another frequently used measure is the Matthews Correlation Coefficient, MCC, [122], which is a discretization of Pearson's correlation coefficient:

$$\mathrm{MCC} = \frac{\mathrm{TP}\cdot\mathrm{TN}+\mathrm{FP}\cdot\mathrm{FN}}{\sqrt{(\mathrm{TP}+\mathrm{FP})(\mathrm{TP}+\mathrm{FN})(\mathrm{TN}+\mathrm{FN})(\mathrm{TN}+\mathrm{FP})}}\,. \tag{9}$$

MCC takes values in the interval $[-1;1]$ and is zero for completely random (uncorrelated) performance. For RNA structure prediction this MCC holds some special properties since there are no more than $N/2$ base pairs on a sequence of length N (unless base triples are allowed, which is not the case for secondary structure prediction). Given that there are $N(N-1)/2$ pairs in total, the TN will always be an order of magnitude larger than TP, FP, and FN. Using this it has been shown [123] that MCC naturally reduces with very high accuracy to the geometric mean of the PPV and the sensitivity:

$$\mathrm{MCC} \approx \frac{\mathrm{TP}}{\sqrt{(\mathrm{TP}+\mathrm{FP})(\mathrm{TP}+\mathrm{FN})}} = \sqrt{\mathrm{PPV}\cdot\mathrm{SEN}} \tag{10}$$

MCC therefore holds a natural interpretation.

When a prediction method depends on internal parameters to provide the cutoff between predicting and not predicting a base pair, e.g., [83, 124], the performance can be measured by the receiver operating characteristic (ROC) curve. Each value of the internal parameter results in values for SEN and PPV and they can be plotted against each other to complete the ROC curve. On an ROC curve one can select the trade-off between SEN and PPV. PPV can also be replaced by the false positive rate $\mathrm{FPR} = \mathrm{FP}/(\mathrm{FP}+\mathrm{TN})$ for types of data where the TN is "balanced" to the rest of the data. A benchmark of the method as a whole is typically found by computing the area under the (ROC) curve (AUC). The same type of measures are also employed for benchmarking RNA–RNA interactions, e.g., [95, 125].

Even though measuring TP, FP, TN, and FN counts the different types of predicted base pairs, this can be done even more conservatively. These four numbers make a 2×2 *confusion matrix*, which can be extended if one also would like to take into account the base pair partner. For example, position and prediction might yield that positions i and j should base pair, but on the known data position i is involved in base pairing, but with another position, say k. Thus, one can include the cases of base pair partner when comparing any two positions in the sequence: (1) the two positions, i and j can pair with each other; (2) i can pair but with another position j' while j can be unpaired; (3) i can pair but with another position j' while j can be paired elsewhere to position i';

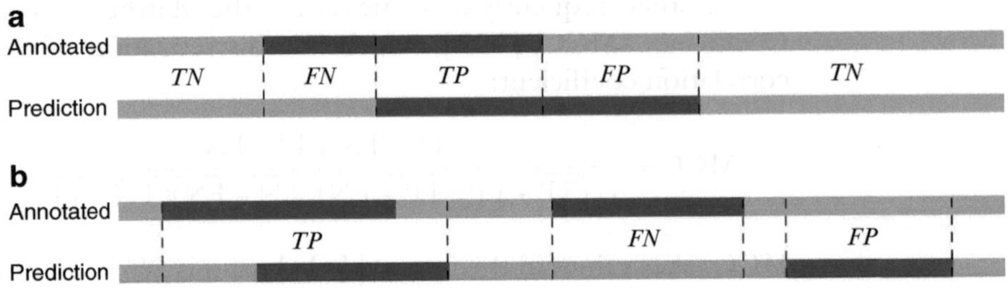

Fig. 11 Benchmarking scenarios for ncRNA gene finding. The *dark regions* are annotated and predicted ncRNA genes, respectively, typically covering an annotated/predicted RNA structure. (**a**) count nucleotide positions and accumulate the numbers TP, FP, TN, and FN. (**b**) Allowing for a minimum overlap between prediction and known gene to be positive prediction. In this case counting false negatives is omitted

(4) i can be unpaired while j can be paired to position i'; (5) both i and j are unpaired [126]. Thus we have a 5×5 confusion matrix extending over what can be handled by MCC. However, MCC has been extended to cope with any size, k confusion matrix, through the extended k-category correlation coefficient R_k [127]. In general one will expect the more conservative benchmark to lead to lower performance, which as also observed in [126], where R_5 was slightly and consistently lower than MCC in the benchmarks carried out.

When employing a benchmark measure the prediction must be compared to known data. Rather than measuring the explicit performance in terms of predicted and non-predicted base pairs, for structure prediction on multiple sequences, measures like number of recovered positions in the (average pairwise) alignments of the sequences as well as the *structural conservation index*, SCI have been employed by comparing predicted to curated structures [115]. The SCI measure is defined from multiple alignment RNA sequences as the folding energy over the multiple alignment divided by the average folding of the individual sequences in the alignment [128]. Note that even though SCI measures the *structural diversity* of the sequences in the alignment, the measure in itself is not a benchmark measure. Benchmarking always requires comparison to known data.

6.2 Benchmarking ncRNA Gene Predictions

Prediction of (structured) ncRNA genes or structured RNA (sometimes elements in UTRs) is faced with the challenge that no genome today can be claimed to be completely annotated for ncRNAs. What one can do is to map known ncRNAs to genomic locations and then count overlap in location. Thus numbers like TP, FP, TN, and FN can be computed, but depending on the type of comparison TN can be omitted as illustrated in Fig. 11. From these numbers the MCC can be computed. Since we presumably only search for ncRNAs in smaller portion of the genome (at least when encountering known cases), one can assume that TN is at least an order of magnitude larger than the other numbers, and thus MMC can still be approximated by the geometric mean of the PPV and the sensitivity. ROC curves have also been used to evaluate prediction of RNA structures on real and random (shuffled multiple alignments), e.g., [129, 130] and as a summary of performance, the AUC has been computed. For such ROC curves SEN and FPR are compared.

7 Perspectives

The area of RNA bioinformatics is in rapid growth and the growing amount of genomic and transcriptomic (such as RNAseq) data will further increase the need for RNA bioinformatics tool, such as RNA folding and RNA structure screens in genomic sequence. There are several interesting subareas in development, a number of which are described or touched upon in this book, and in addition to those, other exciting types of work are emerging including the impact of SNPs on RNA structure, e.g., [110–113] and detection of RNA structural modules [75]. These hold the potential to impact genome analysis, for example in relation to the many (personal) genome projects which are likely to point to genomic variation.

RNA bioinformatics faces new challenges in terms of the many long ncRNAs being (re-)discovered and combining experimental data into the algorithms could push the efforts in this area. RNA systems biology is also an exciting emerging area and is expected to be based on the RNA–RNA and RNA–protein interaction prediction methods. One could anticipate that literature mining will continue to play a bigger role as is the case for discovering protein–protein and protein–chemical associations, e.g., as in STITCH [131].

In the light of the rapid progress in transcriptomic-related areas, including high-throughput versions of RNA structure probing, new challenges will appear and the need for continued development of methods in computational RNA biology and their applications in bioinformatics will further increase.

Acknowledgments

This work is supported by the Danish Council for Independent Research (Technology and Production Sciences), the Danish Council for Strategic Research (Programme Commission on Strategic Growth Technologies), as well as the Danish Center for Scientific Computing.

References

1. Amaral PP, Dinger ME, Mercer TR, John S (2008) The eukaryotic genome as an RNA machine. Science 319(5871):1787–1789
2. Taft RJ, Pang KC, Mercer TR, Dinger M, Mattick JS (2010). Non-coding RNAs: regulators of disease. J Pathol 220(2):126–139
3. Pauli A, Rinn JL, Schier AF (2011) Non-coding RNAs as regulators of embryogenesis. Nat Rev Genet 12(2):136–149
4. Zhou H, Hu H, Lai M (2010) Non-coding RNAs and their epigenetic regulatory mechanisms. Biol Cell 102(12):645–655
5. Clark MB, Johnston RL, Inostroza-Ponta M, Fox AH, Fortini E, Moscato P, Dinger ME, Mattick JS (2012) Genome-wide analysis of long noncoding RNA stability. Genome Res 22(5):885–898
6. Costa FF (2005) Non-coding RNAs: new players in eukaryotic biology. Gene 357(2):83–94
7. Sleutels F, Zwart R, Barlow DP (2002) The non-coding Air RNA is required for silencing autosomal imprinted genes. Nature 415: 810–813
8. Parisien M, Major F (2008) The MC-Fold and MC-Sym pipeline infers RNA structure from sequence data. Nature 452(7183): 51–55
9. Rother M, Rother K, Puton T, Bujnicki JM (2011) ModeRNA: a tool for comparative modeling of RNA 3D structure. Nucleic Acids Res 39(10):4007–4022
10. Kladwang W, VanLang CC, Cordero P, Das R (2011) A two-dimensional mutate-and-map strategy for non-coding RNA structure. Nat Chem 3(12):954–962
11. Berman HM, Westbrook J, Feng Z, Gilliland G, Bhat TN, Weissig H, Shindyalov IN, Bourne PE (2000) The Protein Data Bank. Nucleic Acids Res 28(1):235–242,
12. Tsai MC, Manor O, Wan Y, Mosammaparast N, Wang JK, Lan F, Shi Y, Segal E, Chang HY (2010) Long noncoding RNA as modular scaffold of histone modification complexes. Science 329(5992):689–693
13. Novikova IV, Hennelly SP, Sanbonmatsu KY (2012) Structural architecture of the human long non-coding RNA, steroid receptor RNA activator. Nucleic Acids Res 40(11):5034–5051
14. The FANTOM Consortium and the RIKEN Genome Research Group Phase I & II Team (2002) Analysis of the mouse transcriptome based on functional annotation of 66,770 full-length cDNAs. Nature 420:563–573
15. Ponting CP, Oliver PL, Reik W (2009) Evolution and functions of long noncoding RNAs. Cell 136(4):629–641
16. Jia H, Osak M, Bogu GK, Stanton LW, Johnson R, Lipovich L (2010) Genome-wide computational identification and manual annotation of human long noncoding RNA genes. RNA 16:1478–1487
17. Amaral PP, Clark MB, Gascoigne DK, Dinger ME, Mattick JS (2011) lncRNAdb: a reference database for long noncoding RNAs. Nucleic Acids Res 39:D146–D151
18. Cabili MN, Trapnell C, Goff L, Koziol M, Tazon-Vega B, Regev A, Rinn JL (2011) Integrative annotation of human large intergenic noncoding RNAs reveals global properties and specific subclasses. Genes Dev 25(18):1915–1927
19. Moran VA, Perera RJ, Khalil AM (2012) Emerging functional and mechanistic paradigms of mammalian long non-coding RNAs. Nucleic Acids Res 40(14):6391–6400
20. Ulitsky I, Shkumatava A, Jan CH, Sive H, Bartel DP (2011) Conserved function of lincRNAs in vertebrate embryonic development despite rapid sequence evolution. Cell 147(7):1537–1550
21. Jacob F, Monod J (1961) Genetic regulatory mechanisms in the synthesis of proteins. J Mol Biol 3:318–356
22. International Human Genome Sequencing Consortium (2001) Initial sequencing and analysis of the human genome. Nature 409:860–921
23. Lim LP, Glasner ME, Yekta S, Burge CB, Bartel DP (2003) Vertebrate microRNA genes. Science 299:1540
24. Omer AD, Lowe TM, Russell AG, Ebhardt H, Eddy SR, Dennis PP (2000) Homologs of small nucleolar RNAs in Archaea. Science 288(5465):517–522
25. Ames TD, Rodionov DA, Weinberg Z, Breaker RR (2010) A eubacterial riboswitch class that senses the coenzyme tetrahydrofolate. Chem Biol 17(7):681–685
26. Weinberg Z, Perreault J, Meyer MM, Breaker RR (2009) Exceptional structured noncoding RNAs revealed by bacterial metagenome analysis. Nature 462(7273): 656–659
27. Trausch JJ, Ceres P, Reyes FE, Batey RT (2011) The structure of a tetrahydrofolate-sensing riboswitch reveals two ligand binding sites in a single aptamer. Structure 19(10):1413–1423

28. Hanson RM (2010) Jmol—a paradigm shift in crystallographic visualization. J Appl Crystallogr 43:1250–1260
29. Gardner PP, Daub J, Tate J, Moore BL, Osuch IH, Griffiths-Jones S, Finn RD, Nawrocki EP, Kolbe DL, Eddy SR, Bateman A (2011) Rfam: Wikipedia, clans and the "decimal" release. Nucleic Acids Res 39(Database issue):D141–D145
30. Lorenz R, Bernhart SH, Honer Zu Siederdissen C, Tafer H, Flamm C, Stadler PF, Hofacker IL (2011) ViennaRNA Package 2.0. Algorithms Mol Biol 6:26
31. Piccinelli P, Rosenblad MA, Samuelsson T (2005) Identification and analysis of ribonuclease P and MRP RNA in a broad range of eukaryotes. Nucleic Acids Res 33(14):4485–4495
32. Xie M, Mosig A, Qi X, Li Y, Stadler PF, Chen JJ (2008) Structure and function of the smallest vertebrate telomerase RNA from teleost fish. J Biol Chem 283(4):2049–2059
33. Stadler PF, Chen JJ, Hackermuller J, Hoffmann S, Horn F, Khaitovich P, Kretzschmar AK, Mosig A, Prohaska SJ, Qi X, Schutt K, Ullmann K (2009) Evolution of vault RNAs. Mol Biol Evol 26(9):1975–1991
34. Gorodkin J, Hofacker IL, Torarinsson E, Yao Z, Havgaard JH, Ruzzo WL (2010) De novo prediction of structured RNAs from genomic sequences. Trends Biotechnol 28:9–19
35. Torarinsson E, Sawera M, Havgaard JH, Fredholm M, Gorodkin J (2006) Thousands of corresponding human and mouse genomic regions unalignable in primary sequence contain common RNA structure. Genome Res 16(7):885–889
36. Fujita PA, Rhead B, Zweig AS, Hinrichs AS, Karolchik D, Cline MS, Goldman M, Barber GP, Clawson H, Coelho A, Diekhans M, Dreszer TR, Giardine BM, Harte RA, Hillman-Jackson J, Hsu F, Kirkup V, Kuhn RM, Learned K, Li CH, Meyer LR, Pohl A, Raney BJ, Rosenbloom KR, Smith KE, Haussler D, Kent WJ (2011) The UCSC Genome Browser database: update 2011. Nucleic Acids Res 39(Database issue):D876–D882
37. Gorodkin J, Hofacker IL (2011) From structure prediction to genomic screens for novel non-coding RNAs. PLoS Comput Biol 7:e1002100
38. Altschul SF, Madden TL, Schaffer AA, Zhang J, Zhang Z, Miller W, Lipman DJ (1997) Gapped BLAST and PSI-BLAST: a new generation of protein database search programs. Nucleic Acids Res 25:3389–3402
39. Rivas E, Eddy S (2000) Secondary structure alone is generally not statistically significant for the detection of noncoding RNAs. Bioinformatics 7:583–605
40. Workman C, Krogh A (1999) No evidence that mRNAs have lower folding free energies than random sequences with the same dinucleotide distribution. Nucleic Acids Res 27:4816–4822
41. Washietl S, Hofacker IL (2004) Consensus folding of aligned sequences as a new measure for the detection of functional RNAs by comparative genomics. J Mol Biol 342: 19–30
42. Gesell T, Washietl S (2008) Dinucleotide controlled null models for comparative RNA gene prediction. BMC Bioinformatics 9:248
43. Anandam P, Torarinsson E, Ruzzo WL (2009) Multiperm: shuffling multiple sequence alignments while approximately preserving dinucleotide frequencies. Bioinformatics 25(5):668–669
44. Lu ZJ, Yip KY, Wang G, Shou C, Hillier LW, Khurana E, Agarwal A, Auerbach R, Rozowsky J, Cheng C, Kato M, Miller DM, Slack F, Snyder M, Waterston RH, Reinke V, Gerstein MB (2011) Prediction and characterization of noncoding RNAs in C. elegans by integrating conservation, secondary structure, and high-throughput sequencing and array data. Genome Res 21(2):276–285
45. Underwood JG, Uzilov AV, Katzman S, Onodera CS, Mainzer JE, Mathews DH, Lowe TM, Salama SR, Haussler D (2010) FragSeq: transcriptome-wide RNA structure probing using high-throughput sequencing. Nat Methods 7(12):995–1001
46. Kertesz M, Wan Y, Mazor E, Rinn JL, Nutter RC, Chang HY, Segal E (2010) Genome-wide measurement of RNA secondary structure in yeast. Nature 467(7311):103–107
47. Deigan KE, Li TW, Mathews DH, Weeks KM (2009) Accurate SHAPE-directed RNA structure determination. Proc Natl Acad Sci USA 106(1):97–102
48. Washietl S, Hofacker IL, Stadler PF, Kellis M (2012) RNA folding with soft constraints: reconciliation of probing data and thermodynamic secondary structure prediction. Nucleic Acids Res 40(10):4261–4272
49. Hogeweg P, Hesper B (1984) Energy directed folding of RNA sequences. Nucleic Acids Res 12(1 Pt 1):67–74
50. Le SY, Nussinov R, Maizel JV (1989) Tree graphs of RNA secondary structures and their comparisons. Comput Biomed Res 22(5):461–473
51. Dowell RD, Eddy SR (2004) Evaluation of several lightweight stochastic context-free grammars for RNA secondary structure prediction. BMC Bioinformatics 5:71

52. Nussinov R, Piecznik G, Grigg JR, Kleitman DJ (1978) Algorithms for loop matchings. SIAM J Appl Math 35:62–82
53. Durbin R, Eddy S, Krogh A, Mitchison G (1998) Biological sequence analysis. Cambridge University Press, Cambridge
54. Rivas E, Eddy SR (1999) A dynamic programming algorithm for RNA structure prediction including pseudoknots. J Mol Biol 285(5):2053–2068
55. McCaskill JS (1990) The equilibrium partition function and base pair binding probabilities for RNA secondary structures. Biopolymers 29:1105–1119
56. Flicek P, Amode MR, Barrell D, Beal K, Brent S, Carvalho-Silva D, Clapham P, Coates G, Fairley S, Fitzgerald S, Gil L, Gordon L, Hendrix M, Hourlier T, Johnson N, Kahari AK, Keefe D, Keenan S, Kinsella R, Komorowska M, Koscielny G, Kulesha E, Larsson P, Longden I, McLaren W, Muffato M, Overduin B, Pignatelli M, Pritchard B, Riat HS, Ritchie GR, Ruffier M, Schuster M, Sobral D, Tang YA, Taylor K, Trevanion S, Vandrovcova J, White S, Wilson M, Wilder SP, Aken BL, Birney E, Cunningham F, Dunham I, Durbin R, Fernandez-Suarez XM, Harrow J, Herrero J, Hubbard TJ, Parker A, Proctor G, Spudich G, Vogel J, Yates A, Zadissa A, Searle SM (2012) Ensembl 2012. Nucleic Acids Res 40(Database issue):84–90
57. Pace NR, Thomas BR, Woese CR (1999) Probing RNA structure, function, and history by comparative analysis. In: Gesteland RF, Cech TR, Atkins JF (eds) The RNA world. Cold Spring Harbor, New York, pp 113–141
58. Hertz GZ, Hartzell GW III, Stormo GD (1990) Identification of consensus patterns in unaligned DNA sequences known to be functionally related. CABIOS 6:81–92
59. Lindgreen S, Gardner PP, Krogh A (2006) Measuring covariation in RNA alignments: physical realism improves information measures. Bioinformatics 22:2988–2995. doi:10.1093/bioinformatics/btl514
60. Gorodkin J, Stærfeldt HH, Lund O, Brunak S (1999) Matrixplot: visualizing sequence constraints. Bioinformatics 15:769–770. http://www.cbs.dtu.dk/services/MatrixPlot/
61. Schneider TD, Stephens RM (1990) Sequence logos: a new way to display concensus sequences. Nucleic Acids Res 18:6097–6100
62. Gorodkin J, Heyer LJ, Brunak S, Stormo GD (1997) Displaying the information contents of structural RNA alignments: the structure logos. CABIOS 13:583–586. http://www.cbs.dtu.dk/~gorodkin/appl/slogo.html
63. Chang TH, Horng JT, Huang HD (2008) RNALogo: a new approach to display structural RNA alignment. Nucleic Acids Res 36(Web Server issue):W91–W96
64. Gutell RR, Power A, Hertz GZ, Putz E, Stormo GD (1992) Identifying constraints on the higher-order structure of RNA: continued development and application of comparative sequence analysis methods. Nucleic Acids Res 20:5785–5795
65. Hofacker IL, Fekete M, Stadler PF (2002) Secondary structure prediction for aligned RNA sequences. J Mol Biol 319:1059–1066
66. Menzel P, Seemann SE, Gorodkin J (2012) RIlogo: visualising RNA-RNA interactions. Bioinformatics 28(19):2523–2526. http://rth.dk/resources/rilogo
67. Zuker M (2003) Mfold web server for nucleic acid folding and hybridization prediction. Nucleic Acids Res 31(13):3406–3415
68. Lyngsø RB, Pedersen CN (2000) RNA pseudoknot prediction in energy-based models. J Comput Biol 7(3–4):409–427
69. Ji Y, Xu X, Stormo GD (2004) A graph theoretical approach for predicting common RNA secondary structure motifs including pseudoknots in unaligned sequences. Bioinformatics 20(10):1591–1602
70. Sato K, Kato Y, Hamada M, Akutsu T, Asai K (2011) IPknot: fast and accurate prediction of RNA secondary structures with pseudoknots using integer programming. Bioinformatics 27(13):85–93
71. Reidys CM, Huang FW, Andersen JE, Penner RC, Stadler PF, Nebel ME (2012) Addendum: topology and prediction of RNA pseudoknots. Bioinformatics 28(2):300
72. Mohl M, Salari R, Will S, Backofen R, Sahinalp SC (2010) Sparsification of RNA structure prediction including pseudoknots. Algorithms Mol Biol 5:39
73. Dimitrieva S, Bucher P (2012) Practicality and time complexity of a sparsified RNA folding algorithm. J Bioinform Comput Biol 10(2):1241007
74. Xu X, Ji Y, Stormo GD (2007) RNA Sampler: a new sampling based algorithm for common RNA secondary structure prediction and structural alignment. Bioinformatics 23(15):1883–1891
75. Cruz JA, Westhof E (2011) Sequence-based identification of 3D structural modules in RNA with RMDetect. Nat Methods 8(6):513–521
76. Leontis NB, Lescoute A, Westhof E (2006) The building blocks and motifs of RNA architecture. Curr Opin Struct Biol 16: 279–287
77. zu Siederdissen CH, Bernhart SH, Stadler PF, Hofacker IL (2011) A folding algorithm for extended RNA secondary structures. Bioinformatics 27(13):i129–i136

78. Frellsen J, Moltke I, Thiim M, Mardia KV, Ferkinghoff-Borg J, Hamelryck T (2009) A probabilistic model of RNA conformational space. PLoS Comput Biol 5(6):e1000406
79. Das R, Karanicolas J, Baker D (2010) Atomic accuracy in predicting and designing noncanonical RNA structure. Nat Methods 7(4):291–294
80. Cruz JA, Blanchet MF, Boniecki M, Bujnicki JM, Chen SJ, Cao S, Das R, Ding F, Dokholyan NV, Flores SC, Huang L, Lavender CA, Lisi V, Major F, Mikolajczak K, Patel DJ, Philips A, Puton T, Santalucia J, Sijenyi F, Hermann T, Rother K, Rother M, Serganov A, Skorupski M, Soltysinski T, Sripakdeevong P, Tuszynska I, Weeks KM, Waldsich C, Wildauer M, Leontis NB, Westhof E (2012) RNA-Puzzles: a CASP-like evaluation of RNA three-dimensional structure prediction. RNA 18(4):610–625
81. Rose D, Hiller M, Schutt K, Hackermuller J, Backofen R, Stadler PF (2011) Computational discovery of human coding and noncoding transcripts with conserved splice sites. Bioinformatics 27(14):1894–1900
82. Nawrocki EP, Kolbe DL, Eddy SR (2009) Infernal 1.0: inference of RNA alignments. Bioinformatics 25(10):1335–1337
83. Do CB, Woods DA, Batzoglou S (2006) CONTRAfold: RNA secondary structure prediction without physics-based models. Bioinformatics 22(14):e90–e98
84. Altschul SF, Madden TL, Schäffer AA, Zhang J, Zhang Z, Miller W, Lipman DJ (1997) Gapped blast and psi-blast: a new generation of protein database search programs. Nucleic Acids Res 25:3389–3402
85. Weinberg Z, Ruzzo WL (2006) Sequence-based heuristics for faster annotation of non-coding RNA families. Bioinformatics 22(1):35–39
86. Menzel P, Gorodkin J, Stadler PF (2009) The tedious task of finding homologous noncoding RNA genes. RNA 15(12):2075–2082
87. Lowe TM, Eddy S (1997) tRNAscan-SE: a program for improved detection of transfer RNA genes in genomic sequence. Nucleic Acids Res 25:955–964
88. Hertel J, Stadler PF (2006) Hairpins in a Haystack: recognizing microRNA precursors in comparative genomics data. Bioinformatics 22(14):197–202
89. Bernhart SH, Tafer H, Muckstein U, Flamm C, Stadler PF, Hofacker IL (2006) Partition function and base pairing probabilities of RNA heterodimers. Algorithms Mol Biol 1(1):3
90. Rehmsmeier M, Steffen P, Hochsmann M, Giegerich R (2004) Fast and effective prediction of microrna/target duplexes. RNA 10:1507–1517
91. Tafer H, Hofacker IL (2008) RNAplex: a fast tool for RNA-RNA interaction search. Bioinformatics 24(22):2657–2663
92. Busch A, Richter AS, Backofen R (2008) IntaRNA: efficient prediction of bacterial sRNA targets incorporating target site accessibility and seed regions. Bioinformatics 24(24):2849–2856
93. Chitsaz H, Salari R, Sahinalp SC, Backofen R (2009) A partition function algorithm for interacting nucleic acid strands. Bioinformatics 25(12):i365–i373
94. Seemann SE, Richter AS, Gorodkin J, Backofen R (2010) Hierarchical folding of multiple sequence alignments for the prediction of structures and RNA-RNA interactions. Algorithms Mol Biol 5:22
95. Seemann SE, Richter AS, Gesell T, Backofen R, Gorodkin J (2011) PETcofold: predicting conserved interactions and structures of two multiple alignments of RNA sequences. Bioinformatics 27(2):211–219
96. Li AX, Marz M, Qin J, Reidys CM (2011) RNA-RNA interaction prediction based on multiple sequence alignments. Bioinformatics 27(4):456–463
97. Tafer H, Amman F, Eggenhofer F, Stadler PF, Hofacker IL (2011) Fast accessibility-based prediction of RNA-RNA interactions. Bioinformatics 27(14):1934–1940
98. Poolsap U, Kato Y, Sato K, Akutsu T (2011) Using binding profiles to predict binding sites of target RNAs. J Bioinform Comput Biol 9(6):697–713
99. Lewis BP, Burge CB, Bartel DP (2005) Conserved seed pairing, often flanked by adenosines, indicates that thousands of human genes are microRNA targets. Cell 120(1):15–20
100. Enright AJ, John B, Gaul U, Tuschl T, Sander C, Marks DS (2003) MicroRNA targets in Drosophila. Genome Biol 5(1):R1
101. Krek A, Grun D, Poy MN, Wolf R, Rosenberg L, Epstein EJ, MacMenamin P, da Piedade I, Gunsalus KC, Stoffel M, Rajewsky N (2005) Combinatorial microRNA target predictions. Nat Genet 37(5):495–500
102. Kehr S, Bartschat S, Stadler PF, Tafer H (2011) PLEXY: efficient target prediction for box C/D snoRNAs. Bioinformatics 27(2):279–280
103. Tafer H, Kehr S, Hertel J, Hofacker IL, Stadler PF (2010) RNAsnoop: efficient target

prediction for H/ACA snoRNAs. Bioinformatics 26(5):610–616
104. Puton T, Kozlowski L, Tuszynska I, Rother K, Bujnicki JM (2011) Computational methods for prediction of protein-RNA interactions. J Struct Biol 179(3):261–268
105. Andersen KL, Nielsen H (2012) Experimental identification and analysis of macronuclear non-coding RNAs from the ciliate Tetrahymena thermophila. Nucleic Acids Res 40(3):1267–1281
106. Sarver M, Zirbel CL, Stombaugh J, Mokdad A, Leontis NB (2008) FR3D: finding local and composite recurrent structural motifs in RNA 3D structures. J Math Biol 56 (1–2):215–252
107. Mathews DH, Disney MD, Childs JL, Schroeder SJ, Zuker M, Turner DH (2004) Incorporating chemical modification constraints into a dynamic programming algorithm for prediction of RNA secondary structure. Proc Natl Acad Sci USA 101(19):7287–7292
108. Barash D (2003) Deleterious mutation prediction in the secondary structure of RNAs. Nucleic Acids Res 31:6578–6584
109. Sabarinathan R, Tafer H, Seemann SE, Hofacker IL, Stadler PF, Gorodkin J (2013) RNAsnp: efficient detection of local RNA secondary structure changes induced by SNPs Hum Mutat 34(4):546–556
110. Churkin A, Barash D (2006) `RNAmute`: RNA secondary structure mutation analysis tool. BMC Bioinformatics 7:221
111. Shu W, Bo X, Liu R, Zhao D, Zheng Z, Wang S (2006) RDMAS: a web server for RNA deleterious mutation analysis. BMC Bioinformatics 7:404
112. Waldispühl J, Devadas S, Berger B, Clote P (2008) Efficient algorithms for probing the RNA mutation landscape. PLoS Comput Biol 4:e1000124
113. Halvorsen M, Martin JS, Broadaway S, Laederach A (2010) Disease-associated mutations that alter the RNA structural ensemble. PLoS Genet 6:e1001074
114. Gardner PP, Giegerich R (2004) A comprehensive comparison of comparative RNA structure prediction approaches. BMC Bioinformatics 5:140
115. Gardner PP, Wilm A, Washietl S (2005) A benchmark of multiple sequence alignment programs upon structural RNAs. Nucleic Acids Res 33(8):2433–2439
116. Freyhult EK, Bollback JP, Gardner PP (2007) Exploring genomic dark matter: a critical assessment of the performance of homology search methods on noncoding RNA. Genome Res 17(1):117–125
117. Ellis JC, Brown JW (2009) The RNase P family. RNA Biol 6(4):362–369
118. Andersen ES, Rosenblad MA, Larsen N, Westergaard JC, Burks J, Wower IK, Wower J, Gorodkin J, Samuelsson T, Zwieb C (2006) The tmRDB and SRPDB resources. Nucleic Acids Res 34(Database issue):D163–D168. http://rnp.uthct.edu/rnp/SRPDB/AboutSRPDB.html
119. Brown JW, Birmingham A, Griffiths PE, Jossinet F, Kachouri-Lafond R, Knight R, Lang BF, Leontis N, Steger G, Stombaugh J, Westhof E (2009) The RNA structure alignment ontology. RNA 15(9):1623–1631
120. Widmann J, Stombaugh J, McDonald D, Chocholousova J, Gardner P, Iyer MK, Liu Z, Lozupone CA, Quinn J, Smit S, Wikman S, Zaneveld JR, Knight R (2012) RNASTAR: an RNA STructural Alignment Repository that provides insight into the evolution of natural and artificial RNAs. RNA 18(7):1319–1327
121. Hajiaghayi M, Condon A, Hoos HH (2012) Analysis of energy-based algorithms for RNA secondary structure prediction. BMC Bioinformatics 13:22
122. Matthews BW (1975) Comparison of the predicted and observed secondary structure of T4 phage lysozyme. Biochem Biophys Acta 405:442–451
123. Gorodkin J, Stricklin SL, Stormo GD (2001) Discovering common stem-loop motifs in unaligned RNA sequences. Nucleic Acids Res 29:2135–2144
124. Hamada M, Sato K, Asai K (2010) Prediction of RNA secondary structure by maximizing pseudo-expected accuracy. BMC Bioinformatics 11:586
125. Kato Y, Akutsu T, Seki H (2009) A grammatical approach to RNA-RNA interaction prediction. Pattern Recognit 42:531–538
126. Seemann SE, Gorodkin J, Backofen R (2008) Unifying evolutionary and thermodynamic information for RNA folding of multiple alignments. Nucleic Acids Res 36(20):6355–6362
127. Gorodkin J (2004) Comparing two K-category assignments by a K-category correlation coefficient. Comput Biol Chem 28 (5–6):367–374
128. Washietl S, Hofacker IL, Lukasser M, Hüttenhofer A, Stadler PF (2005) Genome-wide mapping of conserved RNA secondary structure structures predicts thousands of functional non-coding RNAs in human. Nat Biotechnol 23:1383–1390
129. Uzilov AV, Keegan JM, Mathews DH (2006) Detection of non-coding RNAs on the basis of predicted secondary structure formation free energy change. BMC Bioinformatics 7:173

130. Bernhart SH, Hofacker IL, Will S, Gruber AR, Stadler PF (2008) RNAalifold: improved consensus structure prediction for RNA alignments. BMC Bioinformatics 9:474

131. Kuhn M, Szklarczyk D, Franceschini A, von Mering C, Jensen LJ, Bork P (2012) STITCH 3: zooming in on protein-chemical interactions. Nucleic Acids Res 40(Database issue):D876–D880

Chapter 2

The Principles of RNA Structure Architecture

Christian Zwieb

Abstract

Being informational, enzymatic, as well as a nanoscale molecular machine, ribonucleic acid (RNA) permeates all areas of biology and has been exploited in biotechnology as drug and sensor. Here we describe the composition and fundamental properties of RNA and how the single-stranded RNA chains fold and shape certain motifs that are repeatedly observed in different structures. Small and large molecular mass RNA binders are being touched upon, as is the technology for selecting RNA molecules in vitro that bind almost any kind of natural or artificial target. Recognizing the versatility of RNA is expected to foster the development of tools which monitor RNA in the environment, including plants, animals, and patients. Many of the noncoding RNAs are yet to be identified in the rapidly emerging genomes and assigned to functions. It is hoped that these and similar worthwhile goals will be achieved by integrating the efforts of bench and computer scientists.

Key words Ribonucleic acid, Structure, RNA, RNA structure, Base pair, Genome, Motif, Antibiotics, Aptamer, Ribonucleoprotein

1 Introduction

Within the recent years a remarkable expansion has taken place in our understanding of the significance of the role of RNA in all aspects of biology. No longer are RNA molecules only transient intermediates which carry the DNA-encoded information from the nucleus into the cytosol, but as ribozymes they also catalyze biochemical reactions [3, 4]. The discovery of this twofold capacity of RNA to be informational and enzymatic made it possible to select in vitro a wide variety of artificial ribozymes with divergent substrate specificities. Furthermore, it fed the idea of an early "RNA world" where RNA perhaps represented the first primitive form of life [5]. A third attribute of RNA is linked to its ability to function as a nanoscale molecular machine, the ribosome being the prime example. Molecular nanodevices with RNA at their core include, among others, the signal recognition particle (SRP) which binds to ribosomes and directs nascent polypeptides to the

Jan Gorodkin and Walter L. Ruzzo (eds.), *RNA Sequence, Structure, and Function: Computational and Bioinformatic Methods*, Methods in Molecular Biology, vol. 1097, DOI 10.1007/978-1-62703-709-9_2,

cell membrane [6], the transfer-messenger RNP (tmRNP) which rescues immobilized bacterial ribosomes [7], and the spliceosome which removes introns from other RNAs [8]. In the perception of some researchers who work in the areas of biology and biotechnology, there is no end in sight of what RNA might be capable to achieve.

Only about 1.5% of the human genome codes for proteins [9], while the remainder, long being ignored as "junk," has the potential to be transcribed into a bewildering variety of noncoding (nc) RNAs. More than 80% of human disease-associated loci associate with non-protein-coding regions highlighting the physiological importance of the RNA [10]. Technological advances in the analysis of genomes continuously generate formidable amounts of data which require filtering and curation in order to become intelligible [11–13]. Computational tools for identifying ncRNA genes and their processed products are being actively developed and aim to bring order to the massive amount of sequence data. For example, the Rfam database (in its version 10) provides alignments for 1,446 RNA groups or "families" [14], and the catalog of functional RNAs at fRNAdb [15] distinguishes 116 RNAs by name. Nevertheless, with respect to structure and function, RNA molecules can be intrinsically more dynamic than what the databases confine themselves to, leaving much work to be done before we better understand the multifaceted roles of RNA.

This introductory chapter attempts to give a simple account of what RNA is made of. It introduces the names and nomenclature commonly used to describe RNA molecules and provides a small sample of the complexity of this fascinating biopolymer. It is hoped that this effort will encourage investigators to advance RNA research and develop urgently needed computational and biotechnological tools aimed to decipher RNA's sophisticated and dynamic biochemistry as reflected in its structure.

2 RNA Fundamentals

RNA (ribonucleic acid) is a linear polymer of ribonucleosides arranged in sequence referred to as its primary structure. The average size transfer RNAs (tRNAs) contains 73–93 such residues [16], whereas the prototypical bacterial 16S ribosomal RNA (rRNA) of the *Escherichia coli* small ribosomal subunit is composed of 1,542 ribonucleotides [17]. At their size extremes are the 22-residues microRNAs [18] and, for example, the 17 kb Xist RNA [19].

Unless modified, a ribonucleoside is a ribose sugar covalently bound to one of four different nitrogenous bases (Fig. 1). Adenine and guanine (abbreviated as A and G) are derived from a hete-

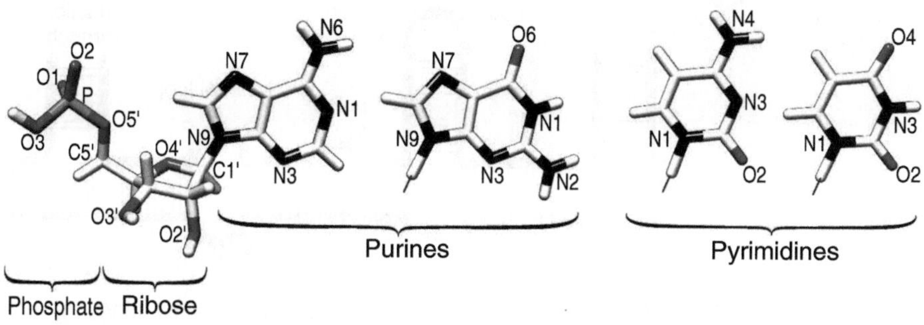

Fig. 1 The prominent building blocks of RNA

rocyclic aromatic purine (abbreviated as R), whereas cytosine (C) and uridine (U) are pyrimidines (abbreviated as Y). The N9 of the purine or the N1 of the pyrimidine base connects via the stereochemically important glycosidic bond with the carbon (C1') of the ribose. A ribonucleotide forms by covalent attachment of one, two, or three phosphates to the ribose 5' carbon. 5'-nucleotide triphosphates (NTPs, where N stands for any of the four bases) are used within the cell to synthesize RNA in the 5'–3' direction. This process releases inorganic pyrophosphate with one phosphate left at each step to join the 3' and 5' positions of neighboring ribose rings. By convention, the sequence of an RNA molecule is written from 5' to 3', left to right starting with the numbering at the 5'-end, typically in groups of three or ten characters.

RNA can be modified after its transcription to include nucleoside methylations or less common bases such as inosine (abbreviated as I), dihydrouridine (D), or pseudouracil (ψ). When residues are removed or inserted by splicing, RNA editing, or other RNA processing steps, the primary structure of the functional RNA molecule is significantly different from the sequence of its gene.

Unlike in DNA, the 2' hydroxyl group of the ribose is capable of forming cyclic phosphate intermediates allowing RNA to be readily hydrolyzed and cleaved by intra- and extracellular RNases. Being able to turn over and reduce its active intracellular pool size allows RNA to regulate vital processes. The 2' hydroxyl group also contributes to the capacity of the RNA to interact in versatile ways with itself or a variety of ligands.

3 The Folded RNA

Although RNA is in general single stranded, the bases have a strong propensity to interact in two principal ways, either perpendicular to their planes (stacking) or hydrogen bonded within the base planes

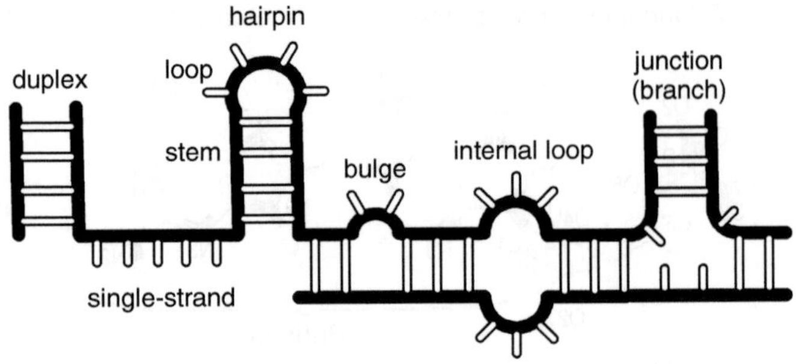

Fig. 2 RNA secondary structure features. Naming conventions for regions in the RNA secondary structure

(pairing). Paired residues are indicated by connecting lines in the RNA secondary structure diagrams composed of stems, bulges, and loops (Fig. 2) characteristic for a particular RNA molecule. Less consistently, stacking may be shown in the diagrams by placing letters closer to each other.

In their elucidation of the structure of the DNA double helix, Watson and Crick proposed two planar purine–pyrimidine base pairs, A–T and G–C, where the bases of each pair are held together by two or three specifically arranged hydrogen bonds [20]. The corresponding pairs in RNA are A–U and G–C, but the non-Watson–Crick G–U wobble pair has approximately the same stability as an A–U [21] and is commonly observed in all the medium-size and larger RNA molecules. The two ribose groups attach to the same side of a base pair and define two types of indentations: a major groove, delineated by N7 of the purine and the C6 of the pyrimidine, and a minor groove with purine N3 and pyrimidine O2 (Fig. 3). The three most frequently used RNA base pairs (A–U, G–C, and G–U) share almost identical overall dimensions and, when placed next to each other, stack continuously to form a rigid A-type helix with 11 bp/turn, a deep narrow major groove, and a relatively shallow minor groove (Fig. 4a, b).

Computational calculations which determine the secondary structure (the base-paired helical regions) of an RNA molecule using energy calculations are readily available but can be unreliable [22, 23]. Considering that biologically active RNAs often bind proteins and other factors that are excluded from the calculation, it is unclear why the structure with the overall low free energy should exist preferentially. A more accurate method observes covariances and compensating base changes (e.g., changes from a G–C to a C–G or A–U pair) in a group of phylogenetically related aligned sequences [24]. A sufficiently large number of compensations strongly support the existence of a base pair because, during evolution, random mutations would not have been corrected to

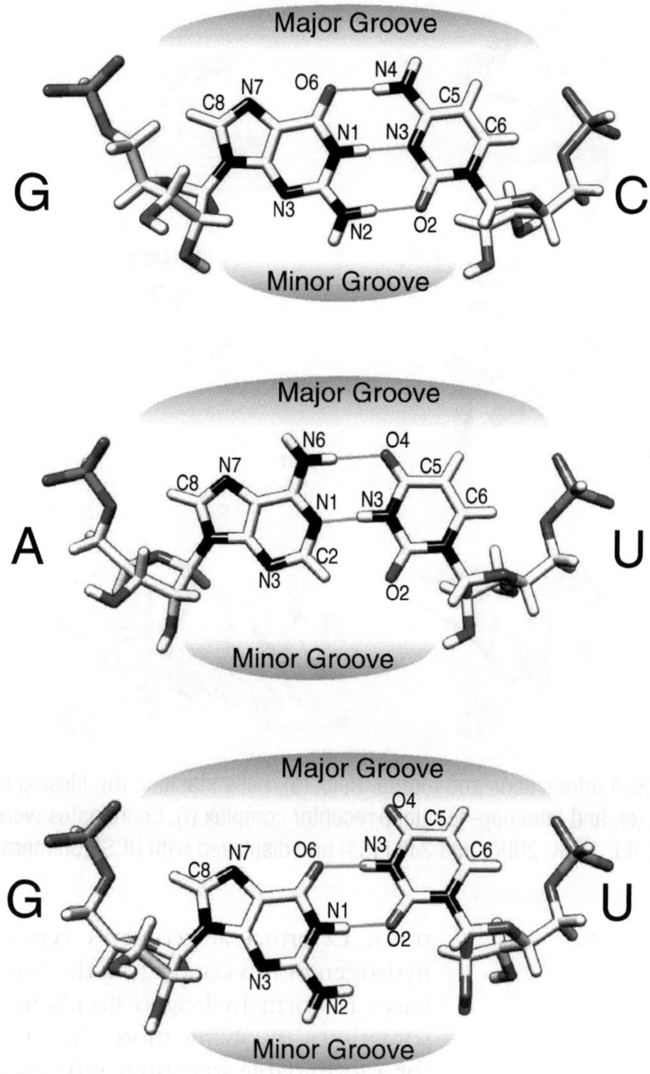

Fig. 3 Watson–Crick and G–U wobble pair geometries

maintain the base pair unless it was required [25]. Semiautomated procedures and tools have been developed to help identify the compensatory base changes and calculate the level of support for each base pair [26].

Canonical A–U and G–C pairs and the G–U wobble pair engage the Watson–Crick edges of the bases to form two or three hydrogen bonds. Planar base interactions can however also occur through the Hoogsteen edge (defined by the purine positions 6, 7, and 8 or the pyrimidine positions 4 and 5) or the sugar edge formed by the 2' hydroxyl group of the ribose with purine positions 2 and 3 or with the pyrimidine oxygen atom at position 2 (Fig. 3). Given that the glycosidic bonds can be oriented either in cis or

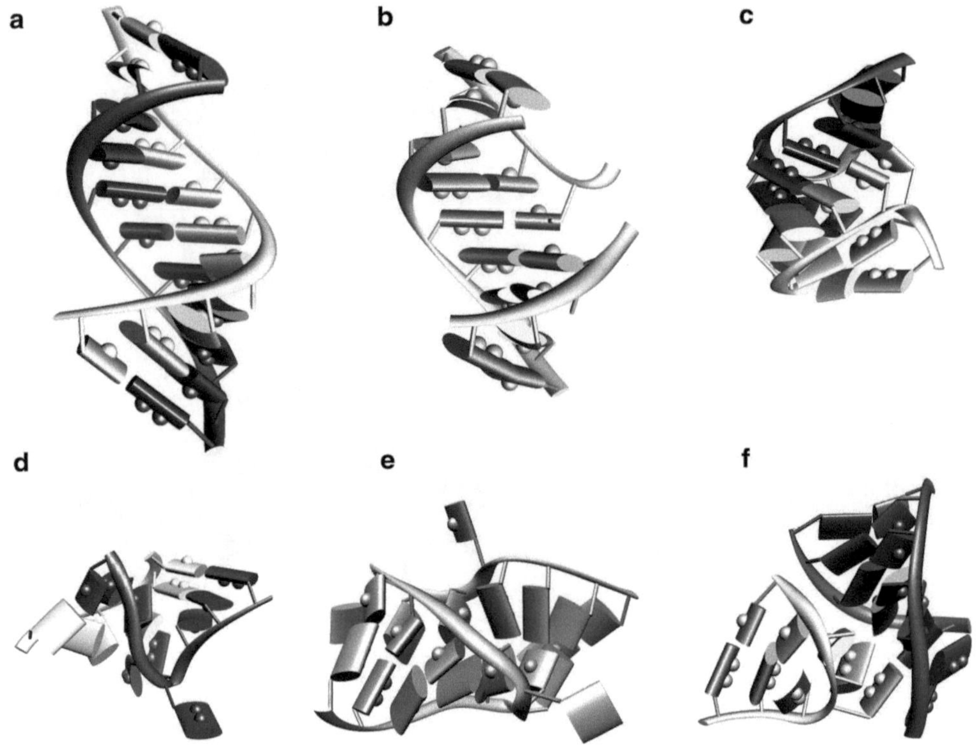

Fig. 4 RNA interactions and motifs. Helix (**a**), helix stacking (**b**), kissing hairpin loops (**c**), kink-turn (**d**), pseudoknot (**e**), and tetraloop–tetraloop receptor complex (**f**). Coordinates were chosen from RCSB PDB files 1KFO, 1EVV, 2JLT, 1FFK, 2RPO, and 2JYJ [54] and displayed with UCSF Chimera [55]

trans, 12 principal geometric types are possible with at least two hydrogen bonds connecting the bases [27]. This ability of the RNA bases to form hydrogen bonds in a multitude of combinations, sometimes involving more than two bases, is largely responsible for a formidable structural and functional variability.

Verification of a predicted RNA secondary structure by carrying out chemical and enzymatic modification experiments is often useful but provides clues only for surface-exposed sites. Watson–Crick and wobble base pairs, even when hidden inside the folded RNA, can be verified by testing the biological activities of molecules with compensatory double mutations which regenerate these pairs. Experimentally derived data guide in the building of three-dimensional models to gain insight into the structure and function of an RNA, an exercise that is often the only option when the high-resolution structure is unavailable [28, 29].

Tertiary interactions in RNA are generally considered to be those which occur between separate regions of the secondary structure. The various ways by which bases stack and form hydrogen bonds provide ample opportunities for secondary structure elements to come together. The first RNA structure determined at atomic resolution was that of yeast $tRNA^{Phe}$ and key to understand some of the architectural principles of an RNA molecule [30, 31]. One such fundamental rule is the ability to preserve its overall three-dimensional shape despite differences in sequence. More recent examples of how unrelated sequences fold unexpectedly the same way are the GUUA and UNCG tetranucleotide (tetra) loops [32]. Another principle is the capacity of RNA strands to adopt a compact and relatively RNase-resistant conformation (*see*, e.g., 1EHZ.pdb [33]).

Subsequent milestones included the deciphering of the molecular structures of the hammerhead ribozyme [34, 35], the P4–P6 domain of the group I intron RNA [36], and the hepatitis delta ribozyme [37]. Solving the structures of ribosomes and its subunits, composed predominantly of RNA, was an astonishing accomplishment [38–40]. Like proteins, RNAs were shown to be able to adopt complicated and yet precise structures. Examples of base triples, adenosine platforms, and ribose zippers were discovered, as were interactions between GNAR tetraloops and their receptors. Drawn in the secondary structures as nondescriptive internal loops, these features were shown to form distinct kinks and helix distortions some suited to bind proteins.

4 RNA Motifs

As more and more structural information became available, the same folding principles were frequently observed in the known RNA structures. It is now clear that, like lego blocks, these structurally strictly defined entities or motifs are reused in combination to generate a rich variety of molecular shapes. As motif examples, the kissing hairpin loops, a kink- or K-turn, a pseudoknot, and the complex between a tetraloop and its receptor are depicted in Fig. 4.

The naming conventions for the different RNA motifs are vague, but classifications according to their structure, functions, and tertiary interaction have been initiated [41, 42] and provide a useful selection of building blocks for constructing "from scratch" biologically meaningful three-dimensional models [28]. Locating RNA motifs in the genomic sequences is possible using computationally intensive pattern matching programs [43]. Progress in this area is desirable in order to identify RNA genes and annotate the genomes.

5 RNA Ligands

Small Molecules. Water and metal or other ions contribute significantly to the folding and conformation of the predominantly negatively charged RNA. Magnesium ions, together with spermidine, have been known to bind in the major groove of $tRNA^{Phe}$ [44]. The relative contributions of these site-bound versus the delocalized ions to RNA folding and stability remain to be explored [45].

RNA-Targeting Drugs. Aminoglycosides, paromomycin, and spectinomycin are among the antibiotics which impede translation by binding to ribosomal RNA [46]. Given the essential role of RNA in biology and the continuing emergence of new RNA families, there appears to be a fertile ground for the discovery of new compounds that inhibit the functions of certain vital RNA molecules. The non-ribosomal ribonucleoproteins (RNPs), such as signal recognition particles and the tmRNPs, are now sufficiently characterized to explore them as promising targets for the development of new antibiotic compounds.

Proteins. We now live in an RNA plus protein world where DNA is being tasked mainly with storing information within the genome. Considering the induced fit-type structural changes commonly observed upon the formation of protein–RNA complexes, proteins and RNA must have experienced a long coevolutionary history. Indeed, the majority of RNAs function with full capacity only when assembled into RNPs. A protein may bind to an RNA molecule or an RNP only temporarily to regulate a biological activity. Being aware of the close structural and functional partnership between RNA and protein is expected to contribute significantly to the genome annotation efforts and an understanding of RNP phylogeny [47].

RNA Aptamers. SELEX (systematic evolution of ligands by exponential enrichment) has been widely used to select in vitro RNA molecules which bind a wide variety of targets, including proteins and antibiotics, with high affinity and specificity [48]. Compared to antibodies, the aptamer approach has the advantage of circumventing the costly use of animals or cultured cells. RNase-resistant aptamers with 2’ modification can be chemically synthesized on a large scale and applied as drugs. Several such aptamers for use as therapeutic molecules are currently in development [49].

6 Prospects

The field of RNA research is rapidly expanding into areas beyond biotechnology, RNA-based therapeutics, or the development of new antibiotics. Crop yields, and ultimately our capacity to respond to challenges in the face of climate change and increasing global populations, are determined by plant noncoding RNAs that control growth, development, and breeding capability [50]. The ribosomal RNAs are now being joined by a host of other noncoding RNA sequences in efforts to better understand phylogeny and the changing diversity and ecology of life. The identification of microRNAs (miRNAs) and Piwi-interacting RNAs (piRNA) [51] in the genomes presents itself as a formidable experimental and computational challenge. The CRISPR RNAs protect bacteria and archaea from phage infections, thereby influencing population dynamics with important ecological consequences [52, 53]. It is almost certain that additional RNAs with unimagined properties and functions will be discovered in the near future. We will be ready to explore.

References

1. Caspersson T, Schultz J (1939) Pentose nucleotides in the cytoplasm of growing tissues. Nature 143:602–603
2. Ochoa S (1959) Enzymatic synthesis of ribonucleic acid. Nobel Lecture
3. Kruger K, Grabowski PJ, Zaug AJ, Sands J, Gottschling DE, Cech TR (1982) Self-splicing RNA: autoexcision and autocyclization of the ribosomal RNA intervening sequence of *Tetrahymena*. Cell 31:147–157
4. Stark BC, Kole R, Bowman EJ, Altman S (1978) Ribonuclease P: an enzyme with an essential RNA component. Proc Natl Acad Sci U S A 75:3717–3721
5. Woese CR (1967) The genetic code: the molecular basis for genetic expression. Harper & Row, New York
6. Walter P, Blobel G (1982) Signal recognition particle contains a 7S RNA essential for protein translocation across the endoplasmic reticulum. Nature 299:691–698
7. Keiler KC, Waller PR, Sauer RT (1996) Role of a peptide tagging system in degradation of proteins synthesized from damaged messenger RNA. Science 271:990–993
8. Lerner MR, Boyle JA, Mount SM, Wolin SL, Steitz JA (1980) Are snRNPs involved in splicing? Nature 283:220–224
9. LanderES, Linton LM, Birren B, Nusbaum C, Zody MC, Baldwin J, Devon K, Dewar K, Doyle M, FitzHugh W, Funke R, Gage D, Harris K, Heaford A, Howland J, Kann L, Lehoczky J, LeVine R, McEwan P, McKernan K, Meldrim J, Mesirov JP, Miranda C, Morris W, Naylor J, Raymond C, Rosetti M, Santos R, Sheridan A, Sougnez C, Stange-Thomann N, Stojanovic N, Subramanian A, Wyman D, Rogers J, Sulston J, Ainscough R, Beck S, Bentley D, Burton J, Clee C, Carter N, Coulson A, Deadman R, Deloukas P, Dunham A, Dunham I, Durbin R, French L, Grafham D, Gregory S, Hubbard T, Humphray S, Hunt A, Jones M, Lloyd C, McMurray A, Matthews L, Mercer S, Milne S, Mullikin JC, Mungall A, Plumb R, Ross M, Shownkeen R, Sims S, Waterston RH, Wilson RK, Hillier LW, McPherson JD, Marra MA, Mardis ER, Fulton LA, Chinwalla AT, Pepin KH, Gish WR, Chissoe SL, Wendl MC, Delehaunty KD, Miner TL, Delehaunty A, Kramer JB, Cook LL, Fulton RS, Johnson DL, Minx PJ, Clifton SW, Hawkins T, Branscomb E, Predki P, Richardson P, Wenning S, Slezak T, Doggett N, Cheng JF, Olsen A, Lucas S, Elkin C, Uberbacher E, Frazier M, Gibbs RA, Muzny DM, Scherer SE, Bouck JB, Sodergren EJ, Worley KC, Rives CM, Gorrell JH, Metzker ML, Naylor SL, Kucherlapati RS, Nelson DL, Weinstock GM, Sakaki Y, Fujiyama A, Hattori M, Yada T, Toyoda A, Itoh T, Kawagoe C, Watanabe H, Totoki Y, Taylor T, Weissenbach J, Heilig R, Saurin W, Artiguenave F, Brottier P, Bruls T, Pelletier E, Robert C, Wincker P, Smith DR, Doucette-Stamm L, Rubenfield M, Weinstock K, Lee HM, Dubois J, Rosenthal A, Platzer M, Nyakatura G, Taudien S, Rump A, Yang H, Yu J, Wang J, Huang G, Gu J, Hood L, Rowen L, Madan A, Qin S, Davis RW, Federspiel NA, Abola AP, Proctor MJ, Myers RM, Schmutz J, Dickson M, Grimwood J, Cox DR, Olson MV, Kaul R, Raymond C, Shimizu N, Kawasaki K, Minoshima S, Evans GA, Athanasiou M, Schultz R, Roe BA, Chen F, Pan H, Ramser J, Lehrach H, Reinhardt R, McCombie WR, de la Bastide M, Dedhia N, Blocker H, Hornischer K, Nordsiek G, Agarwala R, Aravind L, Bailey JA, Bateman A, Batzoglou S, Birney E, Bork P, Brown DG, Burge CB, Cerutti L, Chen HC, Church D, Clamp M, Copley RR, Doerks T, Eddy SR, Eichler EE, Furey TS, Galagan J, Gilbert JG, Harmon C, Hayashizaki Y, Haussler D, Hermjakob H, Hokamp K, Jang W, Johnson LS, Jones TA, Kasif S, Kaspryzk A, Kennedy S, Kent WJ, Kitts P, Koonin EV, Korf I, Kulp D, Lancet D, Lowe TM, McLysaght A, Mikkelsen T, Moran JV, Mulder N, Pollara VJ, Ponting CP, Schuler G, Schultz J, Slater G, Smit AF, Stupka E, Szustakowski J, Thierry-Mieg D, Thierry-Mieg J, Wagner L, Wallis J, Wheeler R,

Williams A, Wolf YI, Wolfe KH, Yang SP, Yeh RF, Collins F, Guyer MS, Peterson J, Felsenfeld A, Wetterstrand KA, Patrinos A, Morgan MJ, de Jong P, Catanese JJ, Osoegawa K, Shizuya H, Choi S, Chen YJ (2001) Initial sequencing and analysis of the human genome. Nature 409:860–921

10. Manolio TA, Collins FS, Cox NJ, Goldstein DB, Hindorff LA, Hunter DJ, McCarthy MI, Ramos EM, Cardon LR, Chakravarti A, Cho JH, Guttmacher AE, Kong A, Kruglyak L, Mardis E, Rotimi CN, Slatkin M, Valle D, Whittemore AS, Boehnke M, Clark AG, Eichler EE, Gibson G, Haines JL, Mackay TF, McCarroll SA, Visscher PM (2009) Finding the missing heritability of complex diseases. Nature 461:747–753
11. Mortazavi A, Williams BA, McCue K, Schaeffer L, Wold B (2008) Mapping and quantifying mammalian transcriptomes by RNA-Seq. Nat Methods 5:621–628
12. Maeda N, Kasukawa T, Oyama R, Gough J, Frith M, Engstrom PG, Lenhard B, Aturaliya RN, Batalov S, Beisel KW, Bult CJ, Fletcher CF, Forrest AR, Furuno M, Hill D, Itoh M, Kanamori-Katayama M, Katayama S, Katoh M, Kawashima T, Quackenbush J, Ravasi T, Ring BZ, Shibata K, Sugiura K, Takenaka Y, Teasdale RD, Wells CA, Zhu Y, Kai C, Kawai J, Hume DA, Carninci P, Hayashizaki Y (2006) Transcript annotation in FANTOM3: mouse gene catalog based on physical cDNAs. PLoS Genet 2:e62
13. Kapranov P, Drenkow J, Cheng J, Long J, Helt G, Dike S, Gingeras TR (2005) Examples of the complex architecture of the human transcriptome revealed by RACE and high-density tiling arrays. Genome Res 15: 987–997
14. Gardner PP, Daub J, Tate JG, Nawrocki EP, Kolbe DL, Lindgreen S, Wilkinson AC, Finn RD, Griffiths-Jones S, Eddy SR, Bateman A (2009) Rfam: updates to the RNA families database. Nucleic Acids Res 37:D136–D140
15. Mituyama T, Yamada K, Hattori E, Okida H, Ono Y, Terai G, Yoshizawa A, Komori T, Asai K (2009) The functional RNA database 3.0: databases to support mining and annotation of functional RNAs. Nucleic Acids Res 37:D89–D92
16. Clark BF (2006) The crystal structure of tRNA. J Biosci 31:453–457
17. Brosius J, Palmer ML, Kennedy PJ, Noller HF (1978) Complete nucleotide sequence of a 16S ribosomal RNA gene from *Escherichia coli.* Proc Natl Acad Sci U S A 75:4801–4805
18. Lund E, Dahlberg JE (2006) Substrate selectivity of exportin 5 and Dicer in the biogenesis of microRNAs. Cold Spring Harb Symp Quant Biol 71:59–66
19. Brown CJ, Hendrich BD, Rupert JL, Lafreniere RG, Xing Y, Lawrence J, Willard HF (1992) The human XIST gene: analysis of a 17 kb inactive X-specific RNA that contains conserved repeats and is highly localized within the nucleus. Cell 71:527–542
20. Watson JD, Crick FH (1953) A structure for deoxyribose nucleic acid. Nature 171:737–738
21. Giese MR, Betschart K, Dale T, Riley CK, Rowan C, Sprouse KJ, Serra MJ (1998) Stability of RNA hairpins closed by wobble base pairs. Biochemistry 37:1094–1100
22. Zuker M (2003) Mfold web server for nucleic acid folding and hybridization prediction. Nucleic Acids Res 31:3406–3415
23. Turner DH, Sugimoto N, Freier SM (1988) RNA structure prediction. Annu Rev Biophys Biophys Chem 17:167–192
24. Fox GE, Woese C (1975) 5S RNA secondary structure. Nature 256:505–507
25. Larsen N, Zwieb C (1991) SRP-RNA sequence alignment and secondary structure. Nucleic Acids Res 19:209–215
26. Andersen ES, Lind-Thomsen A, Knudsen B, Kristensen SE, Havgaard JH, Torarinsson E, Larsen N, Zwieb C, Sestoft P, Kjems J, Gorodkin J (2007) Semiautomated improvement of RNA alignments. RNA 13:1850–1859
27. Leontis NB, Westhof E (2001) Geometric nomenclature and classification of RNA base pairs. RNA 7:499–512
28. Mueller F, Stark H, van Heel M, Rinke-Appel J, Brimacombe R (1997) A new model for the three-dimensional folding of *Escherichia coli* 16 S ribosomal RNA. III. The topography of the functional centre. J Mol Biol 271: 566–587
29. Lavender CA, Ding F, Dokholyan NV, Weeks KM (2010) Robust and generic RNA modeling using inferred constraints: a structure for the hepatitis C virus IRES pseudoknot domain. Biochemistry 49:4931–4933
30. Jovine L, Djordjevic S, Rhodes D (2000) The crystal structure of yeast phenylalanine tRNA at 2.0 A resolution: cleavage by Mg(2+) in 15-year old crystals. J Mol Biol 301:401–414
31. Shi H, Moore PB (2000) The crystal structure of yeast phenylalanine tRNA at 1.93 A resolution: a classic structure revisited. RNA 6:1091–1105
32. Ihle Y, Ohlenschlager O, Hafner S, Duchardt E, Zacharias M, Seitz S, Zell R, Ramachandran R, Gorlach M (2005) A novel cGUUAg tetraloop structure with a conserved yYNMGg-type backbone conformation from cloverleaf 1 of bovine enterovirus 1 RNA. Nucleic Acids Res 33:2003–2011
33. Dutta S, Burkhardt K, Young J, Swaminathan GJ, Matsuura T, Henrick K, Nakamura H, Berman HM (2009) Data deposition and

annotation at the worldwide protein data bank. Mol Biotechnol 42:1–13
34. Pley HW, Flaherty KM, McKay DB (1994) Three-dimensional structure of a hammerhead ribozyme. Nature 372:68–74
35. Scott WG, Finch JT, Klug A (1995) The crystal structure of an all-RNA hammerhead ribozyme: a proposed mechanism for RNA catalytic cleavage. Cell 81:991–1002
36. Cate JH, Gooding AR, Podell E, Zhou K, Golden BL, Kundrot CE, Cech TR, Doudna JA (1996) Crystal structure of a group I ribozyme domain: principles of RNA packing. Science 273:1678–1685
37. Ferre-D'Amare AR, Zhou K, Doudna JA (1998) Crystal structure of a hepatitis delta virus ribozyme. Nature 395:567–574
38. Ban N, Nissen P, Hansen J, Moore PB, Steitz TA (2000) The complete atomic structure of the large ribosomal subunit at 2.4 A resolution. Science 289:905–920
39. Wimberly BT, Brodersen DE, Clemons WM Jr, Morgan-Warren RJ, Carter AP, Vonrhein C, Hartsch T, Ramakrishnan V (2000) Structure of the 30S ribosomal subunit. Nature 407:327–339
40. Yusupov MM, Yusupova GZ, Baucom A, Lieberman K, Earnest TN, Cate JH, Noller HF (2001) Crystal structure of the ribosome at 5.5 A resolution. Science 292:883–896
41. Klosterman PS, Tamura M, Holbrook SR, Brenner SE (2002) SCOR: a structural classification of RNA database. Nucleic Acids Res 30:392–394
42. Hendrix DK, Brenner SE, Holbrook SR (2005) RNA structural motifs: building blocks of a modular biomolecule. Q Rev Biophys 38:221–243
43. Disz T, Akhter S, Cuevas D, Olson R, Overbeek R, Vonstein V, Stevens R, Edwards RA (2010) Accessing the SEED genome databases via web services API: tools for programmers. BMC Bioinformatics 11:319
44. Quigley GJ, Teeter MM, Rich A (1978) Structural analysis of spermine and magnesium ion binding to yeast phenylalanine transfer RNA. Proc Natl Acad Sci U S A 75:64–68
45. Conn GL, Draper DE (1998) RNA structure. Curr Opin Struct Biol 8:278–285
46. Yonath A (2005) Antibiotics targeting ribosomes: resistance, selectivity, synergism and cellular regulation. Annu Rev Biochem 74:649–679
47. Andersen ES, Rosenblad MA, Larsen N, Westergaard JC, Burks J, Wower IK, Wower J, Gorodkin J, Samuelsson T, Zwieb C (2006) The tmRDB and SRPDB resources. Nucleic Acids Res 34:D163–D168
48. Lorsch JR, Szostak JW (1994) In vitro selection of RNA aptamers specific for cyanocobalamin. Biochemistry 33:973–982
49. Keefe AD, Pai S, Ellington A (2010) Aptamers as therapeutics. Nat Rev Drug Discov 9:537–550
50. Auer C, Frederick R (2009) Crop improvement using small RNAs: applications and predictive ecological risk assessments. Trends Biotechnol 27:644–651
51. Girard A, Sachidanandam R, Hannon GJ, Carmell MA (2006) A germline-specific class of small RNAs binds mammalian Piwi proteins. Nature 442:199–202
52. Jansen R, van Embden JD, Gaastra W, Schouls LM (2002) Identification of a novel family of sequence repeats among prokaryotes. OMICS 6:23–33
53. Tang TH, Bachellerie JP, Rozhdestvensky T, Bortolin ML, Huber H, Drungowski M, Elge T, Brosius J, Huttenhofer A (2002) Identification of 86 candidates for small non-messenger RNAs from the archaeon Archaeoglobus fulgidus. Proc Natl Acad Sci U S A 99:7536–7541
54. Berman HM, Westbrook J, Feng Z, Gilliland G, Bhat TN, Weissig H, Shindyalov IN, Bourne PE (2000) The protein data bank. Nucleic Acids Res 28:235–242
55. Pettersen EF, Goddard TD, Huang CC, Couch GS, Greenblatt DM, Meng EC, Ferrin TE (2004) UCSF Chimera—a visualization system for exploratory research and analysis. J Comput Chem 25:1605–1612

Chapter 3

The Determination of RNA Folding Nearest Neighbor Parameters

Mirela Andronescu, Anne Condon, Douglas H. Turner, and David H. Mathews

Abstract

The stability of RNA secondary structure can be predicted using a set of nearest neighbor parameters. These parameters are widely used by algorithms that predict secondary structure. This contribution introduces the UV optical melting experiments that are used to determine the folding stability of short RNA strands. It explains how the nearest neighbor parameters are chosen and how the values are fit to the data. A sample nearest neighbor calculation is provided. The contribution concludes with new methods that use the database of sequences with known structures to determine parameter values.

Key words RNA secondary structure, Optical melting experiment, Gibbs free energy, Nearest neighbor parameters

1 Introduction

This chapter introduces the principles for using nearest neighbor rules to predict RNA conformational stability. It demonstrates how optical melting experiments serve as a foundation for developing an empirical model, called a nearest neighbor model, for predicting the stability of RNA structure. Then, the approach for fitting the model to the experiments is described. Finally, the chapter ends by discussing new frontiers for improving the accuracy of the parameters by using a database of sequences with known structures to refine the values of parameters.

The nearest neighbor rules provide the basis for a number of the algorithms and methods described in other chapters. The nearest neighbor rules are a set of equations and associated parameters that predict the conformational stability for a specific RNA sequence folding into a specific structure. The rules are called nearest neighbor because they employ two assumptions. The first is that the stability of a motif (base pair or loop) depends only on the

Jan Gorodkin and Walter L. Ruzzo (eds.), *RNA Sequence, Structure, and Function: Computational and Bioinformatic Methods*, Methods in Molecular Biology, vol. 1097, DOI 10.1007/978-1-62703-709-9_3,

sequence of that motif and the sequence of the directly adjacent base pairs. The second assumption is that the total stability is the sum of stabilities predicted for each motif.

1.1 Nearest Neighbor Parameters Quantify Stability

Conformational stability is quantified for RNA as a standard state Gibbs free energy change, ΔG°. For an RNA sequence at equilibrium and unimolecular folding, the concentration of folded strands, $[F]$, and the concentration of unfolded strands, $[U]$, are related by an equilibrium constant, K:

$$K = \frac{[F]}{[U]} \tag{1}$$

By definition, the equilibrium constant is related to the ΔG° by

$$K = \mathrm{e}^{-\Delta G^\circ / RT} \tag{2}$$

where R is the gas constant, 1.987 cal/mol/K, and T is the absolute temperature, i.e., the temperature in Kelvins. Note that, by convention, nearest neighbor rules are expressed using the customary unit of calorie, where 1 cal = 4.19 J. The lower the ΔG°, i.e., the larger the magnitude below zero, the more stable the structure as compared to the unfolded state, which is often called a random coil state. The ΔG° is a function of temperature and can be decomposed into two components:

$$\Delta G^\circ = \Delta H^\circ - T\,\Delta S^\circ \tag{3}$$

where ΔH° is the enthalpy change and ΔS° in the entropy change of duplex formation. The enthalpy change is a heat yield, and the entropy change is a measure of the loss of randomness when forming the structure from the unfolded strands. For formation of RNA secondary structure, ΔH° is favorable and ΔS° is unfavorable, i.e., they are both less than zero. Because ΔG° is temperature dependent, it quantifies the stability of the structure at a specific temperature.

When comparing two structures for the same sequence, the ΔG°s determine the relative concentrations of each because the free energy change for each is relative to the same unfolded reference structure. For two structures, 1 and 2,

$$K_1 = \mathrm{e}^{-\Delta G^\circ{}_1 / RT} \text{ and } K_2 = \mathrm{e}^{-\Delta G^\circ{}_2 / RT} \tag{4}$$

Therefore,

$$\frac{K_1}{K_2} = \frac{\mathrm{e}^{-\Delta G^\circ{}_1 / RT}}{\mathrm{e}^{-\Delta G^\circ{}_2 / RT}} = \mathrm{e}^{(\Delta G^\circ{}_2 - \Delta G^\circ{}_1)/RT} \tag{5}$$

and using Eq. 1,

$$\frac{K_1}{K_2} = \frac{[F_1]/[U]}{[F_2]/[U]} = \frac{[F_1]}{[F_2]} \quad (6)$$

meaning that

$$\frac{[F_1]}{[F_2]} = e^{(\Delta G^\circ{}_2 - \Delta G^\circ{}_1)/RT} \quad (7)$$

So, the conformation with lowest folding free energy change is the conformation of highest concentration. This principle forms the basis of predicting RNA secondary structure by finding the lowest free energy structure [1].

For many sequences, there are a large number of possible structures with predicted low free energy change. The thermodynamic equations provided here therefore predict that each of these low free energy structures will be populated to an extent defined by their ΔG° relative to each other. This motivates the prediction of base pair probabilities using partition functions [2] and also the stochastic sampling of structures [3].

1.2 History of Optical Melting Experiments to Determine Nearest Neighbor Parameters

Optical melting measurements leading to parameters for predicting stabilities of folded RNA were first obtained by Olke Uhlenbeck and Frank Martin in 1971 when they were graduate students in the laboratory of Paul Doty [4, 5]. The experiments were made possible by the development of enzymatic synthesis methods [6] that allowed preparation of oligonucleotides, i.e., short strands of RNA. Most sequences were AU rich, so initial experiments were conducted in 1 M NaCl to provide enough stability to allow measurements. The high salt concentration is also expected to minimize length-dependent salt effects on stability [7] so that corrections for oligonucleotide length are not important. As a postdoctoral fellow in the laboratory of Ignacio Tinoco, Jr., Uhlenbeck directed synthesis of additional sequences and collaborated with Tinoco and graduate student Phil Borer to determine six nearest neighbor parameters for duplexes with Watson-Crick pairs [8]. In principle, it is possible to determine 12 parameters for duplexes of all Watson-Crick pairs [9, 10], but few sequences were initially accessible from the enzymatic methods. The limited variety of sequences also restricted studies of loops [11, 12]. New enzymatic [13] and chemical methods [14, 15] allowed synthesis of a wider range of sequences and determination of 11 nearest neighbor parameters for duplexes of all Watson-Crick pairs [16]. Further refinement of synthetic methods [17, 18] allowed preparation of more sequences, including many with loops [19]. Eventually, it was found that 12 parameters were required to provide an optimal fit to data for duplexes with all Watson-Crick pairs [20]. There is one parameter for helix initiation, ten for the possible combinations of Watson-Crick nearest neighbors, and one for terminal AU pairs. The latter accounts for the fact that two duplexes can have the same nearest neighbors but differ by one in the number of AU pairs. Expansion of the database has also been facilitated by computer acquisition and fitting of melting curves [21–23]. The database continues to expand, facilitated by additional advances in synthetic chemistry [24, 25] and "synthesis by purchase order." The underlying approach to analysis, however, is basically the same as that used by the pioneering investigators [8, 12].

2 Experiments

2.1 Overview of Optical Melting

There are many ways to measure the thermodynamics of nucleic acid folding, but optical melting is used most often because it requires less sample than other methods and provides reasonably high throughput. Figure 1 shows a typical optical melting curve. As the temperature is increased, an RNA structure denatures, i.e., becomes unstructured, because the entropy term in Eq. 3 is

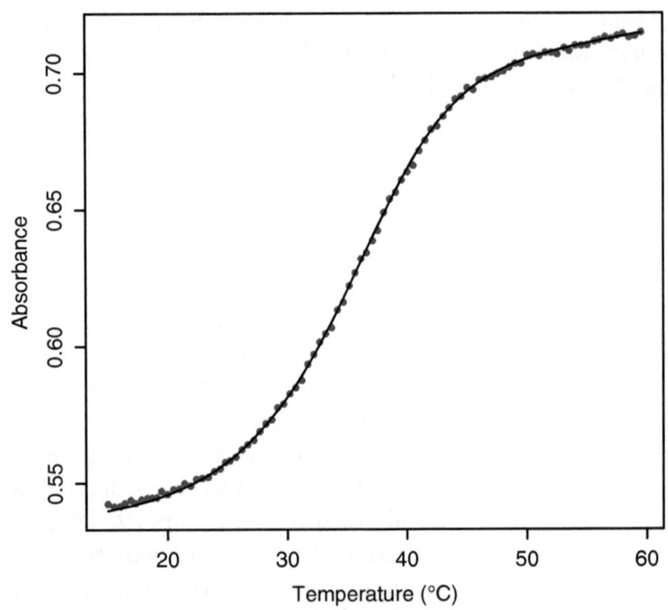

Fig. 1 Optical melting curve, i.e., plot of absorbance vs. temperature, for $(UCUAUAGA)_2$ at a concentration, C_T, of 8.29 μM. For this sequence, absorbance was measured at 260 nm. At low temperature, the RNA is essentially completely folded, while at high temperature it is essentially completely unfolded. Details on making such measurements are described in refs. 26, 27. The points are measurements and the line is the curve fit

unfavorable and the $T\,\Delta S^{\circ}$ increases in magnitude as temperature increases. This section describes how the data are interpreted to determine the stability of RNA structures. For experimental design and experimental procedures, other review articles are available [26, 27].

2.2 Two-State Approximation

Because different states, for example, a specific structure or the random coil state, of the nucleic acid absorb light differently, an optical melting curve contains information about the populations of species as a function of temperature and therefore on the thermodynamics of folding. Usually, however, many approximations are used for extracting this information. For example, in the transition region, there can be fully folded, partially folded, and fully unfolded molecules. The partition function, q, contains information that would allow calculation of the fraction of each species present at a given temperature. In practice, the shape of the curve is almost always fit to a two-state model that contains only fully folded and fully unfolded species. The rationale for this approximation can be seen by considering the conformational partition function, $q_{\mathrm{c}} = -1 + q$ [19, 28]:

$$q_{\mathrm{c}} = -1 + \sum_{j=0}^{n} \mathrm{e}^{-G_j^{\circ}/RT} = \sum_{i=1}^{n} \mathrm{e}^{-\Delta G_i^{\circ}/RT} = \sum_{i=1}^{n} K_i \qquad (8)$$

Here, G_j° is the free energy for forming the jth species. By setting the free energy of the unfolded state to $G_0^{\circ} = 0$ kcal/mol, a sum using free energy changes for folding from the unfolded state can be performed. Then, $\triangle G_i^{\circ}$ is the free energy difference between species i and the unfolded single strand and K_i is the equilibrium constant as defined by Eq. 1. Considering the simple "zipper" model of a completely aligned, fully Watson-Crick base-paired duplex formed from two unfolded strands of different sequence, allowing only one helical region per folded molecule, and assuming the first base pair forms with an equilibrium constant of κ while each additional base pair forms with an equilibrium constant of s give

$$q_{\mathrm{c}} = \kappa \sum_{j=1}^{N} (N - j + 1)\, s^{j-1} = \kappa \left[(N+1) \sum_{j=1}^{N} s^{j-1} - \sum_{j=1}^{N} j s^{j-1} \right] \qquad (9)$$

Here, N is the number of nucleotides in each strand so that $(N - j + 1)$ is the number of species with identical $\triangle G^{\circ}$, i.e., a helix of identical length. In general,

$$\sum_{k=1}^{n} s^{k} = \frac{\left(s^{n+1} - s\right)}{(s-1)} \qquad (10)$$

So,

$$\sum_{j=1}^{N} s^{j-1} = s^{-1}\sum_{j=1}^{N} s^{j} = \frac{\left(s^{N}-1\right)}{(s-1)} \quad (11)$$

Furthermore,

$$\sum_{k=1}^{n} ks^{k} = \frac{s\mathrm{d}\left(\sum_{k=1}^{n} s^{k}\right)}{\mathrm{d}s} = s\frac{\mathrm{d}\left[\left(s^{n+1}-s\right)/(s-1)\right]}{\mathrm{d}s} = \frac{s}{(s-1)^{2}}\left[ns^{n+1}-(n+1)s^{n}+1\right] \quad (12)$$

Thus,

$$\sum_{j=1}^{N} js^{j-1} = s^{-1}\sum_{j=1}^{N} js^{j} = \frac{Ns^{N+1}-(N+1)s^{N}+1}{(s-1)^{2}} \quad (13)$$

Plugging Eqs. 12 and 13 into Eq. 9 yields

$$q_{\mathrm{c}} = \kappa\left[\frac{s^{N+1}-(N+1)s+N}{(s-1)^{2}}\right] \quad (14a)$$

If $s > 1$ and N is "large," then

$$q_{\mathrm{c}} \approx \kappa s^{N-1} \quad (14b)$$

For example, typically $s \approx 30$ at 37 °C and $N \approx 6$ [20], so that q_{c} from Eqs. 14a and 14b differ by less than 7%. The value of q_{c} is thus approximately equal to the equilibrium constant for bimolecular folding, with single strands, A and B, binding to form duplex, $A{\cdot}B$:

$$K = \frac{[A\cdot B]}{[A][B]} \quad (15)$$

where the brackets indicate "concentration of" as in the equilibrium constant for unimolecular folding (Eq. 1). Thus, partially folded states can be neglected.

2.3 Fitting Curves for Non-Self-Complementary Bimolecular Folding

Given the two-state assumption, the experimental curve is fit using nonlinear regression to determine the $\Delta H°$ and $\Delta S°$. Assuming that both the folded and unfolded species increase linearly in absorbance as a function of temperature, then the baselines of absorbance can be expressed as

$$LB(T) = m_{\mathrm{LB}} T + b_{\mathrm{LB}} \quad (16)$$

$$UB(T) = m_{\mathrm{UB}} T + b_{\mathrm{UB}} \quad (17)$$

where the lower baseline, $LB(T)$, describes the absorbance of the folded species; the upper baseline, $UB(T)$, describes the absorbance of the unfolded species; m_{LB} and m_{UB} are slopes; and b_{LB} and b_{UB} are intercepts. $X(T)$, the fraction of strands that are single stranded, is then a function of the measured absorbance, $A(T)$, and the baselines:

$$X(T) = \frac{A(T) - \mathrm{LB}(T)}{\mathrm{UB}(T) - \mathrm{LB}(T)} \quad (18)$$

For the case when two strands interact, if the species A and B were mixed 1:1, as is customary, with a total concentration of C_{T}, then the concentration of the species at a given temperature is

$$[A] = X(T)C_{\mathrm{T}}/2 \quad (19)$$

$$[B] = X(T)C_{\mathrm{T}}/2 \quad (20)$$

$$[A \cdot B] = (1 - X(T))\, C_{\mathrm{T}}/2 \quad (21)$$

Note that the concentration of the duplex species, [A·B], is divided by two because there are two strands in the duplex. According to Eq. 15 and plugging in the expression in Eqs. 19, 20, and 21,

$$K = \frac{(1 - X(T))\, C_{\mathrm{T}}/2}{(X(T)C_{\mathrm{T}}/2)^2} \quad (22)$$

Rearranging Eq. 22 gives

$$K\frac{C_{\mathrm{T}}}{2}X(T)^2 + X(T) - 1 = 0, \quad (23)$$

which can be solved for $X(T)$ using the quadratic equation

$$X(T) = \frac{-1 \pm \sqrt{1 + 2C_{\mathrm{T}}K}}{C_{\mathrm{T}}K} \quad (24)$$

The negative root for Eq. 24 can be neglected because it is not physically meaningful. Equation 18 can be rearranged to show

$$X(T)\,[UB(T) - LB(T)] + LB(T) = A(T) \tag{25}$$

Plugging Eqs. 16, 17, and 24 into Eq. 25 gives the expression

$$\begin{aligned} A(T) &= \frac{-1 \pm \sqrt{1+2C_T K}}{C_T K} \\ &\quad \times [m_{UB} T + b_{UB} - m_{LB} T - b_{LB}] + m_{LB} T + b_{LB} \end{aligned} \tag{26}$$

According to Eqs. 2 and 3, the relationships between free energy and equilibrium constant and enthalpy and entropy (Eq. 26) can be rewritten as

$$\begin{aligned} A(T) &= \frac{-1 \pm \sqrt{1-2C_T e^{[-\Delta H^\circ + T\,\Delta S^\circ]/RT}}}{C_T e^{[-\Delta H^\circ + T\,\Delta S^\circ]/RT}} \\ &\quad \times [m_{UB} T + b_{UB} - m_{LB} T - b_{LB}] + m_{LB} T + b_{LB} \end{aligned} \tag{27}$$

Equation 27 expresses absorbance as a function of six fit parameters, ΔH°, ΔS°, m_{UB}, b_{UB}, m_{LB}, and b_{LB}. This expression facilitates the nonlinear regression of the fit parameters to the experimental curve.

2.4 Assumptions About Baselines and Heat Capacity Change

As shown, the two-state assumption greatly simplifies fitting of melting curves, but additional assumptions were also employed. For example, determining the relative populations of fully folded and unfolded species as a function of temperature requires knowing the absorption of each species as a function of temperature. This was approximated by linear extrapolation of the lower and upper baselines (Eqs. 16 and 17) [22]. It is unlikely, however, that the temperature dependence of each species is linear. For example, the stacking and therefore absorbance of single strands are temperature dependent in a nonlinear way. If the two strands forming a duplex have different sequences, then the absorbance of each strand as a function of temperature can be determined separately and used in the analysis. Most duplexes studied, however, are self-complementary because this requires less synthesis and does not require careful mixing of two strands to give equal concentrations. Studies of hairpin folding also do not allow simple measurement of the temperature dependence of the unpaired state.

Another assumption in analyzing melting curves is that the ΔH° and ΔS° for folding are temperature independent. This is also not exact for many possible reasons [29]. For example, stacking in unfolded strands will provide a temperature-dependent ΔH° and ΔS° of folding. While a temperature dependence for ΔH° and ΔS° could be included when fitting melting curves, simulations indicate that the signal to noise of the data will not allow accurate determination of this temperature dependence, i.e., the heat capacity change, $\Delta C_p{}^\circ$ [30].

While there are many approximations in fitting melting curves, they tend to affect ΔH° and ΔS° in opposite directions [20, 31]. Thus, ΔG° is more accurate than either ΔH° or ΔS°. The effects of approximations are minimized near the melting temperature. Thus, model systems are often designed to have melting temperatures near 37 °C, human body temperature.

2.5 Calculating Melting Temperature

The T_M, or melting temperature, is the temperature at which half the strands are single stranded, i.e., $X(T_M) = 1/2$. According to Eqs. 19, 20, and 21, at the T_M,

$$[A] = C_T/4 \tag{28}$$

$$[B] = C_T/4 \tag{29}$$

$$[A \cdot B] = C_T/4 \tag{30}$$

Then, according to Eqs. 2, 3, and 15, at the T_M,

$$K = \frac{C_T/4}{(C_T/4)^2} = \frac{4}{C_T} = e^{-\Delta G^\circ/RT_M} = e^{-(\Delta H^\circ - T_M \Delta S^\circ)/RT_M} \tag{31}$$

Rearranging, this leads to

$$\ln\left(\frac{4}{C_T}\right) = \frac{-\Delta H^\circ + T_M \Delta S^\circ}{RT_M} \tag{32}$$

$$RT_M \ln\left(\frac{4}{C_T}\right) = -\Delta H^\circ + T_M \Delta S^\circ \tag{33}$$

$$T_M = \frac{-\Delta H^\circ}{R \ln(4/C_T) - \Delta S^\circ} \tag{34}$$

Equation 34 determines the T_M for a non-self-complementary system, from the concentration of strands, the enthalpy, and the entropy change for a system. In the case where the two interacting strands are identical, called self-complementary, the relationship can be shown to be

$$T_M = \frac{-\Delta H^\circ}{R \ln(1/C_T) - \Delta S^\circ} \tag{35}$$

Finally, in the unimolecular case, the T_M is independent of temperature and

$$T_M = \frac{\Delta H^\circ}{\Delta S^\circ} \tag{36}$$

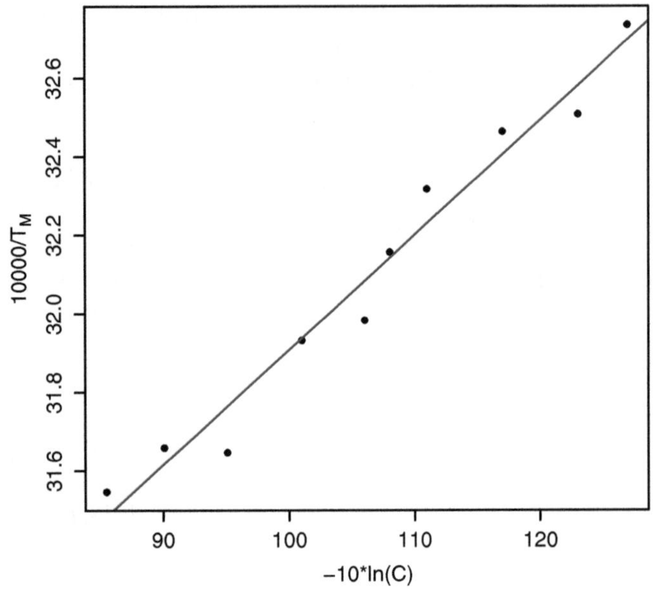

Fig. 2 Plot of T_M^{-1} vs. ln C_T for the series of experiments performed on $(UCUAUAGA)_2$. Note the temperatures in this plot are in Kelvin. Concentrations are expressed in units of molarity for the x-axis

These relationships can be used to calculate the melting temperature from the fit values of $\triangle H^\circ$ and $\triangle S^\circ$.

2.6 Experiments Are Performed at Multiple Concentrations

For a given system studied by optical melting, the strands are melted at multiple concentrations. For unimolecular structures, such as hairpin stem loops, this provides a set of experiments for determining average values of $\triangle H^\circ$ and $\triangle S^\circ$ and also confirms that the structure is giving the expected behavior that the T_M is constant as a function of strand concentration. For non-self-complementary and self-complementary systems, the T_M is a function of temperature, and Eqs. 34 and 35 can be rearranged to show

$$\frac{1}{T_M} = \frac{R}{\Delta H^\circ} \ln\left(\frac{C_T}{a}\right) + \frac{\Delta S^\circ}{\Delta H^\circ} \quad (37)$$

Here, a is one for self-complementary duplexes and four for non-self-complementary duplexes. Thus, $\triangle H^\circ$ and $\triangle S^\circ$ can be obtained by a linear fit to Eq. 37 (van't Hoff analysis) or from averaging $\triangle H^\circ$ and $\triangle S^\circ$ values from fits of individual melting curves. A sample van't Hoff analysis is shown in Fig. 2. A necessary but not sufficient criterion for two-state melting is that the values from the two methods are the same. Usually, a 15% difference is considered acceptable.

There are both random and systematic errors in the measurements described above. Typically, random errors are reported for optical melting studies, but systematic errors are probably more important. On the basis of comparisons of values reported for identical sequences from different laboratories [32–34], systematic errors for $\triangle H^{\circ}$, $\triangle S^{\circ}$, and $\triangle G^{\circ}$ at 37° are estimated as 6%, 6%, and 3%, respectively [20].

As noted above, melting curves for unimolecular transitions are independent of oligonucleotide concentration so that $\triangle H^{\circ}$ and $\triangle S^{\circ}$ can only be obtained from the shapes of curves. Thus, there is no simple necessary test for two-state behavior. At 1 M salt concentration, it is also difficult to keep the melting temperature near 37 °C because of the high local concentration of the nucleotides that will form base pairs. For example, the hairpins GCUUUUGC and GCUUCGGC with only 2 base pairs have melting temperatures of about 30 °C and 50 °C, respectively, in 0.01 M sodium phosphate, pH 7 [35, 36]. Pseudoknots have the additional problem that there are typically several transitions that melt at similar temperatures, making the choice of baselines and deconvolution of transitions difficult.

2.7 Alternatives to Optical Melting

There are alternative methods for measuring thermodynamics that require fewer assumptions than optical melting. The time required for measurements is considerably longer, however. Isothermal titration calorimetry is particularly powerful [37]. In this method, aliquots of one strand are mixed with a second strand at a fixed temperature, and the heat released upon each addition is measured. The sum of the heats provides the $\triangle H^{\circ}$ and the titration curve provides the $\triangle G^{\circ}$ at the temperature of the experiment. Thus, the measurement does not require determination of baselines. Measurements at different temperatures provide the $\triangle C_p{}^{\circ}$.

Differential scanning calorimetry provides another alternative method. In this method, a folded molecule is contained in one cell and the identical buffer solution is contained in another cell [38]. The difference in electrical power required to heat the two cells is measured to provide the $\triangle H^{\circ}$ associated with the unfolding reaction. This method requires determination of baselines.

3 Nearest Neighbor Rules

3.1 Overview

The goal of nearest neighbor development is to devise a set of rules that can predict the stability of a given RNA secondary structure. Rules are then devised for types of motifs, i.e., canonical pairs, hairpin loops, internal loops, multibranch loops, or pseudoknots. As stated in the Introduction (Subheading 1), the rules are based on two assumptions. The first is that the free energy change of a motif depends only on the sequence of that motif and the sequence of the directly adjacent base pairs. The second assumption is that the free energy increments predicted for each motif can be added to get the total free energy change.

3.2 Focus on Folding Free Energy Change at 37 °C

Most of the attention of nearest neighbor parameter development is on the determination of folding free energy changes at 37 °C. There are several reasons for this. A number of organisms, including humans, live at 37 °C or at similar temperatures; therefore, most applications of the rules are for structure formation at 37 °C. The free energy changes are most accurately determined at temperatures close to the melting temperature of the RNA studied. 37 °C is close to the median accessible temperature for most optical melting instruments, and therefore it is a convenient target temperature for the melting of model systems. Finally, using Eq. 3, the relationship between free energy change, enthalpy change, and entropy change, the enthalpy change can be used to adjust the free energy change from 37 °C to other temperatures, and extrapolations of $\pm 20°$ cover most temperatures of interest.

3.3 Developing Rules

Given a set of folding free energy changes determined for model sequences using optical melting (as described above), a set of rules, i.e., equations, need to be devised to account for the observed free energy changes, with as few free parameters as practical [20, 39]. For both helices and loops, the rules need to account for sequence dependence to accurately predict stability. A clear example of the importance of the sequence dependence is seen with experiments on 2×2 internal loops. Optical melting experiments show that tandem AA pairs are 4.3 kcal/mol less stable at 37 °C than tandem GA pairs closed by GC pairs [40]. This is a difference in equilibrium constant of over 1,000-fold.

The nearest neighbor rules are empirical but are generally based on chemical intuition. For example, the free energy changes of terminal mismatches, i.e., a single noncanonical pair, stacking adjacent to a canonical base pair were determined by experiment [41–46]. These free energy changes depend on the sequences of the base pair and the mismatch. These free energy changes also account for much (but not all) of the sequence dependence of hairpin loops, so predictions of hairpin loop stability (for hairpins of four or more unpaired nucleotides) depend on the terminal mismatch parameters [47]. This makes chemical sense because the first mismatch in a hairpin loop is probably similar in structure to a terminal mismatch of the same sequence, so the stabilities should be similar.

Another consideration in the development of nearest neighbor rules is the need to be able to implement them in software. One common use of the nearest neighbor parameters is in dynamic programming algorithms for predicting structure [1, 2, 48, 49]. It is therefore desirable to have nearest neighbor rules that are implemented in dynamic programming algorithms. This may not always be the case; for example, an asymmetric distribution of unpaired nucleotides in a multibranch loop was found by optical melting and fluorescence titration experiments to be destabilizing [50, 51]. Nearest neighbor rules for multibranch loops include this effect, but it is not yet incorporated in dynamic programming algorithms; therefore, this term is ignored in the available programs [47, 52, 53].

3.4 Parameter Fit

Once a set of rules is chosen, parameter values need to be determined for those rules. Depending on the motif, there are different methods for extracting nearest neighbor parameters from the results of optical melting experiments. For example, parameters for dangling ends and terminal mismatches have been determined by measuring duplexes with and without the noncanonical end and subtracting the difference in measured thermodynamics [22, 42–44, 54–57]. Usually, a self-complementary sequence is designed with both ends identical so that the end effect is doubled.

Parameters for canonical nearest neighbor pairs are determined by linear regression to data for many duplexes containing only canonical pairs [20, 39]. Linear regression is ideal because it guarantees that fit values will provide the minimal least square deviation from the data and it provides error estimates. Parameters for loops can be obtained from measurements on molecules containing the loops and then subtracting nearest neighbor parameters for the non-loop portion of the folded molecule. Alternatively, for loops measured in duplexes, a measured value for a molecule containing the same sequence of base pairs but without a loop can be subtracted, and the parameter for the interrupted nearest neighbor added to correct for its absence. Linear regression was then used to fit the loop stabilities for particular types of loops [39, 47]. For hairpin, bulge, and multibranch loops, linear regression revealed useful parameters. For internal loops, linear regression revealed some useful parameters, although the stabilities of internal loops are particularly idiosyncratic [58].

The error estimates from linear regression provide a critical assessment of the quality of the nearest neighbor rules. During rule development, parameters with relatively large errors suggest that the rules need revision to better account for the variation in experimentally determined stabilities. Subheading 4, below, describes how sequences with known structure can be used to inform the fit of parameters.

3.5 Example Nearest Neighbor Calculation

Nearest neighbor parameters have been estimated for RNA base pairs and for loops [20, 47]. Figure 3 shows a sample structure and nearest neighbor prediction of stability at 37 °C. As shown in

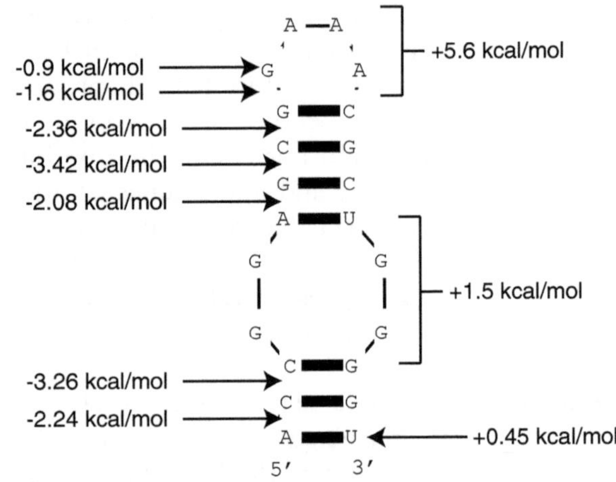

Fig. 3 Sample nearest neighbor calculation. The structure is annotated with the individual nearest neighbor terms for folding free energy change at 37 °C [20, 47]. For helices, the parameters use stacks of adjacent canonical pairs. At the end of a helix, there is a penalty applied for helices that end in AU or GU pairs. The 2 × 2 internal loop free energy change is taken from a table that includes the penalty for the AU pair that closes the loop and ends the helix. The hairpin loop includes a term for the GA noncanonical pair stacking on the adjacent GC pair (–1.6 kcal/mol), a special bonus for having a GA as the first noncanonical pair (–0.9 kcal/mol), and then an initiation penalty for hairpin loops of four unpaired nucleotides (+5.6 kcal/mol). The total free energy change is the sum of the individual terms (–8.3 kcal/mol). As seen in this example, most loop regions are destabilizing, i.e., with free energy change greater than zero. This is because there is an entropic cost for constraining nucleotides in loops that is not completely offset by a favorable enthalpy of interaction between nucleotides

Fig. 3, the parameters for helices are applied to stacks of adjacent base pairs. These rules are directional, so $\begin{array}{c}5'\text{GC}3'\\ \text{CG}\end{array}$ is different from $\begin{array}{c}5'\text{CG}3'\\ \text{GC}\end{array}$. In fact, these cases illustrate the importance of using nearest neighbor rules to estimate the stability of helices. Both $\begin{array}{c}5'\text{GC}3'\\ \text{CG}\end{array}$ and $\begin{array}{c}5'\text{CG}3'\\ \text{GC}\end{array}$ are composed of two GC pairs, but the free energies differ by nearly 1 kcal/mol because of the difference in pair stacking [20].

3.6 Limitations to Nearest Neighbor Parameters

There are some fundamental limitations to nearest neighbor parameterization. The first is that some sequences are known to have non-nearest neighbor effects. For example, the stabilities of single noncanonical pairs (single mismatch) apparently depend on the distance (in base pairs) from the end of a helix [59–61]. The stabilities of bulge loops also depend on their position in a helix [57].

The second limitation is that the parameters also have experimental and systematic errors. The quality of parameters for Watson-Crick helices is excellent, with most errors estimated to be on the order of 0.1 kcal/mol in free energy change at 37 °C per base pair stack [20]. For loop parameters, error estimates are generally about 0.5 kcal/mol per parameter [47]. In equilibrium constant, this is larger than a factor of two. These errors arise from the experimental errors in optical melting and also because the equations do not parameterize every known feature that affects stability. For example, single base pair bulge stabilities are fit neglecting the non-nearest neighbor effects [47].

The third limitation is that there are almost certainly sequences that fold to unusually stable or unstable structures that are currently unknown because they have not been experimentally studied. There are too many possible sequences for most types of loops to be able to study them all by optical melting. Therefore, the nearest neighbor rules are based on incomplete knowledge of sequence-specific stabilities. Ongoing experiments continue to reveal new subtleties in the folding stability of RNA [60–67]. Subheading 4 describes methods that can be used to infer stabilities from the database of known structures.

4 Structure-Informed Parameter Estimation Methods

4.1 Overview

While experimentally determined thermodynamic data are ideal for parameter estimation, such data are in limited supply. It is therefore also valuable to use known reference secondary structures, available in online databases, when estimating parameters. One benefit to the use of reference structures is to avoid overfitting to the limited experimental data available, because the number of available reference structures far exceeds the number of structures for which free energy changes are available and because available reference structures are quite diverse in terms of sequence length and structural features. Another reason to avail of reference structures when estimating parameters is that parameters are often used for secondary structure prediction and structure-informed parameters can have better prediction accuracy than do parameters inferred by linear regression from optical melting data alone.

Mathews et al. used available databases of structures when expanding the nearest neighbor rules and refining parameter values [39]. Since then, advances in the fields of machine learning and algorithm optimization have led to new computational methods and software tools for parameter estimation. These methods have been applied to estimate RNA nearest neighbor parameters from both structural and thermodynamic data, leading to improved prediction accuracy. In this section, we summarize recent progress in structure-informed parameter estimation methods, along with their strengths and limitations.

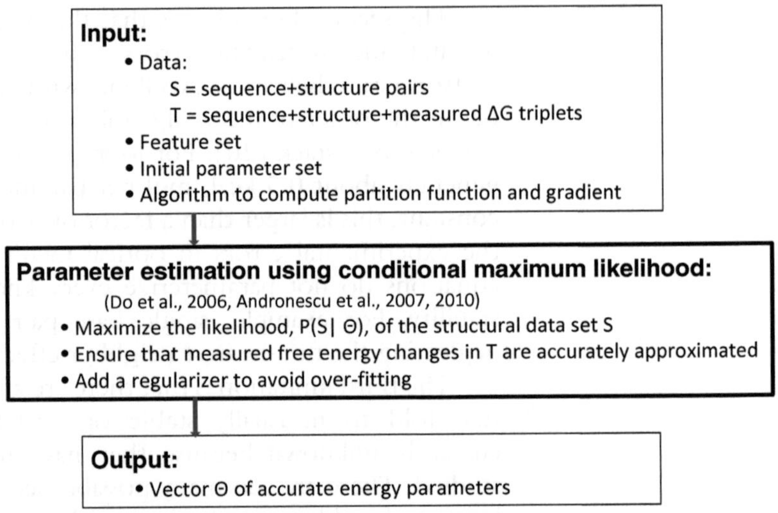

Fig. 4 Schematic of the maximum likelihood parameter estimation method

4.2 Conditional Maximum Likelihood Methods

Suppose we have at hand both a structural set, S, of reference sequence-structure pairs and a thermodynamic set, *T*, of experimentally determined sequence-structure-energy triples, for example, by optical melting. One way to approach parameter estimation is to find the parameters that maximize the likelihood of the reference structures from *S* conditional on the corresponding sequences, as well as accurately approximating the free energy changes of the structures in the thermodynamic set, *T*. Formally, we seek the parameters Θ that represent the mode of the posterior distribution given the structural and thermodynamic data:

$$\Theta = \arg\max_{\Theta} P\left(\Theta \middle| S, T\right) = \arg\max_{\Theta} P\left(S \middle| \Theta\right)\ P\left(T \middle| \Theta\right)\ P(\Theta), \quad (38)$$

where $P(\Theta|S,T)$ is the probability of parameter set Θ given structural set, *S*, and thermodynamic set, *T*. On the right-hand side, $P(S|\Theta)$ is the probability of the structural data set *S* given the parameters Θ, calculated using McCaskill's partition function method [2]. $P(T|\Theta)$ is the probability of the thermodynamic data set, *T*, given the parameters Θ, calculated as a Gaussian distribution with the mean equal to the experimental free energy changes. The last term, $P(\Theta)$, is used to avoid over-fitting to the training sets and is a regularization prior distribution that can be modeled as a Gaussian distribution with mean zero or given by the parameters of Mathews et al. [39]. All terms are convex and the optimization problem can be solved using a standard nonlinear solver. The conditional maximum likelihood approach is illustrated in Fig. 4.

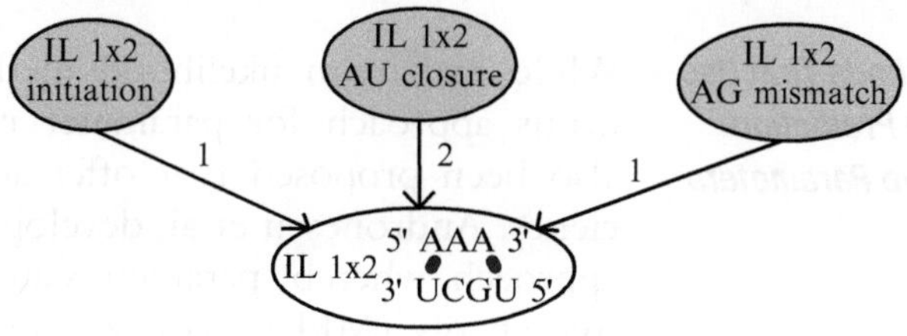

Fig. 5 Example of feature relationship graph for one internal loop 1 × 2. The feature shown in the *bottom node*, which is a 1 × 2 internal loop, has three parents that correspond to components of that loop. The first parent is the initiation for internal loops 1 × 2, the second parent is the AU closure with weight 2 because it appears twice, and the third is an AG mismatch

Do et al. were the first to use a conditional maximum likelihood method for the estimation of 700–900 RNA parameters, but without the thermodynamic set [68]. Andronescu et al. incorporated thermodynamic data and used models with 300–8,000 parameters [69], consistent with the parameters of Xia et al. and Mathews et al. [20, 39, 47]. One challenge in parameter estimation is that some nearest neighbor features, or rules, although based on chemical intuition occur rarely in structures of the data sets, S and T. To address this challenge, Andronescu et al. also leveraged relationships among features so that parameters with reliable estimates can influence parameters that have less support in the data set [70]. To do this, they first modeled feature relationships as a directed acyclic graph (DAG). Nodes of the graph are features, with root nodes (i.e., nodes with no incoming edges) being those features that arise in structures of the thermodynamic data set, T—these are the features whose parameter values can be estimated most reliably. For example, root nodes in the graph include all possible stacked pairs, and small (e.g., 1×1, 1×2, 2×2) internal and bulge loops are also represented as nodes in the graph. A directed edge is placed from one feature to a second if the second, more complex feature incorporates the first. For example, in Fig. 5, the feature corresponding to the depicted 1×2 internal loop includes two closing AU base pairs, and so there is an edge of weight two from the node corresponding to the AU closure feature to the node corresponding to the 1×2 internal loop. The DAGs of Andronescu et al. were designed using extrapolation rules of Mathews et al. and of Znosko and colleagues [39, 61]. Then, when estimating parameters, feature relationships are expressed by setting the prior distribution of a feature's value to be a multivariate Gaussian distribution, conditioned on a linear weighted sum of the values of its parent features in the DAG.

4.3 Methods That Use Structure Prediction to Improve Parameters

While maximum likelihood methods provide a statistically rigorous approach for parameter estimation, other methods have also been proposed that offer advantages of flexibility and efficiency. Andronescu et al. developed a constraint generation (CG) approach, whereby parameters are derived that optimize minimum free energy (MFE) secondary structure prediction accuracy on a reference training set, subject to two types of constraints [70]. Structural constraints specify that the predicted free energy change of the reference structure for a given sequence in the structural set, *S*, should be lower than (or failing that, not much greater than) the predicted free energy changes of alternative structures. Thermodynamic constraints specify that the free energy changes of structures in the thermodynamic data set, *T*, should be close to the experimentally determined values. Since not all possible constraints of the first type can be efficiently enumerated, the optimization process is iterated, and new constraints are added at each iteration that aim to correct mis-predictions of MFE structures obtained with parameters from earlier iterations. At each iteration, a constrained quadratic optimization problem is solved to obtain updated parameters.

Recently, Zakov et al. significantly expanded the number of features or rules, with their largest model having 70,000 features [71]. In their model, features may be context dependent so that the parameter for a CG pair that closes a hairpin can depend both on neighboring bases and on the hairpin length. Moreover, real-valued features, such as the log length of loop regions, are also included. While generalizing the original nearest neighbor rules, the model is still additive. To estimate parameters for their features, Zakov et al. use an iterative approach adapted from prediction frameworks in the natural language processing field: at each step, parameters are updated for features that were over- or underpredicted by the parameters of the previous iteration, with respect to one structure in the structural training data set. The update function employed in this approach is very efficient to compute, since it does not require solving an optimization problem. The method, however, is not currently informed by thermodynamic data, and so the model is not suitable for calculation of free energies or base pair probabilities.

4.4 Database of Known Structures

The structural set used for parameter estimation in these studies [71–73] was derived from a compilation of structures from many publically available databases and is available as the RNA STRAND database of 4,666 secondary structures [74]. Among the families of RNAs that are included in the database are transfer RNAs [75], ribosomal and group I intron RNAs [76], transfer messenger RNAs [77], and ribonuclease P RNAs [78]. The structures included in RNA STRAND are considered reliable because they were determined either by comparative sequence analysis or by

Table 1
Comparison of several parameter sets for RNA

RNA free energy model (kcal/mol)		#Parameters	F-measure on S-test	RMSE on *T*
Mathews-Turner	[39]	363	0.60	1.24
CG*	[70]	363	0.67	0.98
CONTRAFold 2.0	[68]	714	0.69	6.02
BL-FR	[70]	7,726	0.70	1.51
$St^{high}Co^{high}$	[71]	70,000	0.84	N/A

The first row reports on the widely used parameters of Mathews et al. [39]. The CG* parameters were obtained using a variant of the constraint generation method that adopts a loss-augmented max margin approach [70]. The CONTRAFold parameters were obtained using a maximum likelihood approach [68]. The BL-FR parameters were obtained using a maximum likelihood approach with feature relationships [70]. The $St^{high}Co^{high}$ parameters were obtained using the method of Zakov et al. [71]. For each parameter set, the number of features in the corresponding thermodynamic model is reported, along with the F-measure of prediction accuracy on a structural test set S-test with 659 structures and the root-mean-square error (RMSE) of free energy changes on data from 1,291 optical melting experiments.

NMR or X-ray crystallography. The structures in RNA STRAND were processed for use in parameter estimation, for example, by removing noncanonical base pairs and pseudoknots and by breaking long strands into shorter components (no more than 700 nucleotides), and were partitioned into a training set, used for parameter estimation, and a test set, used to assess parameter quality. The resulting structural training set, *S*, contains 2,586 sequence-structure pairs, while the test set, S-test, contains 659 sequence-structure pairs. The thermodynamic set, *T*, used for parameter estimation by Andronescu et al. has 1,291 sequence-structure-energy triples, compiled from over 50 publications.

4.5 Comparison of Methods

Table 1 compares the prediction accuracy of various parameter estimation methods on the structural test set, S-test. Prediction accuracy is given as mean F-measure on the test set, where F-measure is the harmonic mean of sensitivity and positive predictive value, two measures of structure prediction accuracy [70]. F-measure is always between 0 and 1 with 1 meaning perfect predictions. Of the three methods that are informed both by structural and thermodynamic data, the parameter set BL-FR obtained via the maximum likelihood method of Andronescu et al. [70], with feature relationships included, yields the best overall predictions (0.70 F-measure). The parameter set of Zakov et al. yields a significant additional improvement (0.84 F-measure), but the parameters of this model are not able to provide reliable free energy change estimates. The CG* constraint generation method provides the best fit to the thermodynamic data, where fit is measured as root-mean-square error (RMSE) from the experimentally determined free energy changes (the closer to 0, the better).

4.6 Strengths and Limitations of Structure-Informed Parameter Estimation Approaches

The maximum likelihood method with feature relationships provides a statistically rigorous approach to parameter estimation that is informed by both structural and thermodynamic data, obtaining the best accuracy while also achieving a good fit to experimentally determined free energy changes. The current implementation of the method, however, does not handle coaxial stacking features, which are known to contribute significantly to free energy changes [79, 80]. Another challenge in using the method is that an algorithm for calculating the partition function gradients is needed, in addition to an algorithm for partition function calculation. In contrast, the constraint generation approach and the approach of Zakov et al. are easier to adapt to different energy models [71]. All of the methods discussed here rely on the assumption that the reference structures used for training are indeed the minimum free energy (MFE) structures. There is some evidence, such as the success of MFE prediction methods, that RNAs do indeed fold into structures that are good approximations of their MFE structures, but it is also true that co-transcriptional folding and folding kinetics can influence secondary structure [81, 82].

5 Available Parameter Collections

5.1 Experimentally Determined RNA Parameters

Experiments have been performed to determine stabilities for RNA helices, hairpin loops, bulge loops, internal loops, and multibranch loops. Stabilities have been parameterized separately for Watson-Crick helices, GU base pairs, and each type of loop for predicting RNA folding free energy change at 37 °C. The current parameters for Watson–Crick helices were published in 1998 [20]. In 1999, a set of loop and GU pair parameters was assembled that are still in widespread use [39]. In 2004, loop parameters were revised based on new experiments [47]. For the 2004 rules, a set of enthalpy parameters was also assembled, allowing prediction of stability for temperatures other than 37 °C [31].

The 1999 and 2004 thermodynamic parameter sets are available for download from the nearest neighbor database (NNDB) at http://rna.urmc.rochester.edu/NNDB [83]. In addition to providing the parameter values, the rules are summarized and examples are provided for using the parameters. The 2004 set is also incorporated in RNAstructure, which is available at http://rna.urmc.rochester.edu/RNAstructure.html.

5.2 RNA Trained Parameters

The parameters trained by Andronescu et al. [70] using both reference structures and optical melting experiments at 37 °C are available for download as part of the SimFold software package

[84] or as text formatted for the Vienna RNA package [85] or Mfold/Unafold package [86, 87], at http://www.cs.ubc.ca/labs/beta/Projects/RNA-Params.

The parameters trained by Do et al. [68] using reference structures, together with the ContraFold software, are available at http://contra.stanford.edu/contrafold. The parameters trained by Zakov et al. [71] using reference structures, together with the ContextFold software, are available at http://www.cs.bgu.ac.il/~negevcb/contextfold.

5.3 RNA Pseudoknot Parameters

The collections of parameters cited above do not include parameters for predicting pseudoknot stability because few measurements of stability are available [88–93]. In lieu of sufficient experimental data, models for predicting pseudoknot stability have been developed using polymer theory [94–96], lattice models [97, 98], and heuristics [99, 100]. Andronescu et al. [65] used the constraint generation approach (*see* Subheading 4.3) to estimate parameters for pseudoknotted structure prediction, building on the thermodynamic models of Cao and Chen [97, 98] and Dirks and Pierce [100]. These parameters, together with the HotKnots software [73, 101], are available at http://www.cs.ubc.ca/labs/beta/Software/HotKnots.

5.4 Experimentally Determined DNA Parameters

In addition to the RNA nearest neighbor rules for folding, parameter sets have been assembled for DNA. The set in most widespread use was assembled by the SantaLucia laboratory [102]. This set is for predicting folding free energy changes at 37 °C. An alternative set of parameters is available with RNAstructure [103]. It includes parameters for predicting free energy changes at 37 °C and enthalpy changes for extrapolating folding free energy changes to other temperatures.

6 Summary

Determining nearest neighbor parameters for predicting the stability of RNA structures is a paradigm for computational biology. Optical melting and chemical synthesis provide the means to measure the stability of a wide variety of RNA sequences and structures. From these measured stabilities, a set of nearest neighbor parameters were inferred. These parameters are then the underpinnings of the algorithms described in other chapters that predict RNA secondary structure, discover RNA-coding genes in genomes, and design functional sequences.

The parameters are subject to ongoing refinement. These refinements take advantage of new stability data that become available and of new fitting techniques, such as including the set of sequences with known structure to guide fitting.

Acknowledgments

The authors thank Jonathan Chen for preparing Figs. 1 and 2 and Matthew Seetin for helpful comments. The chapter was supported by National Institutes of Health grants R01GM22939 to D.H.T., R01GM076485 to D.H.M., and a grant from the Natural Sciences and Engineering Research Council of Canada to A.C.

References

1. Zuker M, Stiegler P (1981) Optimal computer folding of large RNA sequences using thermodynamics and auxiliary information. Nucleic Acids Res 9:133–148
2. McCaskill JS (1990) The equilibrium partition function and base pair probabilities for RNA secondary structure. Biopolymers 29:1105–1119
3. Ding Y, Lawrence CE (2003) A statistical sampling algorithm for RNA secondary structure prediction. Nucleic Acids Res 31: 7280–7301
4. Uhlenbeck OC, Martin FH, Doty P (1971) Self-complementary oligoribonucleotides: effects of helix defects and guanylic acid-cytidylic acid base pairs. J Mol Biol 57:217–229
5. Martin FH, Uhlenbeck OC, Doty P (1971) Self-complementary oligoribonucleotides: adenylic acid-uridylic acid block copolymers. J Mol Biol 57:201–215
6. Thach RE (1966) Enzymatic synthesis of oligonucleotide of defined sequence. In: Cantoni GL, Davies DR (eds) Procedures in nucleic acid research. Harper and Row, New York, pp 520–524
7. Manning GS (1978) The molecular theory of polyelectrolyte solutions with applications to the electrostatic properties of polynucleotides. Q Rev Biophys 11:179–246
8. Borer PN, Dengler B, Tinoco I Jr, Uhlenbeck OC (1974) Stability of ribonucleic acid double-stranded helices. J Mol Biol 86:843–853
9. Gray DM (1997) Derivation of nearest-neighbor properties from data on nucleic acid oligomers. II. Thermodynamic parameters of DNA.RNA hybrids and DNA duplexes. Biopolymers 42:795–810
10. Gray DM (1997) Derivation of nearest-neighbor properties from data on nucleic acid oligomers. I. Simple sets of independent sequences and the influence of absent nearest neighbors. Biopolymers 42:783–793
11. Gralla J, Crothers DM (1973) Free energy of imperfect nucleic acid helices. 3. Small internal loops resulting from mismatches. J Mol Biol 78:301–319
12. Gralla J, Crothers DM (1973) Free energy of imperfect nucleic acid helices. II. Small hairpin loops. J Mol Biol 73:497–511
13. Uhlenbeck OC, Cameron V (1977) Equimolar addition of oligoribonucleotides with T4 RNA ligase. Nucleic Acids Res 4:85–98
14. England TE, Neilson T (1977) Duplex formation of complementary oligoribonucleotides corresponding to the dihydrouridine loop neck region of several transfer ribonucleic acids. Can J Biochem 55:365–368
15. Kierzek R, Caruthers MH, Longfellow CE, Swinton D, Turner DH, Freier SM (1986) Polymer-supported synthesis and its application to test the nearest-neighbor model for duplex stability. Biochemistry 25: 7840–7846
16. Freier SM, Kierzek R, Jaeger JA, Sugimoto N, Caruthers MH, Neilson T, Turner DH (1986) Improved free-energy parameters for predictions of RNA duplex stability. Proc Natl Acad Sci U S A 83:9373–9377
17. Usman N, Ogilvie KK, Jiang MY, Cedergren RJ (1987) The automated chemical synthesis of long oligoribonucleotides using 2'-O-silylated ribonucleoside 3'-O-phosphoramidites on a controlled-pore glass support: synthesis of a 43-nucleotide sequence similar to the 3'-half molecule of an *Escherichia coli* formylmethionine tRNA. J Am Chem Soc 109:7485–7854
18. Wincott F, DiRenzo A, Shaffer C, Grimm S, Tracz D, Workman C, Sweedler D, Gonzalez C, Scaringe S, Usman N (1995) Synthesis, deprotection, analysis and purification of RNA and ribozymes. Nucleic Acids Res 23:2677–2684
19. Turner DH (2000) Conformational changes. In: Bloomfield V, Crothers D, Tinoco I (eds) Nucleic acids. University Science Books, Sausalito, CA, pp 259–334
20. Xia T, SantaLucia J Jr, Burkard ME, Kierzek R, Schroeder SJ, Jiao X, Cox C, Turner

DH (1998) Thermodynamic parameters for an expanded nearest-neighbor model for formation of RNA duplexes with Watson-Crick pairs. Biochemistry 37:14719–14735
21. McDowell JA, Turner DH (1996) Investigation of the structural basis for thermodynamic stabilities of tandem GU mismatches: solution structure of $(rGAGGUCUC)_2$ by two-dimensional NMR and simulated annealing. Biochemistry 35:14077–14089
22. Petersheim M, Turner DH (1983) Base-stacking and base-pairing contributions to helix stability: thermodynamics of double-helix formation with CCGG, CCGGp, CCGGAp, ACCGGp, CCGGUp, and ACCGGUp. Biochemistry 22:256–263
23. Albergo DD, Marky LA, Breslauer KJ, Turner DH (1981) Thermodynamics of (dG–dC)3 double-helix formation in water and deuterium oxide. Biochemistry 20:1409–1413
24. Hartsel SA, Kitchen DE, Scaringe SA, Marshall WS (2005) RNA oligonucleotide synthesis via 5'-silyl-2'-orthoester chemistry. Methods Mol Biol 288:33–50
25. Scaringe SA, Wincott FE, Caruthers MH (1998) Novel RNA synthesis method using 5'-O-silyl-2'-O-orthoester protecting groups. J Am Chem Soc 120:11820–11821
26. Siegfried NA, Bevilacqua PC (2009) Thinking inside the box: designing, implementing, and interpreting thermodynamic cycles to dissect cooperativity in RNA and DNA folding. Methods Enzymol 455:365–393
27. Schroeder SJ, Turner DH (2009) Optical melting measurements of nucleic acid thermodynamics. Methods Enzymol 468:371–387
28. Cantor CR, Schimmel PR (1980) Biophysical chemistry. W. H. Freeman and Company, New York
29. Mikulecky PJ, Feig AL (2006) Heat capacity changes associated with nucleic acid folding. Biopolymers 82:38–58
30. Chaires JB (1997) Possible origin of differences between van't Hoff and calorimetric enthalpy estimates. Biophys Chem 64: 15–23
31. Lu ZJ, Turner DH, Mathews DH (2006) A set of nearest neighbor parameters for predicting the enthalpy change of RNA secondary structure formation. Nucleic Acids Res 34:4912–4924
32. Breslauer KJ, Frank R, Blocker H, Marky LA (1986) Predicting DNA duplex stability from the base sequence. Proc Natl Acad Sci U S A 83:3746–3750
33. SantaLucia J Jr, Allawi HT, Seneviratne PA (1996) Improved nearest-neighbor parameters for predicting DNA duplex stability. Biochemistry 35:3555–3562
34. Sugimoto N, Nakano S, Yoneyama M, Honda K (1996) Improved thermodynamic parameters and helix initiation factor to predict stability of DNA duplexes. Nucleic Acids Res 24:4501–4505
35. Ma H, Proctor DJ, Kierzek E, Kierzek R, Bevilacqua PC, Gruebele M (2006) Exploring the energy landscape of a small RNA hairpin. J Am Chem Soc 128:1523–1530
36. Proctor DJ, Ma H, Kierzek E, Kierzek R, Gruebele M, Bevilacqua PC (2004) Folding thermodynamics and kinetics of YNMG RNA hairpins: specific incorporation of 8-bromoguanosine leads to stabilization by enhancement of the folding rate. Biochemistry 43:14004–14014
37. Feig AL (2009) Studying RNA–RNA and RNA–protein interactions by isothermal titration calorimetry. Methods Enzymol 468:409–422
38. Breslauer KJ, Freire E, Straume M (1992) Calorimetry: a tool for DNA and ligand-DNA studies. Methods Enzymol 211:533–567
39. Mathews DH, Sabina J, Zuker M, Turner DH (1999) Expanded sequence dependence of thermodynamic parameters provides improved prediction of RNA secondary structure. J Mol Biol 288:911–940
40. Wu M, McDowell JA, Turner DH (1995) A periodic table of symmetric tandem mismatches in RNA. Biochemistry 34:3204–3211
41. Hickey DR, Turner DH (1985) Solvent effects on the stability of A7U7p. Biochemistry 24:2086–2094
42. Freier SM, Kierzek R, Caruthers MH, Neilson T, Turner DH (1986) Free energy contributions of G·U and other terminal mismatches to helix stability. Biochemistry 25:3209–3223
43. Freier SM, Sugimoto N, Sinclair A, Alkema D, Neilson T, Kierzek R, Caruthers MH, Turner DH (1986) Stability of XGCGCp, GCGCYp, and XGCGCYp helixes: an empirical estimate of the energetics of hydrogen bonds in nucleic acids. Biochemistry 25:3214–3219
44. Sugimoto N, Kierzek R, Turner DH (1987) Sequence dependence for the energetics of dangling ends and terminal base pairs in ribonucleic acid. Biochemistry 26:4554–4558
45. Serra MJ, Axenson TJ, Turner DH (1994) A model for the stabilities of RNA hairpins based on a study of the sequence dependence of stability for hairpins of six nucleotides. Biochemistry 33:14289–14296
46. Dale T, Smith R, Serra M (2000) A test of the model to predict unusually stable RNA hairpin loop stability. RNA 6:608–615

47. Mathews DH, Disney MD, Childs JL, Schroeder SJ, Zuker M, Turner DH (2004) Incorporating chemical modification constraints into a dynamic programming algorithm for prediction of RNA secondary structure. Proc Natl Acad Sci U S A 101:7287–7292
48. Hofacker IL, Fekete M, Stadler PF (2002) Secondary structure prediction for aligned RNA sequences. J Mol Biol 319:1059–1066
49. Mathews DH, Turner DH (2002) Dynalign: an algorithm for finding the secondary structure common to two RNA sequences. J Mol Biol 317:191–203
50. Mathews DH, Turner DH (2002) Experimentally derived nearest neighbor parameters for the stability of RNA three- and four-way multibranch loops. Biochemistry 41:869–880
51. Liu B, Diamond JM, Mathews DH, Turner DH (2011) Fluorescence competition and optical melting measurements of RNA three-way multibranch loops provide a revised model for thermodynamic parameters. Biochemistry 50:640–653
52. Mathews DH (2004) Using an RNA secondary structure partition function to determine confidence in base pairs predicted by free energy minimization. RNA 10:1178–1190
53. Mathews DH (2005) Predicting a set of minimal free energy RNA secondary structures common to two sequences. Bioinformatics 21:2246–2253
54. Freier SM, Burger BJ, Alkema D, Neilson T, Turner DH (1983) Effects of 3' dangling end stacking on the stability of GGCC and CCGG double helices. Biochemistry 22:6198–6206
55. Freier SM, Alkema D, Sinclair A, Neilson T, Turner DH (1985) Contributions of dangling end stacking and terminal base-pair formation to the stabilities of XGGCCp, XCCGGp, XGGCCYp, and XCCGGYp helixes. Biochemistry 24:4533–4539
56. Turner DH, Sugimoto N, Freier SM (1988) RNA structure prediction. Ann Rev Biophys Biophys Chem 17:167–192
57. Longfellow CE, Kierzek R, Turner DH (1990) Thermodynamic and spectroscopic study of bulge loops in oligoribonucleotides. Biochemistry 29:278–285
58. Schroeder SJ, Burkard ME, Turner DH (1999) The energetics of small internal loops in RNA. Biopolymers 52:157–167
59. Kierzek R, Burkard ME, Turner DH (1999) Thermodynamics of single mismatches in RNA duplexes. Biochemistry 38:14214–14223
60. Davis AR, Znosko BM (2008) Thermodynamic characterization of naturally occurring RNA single mismatches with G-U nearest neighbors. Biochemistry 47:10178–10187
61. Davis AR, Znosko BM (2007) Thermodynamic characterization of single mismatches found in naturally occurring RNA. Biochemistry 46:13425–13436
62. Miller S, Jones LE, Giovannitti K, Piper D, Serra MJ (2008) Thermodynamic analysis of 5' and 3' single- and 3' double-nucleotide overhangs neighboring wobble terminal base pairs. Nucleic Acids Res 36:5652–5659
63. Clanton-Arrowood K, McGurk J, Schroeder SJ (2008) 3' terminal nucleotides determine thermodynamic stabilities of mismatches at the ends of RNA helices. Biochemistry 47:13418–13427
64. Badhwar J, Karri S, Cass CK, Wunderlich EL, Znosko BM (2007) Thermodynamic characterization of RNA duplexes containing naturally occurring 1×2 nucleotide internal loops. Biochemistry 46:14715– 14724
65. Blose JM, Manni ML, Klapec KA, Stranger-Jones Y, Zyra AC, Sim V, Griffith CA, Long JD, Serra MJ (2007) Non-nearest-neighbor dependence of the stability for RNA bulge loops based on the complete set of group I single-nucleotide bulge loops. Biochemistry 46:15123–15135
66. O'Toole AS, Miller S, Serra MJ (2005) Stability of 3' double nucleotide overhangs that model the 3' ends of siRNA. RNA 11:512–516
67. Vecenie CJ, Morrow CV, Zyra A, Serra MJ (2006) Sequence dependence of the stability of RNA hairpin molecules with six nucleotide loops. Biochemistry 45:1400–1407
68. Do CB, Woods DA, Batzoglou S (2006) CONTRAfold: RNA secondary structure prediction without physics-based models. Bioinformatics 22:e90–e98
69. Andronescu M, Condon A, Hoos HH, Mathews DH, Murphy KP (2007) Efficient parameter estimation for RNA secondary structure prediction. Bioinformatics 23:i19–i28
70. Andronescu M, Condon A, Hoos HH, Mathews DH, Murphy KP (2010) Computational approaches for RNA energy parameter estimation. RNA 16:2304–2318
71. Zakov S, Goldberg Y, Elhadad M, Ziv-Ukelson M (2011) Rich parameterization improves RNA structure prediction. In: Bafna V, Sahinalp, SC (eds) Proceedings of 15th annual international conference on research in computational molecular biology. Springer-Verlag, Berlin, Germany, pp 546–562
72. Do CB, Foo CS, Batzoglou S (2008) A max-margin model for efficient simultaneous alignment and folding of RNA sequences. Bioinformatics 24:i68–i76

73. Andronescu MS, Pop C, Condon AE (2010) Improved free energy parameters for RNA pseudoknotted secondary structure prediction. RNA 16:26–42
74. Andronescu M, Bereg V, Hoos HH, Condon A (2008) RNA STRAND: the RNA secondary structure and statistical analysis database. BMC Bioinformatics 9:340
75. Sprinzl M, Vassilenko KS (2005) Compilation of tRNA sequences and sequences of tRNA genes. Nucleic Acids Res 33:D139–D140
76. Cannone JJ, Subramanian S, Schnare MN, Collett JR, D'Souza LM, Du Y, Feng B, Lin N, Madabusi LV, Muller KM et al (2002) The comparative RNA web (CRW) site: an online database of comparative sequence and structure information for ribosomal, intron, and other RNAs. BMC Bioinformatics 3:2
77. Andersen ES, Rosenblad MA, Larsen N, Westergaard JC, Burks J, Wower IK, Wower J, Gorodkin J, Samuelsson T, Zwieb C (2006) The tmRDB and SRPDB resources. Nucleic Acids Res 34:D163–D168
78. Brown JW (1999) The ribonuclease P database. Nucleic Acids Res 27:314
79. Walter AE, Turner DH, Kim J, Lyttle MH, Müller P, Mathews DH, Zuker M (1994) Coaxial stacking of helixes enhances binding of oligoribonucleotides and improves predictions of RNA folding. Proc Natl Acad Sci U S A 91:9218–9222
80. Kim J, Walter AE, Turner DH (1996) Thermodynamics of coaxially stacked helices with GA and CC mismatches. Biochemistry 35:13753–13761
81. Meyer IM, Miklos I (2004) Co-transcriptional folding is encoded within RNA genes. BMC Mol Biol 5:10
82. Mathews DH, Turner DH (2006) Prediction of RNA secondary structure by free energy minimization. Curr Opin Struct Biol 16:270–278
83. Turner DH, Mathews DH (2010) NNDB: the nearest neighbor parameter database for predicting stability of nucleic acid secondary structure. Nucleic Acids Res 38:D280–D282
84. Andronescu M, Zhang ZC, Condon A (2005) Secondary structure prediction of interacting RNA molecules. J Mol Biol 345:987–1001
85. Hofacker IL, Fontana W, Stadler PF, Bonhoeffer LS, Tacker M, Schuster P (1994) Fast folding and comparison of RNA secondary structures. Monatsh Chem 125: 167–168
86. Zuker M, Mathews DH, Turner DH (1999) Algorithms and thermodynamics for RNA secondary structure prediction: a practical guide. In: Barciszewski J, Clark BFC (eds) RNA biochemistry and biotechnology. Kluwer Academic Publishers, Boston, pp 11–43
87. Markham NR, Zuker M (2008) UNAFold: software for nucleic acid folding and hybridization. Methods Mol Biol 453:3–31
88. Theimer CA, Blois CA, Feigon J (2005) Structure of the human telomerase RNA pseudoknot reveals conserved tertiary interactions essential for function. Mol Cell 17:671–682
89. Nixon PL, Giedroc DP (2000) Energetics of a strongly pH dependent RNA tertiary structure in a frameshifting pseudoknot. J Mol Biol 296:659–671
90. Theimer CA, Finger LD, Trantirek L, Feigon J (2003) Mutations linked to dyskeratosis congenita cause changes in the structural equilibrium in telomerase RNA. Proc Natl Acad Sci U S A 100:449–454
91. Soto AM, Misra V, Draper DE (2007) Tertiary structure of an RNA pseudoknot is stabilized by "diffuse" Mg2+ ions. Biochemistry 46:2973–2983
92. Liu B, Shankar N, Turner DH (2010) Fluorescence competition assay measurements of free energy changes for RNA pseudoknots. Biochemistry 49:623–634
93. Wyatt JR, Puglisi JD, Tinoco I Jr (1990) RNA pseudoknots, stability and loop size requirements. J Mol Biol 214:455–470
94. Aalberts DP, Hodas NO (2005) Asymmetry in RNA pseudoknots: observation and theory. Nucleic Acids Res 33:2210–2214
95. Gultyaev AP, van Batenburg FHD, Pleij CWA (1999) An approximation of loop free energy values of RNA H-pseudoknots. RNA 5:609–617
96. Xayaphoummine A, Bucher T, Thalmann F, Isambert H (2003) Prediction and statistics of pseudoknots in RNA structures using exactly clustered stochastic simulations. Proc Natl Acad Sci U S A 100:15310–15315
97. Cao S, Chen SJ (2006) Predicting RNA pseudoknot folding thermodynamics. Nucleic Acids Res 34:2634–2652
98. Cao S, Chen SJ (2009) Predicting structures and stabilities for H-type pseudoknots with interhelix loops. RNA 15:696–706
99. Rivas E, Eddy SR (1999) A dynamic programming algorithm for RNA structure prediction including pseudoknots. J Mol Biol 285:2053–2068
100. Dirks RM, Pierce NA (2003) A partition function algorithm for nucleic acid secondary structure including pseudoknots. J Comput Chem 24:1664–1677
101. Ren J, Rastegari B, Condon A, Hoos HH (2005) HotKnots: heuristic prediction of RNA secondary structures including pseudoknots. RNA 11:1494–1504

102. SantaLucia J, Hicks D (2004) The thermodynamics of DNA structural motifs. Annu Rev Biophys Biomol Struct 33: 415–440

103. Reuter JS, Mathews DH (2010) RNAstructure: software for RNA secondary structure prediction and analysis. BMC Bioinformatics 11:129

Chapter 4

Energy-Directed RNA Structure Prediction

Ivo L. Hofacker

Abstract

In this chapter we present the classic dynamic programming algorithms for RNA structure prediction by energy minimization, as well as variations of this approach that allow to compute suboptimal foldings, or even the partition function over all possible secondary structures. The latter are essential in order to deal with the inaccuracy of minimum free energy (MFE) structure prediction, and can be used, for example, to derive reliability measures that assign a confidence value to all or part of a predicted structure. In addition, we discuss recently proposed alternatives to the MFE criterion such as the use of maximum expected accuracy (MEA) or centroid structures. The dynamic programming algorithms implicitly assume that the RNA molecule is in thermodynamic equilibrium. However, especially for long RNAs, this need not be the case. In the last section we therefore discuss approaches for predicting RNA folding kinetics and co-transcriptional folding.

Key words Secondary structure prediction, Minimum free energy structure, Partition function, Pair probabilities, Suboptimal structures, Reliability, Maximum expected accuracy, Folding kinetics

1 Introduction

RNA bioinformatics is a *structural* bioinformatics in that for almost all tasks both sequence and structure of the molecule need to be taken into account. The problem of predicting an RNA's structure from its sequence is therefore the most basic task in RNA bioinformatics. It is especially important since known (experimentally verified) RNA structures are in short supply, while sequences are plentiful.

In most cases we content ourselves with prediction of secondary structure, which can be accomplished by efficient and reasonably accurate algorithms. Moreover, secondary structure has proven to be very useful for understanding the biological function of RNAs. While prediction of the full tertiary structure might be desirable, it is still out of reach for most applications. This chapter will therefore deal with secondary structure prediction only, while

Jan Gorodkin and Walter L. Ruzzo (eds.), *RNA Sequence, Structure, and Function: Computational and Bioinformatic Methods*, Methods in Molecular Biology, vol. 1097, DOI 10.1007/978-1-62703-709-9_4,

progress in tertiary structure prediction will be presented in a later chapter. In addition we only consider pseudo-knot free secondary structures.

2 The Minimum Free Energy Structure

From a biophysical view the natural way to approach structure prediction is to search for the ground state structure, i.e. the structure with minimal free energy. The previous chapter has outlined the energy model that can be used to assign a free energy to every secondary structure. The MFE prediction problem is thus simply to find the structure of minimal free energy among all possible structures. The number of possible structures, however, grows exponentially with the length of the sequence. For random sequences with equal A, C, U, G content the number of pseudo-knot free structures S_n grows asymptotically as $S_n \sim n^{-\frac{3}{2}} 1.85^n$. Thus, the *E. Coli* tRNA^{phe} with a length of 76nt can form about $2.8 \cdot 10^{17}$ different secondary structures.

3 Dynamic Programming Solution of the Folding Problem

In spite of the huge number of possible structures, the folding problem can be solved efficiently using a technique called dynamic programming. In general, dynamic programming involves building the solution of a larger problem from the optimal solutions of smaller subproblems. Ruth Nussinov was the first to apply this principle to RNA secondary structure prediction [1]. The Nussinov algorithm, outlined in the introduction chapter, precedes the current loop-based energy models and simply tries to maximize the number of base pairs in a structure.

For loop-based energy models the decomposition (grammar) has to be slightly more complicated, since for each base pair we have to distinguish which type of loop is closed by the pair. The decomposition shown in Fig. 1 is unambiguous in the sense that each structure can be decomposed in only one way. It starts out as the Nussinov case (*see* Chapter 1) by noting that any structure either starts with an unpaired nucleotide or a substructure enclosed in a pair. For closed structures (C), we have to distinguish whether the enclosing pair closes a hairpin loop, interior loop, or multi-loop. Multi-loops consist of at least two components (M and M^1). To make the decomposition unambiguous we split multi-loops into their final component (M^1), consisting of a close substructure and an unpaired tail, while the M stands for a substructure within a multi-loop consisting of one or more components. Note that the

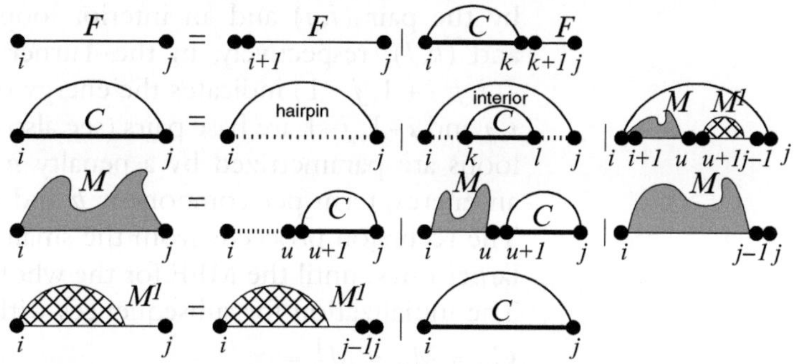

Fig. 1 Recursive decomposition of secondary structures. The decomposition as shown is used in most folding routines of the Vienna RNA package

distinction between M and M^1 can be dropped if ambiguity is not an issue as in MFE folding. Moreover, the decomposition of multi-loops into their constituent components requires that multi-loop energies grow linearly with the length and number of components.

The above decomposition scheme immediately translates into a set of recursions (1) to compute the structure with minimum free energy (MFE). We denote by F_{ij} the minimum free energy on the subsequence $x[i \dots j]$, while C_{ij} denotes the MFE over closed structures only. M_{ij} is the optimal free energy for a substructure that is part of a multi-loop, and M^1_{ij} for a substructure in a multi-loop that starts with a closed structure followed by zero or more unpaired bases.

$$
\begin{aligned}
F_{ij} &= \min\left\{F_{i+1,j},\ \min_{i<k\leq j} C_{ik} + F_{k+1,j}\right\} \\
C_{ij} &= \min\left\{\mathcal{H}(i,j),\ \min_{i<k<l<j} C_{kl} + \mathcal{I}(i,j;k,l),\right. \\
&\qquad \left.\min_{i<u<j} M_{i+1,u} + M^1_{u+1,j-1} + a\right\} \\
M_{ij} &= \min\left\{\min_{i<u<j} (u-i+1)c + C_{u+1,j} + b,\right. \\
&\qquad \left.\min_{i<u<j} M_{i,u} + C_{u+1,j} + b,\ M_{i,j-1} + c\right\} \\
M^1_{ij} &= \min\left\{M^1_{i,j-1} + c,\ C_{ij} + b\right\}
\end{aligned} \tag{1}
$$

The recursions assume there are functions $\mathcal{H}(i,j)$ and $\mathcal{I}(i,j;k,l)$ that look up the energy of a hairpin loop closed by the pair (i, j) and an interior loop closed by the pairs (i, j) and (k, l), respectively, in the Turner energy model. Note that $I(i, j; i+1, j-1)$ indicates the energy of a stacked pair when both i, j and $i+1, j-1$ are base pairs (see also Fig. 3 of chapter 3). Multi-loops are parametrized by a penalty a for forming a multi-loop, an energy term per component b and per unpaired nucleotide c. The recursion proceeds from the smallest subsequences ($i = j$) to larger ones, until the MFE for the whole sequence $F_{1,n}$ is reached. The initialization for subsequences with $j - i < 4$ sets $F_{ij} = 0$ and $C_{ij} = M_{ij} = M^1_{ij} = \infty$.

In total the above recursions use four tables where the Nussinov algorithms needs only one. Space requirements therefore grow quadratically as before. The decomposition of interior loops takes $\mathcal{O}(n^4)$ steps, in principle. However, most implementations restrict the maximum size of interior loops, typically to 30 nt. With this restriction the recursion is dominated by the multi-loop decomposition and the time complexity becomes $\mathcal{O}(n^3)$ as in the Nussinov algorithm. As is usual dynamic programming algorithms an additional backtracing phase is needed to determine the optimal structure.

The recursions above can be extended to account for various variations of the folding problem, such as secondary structures with a minimal helix length [2], folding of circular RNAs [3], computing of hybridization structures through virtual concatenation [4], or consensus structure prediction [5, 6].

4 Prediction Accuracy

The dynamic programming algorithms outlined above provide an exact solution for the mathematical problem of finding the optimal structure given our energy model. This does of course not ensure predictions will be correct. Inaccuracies in the energy parameters, limitations of the model (such as excluding pseudo-knots), interaction with co-factors, and kinetic folding effects can lead to predictions that differ significantly from the biologically functional structure. The exponential growth in the number of structures implies that these problems become more pronounced as sequence length grows.

The accuracy of structure prediction can be measured using test sets of RNAs with known structure, derived from either experiment or comparative structure analysis. Prediction accuracy based on the Turner energy model was reported to be close to 70% in [7, 8], while a study based on ribosomal RNAs [9] yielded significantly lower accuracies of only 20–60%. Presumably this is since the first studies used only short RNAs up to 700nt or split larger RNAs into domains, while the latter used full length 16S and 23S ribosomal RNAs. In addition [7, 8] used a relaxed accuracy definition in which base pair shifted by one nucleotide compared to the reference was still counted as correct.

Improvements of the energy parameters are unlikely to fundamentally change this situation dramatically. Moreover, predicted structures can be highly useful even without being perfectly accurate. It is clear, however, that it is unwise to rely on a single predicted structure. The main avenues to cope with limited prediction accuracy are to (i) consider alternative structures or (ii) use reliability information to estimate how accurate the MFE structure might be and which regions are likely to be predicted correctly.

5 The Partition Function

Even in the absence of errors in the energy model an RNA molecule is not expected to form just a single structure. Statistical mechanics tells us that in thermodynamic equilibrium structures will be populated according to Boltzmann's law, which states that the probability to find a molecule in structure s is given by

$$p(S) = \frac{1}{Z}\exp(-E(s)/kT), \qquad Z = \sum_s \exp(-E(s)/kT)$$

where Z called the partition function. The partition function is the central quantity of statistical mechanics, since the other thermodynamic variables such as free energy, entropy, or specific heat can be computed from it.

The above equation is of course useless for *computing* the partition function, since the sum involves an exponential number of terms. However, as shown by John McCaskill [10] the partition function can be calculated following the dynamic programming same scheme used for MFE structure prediction. Essentially, one has to note that whenever a structure can be decomposed into substructures the partition function becomes the product of the partition function over the substructures. Thus, an algorithm for the partition function can be obtained simply by replacing minimum operations by sums, additions by products, and energy values by their corresponding Boltzmann factors. Using the Z, Z^B, Z^M, and Z^{M1} as the partition functions corresponding to F, C, M, and M^1, we obtain the recursions

$$
\begin{aligned}
Z_{ij} &= Z_{i+1,j} + \sum_{i<k\leq j} Z^B_{ik} Z_{k+1,j} \\
Z^B_{ij} &= e^{-\beta\mathscr{H}(i,j)} + \sum_{i<k<l<j} Z^B_{kl} e^{-\beta\mathscr{I}(i,j;k,l)} + \sum_{i<u<j} Z^M_{i+1,u} Z^{M1}_{u+1,j-1} e^{-\beta a} \\
Z^M_{ij} &= \sum_{i<u<j} e^{-\beta(u-i+1)c} Z^M_{u+1,j} + \sum_{i<u<j} Z^M_{i,u} Z^B_{u+1,j} e^{-\beta b} + Z^M_{i,j-1} e^{-\beta c} \\
Z^{M1}_{ij} &= Z^{M1}_{i,j-1} e^{-\beta c} + Z^B_{ij} e^{-\beta b} \\
Z_{ii} &= 1,\ Z^B_{ii} = Z^M_{ii} = Z^{M1}_{ii} = 0,
\end{aligned} \tag{2}
$$

where $\beta = \frac{1}{RT}$ with T the absolute temperature and R the gas constant. The approach can be used to compute not only the overall partition function but also the equilibrium probability that a particular base pair will be formed. To do this one computes not only the partition function Z^B_{ij} over all structures inside a pair (i, j) but also $\hat{Z}^B_{ij}$ over all structures outside of (i, j) (i.e., on the subsequence $x_1 \ldots x_i, x_j \ldots x_n$). We then get the probability p_{ij} for the formation of base pair (i, j) as

$$
p_{ij} = \frac{Z^B_{ij} \cdot \hat{Z}^B_{ij}}{Z} \tag{3}
$$

Computation of $\hat{Z}^B$ proceeds more or less analogously to the computation of Z^B, except that we proceed from larger to smaller subsequences. Moreover, we can reuse the Z^M and Z^{M1} tables from the forward recursion. For the technically interested, we list the resulting full recursion below:

$$
\begin{aligned}
\hat{Z}^B_{ij} = Z_{1,i-1} Z_{j+1,n} \sum_{k<i;j>l} \hat{Z}^B_{kl} \Big\{ & e^{-\mathscr{I}(k,l;i,j)} + Z^M_{k+1,i-1} Z^M_{j+1,l-1} \\
& + Z^{M2}_{j+1,l-1}) e^{-(a+(i-k-1)c)} + Z^{M2}_{k+1,i-1}) e^{-(a+(l-j-1)c)} \Big\}
\end{aligned} \tag{4}
$$

The first term corresponds to the case where there is no pair outside of (i, j), while for the remaining terms there exists an enclosing base pair (k, l) with $k < i < j < l$. That enclosing pair either closes an interior loop or multi-loop (the last three terms). For the multi-loop case we have to distinguish whether there are additional multi-loop components both left and right of (i, j), whether (i, j) is the leftmost pair, or rightmost pair in

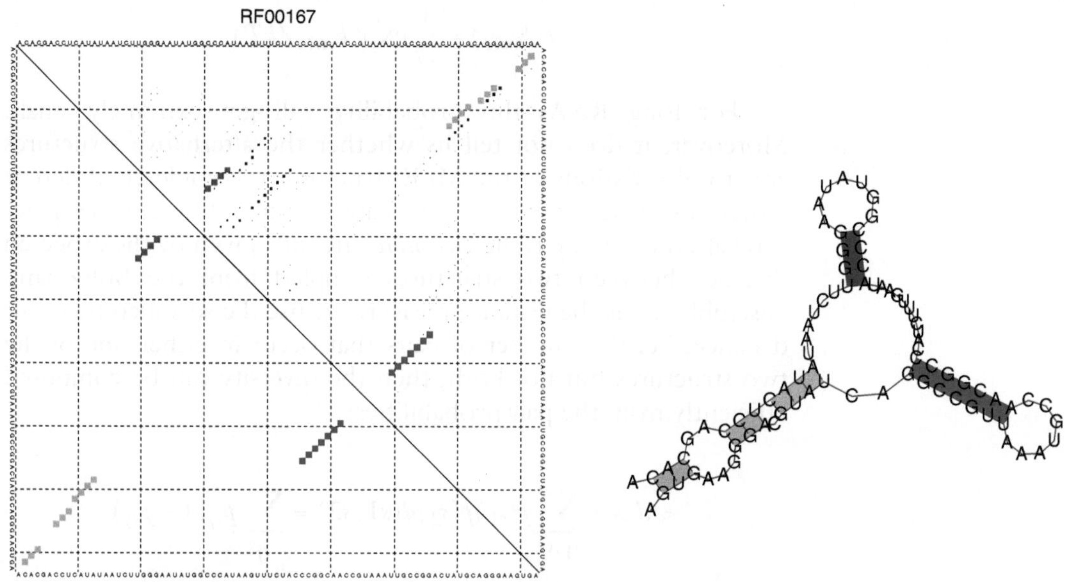

Fig. 2 Dot plot showing the pair probabilities (*left*) and MFE structure (*right*) of the purine riboswitch from *B. subtilis* (Rfam RF00167). Base pairs present in the MFE structure are colored. The lower part of the dot plot shows only the MFE structure. The dot plot shows an overall well-defined structure, but several pairs from alternative foldings are visible in the vicinity of the blue helix

the multi-loop. For convenience we have used Z_{kl}^{M2} to denote the partition function over substructures of a multi-loop containing at least two components, $Z_{kl}^{M2} = \sum_u Z_{ku}^{M} Z_{u+1,l}^{M1}$. As written above the above recursions seem to require $\mathcal{O}(n^4)$ steps. To reduce this to $\mathcal{O}(n^3)$ we can split the multi-loop part into two loops each requiring $\mathcal{O}(n^3)$ steps, e.g. by tabulating $\sum_k \hat{Z}_{kl}^{B} Z_{k+1,i-1}^{M}$ which then can be reused for different values of j.

Pair probabilities can be visualized in the so-called *dot plots*, *see* Fig. 2. They are a highly useful concept since they faithfully represent the complete ensemble of structures in thermodynamic equilibrium.

6 Reliability Measures and Visualization

Partition functions and pair probabilities are the ideal starting point for *reliability* measures to tell us how much confidence we can put into a structure prediction. From the partition function we obtain, for example, immediately the Boltzmann probability of the MFE structure

$$p(S_{\mathrm{mfe}}) = \frac{1}{Z}\exp(E_{\mathrm{mfe}}/kT)$$

For long RNAs this probability will get extremely small. Moreover, it does not tell us whether the alternative structures are trivial variations of the MFE structure or completely different structures. A better measure that captures how different the structural alternatives are is the *ensemble diversity*, given by the expected distance between two structures sampled from the Boltzmann ensemble. If, as the distance measure we use the so-called base pair distance, i.e. the number of pairs that occur in either one of the two structures but not both, then the diversity can be computed efficiently from the pair probabilities:

$$< d > = \sum_{s_1,s_2} p(s_1)p(s_2)d(s1,s2) = \sum_{ij} p_{ij}(1-p_{ij})$$

Programs such as RNAfold from the Vienna RNA package output the Boltzmann probability of the MFE structure and ensemble diversity whenever partition function folding is switched on.

Apart from the *global* reliability measures, we are also interested in local measures that help us identify regions where we can trust the local structure. The `Mfold` program computes a *P-num*(k) value [11] which counts the number of different pairings of position k in a set of suboptimal structures. When using pair probabilities, one can define a *positional entropy*

$$S_k = -\sum_i p_{ik} \ln p_{ik}$$

where p_{ii} is defined as the probability that i does not pair $p_{ii} = 1 - \sum_{j \neq i} p_{ij}$. Well-defined regions should have low *P-num*(k) values, low positional entropy, and high pair probabilities.

Local reliability measure can be conveniently used to annotate structure drawings, *see* Fig. 3 for an example.

7 Suboptimal Folding

Pair probabilities can describe the distribution of all structures in the thermodynamic ensemble but are not always as convenient to as concrete structures. An alternative approach is therefore to predict additional suboptimal structure in order to cover the space of plausible foldings. This can be done in a variety of ways, depending on how many structures are desired and how they are sampled.

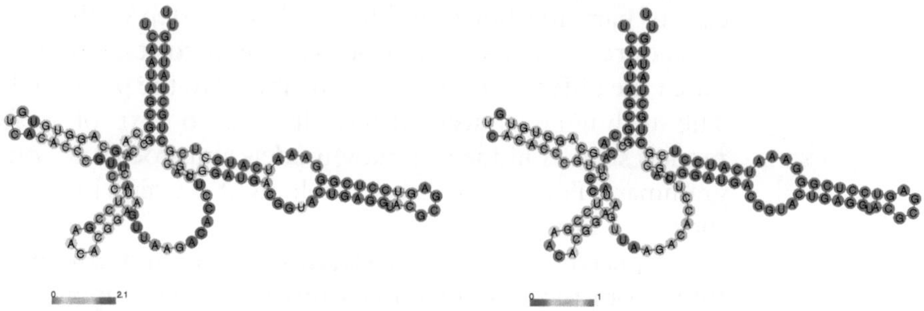

Fig. 3 Predicted MFE structure for 5S rRNA from *M. hungatii* with reliability annotation using positional entropy (*left*) or pair probabilities (*right*). The two left arm show high entropy and low pair probabilities and are in fact mis-predicted. The other, correctly predicted, helices have low entropy and pair probabilities close to 1

7.1 Zuker Suboptimals

The earliest approach to predict suboptimal structures goes back to Michael Zuker [12] and is used in the `mfold` program. In a manner similar to the inside–outside computation of pair probabilities, it computes for each possible base pair (i,j) the optimal folding energy given that this pair is present. It is then straightforward to compute the corresponding optimal structure. Since there are less than $n^2/2$ possible base pairs, it is clear that this approach can only explore a small subset of possible structures. The idea is to present the user with a very small number of structures that are nevertheless representative. In fact `mfold` employs additional filters to remove similar structures and further pare down the resulting list.

While the approach often works well at selecting suitable representatives, it is unavoidable that it will occasionally miss important structures. The method can, for example, not produce structures where more than one substructure is not part of the MFE.

7.2 Complete Suboptimal Folding

It is also possible to extend the normal backtracing procedure such that *all* structures with some energy increment from the MFE are produced. This approach has been implemented in RNAsubopt [13] following ideas from [14].

Of course, even for a fixed energy interval, the number of structures produced by the algorithm will grow exponentially. For example, the 5S RNA from Fig. 3 with 126 nt produces over 8,000 structures within 5 Kcal/mol of the MFE and over 600,000 within 10 kcal/mol. While these numbers are too large for manual inspection of each structure, they can be the starting point for further (automated) analyses. The `barriers` program [15], for example, uses the RNAsubopt output to analyze the energy landscape and predict folding kinetics.

7.3 Stochastic Backtracking

The third popular strategy for producing suboptimal structures is stochastic sampling from the Boltzmann ensemble. Stochastic sampling is the equivalent to backtracing the MFE structure in the case of partition function folding. At each step of the backtracing procedure a stochastic decision is made in correspondence to how much the different alternative contributes to the partition function. The technique is well known in the context of probabilistic models such as hidden Markov models and stochastic context free grammars. For energy directed folding it was first implemented in the `Sfold` program [16].

Repeated stochastic backtracing results in a sample of structures from the Boltzmann ensemble, and most applications have used a sample size of 1,000 structures. One of the attractions of sampling is that it allows in principle to compute any desired thermodynamic average. For any quantity A that can be computed from a secondary structure s we have

$$< A > = \sum_s p(s)A(s) \approx \frac{1}{n} \sum_{s \in \text{sample}} A(s)$$

8 Alternatives to the MFE Criterion

Even if one wants the convenience of a single predicted structure, it is unclear whether the MFE structure is always the optimal choice. The MFE structure is of course the structure with highest probability in the equilibrium ensemble, and therefore can be interpreted as the structure most likely to be correct. Note, however, that the probability $p(\S_{\text{mfe}})$ will still become very small long sequences. It is therefore legitimate to ask whether other criteria than minimal energy could produce a better representative structure.

One approach would be to argue that the best representative for the equilibrium ensemble of structures should be the centroid structure of the ensemble, where the centroid Ψ_c is the structure which minimizes the Boltzmann weighted distance to all other structures

$$< d(\Psi) > = \sum_s p(s)d(\Psi, s) \tag{5}$$

Usually the base pair distance is used as the underlying distance measure d between structures. It turns out that for this choice the centroid structure can be computed trivially from the pair probabilities: It is simply the structure containing all pairs (i, j) with $p_{i,j} > \frac{1}{2}$. Even more restrictive versions can be built by increasing the threshold beyond $\frac{1}{2}$, following the observation [17] that pairs with high probability are seldom mis-predicted.

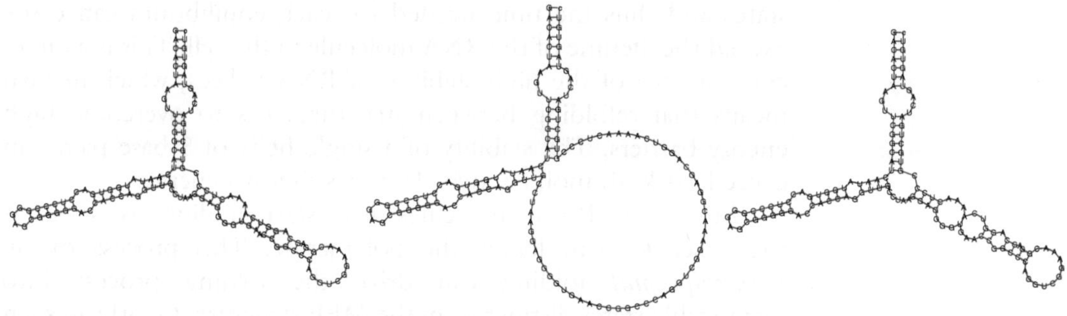

Fig. 4 Structure of the 7SL RNA from human signal recognition particle, comparing MFE, centroid structure, and MEA structure as predicted by RNAfold. The upper and left arm are correctly predicted in all cases, while the lower right arm is shifted compared to the crystal structure in the MFE and MEA structures. Since the pairs in this arm have low probability, the centroid structure leaves that whole region unpaired

Another possibility is to argue that it is better to ask for the structure expected to contain largest number of correct pairs, rather than the one most likely to be perfectly correct. This approach is known as maximum expected accuracy (MEA) folding and has recently been proposed as an alternative to MFE folding [18, 19].

In the simplest case, we might measure accuracy as the number of correct base pairs in our predicted structure. Assuming that the base pair probability p_{ij} is a good measure for the probability that pair (i, j) is correct, then the expected accuracy of a structure is given by

$$\langle A(s) \rangle = \sum_{(i,j) \in s} p_{ij} \qquad (6)$$

The MEA structure is simply the structure maximizing the sum of pair probabilities and can be computed simply using the Nussinov algorithm.

While MEA structures often do perform better in benchmarks than MFE structures, one should be aware that the MEA or centroid structure can well be a very unlikely structure. In particular, the MEA structure can be a mixture of mutually exclusive solutions, while the centroid structure will often be the open chain conformation (i.e., the structure containing no pairs). An example comparing MFE, MEA, and centroid structure prediction is given in Fig. 4.

9 RNA Folding Kinetics

The dynamic programming approaches to RNA folding presented above implicitly assume that RNA molecules are in thermodynamic equilibrium. However, RNAs may exhibit long-lived metastable states and thus the time needed to reach equilibrium can easily exceed the lifetime of the RNA molecule in the cell. This is a direct consequence of the high stability of RNA helices, which in turn means that refolding between structures has to overcome high energy barriers. The stability of a single helix of 5 base pairs can exceed 10 kcal/mol, i.e. over 16 times thermal energy.

Moreover, RNA molecules will start folding as soon as the nascent chain leaves the polymerase. This process of *co-transcriptional* folding can drive the folding process into meta-stable states distinct from the MFE structure. Clearly, in such cases the native structure of the molecule can only be predicted by methods that take the kinetics of co-transcriptional folding into account. Nevertheless, methods that consider RNA folding kinetics are still relatively scarce. Below we briefly describe the most prominent approaches, for a more detailed review see [20].

The most straightforward approaches to kinetic folding try to directly simulate the folding process. Since RNA secondary structures are discrete this usually amounts to a Monte-Carlo simulation rather than the molecular dynamics simulations popular for protein tertiary structures. Two of the most widely used methods of this type are the `Kinefold` of Hervé Isambert [21] and the `Kinfold` program [22] in the Vienna RNA package. The main difference between the two approaches lies in the *move-set* used. `Kinfold` performs a fine-grained simulation using opening and closing of single base pairs as the move-set. `Kinefold`, on the other hand, opens or closes entire helices. Estimating proper rates for these moves is much more difficult and error prone for helix moves, but the larger moves allow simulation of much longer molecules. Moreover, `Kinefold` allows the formation of pseudo-knots. Both programs can either start from a full-length molecule or simulate folding during transcription.

Simulation approaches tend to be time-consuming since they require a large number of simulated trajectories for statistical analysis. An alternative is the direct analysis of folding energy landscapes. Here one is primarily interested in structures that are local energy minima, i.e. meta-stable states, and the energy barriers separating them. Energy barriers are of interest, since they can be readily converted into an estimate for refolding times. The `barriers` program [15] analyzes the output of RNAsubopt to identify local minima and energy barriers and presents the results in form of a *barrier tree*. The leafs of the tree correspond to local minima while the internal nodes represent saddle points connecting the minima, *see* Fig. 5 for an example. Since the number of structures that have to be considered grows exponentially with sequence length, the algorithm is usable only for RNAs up to about 100nt.

The barrier tree can also be used as the basis for a *coarse grained* calculation of folding dynamics. Since the number of states (local

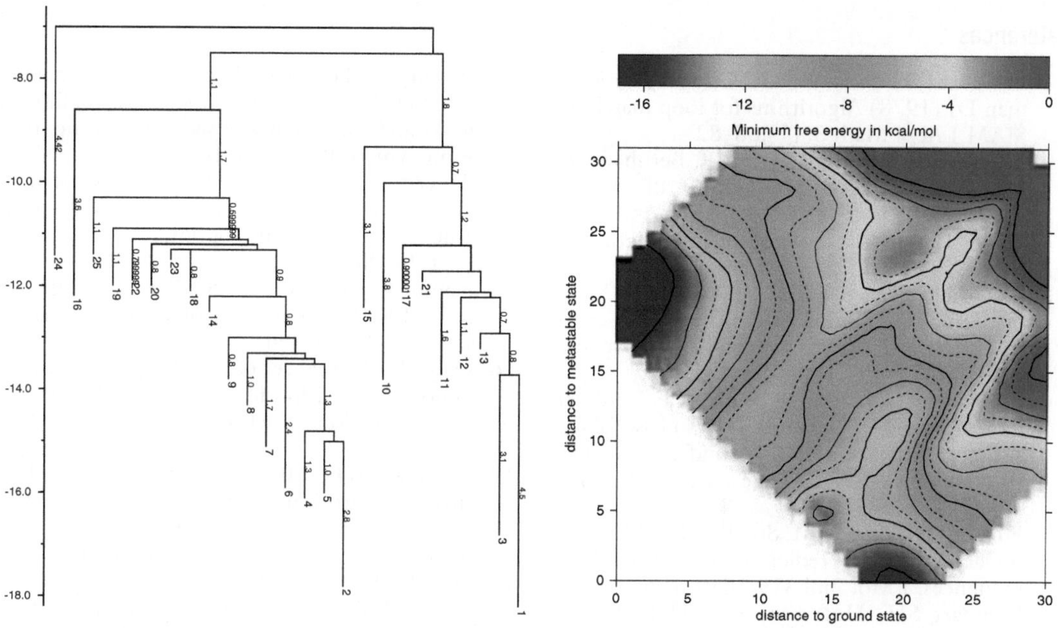

Fig. 5 Barrier tree and 2D projection of the energy landscape for an artificial RNA switch

minima) is small, the folding dynamics can be computed directly by numerically integrating the master equation of the Markov process using the `treekin` program [23]. Landscape analysis and coarse grained dynamics calculations can be performed online on the Vienna RNA websuite [24].

A recent approach that overcomes the length limitations of `barriers` uses classified dynamic programming to compute 2D projections of the energy landscape. Suppose we have already identified the MFE structure as well as one alternative folding of an RNA. All possible structures of the RNA can now be classified by their distances κ, λ from the two reference structures. It is possible to extend the folding recursions (1) to keep track of these two distances and thus compute the MFE and/or partition function over all structures with given values of κ, λ. This increases space and time complexity to $\mathcal{O}(n^4)$ and $\mathcal{O}(n^7)$. As implemented in the RNA2Dfold program of the Vienna package [25], the approach is usable for sequence up to 500nt which covers most biologically interesting cases such as riboswitches.

The resulting projection of the landscape onto the plane spanned by the two distances κ, λ is well suited for visualization and gives qualitative information about the energy barrier separating the two reference states and likely refolding paths. For co-transcriptional folding, the algorithm can compute the 2D landscapes for all subsequence $x[1 \ldots k]$ at no extra cost. The series of 2D projections thus obtained can be visualized as an animation.

References

1. Nussinov R, Piecznik G, Griggs JR, Kleitman DJ (1978) Algorithms for loop matching. SIAM J Appl Math 35(1):68–82
2. Bompfünewerer AF, Backofen R, Bernhart SH, Hertel J, Hofacker IL, Stadler PF, Will S (2008) Variations on RNA folding and alignment: Lessons from Benasque. J Math Biol 56:119–144
3. Hofacker IL, Stadler PF (2006) Memory efficient folding algorithms for circular RNA secondary structures. Bioinformatics 22(10):1172–1176
4. Hofacker IL, Fontana W, Stadler PF, Bonhoeffer S, Tacker M, Schuster P (1994) Fast folding and comparison of RNA secondary structures (the Vienna RNA Package). Monatsh Chem 125(2):167–188
5. Hofacker IL, Fekete M, Stadler PF (2002) Secondary structure prediction for aligned RNA sequences. J Mol Biol 319:1059–1066
6. Bernhart SH, Hofacker IL, Will S, Gruber AR, Stadler PF (2008) RNAalifold: improved consensus structure prediction for RNA alignments. BMC Bioinformatics 9:474
7. Mathews DH, Sabina J, Zuker M, Turner H (1999) Expanded sequence dependence of thermodynamic parameters provides robust prediction of RNA secondary structure. J Mol Biol 288:911–940
8. Mathews DH, Disney MD, Childs JL, Schroeder SJ, Zuker M, Turner DH (2004) Incorporating chemical modification constraints into a dynamic programming algorithm for prediction of RNA secondary structure. Proc Natl Acad Sci USA 101:7287–7292
9. Doshi K, Cannone J, Cobaugh C, Gutell R (2004) Evaluation of the suitability of free-energy minimization using nearest-neighbor energy parameters for RNA secondary structure prediction. BMC Bioinformatics 5(1):105
10. McCaskill JS (1990) The equilibrium partition function and base pair binding probabilities for RNA secondary structure. Biopolymers 29:1105–1119
11. Zuker M, Jacobson AB (1995) "Well-determined" regions in RNA secondary structure prediction: analysis of small subunit ribosomal RNA. Nuclic Acids Res 23:2791–2798
12. Zuker M (1989) On finding all suboptimal foldings of an RNA molecule. Science 244(4900):48–52
13. Wuchty S, Fontana W, Hofacker IL, Schuster P (1999) Complete suboptimal folding of RNA and the stability of secondary structures. Biopolymers 49(2):145–165
14. Waterman MS, Byers TH (1985) A dynamic programming algorithm to find all solutions in a neighborhood of the optimum. Math Biosci 77:179–188
15. Flamm C, Hofacker IL, Stadler PF, Wolfinger MT (2002) Barrier trees of degenerate landscapes. Z Phys Chem 216:155–173
16. Ding Y, Lawrence CE (2003) A statistical sampling algorithm for RNA secondary structure prediction. Nucleic Acids Res 31:7280–7301
17. Mathews DH (2004) Using an RNA secondary structure partition function to determine confidence in base pairs predicted by free energy minimization. RNA 10(8):1178–1190
18. Do CB, Woods DA, Batzoglou S (2006) CONTRAfold: RNA secondary structure prediction without physics-based models. Bioinformatics 22(14):e90–e98
19. Kiryu H, Kin T, Asai K (2007) Robust prediction of consensus secondary structures using averaged base pairing probability matrices. Bioinformatics 23(4):434–441
20. Flamm C, Hofacker IL (2008) Beyond energy minimization: Approaches to the kinetic folding of RNA. Monatsh f Chemie 139(4):447–457
21. Isambert H, Siggia ED (2000) Modeling RNA folding paths with pseudoknots: application to hepatitis delta virus ribozyme. Proc Natl Acad Sci USA 97(12):6515–6520
22. Flamm C, Fontana W, Hofacker IL, Schuster P (2000) RNA folding kinetics at elementary step resolution. RNA 6:325–338
23. Wolfinger MT, Andreas Svrcek-Seiler W, Flamm C, Hofacker IL, Stadler PF (2004) Efficient folding dynamics of RNA secondary structures. J Phys A Math Gen 37: 4731–4741
24. Gruber AR, Lorenz R, Bernhart SH, Neuböck R, Hofacker IL (2008) The Vienna RNA websuite. Nuclic Acids Res 36:W70–W74
25. Lorenz R, Flamm C, Hofacker IL (2009) 2D projections of RNA folding landscapes. In: Grosse I, Neumann S, Posch S, Schreiber F, Stadler PF, (eds) German conference on bioinformatics 2009, vol 157 of Lecture notes in informatics, pp 11–20, Bonn. Gesellschaft f Informatik

Chapter 5

Introduction to Stochastic Context Free Grammars

Robert Giegerich

Abstract

Stochastic context free grammars are a formalism which plays a prominent role in RNA secondary structure analysis. This chapter provides the theoretical background on stochastic context free grammars. We recall the general definitions and study the basic properties, virtues, and shortcomings of stochastic context free grammars. We then introduce two ways in which they are used in RNA secondary structure analysis, secondary structure prediction and RNA family modeling. This prepares for the discussion of applications of stochastic context free grammars in the chapters on RFAM (6), *Pfold* (8), and INFERNAL (9).

Key words RNA structure prediction, Stochastic grammars, RNA family models, Ambiguity

1 Stochastic Context Free Grammars: Definitions

Stochastic context free grammars(SCFGs) were introduced in bioinformatics for the purpose of modeling RNA secondary structure, the original early references being [1, 2]. In computer science, stochastic grammars have a longer tradition and were studied and used mainly in the field of natural language processing. Early references to that literature are [3, 4]. Both lines of work have evolved quite independently, and the terminology is not always the same. We will adhere to the terminology and notational conventions used in bioinformatics. Let us start with a one-sentence summary of what SCFGs are.

For our readers who are familiar with stochastic modeling and have worked with Hidden Markov Models (HMMs):

> SCFGs are HMMs where the linear path of state transitions is replaced by a tree of states.

For our readers with a background in computer science and formal language theory:

> SCFGs are context free grammars augmented with a probabilistic scoring scheme.

Jan Gorodkin and Walter L. Ruzzo (eds.), *RNA Sequence, Structure, and Function: Computational and Bioinformatic Methods*, Methods in Molecular Biology, vol. 1097, DOI 10.1007/978-1-62703-709-9_5,

For our readers with a background of neither type:

Don't worry, this chapter assumes neither kind of previous experience, but starts from first principles.

1.1 Context Free Grammars

1.1.1 Grammars, Languages, and Derivations

Let $\mathcal{A}$ be a finite set of symbols, called the *alphabet*. $\mathcal{A}^*$ denotes the set of all finite strings of symbols from $\mathcal{A}$, including the empty string, which has no symbols at all. To make it visible, we write ϵ for the empty string. $|x|$ denotes the length of the string x. x^{-1} denotes the reverse of x. A *formal language* L is simply a subset of $\mathcal{A}^*$. Elements of L are called "words" in formal language theory, "phrases" or "sentences" in linguistics, or "sequences" in bioinformatics.

Formal languages can be described in many ways—the most popular one is the use of grammars.

Definition 1: A *context free grammar* G is a formal system that generates a language $L(G) \subseteq \mathcal{A}^*$. It uses a set V of *non-terminal symbols*, one of which is designated as the *axiom*, and a set of *derivation rules* (also called *productions*) that have the form $X \to \alpha$, where $X \in V$ and $\alpha \in (V \cup \mathcal{A})^*$.

We shall use uppercase letters for non-terminal symbols and lowercase for symbols from $\mathcal{A}$. The productions serve to derive the words of the language, starting from the axiom. This will be defined more precisely in a moment. In contrast to the nonterminal symbols from V, the symbols from $\mathcal{A}$ are called terminal symbols, because once generated in the course of a derivation, they are never replaced.

It is customary to collect all rules for the same non-terminal symbol in a rule with a single left hand side, and several alternatives on the right, separated by a '|'. Hence, $\{A \to \alpha_1,\ A \to \alpha_2,\ A \to \alpha_3\}$ becomes $\{A \to \alpha_1 \mid \alpha_2 \mid \alpha_3\}$.

Figure 1 shows three simple grammars. There, we have associated names with all productions, which is not foreseen in the standard definition of CFGs, but will be convenient for later reference.

Next we make precise the notion of a derivation:

Definition 2: A derivation of a word $w \in L(G)$ starts with the axiom symbol, and in each step, replaces one of the non-terminal symbols in the emerging string according to one of the productions: When a derivation has already produced (say) xXy, and $X \to \alpha$ is a production of G, we may rewrite xXy to $x\alpha y$.

A derivation can be represented uniquely in the form of a tree. Figure 2 shows several derivation trees for the grammars of Fig. 1. The inner nodes of the tree are labeled with the production names. Terminal symbols are leaf nodes in the derivation tree, and considering just these leaves in left-to-right order, we obtain the string $w \in L(G)$ produced by the derivation.

Grammar *DotPar*.
$\mathcal{A} = \{(,),.\}$, $V = \{S\}$, axiom is S.

production rule	rule name
$S \to \varepsilon$	*end*
$\mid .S$	*dot*
$\mid (S)S$	*pairsplit*

Grammar *Pali*.
$\mathcal{A} = \{a, b\}$, $V = \{P, T\}$, axiom is P.

production rule	rule name
$P \to T$	*turn*
$\mid xPy$	$pair_{xy}$, for $x, y \in \mathcal{A}$ and $x = y$
$T \to \varepsilon$	*end*
$\mid xT$	add_x, for $x \in \mathcal{A}$

Grammar *Ali*.
$\mathcal{A} = \{a, c, g, t, \$\}$, $V = \{A\}$, axiom is A.

production rule	rule name	
$A \to xAy$	$match_{xy}$,	for $x, y \in \mathcal{A} \setminus \{\$\}$ and $x = y$
$\mid xAy$	rep_{xy},	for $x, y \in \mathcal{A} \setminus \{\$\}$ and $x \neq y$
$\mid xA$	del_x,	for $x \in \mathcal{A} \setminus \{\$\}$
$\mid Ax$	ins_x,	for $x \in \mathcal{A} \setminus \{\$\}$
$\mid \$$	*end*	

Fig. 1 Three context free grammars. *DotPar* describes RNA structure in dot-parenthesis notation, *Pali* describes palindromes, and *Ali* DNA sequence alignments, allowing for simple insertions and deletions, matches and replacements. Production rules have been named for later reference. Grammars *Pali* and *Ali* use abbreviation. For example in grammar *Ali*, the rule $A \to xAy$ named $match_{xy}$ actually stands for four distinct rules for the indicated choices of x and y with $x = y$, and the rule rep_{xy} has 12 variants.

We can also write derivations as formulas: Consider grammar *Ali* with its derivation rules *match, del, ins, rep, end*. Interpreting the production names as functions which actually construct trees, we can write a derivation tree as an expression, such as $match(c, del(a, ins(rep(t, end(\$), g), g)), c)$ for the *Ali*-tree in Fig. 2.

Which are the languages described by the grammars in Fig. 1?

- Grammar *DotPar* allows to derive all strings which denote RNA secondary structures in the familiar dot-bracket notation, where a dot means an unpaired base and matching parentheses denote two bases forming a base pair.
- Grammar *Pali* allows to derive palindromic strings of the form $p = uvu^{-1}$, which allows for a "turn" v of any length in the middle, where the characters do not have to match up.
- Grammar *Ali* allows to derive two strings from the DNA alphabet, separated by a '$.' Thus, $L(Ali)$ by itself does not appear very interesting and could be described by an even simpler grammar. But consider *how Ali* derives these strings: A derivation of $x\$y$ actually models an alignment of x and y^{-1}, allowing for base matches and replacements, insertions and deletions. This is made apparent by the production names we have chosen.

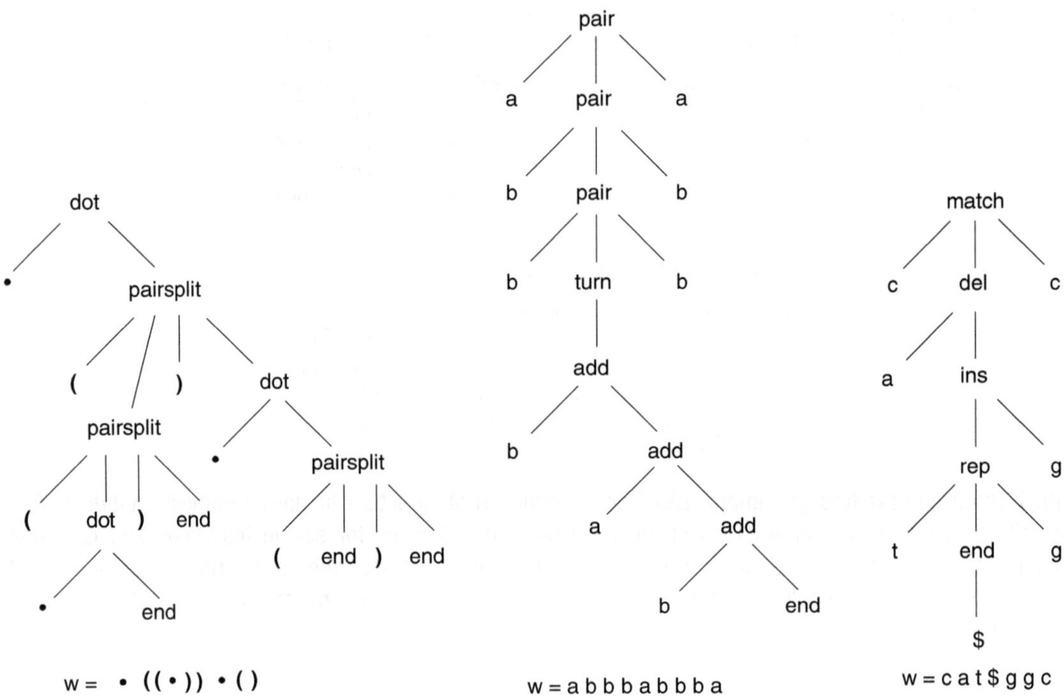

Fig. 2 Three derivation trees for the grammars of Fig. 1. The *DotPar* tree (*left*) makes explicit the base pairing in the RNA structure representation w = ".((.)).()": matching parenthesis, denoting base pairs, are jointly produced by an application of the *pairsplit* rule. This is the only derivation tree for this w. The *Pali* tree (*middle*) indicates a palindromic structure of w = "abbbabbba, " where the innermost three letters are considered part of the turn—this is one out of five different derivation trees for this string. The *Ali* tree describes one particular alignment (out of many) for cat and cgg, written as input w = cat$ggc. This particular alignment uses each edit operation once.

1.1.2 Grammar Normal Forms

The standard definition allows CFGs to contain certain elements of "junk." (i) There could be non-terminal symbols which are useless because they can never be reached in a derivation starting from the axiom. (ii) A rule like $D \rightarrow BD$, when other alternative rules for D are lacking, would allow for endless, unproductive derivations, which never get rid of the D non-terminal symbol. And (iii) there could be unproductive derivation cycles $C \rightarrow B \rightarrow D \rightarrow C$ that do not produce any terminal symbols. This would blow up, without need, the number of possible derivations for any string that is derived using C. We call a grammar *clean* when it does not have any of these features. Grammars *DotPar*, *Pali*, and *Ali* are clean. Our discussion in the sequel will tacitly assume that grammars are clean.

Even with clean grammars, there are many ways to describe the same formal language with different grammars, i.e. $L(G_1) = L(G_2)$. This is important, as some grammars are more convenient algorithmically than others. In particular, there is always the following normal form, which has especially simple production rules:

Definition 3: A grammar is in *Chomsky Normal Form*,[1] if each right-hand side holds no more than two symbols from $V \cup \mathcal{A}$.

Chomsky Normal Form of a grammar can always be achieved without changing the language. For example, the rule

$$S_1 \to aS_2bS_1S_3c$$

can be replaced by

$$S_1 \to aA,\ A \to S_2B,\ B \to bC,\ C \to S_1D,\ D \to S_3c$$

where A, B, C, D are new non-terminal symbols. Such a transformation of the productions does not affect the language $L(G)$, but makes the grammar more handsome for the issue we study next: parsing.

1.1.3 Parsing and Ambiguity

Given $w \in \mathcal{A}^*$, we want to solve the *word problem*, also known as *syntax checking*: Is $w \in L(G)$? This question is answered by constructing a derivation tree for x, or by showing that no such tree exists.

Definition 4: The construction of a derivation tree for given $w \in \mathcal{A}^*$ is called *parsing*, and the derivation tree (if any) is called a *parse tree* in this context. A CFG G is syntactically *ambiguous*, if there is more than one parse tree for some w. By $T_G(w)$ we denote the set of all parse trees for w.

Syntactic ambiguity of context free grammars is sometimes welcome, sometimes a nuisance. For example, when grammars are used to define the syntax of programming languages, ambiguity is to be avoided. If (say) a Java program could be parsed in two ways, machine code implementing different algorithms might be generated from it! Therefore, programming language research mainly studies non-ambiguous grammars, which also allow for very efficient parsing algorithms. An algorithm for checking ambiguity of grammars was intensively sought for—until this problem was proved to be algorithmically undecidable in the 1960s [5].

The situation is completely different in natural language processing, as well as in bioinformatics. Syntactic correctness, $w \in L(G)$, is taken for granted. Here, grammars are typically ambiguous. We have a large number of parse trees for any given word, and we want to select from them a most plausible one, based on a suitable scoring scheme and objective function. Hence, with ambiguous grammars, parsing naturally generalizes to an optimization problem. Adding a probabilistic scoring scheme to CFGs will take us to SCFGs in the next section. Scoring instead with a thermodynamic energy model takes us to RNA folding, as discussed in the Chapter 4 in this book.

[1] Our definition is a bit more relaxed than the one found in the formal language literature.

Considering our example grammars, we note the following:

- Grammar *DotPar* is non-ambiguous. This is easy to check: Given any word $w \in L(DotPar)$, generate a derivation for it, producing w from left to right. In each step, exactly one of the three rules applies, so there are never two different derivation trees for any w.
- With grammar *Pali*, when deriving, for example, *abbbabbba*, we find that there are five different derivations, depending whether we choose $w = uvu^{-1}$ with $u = abbb$ and $v = a$, $u = abb$ and $v = bab$, $u = ab$ and $v = bbabb$, $u = a$ and $v = bbbabbb$, or $u = \varepsilon$ and $v = abbbabbba$. You may argue that the first choice is the "most palindromic" one, but this is sort of your personal opinion. The grammar, per se, does not make such preference.
- Grammar *Ali*, finally, has as many derivations for $x\$y$ as there are alignments of x and y^{-1}. No preference is expressed within the grammar itself, but of course, we can add a suitable scoring scheme and search for an optimal alignment.

1.1.4 CYK Parsing

A simple parsing algorithm for ambiguous CFGs, which finds all parses for a word, was independently suggested by Cocke, Younger, and Kasami in the 1960s. It is commonly refered to as the CYK-algorithm in bioinformatics.[2]

To decide if (and how) a string w can be parsed according to a particular production, CYK splits w into as many parts as there are symbols on the right-hand side. The split is done in all possible ways, and the i-th substring in a split is derived (if possible) from the corresponding i-th symbol in the right-hand side. Using our production names as tree construction functions, the logic of a CYK algorithm for a grammar is easily written as a recursive function: $parses_S(w)$ constructs all derivation trees for a word w, starting from non-terminal symbol S. For grammar *DotPar* this leads to the recurrence

$$\begin{aligned} parses_S(w) = {} & \{end \mid w = \varepsilon\} \\ & \cup \{dot('.', x) \mid w = '.'v,\ x \in parses_S(v)\} \\ & \cup \{pairsplit('(', x, ')', y) \mid w = '('u')'v,\ x \in parses_S(u), \\ & \quad y \in parses_S(v)\} \end{aligned}$$

[2] Amusingly, the natural language processing community consistently refers to it as the CKY algorithm. The reason for such confusion is that there is no joint paper by these three authors. Reference [6] tells the story behind CYK. This classical textbook presents CYK because of its "intuitive simplicity," but remains "doubtful, however, that it will find practical use." Those were the days when a state-of-the-art computer had 65K bytes of memory.

In the *pairsplit* case, the split of w must be done in all possible ways, but only the split which chooses the matching '(' and ')' will be successful and deliver a parse tree. In general, a CYK parser has a parser $parse_X$ for each $X \in V$, and the parser has as many cases to distinguish as there are alternative rules for X. The axiom parser called with input w produces all parse trees for w, i.e. $T_G(w)$.

Here, it is important that the grammar is clean, otherwise the recursion in the parser may not terminate. Our $parse_S$ function constructs a derivation in a top-down fashion. The order of computation may also be reversed, computing parses proceeding from smaller to larger subwords. This is normally done in CYK parsing. To achieve good efficiency, CYK parsers tabulate their results for each subword on which they are called, to avoid recomputation in the case of another call on the same subword. When $|w| = n$, there are at most $O(n^2)$ subwords. If the grammar is in Chomsky Normal Form, a right-hand side requires no more than $|w|$ different splits, and the CYK algorithm runs in $O(n^3)$ time and $O(n^2)$ space. However, the idea behind CYK is not restricted to grammars in Chomsky Normal Form. A production of form $A \rightarrow BCD$, for example, will increase runtime to $O(n^4)$. Two extra, nested loops are required to iterate over all internal splits between BC and CD.

In the case of *DotPar*, since the grammar is non-ambiguous, at most one of the sets on the right-hand side of $parse_S$ contains a derivation tree. For ambiguous grammars, several sets can hold multiple trees, and with productions like $A \rightarrow BC$, the numbers of parses for B and C multiply when considering all parses for A. In general, with an ambiguous grammar, the number of parses grows exponentially with the length of the input string.

1.2 Stochastic CFGs

To make a rational choice between multiple parses, a CYK parser is typically equipped with a scoring scheme. Parse trees are scored and selected based on their scores on-the-fly—this combination of parsing, scoring, and choice is commonly known as dynamic programming [7]. If the scoring is based on a probabilistic model, this brings us to SCFGs.

Definition 5: A *stochastic* context free grammar G is a CFG which associates with each production rule r a transition probability π_r. For all $A \in V$, with $A \rightarrow \alpha_1 \mid \ldots \mid \alpha_k$, with these alternatives named $r_1, \ldots, r_k$, $\sum_{i=1}^{k} \pi_{r_i} = 1$ must hold. The *probability* $P_G(t)$ *of a derivation* (or parse tree) t is the product of the π_{r_i} for all uses of productions r_i in t. The *word probability* w assigned by grammar G is defined as $P_G(w) = \sum_{t \in T_G(w)} P_G(t)$.

Note that by this definition, $P_G(w) = 0$ if $w \notin L(G)$. Different grammars for the same language, naturally, assign different probabilities to words and parses. When the grammar is not relevant or clear from the context, we omit the G with $P_G(w)$ and $P_G(t)$. $P(t)$ is sometimes called the *joint probability* of parse t and word w, to emphasize that the tree t includes w as its string of leaves. In Fig. 3

Probabilites assigned to *DotPar*

$$\pi_{\text{end}} = 0.1$$
$$\pi_{\text{dot}} = 0.5$$
$$\pi_{\text{pairsplit}} = 0.4$$

Probabilites assigned to *Pali*

$$\pi_{\text{turn}} = 0.1$$
$$\pi_{\text{pair } aa} = 0.7$$
$$\pi_{\text{pair } bb} = 0.2$$
$$\pi_{\text{end}} = 0.1$$
$$\pi_{\text{add } a} = 0.4$$
$$\pi_{\text{add } b} = 0.6$$

Probabilites assigned to *Ali*

$\pi_{\text{match } xx}$	$= 0.2$	for 4 choices of x
$\pi_{\text{replace } xy}$	$= 0.033$	for 12 choices of x, y
$\pi_{\text{del } x}$	$= 0.025$	for 4 choices of x
$\pi_{\text{ins } x}$	$= 0.025$	for 4 choices of x
π_{end}	$= 0.0004$	

Fig. 3 Parameters assigned to grammars *DotPar*, *Pali*, and *Ali*. With these assignments, the trees in Fig. 2 have probabilities $\pi^4_{\text{end}} \cdot \pi^3_{\text{dot}} \cdot \pi^3_{\text{pairsplit}} = 8 \cdot 10^{-7}$, $\pi^1_{\text{pair aa}} \cdot \pi^2_{\text{pair bb}} \cdot \pi^1_{\text{turn}} \cdot \pi^1_{\text{add a}} \cdot \pi^2_{\text{add b}} \cdot \pi^1_{\text{end}} = 0.4032 \cdot 10^{-3}$, and $\pi_{\text{match cc}} \cdot \pi_{\text{del a}} \cdot \pi_{\text{ins g}} \cdot \pi_{\text{rep tg}} \cdot \pi_{\text{end}} = 1.65 \cdot 10^{-9}$, respectively.

we associate probabilities with the rules of grammars *DotPar*, *Pali*, and *Ali*.

The relationship between parses and their probability can be expressed in a neat way by considering a parse tree t as a formula, say $t = dot(a, pairsplit(c, end, g, end))$. By reinterpreting the production names as scoring functions

$$\begin{aligned} dot(x, s) &= \pi_{\text{dot}} \cdot s \\ pairsplit(x, s, y, s') &= \pi_{\text{pairsplit}} \cdot s \cdot s' \\ end() &= \pi_{\text{end}} \end{aligned}$$

we obtain

$$\begin{aligned} P(t) &= dot('.', pairsplit('(', end, ')', end)) \\ &= \pi_{\text{dot}} \cdot (\pi_{\text{pairsplit}} \cdot \pi_{\text{end}} \cdot \pi_{\text{end}}) = 0.002 \end{aligned}$$

under the above assignment of probabilities.

Under mild conditions, an SCFG defines a probability distribution on $L(G)$, i.e. $\sum_{w \in L(G)} P_G(w) = 1$. Here, it is important that the grammar is clean. In a clean grammar, the sum defining $P_G(w)$ is finite, and no probability mass gets lost by entering unproductive derivations.

A probability measure defined in this way has a number of properties one should be aware of. For example, since multiplication is a commutative and associative operation, $P_G(t)$ can always be re-factored as $\pi^{n_1}_{r_1} \cdot \pi^{n_2}_{r_2} \cdot \ldots \cdot \pi^{n_k}_{r_k}$, where n_i denotes the number of times rule r_i is applied in t. Hence, all derivations which use the same rule the same number of times have the same probability, irrespective of their arrangement in the derivation.

Looking more closely at *DotPar*, we find that a derivation tree with k *pairsplit*-nodes must have $k + 1$ *end*-nodes. Hence, for a structure w with k base pairs and l unpaired bases, we have $P_{\text{DotPar}}(w) = \pi_{\text{dot}}^{l} \cdot (\pi_{\text{pairsplit}} \cdot \pi_{\text{end}})^k \cdot \pi_{\text{end}}$. This stochastic model simply scores the number of base pairs against the number of unpaired bases! It is insensitive to their arrangement, which makes it certainly a rather crude structure model. Larger grammars not only can capture more structural features but also have more parameters and require more data to derive concrete parameters from. This balance must be carefully chosen. Grammar *DotPar* is an extreme choice: it is wellsuited for expository purposes, because it is so small, but not useful for practical modeling.

Another property of this type of model is that, since all rule probabilities multiply over a derivation, and all $\pi_r < 1$, longer words tend to have small probability. In fact, for any $\varepsilon > 0$, there is a limit N such that for all "long" words with $|w| > N$, we have $P(w) < \varepsilon$. See [3] for a detailed discussion of such issues. Should we intend to study probabilities assigned to words of varying length, we should normalize scores with respect to word length. This can be achieved, for example, by dividing the probability $P_G(w)$ by the probability of generating the *sequence* w from a background distribution.

Definition 6: Three important algorithms are encountered with SCFGs:

- The most-likely-parse algorithm or "Viterbi-Algorithm": Given w, compute $p^* = \max_{t \in T_G(w)} P(t)$, and also some or all t^* such that $P(t^*) = p^*$. In words: t^* is the most likely parse, and p^* is its probability.
- The "Inside-Algorithm": Given w, compute the word probability $P(w)$, as defined above as the sum over all parse trees of w.
- The "Outside Algorithm": Given $w = xyz$, compute $\sum_{t \in T_G(w)} P(t[y])$ where $t[y]$ is a derivation tree for xyz *excluding* the subtree which derives subword y.

The name "Viterbi" for the most-likely-parse algorithm is borrowed from HMM terminology. Both the Viterbi and the Inside algorithms are commonly based on a CYK parsing algorithm, equipped with a different handling of multiple parses. The Viterbi algorithm only pursues the most likely parse at each point, while the Inside algorithm takes the probability sum over all parses. In the bioinformatics literature, one occasionally finds the name CYK used with the meaning of most-likely-parse. At least for the present chapter, we avoid this confusion. The outside algorithm is relevant to parameter training and uses a recursion scheme different from CYK [8].

An SCFG G can always brought into Chomsky Normal Form while preserving the probability distribution it defines on $L(G)$. Let there be a rule named r, which is not in Chomsky Normal Form, say

$$r: \quad A \rightarrow BCD \quad : \pi_r$$

We can savely replace it by two rules

$$r_1: \quad A \rightarrow BX \quad : \pi_{r_1} = \pi_r$$

$$r_2: \quad X \rightarrow CD \quad : \pi_{r_2} = 1$$

(Remember that X is a new non-terminal symbol with no other rules.) While the original rule would cause a runtime of $O(n^4)$ of the CYK parser, the transformed grammar only requires $O(n^3)$. Since this transformation is always possible,[3] feel free to use the most natural form when you design an SCFG, and use Chomsky Normal Form only for the implementation.

1.3 Connecting SCFG to HMM Terminology

Hidden Markov Models (HMMs) are stochastic models based on state transitions. Each (hidden) state from a state set S emits an observable symbol from alphabet $\mathcal{A}$ and enters a subsequent state. A series of transitions, also called a state path, thus emits a sequence of symbols. Emissions and transitions are made with a certain probability, and these probabilities multiply along a state path. HMMs are often depicted as state transition diagrams, or as transition probability matrices (assuming any state can transit into any state) and emission probability vectors (assuming any state can emit any $a \in \mathcal{A}$). However, the grammar view does also apply: A combination of emission and transition can be written as a rule

$$S_1 \rightarrow aS_2$$

indicating that a symbol a is emitted upon transit from state S_1 to S_2. This resembles a production rule, where states become the non-terminal symbols. In fact, this rule has a very simple form, as it holds only one non-terminal symbol on the right-hand side. Grammars under this restriction are called *regular grammars* in formal language theory, and from this point of view, HMMs are SCFGs where the underlying grammar is regular. In the frequent case where any state can transit into any state and emit any symbol, albeit with different probability, the language generated by this grammar is trivially $\mathcal{A}^*$. This is why formal language terminology is not very useful with HMMs.

[3] When other types of scoring schemes are associated with G, such an efficiency improving transformation may not be possible. This has been called the *yield parsing paradox* in dynamic programming [7].

The generalization towards SCFGs, formulated in HMM terminology, allows each state to generate not a single successor state, but any (fixed) number of immediate successors. For example, a transition could be

$$S_1 \rightarrow aS_2 bS_1 cS_3 d$$

with several symbols a, b, c, d emitted simultaneously and successor states S_2, S_1, S_3 created. The stochastic process branches after such a transition, and further transitions from the generated states S_1, S_2, S_3 proceed independently. The view of a state "path" breaks down, as the transitions branch off into a tree. Transition diagrams become unreadable, and transition matrices are no longer convenient. This is why the HMM terminology is not very useful with SCFGs.

Considering the HMM tradition of separating state transition and symbol emission, we note that SCFGs, per se, do not need to distinguish between emission and transition probabilities. Nor are they restricted to emit one symbol at a time. However, we can make a grammar more HMM-like by rewriting it a bit (without changing the language). We can transform it such that some non-terminal symbols produce only terminal symbols—the "emissions," and others only produce non-terminal symbols—the "transitions," and re-factor the probabilities accordingly. The above rules

$$S_1 \rightarrow aS_2 \text{ for } a \in \mathscr{A}$$

would be rewritten to

$$S_1 \rightarrow AS_2$$

$$A \rightarrow a \mid b \mid c \mid \ldots$$

and the original probabilities re-assigned as a transition probability with the first rule, and as emission probabilities with the second. But note that any probability for two (or more) symbols emitted jointly as in $S \rightarrow aSb$ must remain associated with the production that does the joint emission.

The HMM counterparts of the most-likely-parse, Inside and Outside algorithms are the Viterbi, Forward and Backward algorithms in HMM terminology. In fact, when the grammar underlying our SCFG is a regular grammar, practically the same computation is performed.

1.4 Semantics of Stochastic Models

So far, we have discussed SCFGs purely as a formal device. An SCFG G assigns a probability $P_G(w)$ to a word $w \in L(G)$, and a probability $P_G(t)$ to each of its derivations t. When it comes to using stochastic models for the analysis of real-world objects or phenomena, it must be specified what the computed probabilities refer to in reality. The stochastic model must be given a semantics. We will do this twice in the second part of this chapter, where we use SCFGs to derive a most likely structure for a single RNA sequence, and to compute the likelihood of the sequence belonging to a particular RNA family for which an SCFG family model has been created.

2 Analyzing RNA Secondary Structure with SCFGs

The use of SCFGs for RNA structure analysis has two forms: An SCFG can be used to assign a structure to an RNA sequence, and it can be used to build a model of an RNA sequence family with a conserved structure. The present section introduces both kinds of use and then links to the applied chapters of this book where these scenarios are treated in detail.

2.1 SCFGs Modeling Secondary Structure of a Single RNA Sequence

2.1.1 SCFG Parses Indicating RNA Secondary Structure

For secondary structure prediction, we use grammars which generate RNA *sequences*, and whose parses designate potential secondary *structures* for these sequences. Let $F(w)$ be the folding space of sequence w, i.e. the set of all structures w can fold into according to the chosen rules for base pairing, and let there be a semantic mapping $\mu : T_G(w) \to F(w)$, which relates parses to structures. Most desirable, this mapping is bijective: Surjectivity of μ ensures that the CYK parser actually evaluates all possible foldings. Injectivity of μ guarantees that the most likely parse also denotes the most likely structure under the given parameters.

Recall grammar *DotPar*. It generates dot-bracket strings, each of which denotes an RNA secondary structure, independent of a particular sequence of bases. As we have seen, each dot-bracket string has a unique parse tree. If we replace the terminal alphabet and make the grammar generate RNA sequences instead, each parse with this grammar will indicate a secondary structure for that sequence. In this way, we obtain the grammar *MiniRNA*—*see* Fig. 4.

Grammar *MiniRNA*.
$\mathcal{A} = \{a, c, g, u\}$, $V = \{S\}$, axiom is S.

production rule	rule name
$S \to \varepsilon$	*end*
$\mid xS$	dot_x for $x \in \mathcal{A}$
$\mid xSyS$	$pairsplit_{xy}$ for $x, y \in \mathcal{A}$

Fig. 4 Grammar *MiniRNA* is modeled after grammar *DotPar*. Note that rule *pairsplit* generates arbitrary pairs of bases. Stochastic parameters trained from real data will associate high probabilities with canonical base pairs (a–u, c–g, g–u), but allow nonstandard base pairs with a small probability

Grammar *MiniRNA* is shaped after *DotPar*, using isomorphic rules, but deriving an RNA sequence rather than a dot-bracked string. For a parse tree $t \in T_{\mathrm{MiniRNA}}(w)$, let $\hat{t}$ be the isomorphic tree of grammar *DotPar*. $\hat{t}$ derives a dot-bracket string s, and this s is the structure assigned to w by t—we define $\mu(t) = s$. Since we already know that any dot-bracket string s has only one parse tree in *DotPar*, in our case $\hat{t}$, the parse tree t is the only parse tree of grammar *MiniRNA* with $\mu(t) = s$. This means that with *MiniRNA*, the most likely parse tree t^* indicates the most likely structure.

2.1.2 Semantic (Non-)Ambiguity

What if μ is not injective, i.e. several parse trees indicate the same secondary structure? This situation is called *semantic ambiguity.* This is a subtle pitfall with SCFGs, which, when ignored, will make your SCFG modeling ill-defined. Let us cast this in general terms, following [9].

Definition 7: Let G be an SCFG and $\mathcal{M}$ a set of objects of interest, called meanings. A *semantics* for G is a mapping $\mu : \{T_G(w) \mid w \in L(G)\} \to \mathcal{M}$. The probability of object $m \in \mathcal{M}$ given w is defined as $P(m) = \sum\{P(t) \mid t \in T_G(w), \mu(t) = m\}$. G is *semantically ambiguous*, if there is a $w \in L(G)$ with $t, t' \in T_G(w)$ such that $t \neq t'$ and $\mu(t) = \mu(t')$.

With semantic ambiguity, the probability of any object m is distributed over all the parses t where $\mu(t) = m$. When interested in the most likely object m^*, it does not help to compute the most likely parse t^*, since generally, $\mu(t^*) \neq m^*$. $P(m^*) > P(t^*)$ can occur, with several parses t' contributing to $P(m^*)$, all having $p(t') < p(t^*)$. In fact, mistaking $\mu(t^*)$ for m^* can be strongly misleading, as has been evaluated empirically in [10]. The Viterbi algorithm is not meaningful with a grammar that is semantically ambiguous.

Semantic ambiguity is discussed in HMM literature under the name "path labeling problem," where a path labeling corresponds to our meaning function μ, and maps the state paths of an HMM to more abstract objects of interest. In [11] it was shown that the problem of computing the optimal path labeling (or in our terms: the optimal meaning) is NP-hard in general. As SCFGs properly subsume HMMs, this also applies to SCFGs, *in general.* It has not been shown whether this holds *in particular* for SCFGs with μ mapping parses to structures, but can be taken as a warning that computing $P(m^*)$ from a semantically ambiguous SCFG may be intractable. Therefore, it seems advisable to avoid semantic ambiguity altogether.

As with syntactic grammar ambiguity, there is no algorithm which, given a grammar G and a meaning function μ, can decide whether G is semantically ambiguous with respect to μ. This problem was shown to be undecidable in [12]. However, we shall point out a pragmatic approach to semantic ambiguity checking below.

Grammar *RNAFeatures*.
$\mathcal{A} = \{a, c, g, u\}$,
$V = \{Externalloop, \ldots, MLComponents\}$, axiom is *ExternalLoop*.

production rule		rule name	
ExternalLoop	$\rightarrow \varepsilon$	el_1	
	$\mid$ *x ExternalLoop*	$el_{2,x}$	for $x \in \mathcal{A}$
	$\mid$ *Stack ExternalLoop*	el_3	
Stack	$\rightarrow$ *x Stack y*	$st_{1,xy}$	for $x, y \in \mathcal{A}$
	$\mid$ *x Weak y*	$st_{2,xy}$	for $x, y \in \mathcal{A}$
Weak	$\rightarrow$ *HairpinLoop*	wk_1	
	$\mid$ *InternalLoop*	wk_2	
	$\mid$ *BulgeLeft*	wk_3	
	$\mid$ *BulgeRight*	wk_4	
	$\mid$ *MultiLoop*	wk_5	
HairpinLoop	$\rightarrow$ *x SingleStrand y*	hl_{xy}	for $x, y \in \mathcal{A}$
InternalLoop	$\rightarrow$ *x SingleStrand Stack SingleStrand y*	il_{xy}	for $x, y \in \mathcal{A}$
BulgeLeft	$\rightarrow$ *x SingleStrand Stack y*	bl_{xy}	for $x, y \in \mathcal{A}$
BulgeRight	$\rightarrow$ *x Stack SingleStrand y*	br_{xy}	for $x, y \in \mathcal{A}$
MultiLoop	$\rightarrow$ *x ML_Component ML_Components y*	ml_{xy}	for $x, y \in \mathcal{A}$
ML_Component	$\rightarrow$ *SingleStrand Stack*	co_1	
	$\mid$ *Stack*	co_2	
ML_Components	$\rightarrow$ *ML_Component ML_Components*	cs_1	
	$\mid$ *ML_Component SingleStrand*	cs_2	
	$\mid$ *ML_Component*	cs_3	
SingleStrand	$\rightarrow$ *x SingleStrand*	$ss_{1,x}$	for $x \in \mathcal{A}$
	$\mid$ x	$ss_{2,x}$	for $x \in \mathcal{A}$

Fig. 5 Grammar *RNAFeatures*, which explicitly identifies structural components

2.1.3 Grammar Design

Grammar *MiniRNA* cannot be expected to be a very useful modeling device. Remember what we observed about *DotPar*: it merely weights base pairs against unpaired bases. *MiniRNA* has different parameters for each base or base pair, but else has little extra distinctive power. For example, all unpaired residues are treated alike, no matter whether they reside in a hairpin loop, in a bulge, or in the external loop. All C–G pairs contribute the same score, independent of their structural context, and so on. Let us create a grammar which gives us more control.

Grammar *RNAFeatures* is designed to explicitly designate the different structural features which humans refer to when speaking about secondary structure. *See* Fig. 5.

Grammar *RNAFeatures* uses more non-terminal symbols and rules than *MiniRNA* to explicitly designate multiloops, bulges, internal loops, and so on. If a structure contains two multiloops, it is because the rule named ml_{xy} is used twice, and the same holds for internal loops, bulges, and hairpin loops. This allows the respective probabilities to reflect the statistics of structural features. *RNAFeatures* also enforces the convention that a structure should not have "lonely pairs" (i.e., base pairs not stacked onto an adjacent pair). This is achieved by the use of two non-terminal symbols: *Weak* derives substructures "weakly" closed by a single base pair, while *Stack* derives substructures closed by two or more base pairs. Since *Weak* substructures can only be embedded in larger parse trees via the rule *Stack* $\rightarrow$ *x Weak y*, there is no way that a structure derived from the axiom *ExternalLoop* can have an isolated base pair. (If you prefer to allow lonely pairs, just identify *Weak* and *Stack* and remove the redundant production rule that results from this merge.)

Another interesting point is the handling of multiloops. First of all, substructures inside a multiloop are produced by different rules than substructures in the external loop. All unpaired bases in the external loop are produced via rules $el_{2,x}$, while all others stem from rules $ss_{1,x}$ and $ss_{2,x}$. This allows to assign independent probabilities to them. Second, care has been taken that inside the multiloop, there are at least two closed substructures. This is important, since a multiloop with a single stem inside would rather be considered an internal loop, which is already modeled by the appropriate rule. Ignoring this fact would make the grammar semantically ambiguous. Third, the grammar takes care not to derive two adjacent *SingleStrand* non-terminal symbols. This would happen, for example, if the rule for *ML_Component* was written in the more natural, symmetric way, as in *ML_Component* $\rightarrow$ *SingleStrand Stack Singlestrand*. With such a rule, two helices branching from a multiloop would lead to a derivation via *SingleStrand Stack SingleStrand SingleStrand Stack SingleStrand*, and unpaired bases between two stacks could be ambiguously derived from the two adjacent *SingleStrand* non-terminal symbols.

We leave it to the reader to define the semantic mapping μ and show that this grammar is semantically non-ambiguous.

2.1.4 Grammar Design Trade-Offs

Let us consider the trade-offs when designing a (stochastic) grammar modeling RNA structure. Grammar *MiniRNA* has only three rules and only 20 parameters to be trained from the data—π_{end}, 4 for π_{dot_x}, and 16 for $\pi_{\text{pairsplit}_{xy}}$, which makes 21–1, since they must sum up to 1. In particular, the *pairsplit* rule combines both the generation of a base pair and a potential branch in the structure. Since continuous stacking pairs are much more frequent than branching structures, this is an unfortunate coupling of situations. They should rather be modeled independently.

Grammar *MiniRNA* has been reported to have "abysmal" practical performance [10] in modeling RNA structure.

Grammar *RNAFeatures*, on the other hand, has more than 120 parameters. Each structural feature is governed by an extra parameter for its special rule. If structures in our training data have lots of bulges, but few internal loops, the grammar will be trained to reflect this. Even the base pairs enclosing an internal, bulge or multiloop are assigned separate parameters, which may come out different in training from data. In practice, one may want to tie some of these parameters to each other, to avoid overfitting. Grammars that constitute a compromise between *MiniRNA* and *RNAFeatures* are used, for example, in *Pfold* [13] and *Infernal* [14].

A stochastic model will always reflect the statistical properties of the training data set, including structural feature frequency, sequence length, and base composition. There are two sides to this coin:

- A general model requires a large data set, "typical" of all RNA structures. This is hard to achieve. An extreme case is a grammar presented in [15], which reaches the sophistication of the thermodynamic parameter space. This grammar requires 17 non-terminal symbols and 41 rules, and thousands of parameters.
- If you use training data from a specific RNA family, the model can adapt to family properties to a certain extent, again in terms of base composition, sequence length, and frequency of structural features. It will not be able to capture the specific arrangement of structural components. This is why we must move on from general RNA folding grammars to grammars specialized to specific structures for RNA family modeling. We describe this route in more detail in Subheading 2.2.

2.1.5 Avoiding Semantic Ambiguity

Should you design a novel grammar, here is some guidance for avoiding semantic ambiguity.

- Stay away from rules like $S \rightarrow S\ S$, as it generates notorious ambiguity. Distinguishing multiple occurrences of S by superscripts, $S^1 S^2 S^3$ can be derived in two ways, $S \rightarrow S^1 S \rightarrow S^1 S^2 S^3$ and $S \rightarrow SS^3 \rightarrow S^1 S^2 S^3$.
- Use two non-terminal symbols instead to generate sequences of substructures, e.g. $T \rightarrow S\ T \mid S$.
- Finally, avoid rules like $S \rightarrow aS \mid Sa \mid T$, as the phrase aTa has two derivations.

In all of these cases, the competing derivations make no difference for the assigned structure under the semantic mapping μ and lead to semantic ambiguity.

Still, avoiding ambiguity can be subtle, and can lead to grammars with more rules and hence, more parameters to train. A simple check against semantic ambiguity is the following. Modify your scoring functions such that they score each candidate by 1 and sum up the scores as in the Inside algorithm. This gives you the size of your grammar's search space, $|T_G(w)|$, for any test sequence w. Compare this to $|T_{\text{MiniRNA}}(w)|$, which should give the same result. (Instead of MiniRNA, you can use any other grammar whose semantic non-ambiguity you trust in, but make sure it uses the same conventions on lonely pairs, minimal hairpin loop sizes, etc. as your grammar G does.) The number of structures is very large (use long integers!), and if the numbers coincide, there is strong evidence that the search space of G does not have redundant candidates. Still, it is only a test.

In spite of the general undecidability of semantic ambiguity checking, there is a quite powerful method to do it for grammars modeling RNA structure [16]: Given grammar G, replace each rule generating unpaired bases $x \in \{a, c, g, u\}$ by a single rule generating a '.' instead, and each rule generating possible base pairs by a single rule generating '(' and ')' instead. This turns the grammar into a grammar $\hat{G}$ which derives dot-bracket strings, i.e. $L(\hat{G}) = L(DotPar)$. In fact, $\widehat{MiniRNA} = DotPar$. Then, submit the grammar $\hat{G}$ to a syntactic ambiguity checker for context free grammars, such as the ACLA server described in [17].[4] If ACLA proves that $\hat{G}$ is syntactically non-ambiguous, then your grammar G is semantically non-ambiguous.

2.1.6 Implementing Your SCFG

When it comes to implementing your own SCFG, you must produce code for the Viterbi and/or Inside algorithms. Chapter 8 in this book provides some guidance by specifying low-level pseudocode for these algorithms, which needs to be adapted to your grammar. However, hand-programming dynamic programming recurrences is error-prone and their debugging is tedious.

The most convenient way to develop your own CSFG models presently may be the TORNADO system [18]. It provides a language to describe SCFGs annotated with details about distributions for emission probabilities, context dependence, tied parameters, and so on. The TORNADO compiler generates an SCFG parser, and the system also provides methods for parameter training. In [18], Rivas et al. have evaluated various grammars and give advice on how to choose data sets to safeguard against overfitting. Possibly, you can reuse one of the grammars already reported there, some of which have been shown to be semantically non-ambiguous.

[4]The present URL of this tool is http://www.brics.dk/grammar/.

The recent Bellman's GAP programming system [19] supports dynamic programming over sequence data in general, which subsumes the implementation of CYK-type algorithms. It allows you to specify grammars and one or more scoring schemes separately and generates efficient code from these declarative constituents. This by-passes all subscript-fiddling on your side, gives you stochastic sampling and full backtrace for free, and the resulting programs are easy to modify when your ideas about the grammar or the scoring scheme evolve. However, Bellman's GAP does not (yet) provide parameter training methods. The system eases the combined use of scoring schemes of any type. For example, it allows you to combine stochastic scoring with abstract shape analysis (*see* Chapter 11).

2.2 SCFGs Modeling Structural RNA Families

A structural family of RNAs is defined as a set of RNA sequences which fold into a consensus structure, either with good free energy in the thermodynamic model, or with high probability in a stochastic model. However, the consensus structure need not be realized exactly by a sequence in order to fit the model. The sequence may have more or fewer residues than the consensus, and it should achieve a good number of the base pairs in the consensus, but not necessarily all of them.

Definition 8: A *family model grammar* is an SCFG whose parse trees encode alignments of a *query* sequence to a *consensus* structure, allowing for insertions and deletions.

Parameters for a family model grammar M are typically derived from a set of aligned "seed" sequences, and an explicitly given consensus structure. As before, M assigns a probability $P_M(q)$ to any query sequence q. When this probability exceeds a model-specific threshold, q is accepted as a member of the family modeled by M.

To construct a family model M, we can build on our grammars introduced earlier. Since *DotPar* encodes structures, *MiniRNA* assigns structures to sequences, and *Ali* models sequence alignments, a suitable combination of these three grammars will give an SCFG that serves as a family model. We will describe the construction of the family model grammar using a small example. Let our consensus structure be c =" . ((.)) . ( ) , " shown as a *DotPar* tree in Fig. 2.

A family model does not allow the query to fold into arbitrary structures. Only structures are allowed which are formed by a subset of the base pairs in the consensus. Hence, the model specializes grammar *MiniRNA* with respect to c by using several copies of its rules and non-terminal symbols, indexed by the positions in the consensus to which they correspond. Let us number the positions in c as

$$\cdot_1(_2(_3\cdot_4)_5)_6\cdot_7(_8)_9$$

Grammar *StrictConsensus*.
$\mathcal{A} = \{a,c,g,u\}$, $V = \{S_1,...,S_{10}\}$, axiom is S_1.

production rule	rule name (position subscripts omitted)
$S_1 \to xS_2$	dot_x for $x \in \mathcal{A}$
$S_2 \to xS_3yS_7$	$pairsplit_{xy}$ for $x,y \in \mathcal{A}$
$S_3 \to xS_4yS_6$	$pairsplit_{xy}$ for $x,y \in \mathcal{A}$
$S_4 \to xS_5$	dot_x for $x \in \mathcal{A}$
$S_5 \to \varepsilon$	*end*
$S_6 \to \varepsilon$	*end*
$S_7 \to xS_8$	dot_x for $x \in \mathcal{A}$
$S_8 \to xS_9yS_{10}$	$pairsplit_{xy}$ for $x,y \in \mathcal{A}$
$S_9 \to \varepsilon$	*end*
$S_{10} \to \varepsilon$	*end*

Fig. 6 Grammar *StrictConsensus* obtained specializing *MiniRNA* to c = ".((.)).()"

```
.((.)).(--)      .((.)).-()      .((.)).-()
-cguc-aucca      c-uca-cuua      cu-c-acuua
    (1)              (2)             (3)
 cgucaucca        cucacuua        cucacuua
 .(.).(..)        .(.)..()        .(.)..()
```

Fig. 7 *Top row*: Three alignments of a query sequence to the family consensus structure. *Bottom row*: The structure assigned to the query by the respective alignment. Due to deletions and insertions in the alignments, all assigned structures in (1)–(3) differ slightly from the consensus

Specializing *MiniRNA* to c will yield, as an intermediate step, a grammar *StrictConsensus* (*see* Fig. 6), which folds each sequence *strictly* into the consensus, with no base pairs omitted or bases inserted. We proceed as follows:

The length of c is 9. We make 10 copies of *MiniRNA*, renaming the non-terminal symbol S into $S_1, \ldots, S_{10}$. For each position, we retain the rule alternative required to derive c in that place and delete the other two alternatives. We start with $S_1 \to xS_2$, since the position 1 in c is unpaired. We continue with $S_2 \to xS_3yS_7$, as x and y correspond to the matching brackets at consensus positions 2 and 6. Similarly, we get $S_3 \to xS_4yS_6$. For S_6, we create the rule $S_6 \to \varepsilon$, since the base at position 6 is already generated as y by S_2. And so on.

The grammar *StrictConsensus* is most rigid. It will parse any query sequence of length $|c|$ into the consensus structure and find no parse for any other sequence. What we want to compute are alignments like those shown in Fig. 7.

In order to allow for deletions and insertions with respect to the consensus, let us incorporate what we have learnt from grammar *Ali*. *Ali* aligns two sequences—here we want to align the query sequence to the consensus. As the consensus is already encoded in the grammar *StrictConsensus*, we do not have to treat the consensus as a second sequence. But we have to allow (1) for

Grammar *FamilyModel*.
$x, y \in \mathscr{A} = \{a, c, g, u\}$, $V = \{S_1, ..., S_{10}\}$, axiom is S_1.

production rule	rule names omitted
$S_1 \rightarrow xS_1 \mid xS_2$	$\mid S_2$
$S_2 \rightarrow xS_2 \mid xS_3yS_7$	$\mid S_3yS_7 \mid xS_3S_7 \mid S_3S_7$
$S_3 \rightarrow xS_3 \mid xS_4yS_6$	$\mid S_4yS_6 \mid xS_4S_6 \mid S_4S_6$
$S_4 \rightarrow xS_4 \mid xS_5$	$\mid S_5$
$S_5 \rightarrow xS_5 \mid \varepsilon$	
$S_6 \rightarrow xS_6 \mid \varepsilon$	
$S_7 \rightarrow xS_7 \mid xS_8$	$\mid S_8$
$S_8 \rightarrow xS_8 \mid xS_9yS_{10}$	$\mid S_9yS_{10} \mid xS_9S_{10} \mid S_9S_{10}$
$S_9 \rightarrow xS_9 \mid \varepsilon$	
$S_{10} \rightarrow xS_{10} \mid \varepsilon$	

Fig. 8 Grammar *FamilyModel* obtained by extending *StrictConsensus* with rules for insertions and deletions

residues in the query, which are not aligned to residues in the model, and (2) for positions in the consensus structure, which are not matched by bases in the query. Point (1) is handled by adding a rule alternative $S_i \rightarrow xS_i$ for $x \in \mathscr{A}$. Point (2) requires, for each rule alternative which generates query residues, an alternative to make the same transitions without generating a symbol. For $S_i \rightarrow xS_{i+1}$, we add the alternative $S_i \rightarrow S_{i+1}$—without the x, which means that no symbol in the query corresponds to position i in the consensus. For $S_i \rightarrow xS_{i+1}yS_k$, we add the alternatives $S_i \rightarrow S_{i+1}yS_k \mid xS_{i+1}S_k \mid S_{i+1}S_k$, which allow the left, the right, and both partners of a consensus base pair to be missing in the query. This completes our construction of the family model grammar; the result is shown in Fig. 8.

We have used grammar *MiniRNA* as the prototype grammar. It was specialized with respect to a consensus structure, and then extended to provide for query-consensus alignments. It can be shown that, when proceeding in this systematic fashion from a prototype grammar which is semantically non-ambiguous, this property also holds for the derived family model grammar: Each parse tree $t \in T_{\text{FamilyModel}}(w)$ uniquely encodes a query/consensus alignment. It is not strictly necessary to proceed this way: *Infernal*, for example, starts from an ambiguous prototype grammar, but still ensures non-ambiguity of the family model by a more refined generation process.

What can we compute once we trained the parameters for our family model grammar? The Viterbi algorithm computes the most likely alignment of the query to the model. This alignment can be used, for example, to extend the seed sequence alignment, from which the model was generated, by adding a high scoring query as a bona fide family member. The Inside algorithm computes the overall probability of the query with respect to the model, $\sum\{P(t) | t \in T_{\text{FamilyModel}}(w)\}$.

Another interesting piece of information would be (the probability of) the most likely structure assigned to the query by a family model grammar M. All base pairs produced by the rules $S_i \rightarrow xS_jyS_k$ make up the secondary structure assigned to w, irrespective of how gaps and unpaired bases are placed. Here, the semantic mapping μ maps query/consensus alignments to structures of the query. In Fig. 7, the same structure is assigned to the query in case (2) and (3), by *different* alignments to the consensus. Hence, the most likely structure would be the query structure s^* which maximizes $\sum\{P_M(t)|\mu(t) = s\}$. However, at the point of this writing, it is not known how to compute this information efficiently.

3 Further Reading

Applications of stochastic context free grammars are treated in three further chapters of this book:

- An introduction to RNA databases, by Marc Hoeppner, Lars Barquist, and Paul Gardner (see Chapter 6),
- SCFGs for RNA structure prediciton, by Zsuzsanna Skosd, Ebbe S. Andersen, and Rune Lyngsoe (see Chapter 8),
- Annotating ncRNAs in genomes with INFERNAL, by Eric Nawrocki (see Chapter 9).

These chapters cover the most important uses of SCFGs in RNA structure analysis at the time of this writing. For further reading, please consult the literature given therein.

Acknowledgements

Thanks go to Jan Reinkensmeier for a careful reading of this manuscript.

References

1. Eddy SR, Durbin R (1994) RNA sequence anaysis using covariance models. Nucleic Acids Res 22(11):2079–2088
2. Sakakibara Y, Brown M, Hughey R, Mian IS, Sjölander K, Underwood RC, Haussler D (1994) Stochastic context-free grammars for tRNA modeling. Nucleic Acids Res 22(23):5112–5120. URL http://www.hubmed.org/display.cgi?uids=7800507
3. Booth TL, Thompson RA (1973) Applying probability measures to abstract languages. IEEE Trans Comput 22(5):442–450
4. Baker JK (1979) Trainable grammars for speech recognition. J Acoust Soc Am 54–550
5. Hopcroft JE, Ullman JD (1969) Formal languages and their relation to automata. Addison-Wesely, Reading, MA
6. Aho AV, Ullman JD (1973) The theory of parsing, translation and compiling. Prentice-Hall, Englewood Cliffs, NJ. I and II.
7. Giegerich R, Meyer C, Steffen P (2004) A discipline of dynamic programming over sequence data. Sci Comput Program 51(3): 215–263

8. Durbin R, Eddy S, Krogh A, Mitchison G (1998) Biological sequence analysis, 2006 edn. Cambridge University Press, Cambridge
9. Giegerich R (2000) Explaining and controlling ambiguity in dynamic programming. In: Proceedings of combinatorial pattern matching, vol 1848 of Lecture notes in computer science, pp 46–59. Springer, New York
10. Dowell RD, Eddy SR (2004) Evaluation of several lightweight stochastic context-free grammars for RNA secondary structure prediction. BMC Bioinformatics 5:71–71. doi: 10.1186/1471-2105-5-71. URL http://www.hubmed.org/display.cgi?uids=15180907
11. Brejová B, Brown DG, Vinař T (2007) The most probable annotation problem in HMMs and its application to bioinformatics. J Comput Syst Sci 73(7):1060–1077
12. Reeder J, Steffen P, Giegerich R (2005) Effective ambiguity checking in biosequence analysis. BMC Bioinformatics 6(153). URL http://www.biomedcentral.com/1471-2105/6/153
13. Knudsen B, Hein J (2003) Pfold: RNA secondary structure prediction using stochastic context-free grammars. Nucleic Acids Res 31(13):3423–3428. URL http://www.hubmed.org/display.cgi?uids=12824339
14. Nawrocki EP, Kolbe DL, Eddy SR (2009) Infernal 1.0: inference of RNA alignments. Bioinformatics 25(10):1335–1337. doi: 10.1093/bioinformatics/btp157. URL http://www.hubmed.org/display.cgi?uids=19307242
15. Nebel M, Scheid A (2011) Analysis of the free energy in a stochastic RNA secondary structure model. IEEE/ACM Trans Comput Biol Bioinformatics 8(6):1468–1482
16. Giegerich R, Höner zu Siederdissen C (2011) Semantics and ambiguity of stochastic rna family models. IEEE/ACM Trans Comput Biol Bioinformatics 8(2):499–516. ISSN 1545-5963. doi: http://doi.ieeecomputersociety.org/10.1109/TCBB.2010.12
17. Braband C, Giegerich R, Møller A (2010) Analyzing ambiguity of context-free grammars. Sci Comput Program 75(3):176–191. Earlier version in Proc. 12th International Conference on Implementation and Application of Automata, CIAA '07, Springer LNCS vol. 4783
18. Rivas E, Lang R, Eddy S (2012) A range of complex probabilistic models for RNA secondary structure prediction that includes the nearest-neighbor model and more. RNA 18:193–212
19. Sauthoff G, Janssen S, Giegerich R (2011) Bellman's GAP - a declarative language for dynamic programming. In: Schneider-Kamp (ed) Principles and practice of declarative programming. ACM Press, New York, NY, pp 29–40

Chapter 6

An Introduction to RNA Databases

Marc P. Hoeppner, Lars E. Barquist, and Paul P. Gardner

Abstract

We present an introduction to RNA databases. The history and technology behind RNA databases are briefly discussed. We examine differing methods of data collection and curation and discuss their impact on both the scope and accuracy of the resulting databases. Finally, we demonstrate these principles through detailed examination of four leading RNA databases: Noncode, miRBase, Rfam, and SILVA.

Key words ncRNA, Database, Alignment database, Sequence database, SILVA, Rfam, Noncode, miRBase

1 Introduction

The introduction of targeted molecular and bioinformatic approaches [1, 2] and the availability of affordable sequencing technologies have lead to a glut of novel ncRNA sequences (Fig. 1). NcRNAs have been shown to be involved in a diverse array of cellular processes, from long-known roles in the translational process to more recently discovered functions in the regulation of gene expression and genomic defense [3, 4]. Databases provide a central resource for researchers to obtain and deposit this information in the form of sequences and descriptive metadata.

The purpose of this chapter is to introduce the reader to the state of the art of RNA databases. The first section of this chapter will provide a brief history of RNA databases which will put current databases in context. The second section covers approaches to data collection and curation and how these differing approaches affect the utility of data served. Finally, we illustrate these points by exploring a few exemplar databases in depth: Noncode, miRBase, Rfam, and SILVA.

Jan Gorodkin and Walter L. Ruzzo (eds.), *RNA Sequence, Structure, and Function: Computational and Bioinformatic Methods*, Methods in Molecular Biology, vol. 1097, DOI 10.1007/978-1-62703-709-9_6,

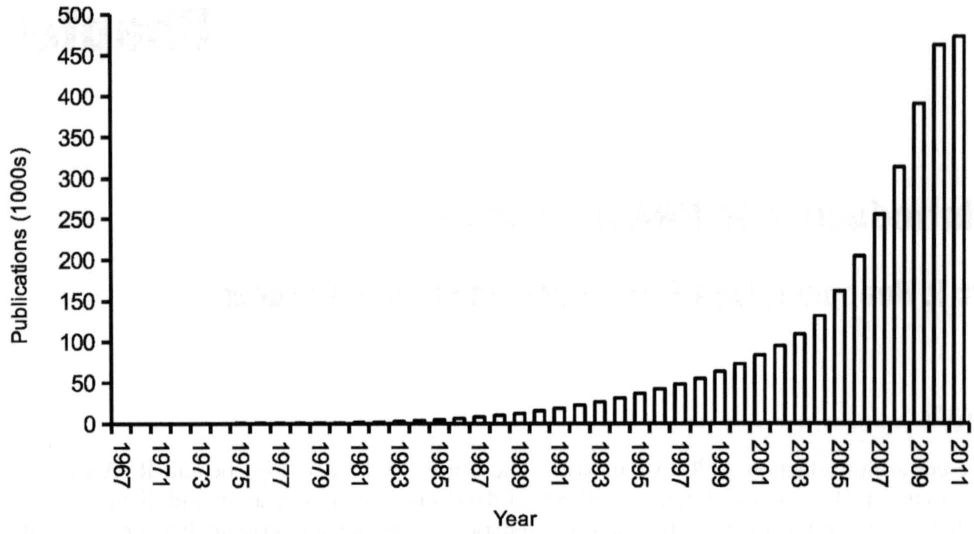

Fig. 1 The expanding RNA world. The expanding picture of noncoding RNAs as established by the number of publications dedicated to the subject matter and listed in the PubMed database (identified by the tag "npcRNA")

2 RNA Databases: A Historical Perspective

The development of modern RNA databases reflects both our growing knowledge of RNA biology and the development of modern information technologies. The earliest databases focused on long-known classes of RNA molecules. For instance, the original release of the Sprinzl tRNA database was distributed as the text of a journal article [5]. Other early databases include the Signal Recognition Particle Database [6] and the Ribosomal RNA Database project [7], both distributed as "flat" text files over FTP.

The growing number of sequenced RNAs drove a shift in database technology. While the original Sprinzl tRNA database contained approximately 700 sequences, its modern inheritor contains more than 12,000 tRNA genes [8], and the current release of the Rfam database contains over one million sequences computationally identified as encoding tRNAs. This amount of data would be impractical to store and search as flat files. The solution to this problem that has been adopted is the use of relational databases accessible over the World Wide Web.

The strength of relational databases lies in the ability to have one data repository from which multiple outputs can be generated dynamically. Any updates to the database will affect all output. In contrast, flat files need to be updated individually. More specifically, a relational database acts as a server which organizes data and provides it in an interactive fashion to client applications. All output draws from the same source, making data maintenance considerably easier. In addition, the way data is organized allows for

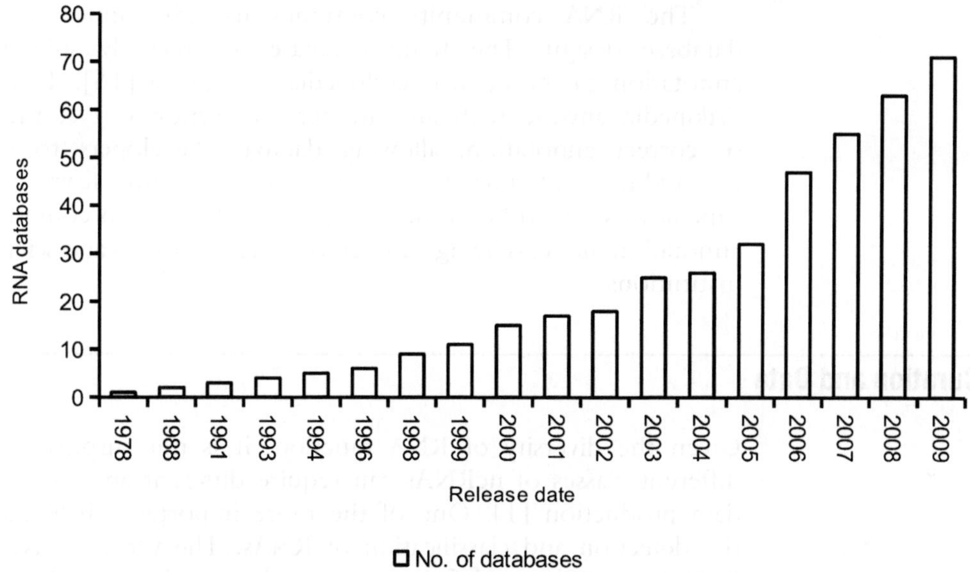

Fig. 2 Number of RNA databases. The growth in available data and general interest in the various classes of RNAs is mirrored in the increase of the number of RNA databases (based on data from http://www.oxfordjournals.org/nar/database/a/)

the use of complex Boolean queries to retrieve specific information of interest. An example would be collecting all human ncRNAs that belong to a specific class, are between 50 and 60 nucleotides long, and were published after 2004. This would require query-specific scripting with flat file data and would then require constant updating of a multitude of files as new information becomes available. In a relational database, we would simply select the relevant entries from our tables, and any changes in the underlying data set could be captured by rerunning our query.

At the time of writing, there are at least 72 active databases dedicated to RNA [9] (http://www.oxfodjournals.org/nar/database/a) (Fig. 2). Many ncRNAs have inspired their own specialist databases as their functional importance has grown clear. Examples include bacterial small RNAs [10] and eukaryote microRNAs [11], both of which function in the regulation of gene expression. Other abundant classes are small interfering RNAs (siRNA) [12] and small nucleolar RNAs [13]. The latter have been found in both archaea and eukaryotes and have spawned several databases over the years, focused on the inventory of single species or on collecting information from a broad taxonomic range.

In contrast to these "specialist" databases and as a consequence of the widening spectrum of known RNA classes, the need for "generalist databases" has emerged. Examples include the noncoding sequence database Noncode [14] and the RNA family database Rfam [15]. Focusing on information relevant only to a narrow functional class of RNA, these databases aim to provide broad information about all RNAs.

The RNA community continues to drive innovation in database design. The Rfam database has recently shifted its annotation to the open encyclopedia, Wikipedia [16]. Through Wikipedia anyone with an Internet connection can contribute or correct annotation, allowing database developers to focus on adding and improving data sources. Early fears about vandalism seem to be unfounded [15], and the open community annotation model is being adopted by other databases and scientific institutions.

3 Curation and Data

Given the diversity of RNA function, it is not surprising that different classes of ncRNAs can require different approaches for data production [1]. One of the more important challenges is the detection and classification of RNAs. The various classes of RNA have different defining features that need to be taken into consideration. An example of this is microRNAs (miRNAs). Given the relatively short sequence of a mature miRNA (~22 bp), testing for the presence of their characteristic stem-loop structure is an essential step in the prediction process to decrease the chance of false annotations. In contrast, small nucleolar RNAs possess well-conserved sequence motifs in addition to structural features. Similar requirements exist for many other classes as well. Consequently, detection algorithm choice depends on the class of RNA under investigation.

Another distinguishing factor across databases is scope. Some databases focus on a particular type of ncRNA (specialist databases), while others provide access to sequences from a broad range of ncRNA classes (generalist databases). Similarly, sequences may be presented individually (sequence databases) or grouped based on common function and inferred ancestry (alignment databases). Some databases rely on data manually curated by domain experts in addition to or instead of computational predictions. In the following, we will discuss the various approaches used in RNA databases and then present some specific examples that demonstrate their application (Fig. 3).

3.1 Manual Versus Automated Annotation

One of the main differences between RNA databases is whether the data is the result of experimental discovery and verification or derived from automatic, computational scans (discussed in other chapters). The former is often used in specialist databases, whereas the latter generally finds application in genome-scale annotation processes.

Automated annotation has a clear advantage in that it can be used to quickly identify potential RNAs in very large data sets. Relevant tools range from sequence similarity search methods

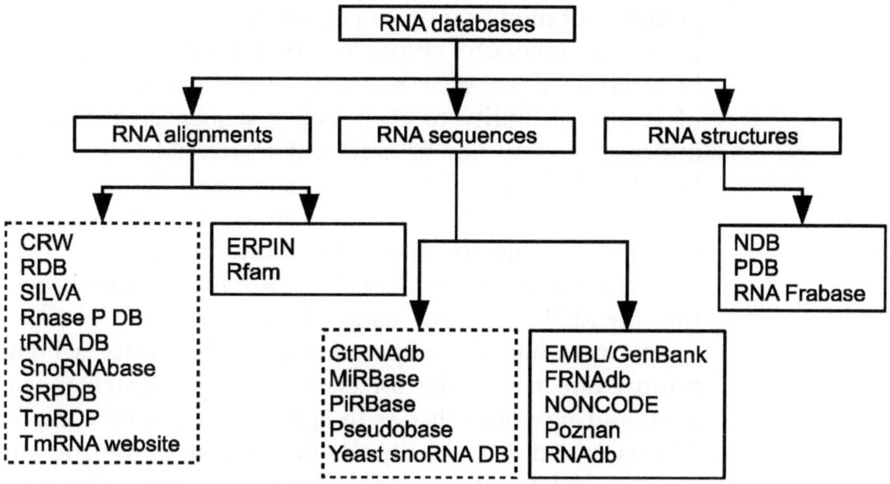

Fig. 3 RNA databases can be broadly organized into alignment, sequence, and structure databases. These are further grouped into class-specific databases (*dotted line*) and general databases (*solid line*). An index of RNA databases is maintained at http://www.oxfodjournals.org/nar/database/a

such as BLAST [17, 18] to complex probabilistic models [19]. Rather than relying entirely on experimentally confirmed results, automated annotation uses some criterion of sequence and/or structural similarity to define thresholds for the inclusion of new sequences (e.g., covariance models as implemented in Infernal [20]). This "thresholding" allows for greater transparency and makes it possible for researchers to apply the same method to their own data. Likewise, new insights into the sequence or structure of a given RNA can be easily incorporated into the annotation process without requiring a time-consuming manual reevaluation of all data.

Despite technological advances, computational predictions still come with a number of caveats. First, they are only as good as the information they are built on—such as a seed alignment for covariance models. While manual curation of data is potentially subjective and can result in the occasional false annotation, any error in the automated annotation process will affect all downstream predictions and could be more severe. However, these mistakes can also be more easily rectified by adjusting the relevant parameters and rerunning the analysis.

Second, the degree of sequence divergence of ncRNAs remains a complicated issue and can be highly variable across classes and lineages. Some ncRNAs, such as tRNAs, have relatively well-conserved sequences and have been predicted with a low error rate across all domains of life [21]. However, many other ncRNAs, such as small nucleolar RNAs (snoRNAs), can exhibit a high degree of sequence plasticity [22]. SnoRNAs are involved in the guidance of modifications on other RNAs (mostly rRNA) through the interaction with a conserved protein complex. As their function is determined by their secondary structure and a small stretch of complementarity to their target sequence, primary sequence information may not be sufficient for reliable identification over larger evolutionary distances. In such cases, finding genuine RNA genes without producing many false-positives is challenging, and experimental validation is of crucial importance.

Whether to prefer manual or automatic curation depends on the intended use of the data. Manual curation will provide high specificity, while automatic curation will produce a higher false-positive rate in exchange for a faster and more systematic detection. It should be noted that both approaches can be complimentary. Manual curation is usually the first step in automatic annotation pipelines, while automatic annotation in turn recovers many candidate sequences later subjected to manual inspection and curation. In any case, information from databases cannot replace a solid understanding of the biology of the ncRNAs under investigation and scrutiny of the information is always advisable.

3.2 Sequence Versus Alignment Databases

A large fraction of RNA databases can be further divided into sequence databases and alignment databases. Sequence databases, such as GenBank [23], are primarily designed as repositories. Their primary purpose is to store individual sequences, usually complemented by cross-references to publications and other databases. This approach makes it easier for both developers and external contributors to add to the database, as little specialized analysis beyond sequence discovery is necessary for a new entry.

However, sequence databases generally contain no detailed information on individual ncRNAs, their relationships to each other, or a robust nomenclature. The lack of explicit homology information or a stringent nomenclature may negatively impact the value of such resources to certain users. For example, discovering all of the U3 snoRNA sequences in GenBank presents a considerable challenge. Some of the relevant entries may contain helpful information in their descriptions, but many others will not. Even more are probably not annotated at all.

In contrast, to better capture the diversity of an ncRNA, some of the more specialized projects (e.g., miRBase) bin ncRNAs into families based on their expected common ancestry using similarity in structure and sequence. These sequences can then be aligned, producing an estimate of the diversity of each nucleotide. The alignments may subsequently be employed to, e.g., train covariance models and thus greatly improving our ability to identify homologs across genomes. On the downside, this process is time-consuming as each sequence and alignment need to be manually curated to ensure optimal sensitivity. As such, expansion of these databases may be limited by the availability of expert curators able to perform such tasks.

4 RNA Databases: Examples and Practical Application

In the previous section, we defined RNA databases in terms of their curation style and data types. These differences have implications for when and how databases should be used. In the following, we will explore these implications by the example of a few leading databases covering the breadth of methods and scopes.

4.1 General Sequence Databases: Noncode

Noncode, established in 2005, is among the most exhaustive general databases on the topic of ncRNAs [14]. At the time of writing, it holds information on 112 distinct classes from over 800 species. Noncode is a general sequence database and as such focuses on presenting individual sequences with relevant metadata. In contrast to many other databases, candidate genes are derived primarily from experimental data—over 80% of sequences. This is an impressive feat when considering that there are over 200,000 sequence entries in Noncode (release 2.0, 2007).

Data production in Noncode is a stepwise, semiautomated process. Initially, a set of broad keywords is used to identify putative ncRNAs in the literature, and new queries are iteratively added to the original list as they are recovered during this search. Based on the results of this process, the GenBank database is then automatically filtered for candidate entries. All data is manually vetted by reference to relevant publications to ensure that they constitute genuine RNAs and to gather additional information relevant to their biological role. Sequences are checked for redundancy before being added to the final data set (Fig. 4).

A useful innovation in Noncode is the introduction of a "process function classification" (PFC). PFC is a vocabulary that describes the cellular functions an ncRNA takes part in, similar to GO terms [24]. This system was introduced in an attempt to systematize RNAs functional nomenclature and allows quick access to particular functional classes. Other useful parameters used to describe RNAs include the molecular mechanism of their function, their subcellular location, or their cellular role. Noncode can also report ncRNAs based on their organismal range.

Noncode features a Boolean search engine that allows users to perform complex queries against the data set, enabling efficient data mining—with certain limitations. Among these, it relies on existing annotation and so does not contain unannotated homologous sequences, though a BLAST sequence search is available for specific sequence queries. Additionally, while all sequences are available as a bulk FASTA-format download, the metadata is not.

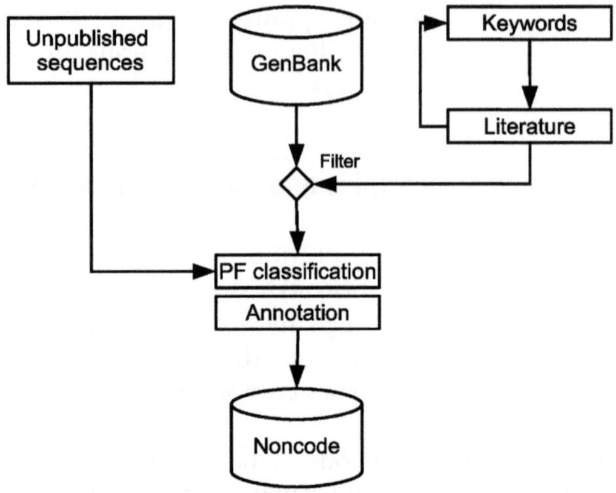

Fig. 4 The Noncode annotation pipeline compiles experimentally verified ncRNAs from the literature and the nucleotide database GenBank using a set of relevant keywords. In addition, the authors also include unpublished RNAs studied in their lab. The resulting data is subjected to manual inspection and annotation (such as the Noncode-specific process function classification) prior to inclusion in the database

This limits the prospects for performing bespoke analysis of the Noncode data set.

At the time of writing (March 2011), updates to Noncode have been sparse (2005 and 2007) and so may not reflect the current state of ncRNA research. A contributing factor here may be the presumably time-consuming manual vetting of data as part of the production pipeline.

4.2 Specialized Sequence Databases: miRBase

The microRNA (miRNA) database miRBase was first released in 2002 as the "microRNA Registry" [25]. It is currently the most complete resource for information on miRNAs, a diverse group of eukaryote RNAs involved in the regulation of gene expression [26]. The primary goal of miRBase is to collect published, experimentally verified miRNA sequences and provide researchers with a consistent nomenclature. As of March 2011, the database has seen 16 major releases and contains entries for over 17,000 distinct mature miRNA and their sequences in over 140 species (Fig. 5).

Conceptually, miRBase can be divided into three parts (Fig. 6). The first is the registry, in which novel miRNAs are included and given a unique id and stable accession number. In order to be considered for inclusion into miRBase, sequences must conform to a set of quality criteria and be either derived from experimental studies or show clear homology to existing entries [27]. De novo computational predictions are not part of the miRBase data set. Submitted sequences are manually inspected and integrated into

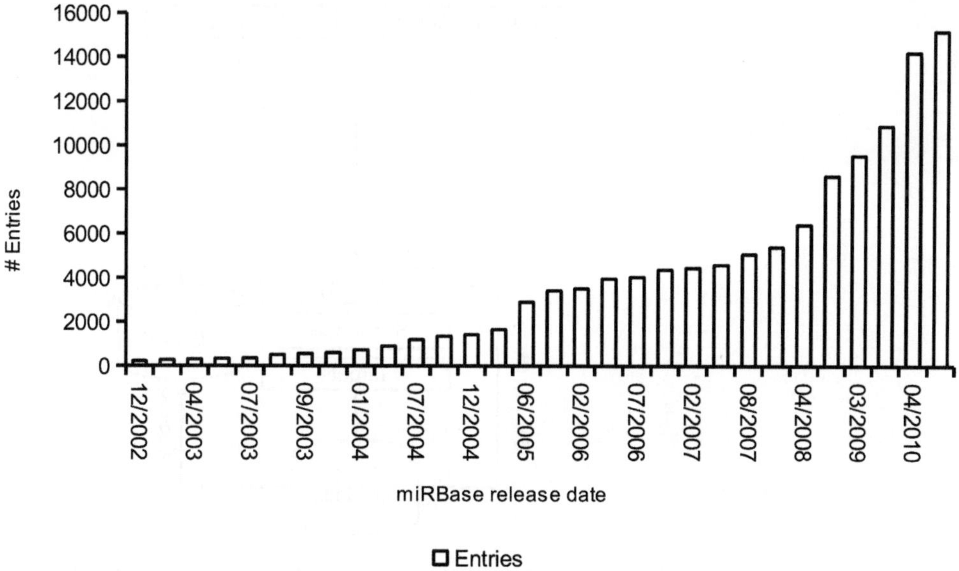

Fig. 5 The miRBase database was first released in 2002 and as of early 2011 has seen 16 major releases. During this time it has grown significantly, mirroring closely the growth of the RNA field and the increasing amount of sequence data

the database, but only after the work describing the new data has been accepted for publication in a peer-reviewed journal. This serves to maintain a high standard and clean data set with few false annotations. Where applicable, entries are also assigned to a higher-order family which group miRNAs across species based on common ancestry. Owing to the increasing amount of available sequencing data, miRBase has recently integrated new procedures to recover miRNAs from the Gene Expression Omnibus database [28] using, among other things, significant sequence similarity to existing entries and characteristic expression profiles as criteria to identify genuine miRNAs.

The second part of miRBase focuses on miRNA sequences. Individual miRNAs are annotated in depth with information such as genome coordinates and genomic context (intronic, intergenic, clustered, or singleton). Relevant metadata is extracted from the literature, including details on experimental procedures used to identify and characterize the respective genes. Finally, miRBase links individual miRNA entries to external databases that focus on automated prediction of possible mRNA targets, such as "microCosm" which was originally developed as part of miRBase.

MiRBase offers a range of options for users to interact with its data. Most importantly, it can be searched for a set of predefined criteria including species, genomic location, and expression patterns. While a free-form search is possible, it does not support complex Boolean queries. In addition, BLASTN and SSEARCH

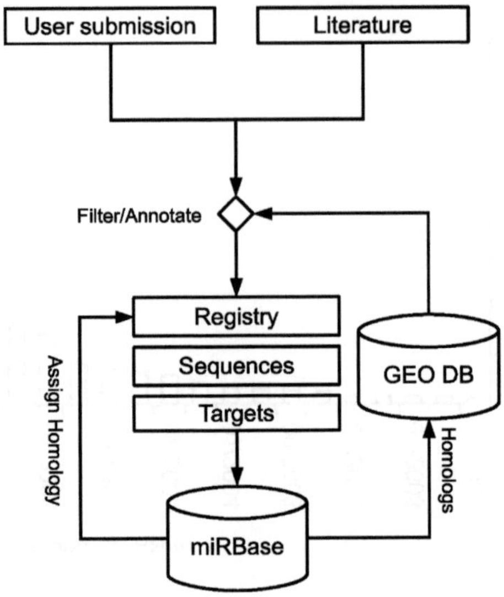

Fig. 6 Data in miRBase stems primarily from experimental evidence. After submission, new miRNAs are given an entry in the registry, annotated, and linked to external sources for target prediction. In an attempt to capitalize on the increasing amount of sequencing data, miRBase more recently expanded its pipeline to identify expressed miRNA candidates in the GEO database, based on expected homology to existing entries

can be used to find known miRNAs similar to a query sequence. The wealth of metadata makes miRBase particularly useful for bioinformaticians. Not only is it possible to download all the sequences, but the entire database is freely available as well. This opens up the possibility of designing custom data mining pipelines which incorporate all of the information contained in miRBase.

4.3 General Alignment Databases: Rfam

The RNA family database, Rfam, is the largest general alignment database currently available [15, 29]. As of release 10, it provides over 3 million annotations for 1446 distinct ncRNAs across the entire EMBL nucleotide database. Rfam defines all sequences that align to a covariance model constructed from a "seed" alignment within certain sequence and structural similarity criteria as a family. These "seed" models are constructed from two or more representative sequences and are manually curated, using published data. The thresholds of Rfam covariance models are individually adjusted to account for the varying degree of sequence plasticity specific to each family, reducing the fraction of false annotations. A possible shortcoming of this approach is that distantly related RNAs may not be captured by a model. To address this issue, Rfam has introduced clans, which are a higher-order grouping of families based on expected common descent using information from the

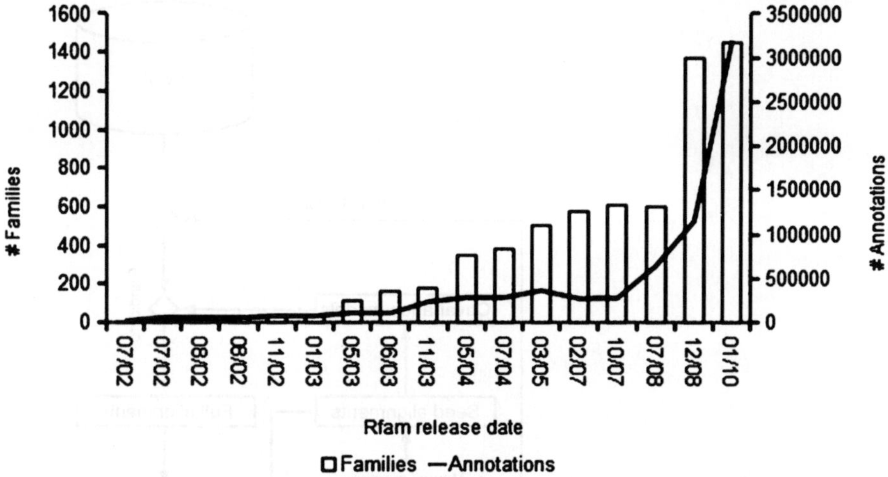

Fig. 7 Release 10.0 of Rfam contains information on 1446 ncRNA families, yielding more than 3 million annotations across the EMBL nucleotide database. Each family is defined as sequences aligning to a covariance model, based on manually curated seed alignments from published ncRNA data

literature and measures of similarity between families [16]. Clans and families provide a robust nomenclature to identify homologous RNAs across genomes (Fig. 7).

The Rfam annotation pipeline consists of a two-step process (Fig. 8). First, verified RNAs from the literature or external databases are used to create seed alignments. Sequences from these families are then used as queries for WU-BLAST searches against the EMBL nucleotide database to identify candidate RNAs. This step is made necessary by the comparatively high computational requirements of the covariance model search but may be replaced by accelerated profile hidden Markov models in the near future [30, 31]. All candidate sequences are subsequently subjected to more rigorous covariance model searches [20]. These models are calibrated so as to capture the suspected range of an ncRNA family while providing a low false-positive rate.

The tools and information provided by Rfam are diverse. Full alignments of all annotations as well as evolutionary trees are available to provide users with insight into the diversity and phylogenetic distribution of each family. In addition, users can search sequences against the Rfam database to identify putative ncRNAs. In addition, the Rfam database and the annotation pipeline are available for download from the website. Rfam has also found an application in, for example, the Ensembl database [32] where it is used to provide annotations of ncRNAs in a range of different genomes. Rfam annotations are provided through Wikipedia, allowing researchers to rapidly update and correct annotations as new information becomes available.

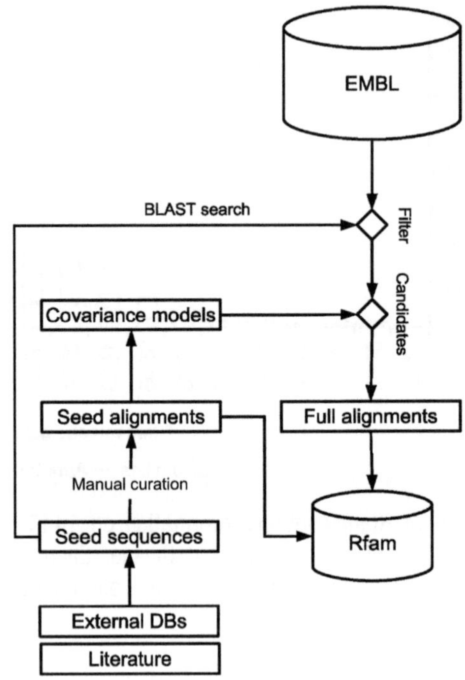

Fig. 8 The Rfam pipeline. The Rfam pipeline consists of two major steps. First, experimentally verified sequences are grouped based on their expected common ancestry to create seed alignments. Sequences from these alignments are then used in a traditional BLAST search against the EMBL database to identify putative ncRNA genes. To further increase the confidence in these predictions, all candidate hits are analyzed with manually curated covariance models. Sequences that pass both these tests are included in Rfam

The use of computational predictions means that Rfam provides ncRNA annotations for many genomes that have not yet been studied in detail. However, there are limitations to this approach. Despite rigorous examination and thresholding of families, the Rfam pipeline can produce false-positives. Secondly, Rfam is not exhaustive but features only a limited number of ncRNAs. Newly discovered ncRNAs need to be manually curated before being included in the database, which requires time. In an effort to speed up this process, Rfam has introduced a special publication track in collaboration with the journal RNA Biology where researchers can publish alignments and Wikipedia annotations of the ncRNAs they work on, which are then included in the database.

4.4 Specialized Alignment Databases: SILVA

Ribosomal RNAs (rRNAs) were among the first RNAs to be cataloged in databases, and several projects have done this over the years. This continued interest is partially due to the fact that rRNAs are widely used as phylogenetic markers. This has special relevance

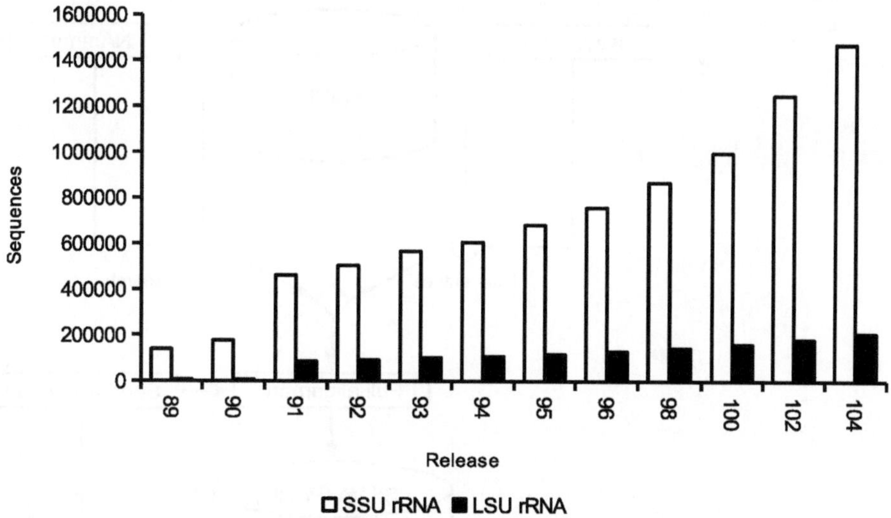

Fig. 9 Since its initial release in February 2007 (89, based on EMBL 89), data for both small subunit rRNA (SSU) and large subunit rRNA (LSU) has increased markedly over the course of only 3 years. As of release 104, SILVA contains information on over 1.4 million SSU sequences and more than 200.000 LSU sequences

to the emerging field of metagenomics, where the taxonomic diversity of a sample can be difficult to determine.

At the time of writing, the most complete rRNA database is SILVA, which was originally released in 2007 [33]. The release 104 (October 2010) holds information on almost 1.5 million small subunit (SSU) and over 200.000 large subunit (LSU) rRNA entries from all domains of life. The SILVA release cycle is synchronized with the EMBL nucleotide database, from which it draws sequence data. As a result, it is updated on a regular basis (Fig. 9).

SILVA's automated production pipeline is based on a set of keywords that are used to identify putative rRNAs in the EMBL database, given existing annotations or descriptions. To account for the increasing number of unannotated rRNAs produced by large-scale sequencing projects, EMBL sequences are scanned using hidden Markov models to identify additional candidates [34, 35]. Data retrieved in these ways are subsequently filtered based on a set of stringent criteria aimed at identifying genuine rRNA genes, including a minimum size requirement and a maximum number of allowed ambiguous positions. All rRNAs which pass these filters are then aligned against aforementioned seed alignments and stored in the database. In addition to this primary, comprehensive data set (referred to as "Parc"), SILVA compiles the "Ref" data set, a subset of "Parc" comprised of high-quality, full- or nearly full-length sequences (Fig. 10).

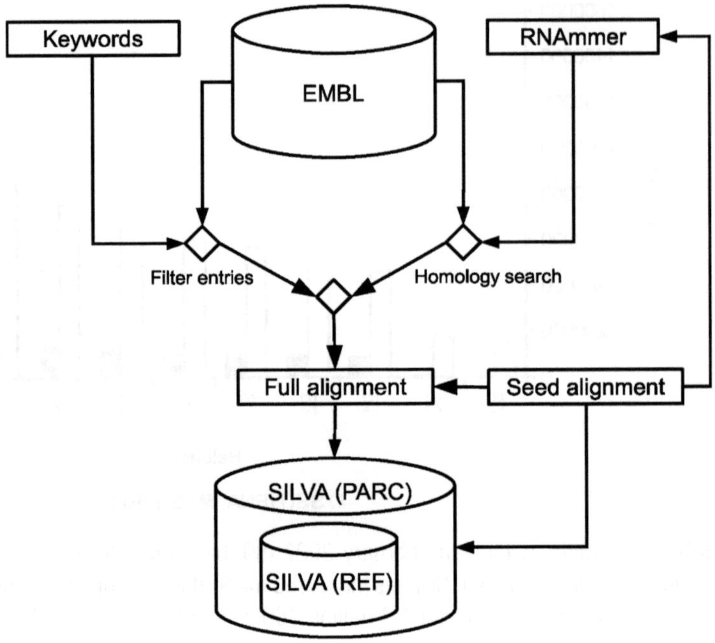

Fig. 10 SILVA data production. The SILVA rRNA database is built in three automated steps. Starting with a set of keywords, putative ribosomal RNAs are retrieved from the EMBL nucleotide database. rRNA candidates are identified by alignment to a curated profile hidden Markov model. The resulting data then has to meet a number of criteria, including a minimum length of 300 bases and a maximum of 2% ambiguities, for inclusion in the database. The comprehensive data set of SSU and LSU sequences is referred to as "Parc," from which a subset of full-length, high-quality sequences is created ("Ref")

SILVA provides users with a range of tools. Most importantly, annotations can be searched using numerous criteria, including organism names, accession numbers, or related publications. All retrieved hits can be downloaded in either FASTA or the ARB format. Another helpful feature is the use of ontologies, such as the environmental or the aforementioned taxonomic affiliation, to further characterize sequences. Finally, user-submitted queries can be aligned against the respective seed alignments, or subsets thereof, to create quality alignments from their own data.

Data from SILVA is free for academic use and prefiltered data sets are available for download.

4.5 Summary (Table 1)

Table 1
Summary and comparison of major RNA databases

	Noncode	Rfam	miRBase	SILVA
DB type	General sequence DB	General alignment DB	Specialist sequence DB	Specialist alignment DB
Manual annotation	Yes (based on published data)	Yes (seed alignments and Wikipedia)	Yes	Yes (seed alignments)
Comp. annotation	No	Yes	Yes (homologs only)	Yes
Data source	GenBank/ literature	EMBL	User submission	EMBL
Nomenclature	No	Yes	Yes	NA
Data download	Sequence files	Everything	Everything	Sequence files
Release cycle	Irregular	~1–2/year	~1–2/year	Synced with EMBL

5 Closing Remarks

Two major trends have been driving the development of ncRNA databases: an increasing appreciation of the importance of ncRNA genes, particularly in the face of high-throughput sequencing technologies, and the development of faster, more accurate computational tools for identifying ncRNA sequences. We have presented four databases here with very different approaches to making sense of increasingly large data sets. We believe these approaches are complimentary, as these diverse molecules demand diverse approaches to their characterization. It is important that users be aware of the potential for false-positives, particularly in computationally produced predictions. In these cases, cross-validation from multiple sources can provide higher certainty in an annotation. Finally, at the time of writing this chapter, a proposal has been made to produce a major new sequence-based resource for RNA called "RNAcentral" [36]. The goals of this resource should help address many of the concerns raised above.

References

1. Griffiths-Jones S (2007) Annotating noncoding RNA genes. Annu Rev Genomics Hum Genet 8:279–298
2. Hüttenhofer A, Brosius J, Bachellerie JP (2002) RNomics: identification and function of small, non-messenger RNAs. Curr Opin Chem Biol 6:835–843
3. Mattick JS, Makunin IV (2006) Non-coding RNA. Hum Mol Genet 15:R17–R29
4. Eddy SR (2001) Non-coding RNA genes and the modern RNA world. Nat Rev Genet 2(12):919–929
5. Sprinzl M, Vorderwülbecke T, Hartmann T (1985) Compilation of sequences

of tRNA genes. Nucleic Acids Res 13: r51–r104
6. Zwieb C, Larsen N (1992) The signal recognition particle (SRP) database. Nucleic Acids Res 20:2207
7. Olsen GJ, Larsen N, Woese CR (1991) The ribosomal RNA database project. Nucleic Acids Res 19:2017–2021
8. Jühling F, Mörl M, Hartmann RK, Sprinzl M, Stadler PF, Pütz J (2009) tRNAdb 2009: compilation of tRNA sequences and tRNA genes. Nucleic Acids Res 37(Database issue):D159–D162
9. Galperin MY, Cochrane GR (2011) The 2011 nucleic acids research database issue and the online molecular biology database collection. Nucleic Acids Res 39(Database issue):D1–D6
10. Huang H-Y, Chang H-Y, Chou C-H, Tseng C-P, Ho S-Y, Yang C-D et al (2009) sRNAMap: genomic maps for small non-coding RNAs, their regulators and their targets in microbial genomes. Nucleic Acids Res 37(Database issue):D150–D154
11. Griffiths-Jones S, Grocock RJ, van Dongen S, Bateman A, Enright AJ (2006) miRBase: microRNA sequences, targets and gene nomenclature. Nucleic Acids Res 34(Database issue):D140–D144
12. Chalk AM, Warfinge RE, Georgii-Hemming P, Sonnhammer ELL (2005) siRNAdb: a database of siRNA sequences. Nucleic Acids Res 33(Database issue):D131–D134
13. Lestrade L, Weber MJ (2006) snoRNA-LBME-db, a comprehensive database of human H/ACA and C/D box snoRNAs. Nucleic Acids Res 34(Database issue):D158–D162
14. Liu C, Bai B, Skogerbø G, Cai L, Deng W, Zhang Y et al (2005) NONCODE: an integrated knowledge database of non-coding RNAs. Nucleic Acids Res 33(Database issue):D112–D115
15. Gardner PP, Daub J, Tate J, Moore BL, Osuch IH, Griffiths-Jones S et al (2011) Rfam: Wikipedia, clans and the "decimal" release. Nucleic Acids Res 39(Database issue):D141–D145
16. Daub J, Gardner PP, Tate J, Ramsköld D, Manske M, Scott WG et al (2008) The RNA WikiProject: community annotation of RNA families. RNA 14(12):2462–2464
17. Altschul SF, Gish W, Miller W, Myers EW, Lipman DJ (1990) Basic local alignment search tool. J Mol Biol 215(3):403–410
18. Gardner PP, Wilm A, Washietl S (2005) A benchmark of multiple sequence alignment programs upon structural RNAs. Nucleic Acids Res 33(8):2433–2439
19. Eddy SR, Durbin R (1994) RNA sequence analysis using covariance models. Nucleic Acids Res 22(11):2079–2088
20. Nawrocki EP, Kolbe DL, Eddy SR (2009) Infernal 1.0: inference of RNA alignments. Bioinformatics 25(10):1335–1337
21. Lowe TM, Eddy SR (1997) tRNAscan-SE: a program for improved detection of transfer RNA genes in genomic sequence. Nucleic Acids Res 25(5):955–964
22. Gardner P, Bateman A, Poole A (2010) SnoPatrol: how many snoRNA genes are there? J Biol 9(1):4
23. Benson DA, Karsch-Mizrachi I, Lipman DJ, Ostell J, Sayers EW (2009) GenBank. Nucleic Acids Res 37(Database issue):D26–D31
24. Ashburner M, Ball CA, Blake JA, Botstein D, Butler H, Cherry JM et al (2000) Gene ontology: tool for the unification of biology. The Gene Ontology Consortium. Nat Genet 25(1):25–29
25. Griffiths-Jones S (2004) The microRNA Registry. Nucleic Acids Res 32(Database issue):D109–D111
26. Bartel DP (2009) MicroRNAs: target recognition and regulatory functions. Cell 136(2):215–233
27. Ambros V, Bartel B, Bartel DP, Burge CB, Carrington JC, Chen X et al (2003) A uniform system for microRNA annotation. RNA 9(3):277–279
28. Barrett T, Troup DB, Wilhite SE, Ledoux P, Evangelista C, Kim IF et al (2010) NCBI GEO: archive for functional genomics data sets–10 years on. Nucleic Acids Res 39(Database issue):D1005–D1010
29. Griffiths-Jones S, Bateman A, Marshall M, Khanna A, Eddy SR (2003) Rfam: an RNA family database. Nucleic Acids Res 31(1):439–441
30. Eddy SR (2008) A probabilistic model of local sequence alignment that simplifies statistical significance estimation. PLoS Comput Biol 4(5):e1000069
31. Eddy SR (2009) A new generation of homology search tools based on probabilistic inference. Genome Inform 23(1):205–211
32. Flicek P, Amode MR, Barrell D, Beal K, Brent S, Chen Y et al (2011) Ensembl. Nucleic Acids Res 39(Database issue):D800–D806
33. Pruesse E, Quast C, Knittel K, Fuchs BM, Ludwig W, Peplies J et al (2007) SILVA: a comprehensive online resource for quality checked and aligned ribosomal RNA sequence data compatible with ARB. Nucleic Acids Res 35(21):7188–7196
34. Lagesen K, Hallin P, Rødland EA, Staerfeldt H-H, Rognes T, Ussery DW (2007) RNAmmer: consistent and rapid annotation of ribosomal RNA genes. Nucleic Acids Res 35(9):3100–3108
35. Pruesse E, Quast C, Knittel K, Fuchs BM, Ludwig W, Peplies J et al

(2010) Release information: SILVA 104. SILVA: comprehensive ribosomal RNA database.http://www.arb-silva.de/documentation/background/release-104/. Accessed 7 Apr 2010

36. Bateman A, Agrawal S, Birney E, Bruford EA, Bujnicki JM, Cochrane G, Cole JR et al (2011) RNA central: a vision for an international database of RNA sequences. RNA 17(11):1941–1946

Chapter 7

Energy-Based RNA Consensus Secondary Structure Prediction in Multiple Sequence Alignments

Stefan Washietl, Stephan H. Bernhart, and Manolis Kellis

Abstract

Many biologically important RNA structures are conserved in evolution leading to characteristic mutational patterns. RNAalifold is a widely used program to predict consensus secondary structures in multiple alignments by combining evolutionary information with traditional energy-based RNA folding algorithms. Here we describe the theory and applications of the RNAalifold algorithm. Consensus secondary structure prediction not only leads to significantly more accurate structure models, but it also allows to study structural conservation of functional RNAs.

Key words RNA structure, Consensus structure, Structure prediction, Functional RNA

1 Introduction

1.1 Conserved RNA Secondary Structures in Functional RNAs

As for most biological macromolecules, there is a close connection between the function of an RNA molecule and its structure. Therefore, accurate structure predictions can help to better understand RNA functions. This applies to the whole functional spectrum of RNAs, which reflects a diverse structural spectrum ranging from simple hairpin structures up to the intricate multi-loop and pseudoknot structures that build up complex molecular machines such as the ribosome.

Often, evolutionary processes lead to distinct and characteristic evolutionary signatures for different types of functional molecules. For structural RNA molecules, it is the structure and not necessarily their primary DNA sequence which is selectively maintained during evolution. Because of the simple rules that govern their secondary structure, structural RNAs provide exceptionally clear patterns of selection with base pairing patterns directly reflecting structural conservation. With the exception of nonstandard base pairs, RNA secondary structures are generally formed by the Watson/Crick base pairs A·U and G·C as well as G·U pairs.

Jan Gorodkin and Walter L. Ruzzo (eds.), *RNA Sequence, Structure, and Function: Computational and Bioinformatic Methods*, Methods in Molecular Biology, vol. 1097, DOI 10.1007/978-1-62703-709-9_7,

Two nucleotides that form a base pair may be changed by mutations but preserve the propensity to form a valid base pair. We distinguish consistent mutations and compensatory mutations. A consistent mutation changes one base (e.g., A·U ↔ G·U) while compensatory mutations change both bases in the base pair (e.g., A·U ↔ G·C or C·G ↔ G·C).

These mutational patterns have some concrete, practical implications. Interpreted as "evolutionary signatures" they can help to decide if some RNA forms a functional structure or if some region of genomic DNA is likely to be expressed into a functional RNA [1]. Another very useful aspect is the fact that the additional evolutionary information from homologous sequences greatly enhances our ability to predict accurate secondary structure models. The combination of this information with traditional thermodynamic folding algorithms will be the topic of this chapter.

1.2 Strategies to Predict Consensus Structures

Assuming several homologous RNA sequences form the same or very similar structures, our goal is to predict the *consensus structure* common to these sequences. Analyzing evolutionarily conserved structures generally requires a direct or indirect estimation of the underlying sequence alignment, which can itself depend on the structure or not.

There are three main strategies to consensus structure prediction, which differ in how structure prediction and sequence alignment are combined [2]: (1) The alignment is created first based on the sequence alone and the structure is predicted afterward, (2) The structure is predicted first and the structures are aligned afterwords, (3) sequence alignment and structure predictions are performed simultaneously.

In this chapter, we focus on the first approach which is conceptually the simplest. We are given a sequence alignment for which we want to predict a consensus structure. Clearly, this approach can only be successful if the alignment reflects the structural properties. Extensive benchmarks [3, 4] showed that if sequences are similar enough (more than ≈70% sequence identity) this requirement is usually met. It also does not matter which of the many sequence alignment programs are used because on high-sequence similarity data these programs agree well. So, for the purpose of getting high-quality alignments, high-sequence similarities are desirable. In contrast, for the purpose of structure prediction the opposite is true. It is only possible to efficiently exploit the evolutionary signal if there is enough evolutionary divergence in the sequences compared. Fortunately, in many practical applications we have access to data sets at reasonable range of sequence similarity, i.e. similar enough to be reliably aligned and diverse enough to infer secondary structures from mutation patterns.

Strategy (2) is rarely used in practice. One program of this class is `RNAforester`, which uses a tree alignment algorithm to align sequences based on their pre-calculated structure [5].

Strategy (3) solves the problem in the most rigorous way and is clearly the most appealing. Algorithms following this approach are usually variants of the well-known Sankoff algorithm [6]. The main drawback of these algorithms is their computational complexity in terms of both CPU time and memory. However, more recent variants of the Sankoff algorithm introduced several heuristics that improved the performance considerably. Some of these algorithms are discussed in Chapters 13–15.

Here, we concentrate on the case of a fixed alignment. In some rare cases there is enough evolutionary information that allows to predict an accurate structure exclusively from the analysis of the mutational patterns. Such covariation analyses typically use a mutual information score to find columns that show highly correlated mutation patterns. For example, this method led to surprisingly accurate structures for rRNAs already 30 years ago [7].

In practice, however, most data sets do not contain enough evolutionary information to predict reliable structures based on this information alone. In that case, thermodynamic folding algorithms might give better results. An obvious solution is to combine both strategies. This approach was first taken 2002 when `RNAalifold` [8] was presented as a tool to fold aligned sequences by extending Zuker's algorithm for folding single sequences. `RNAalifold` is now part of the widely used Vienna RNA packages and is routinely used to predict consensus structures by many researches.

2 The RNAalifold Algorithm

2.1 Averaged Energy Minimization for Multiple Alignments

`RNAalifold` extends the RNA structure prediction algorithms based on the nearest neighbor energy model in two ways: It averages the energy contributions over the sequences in the alignments and incorporates phylogenetic information as "pseudo-energies" into the energy model.

First, we want to recall the algorithm to predict the minimum free energy as introduced by Zuker and Stiegler [9] (refer to Chapter 4 for details). The recursions to calculate the minimum free energy F on a subsequence from i to j can be written as follows:

$$F(i,j) = \min\{F(i+1,j), \min_{i<u\leq j} C(i,u) + F(u+1,j)\} \qquad (1)$$

$$C(i,j) = \min\{\mathscr{H}(i,j), \min_{i<k<l<j}\{C(k,l) + \mathscr{I}(ij,kl)\},$$

$$\min_{i<u<j} M(i+1,u)M^1(u+1,j-1) + a\}$$

$$M(i,j) = \min\{M(i+1,j) + c, \min_{i<u\leq j} C(i,u) + b + (j-u)c,$$

$$\min_{i<u<j} C(i,u) + b + M(u+1,j)\}$$

$$M^1(i,j) = \min\{M^1(i,j-1) + c, C(i,j)\}$$

$C(i,j)$ is the minimum free energy given i,j form a base pair. The technical details of the multi-loop matrices M and M^1 will not concern us here. Relevant to understand how `RNAalifold` extends the single sequence folding are the terms $\mathcal{H}(i,j)$ (the free energy of a hairpin between bases i and j) and $\mathcal{I}(ij,kl)$ (the free energy of an interior loop between base pairs i,j and k,l).

In `RNAalifold`, the free energy contributions from $\mathcal{H}(i,j)$ and $\mathcal{I}(ij,kl)$ are replaced by the mean contributions over all m sequences s of the alignment $\mathcal{A}$. The recursion for matrix C in the case of a multiple alignment thus reads:

$$C^{\mathcal{A}}(i,j) = \min \begin{cases} \frac{1}{m}\sum_{s\in\mathcal{A}} \mathcal{H}(i,j,s), \\ \min_{i<k<l<j}\{C^{\mathcal{A}}(k,l) + \frac{1}{m}\sum_{s\in\mathcal{A}} \mathcal{I}(ij,kl,s)\}, \\ \min_{i<u<j} M(i+1,u)M^1(u+1,j-1) + a \end{cases} \tag{2}$$

Here, the indices i, j, k, and l correspond to columns of the alignment $\mathcal{A}$. While this step is straightforward, some problems arise. One of them is the presence of gaps in the alignments. Clearly, a gap character is not considered in the standard energy model but needs to be dealt with during the energy evaluation step. It can cause problems, for example, when evaluating loop energies that depend on the loop length or assigning dangling energies to base pairs. As a solution, energy contributions are generally computed from the *gap free* sequences s. To this end, a map from the alignment column to sequence index is computed for every sequence s before the recursion is started.

Another problem may arise when columns i and j cannot pair for some sequences of the alignment—either because the bases do not form a valid base pair or there are gaps at one or both of these positions. All these cases are treated as "nonstandard" base pairs which are considered in the standard thermodynamic model. As a consequence, these positions are assigned unfavorable energies and do not require special treatment in the algorithm.

2.2 A Simple Covariance Score

The second extension to single sequence structure prediction is the introduction of covariance terms that take into account the phylogenetic signal of a conserved structure. `RNAalifold` uses two distinct approaches for this.

The simpler one calculates the sum of the hamming distances of all sequence pairs s_α and s_β at columns i and j if they can form a valid base pair between i and j. More formally, with h being the hamming distance, $\mathscr{B}$=AU,CG,GC,GU,UA,UG the set of allowed base pairs, the conservation score γ' between two columns i and j can be written as:

$$\gamma'(i,j) = \frac{1}{2} \sum_{s_\alpha, s_\beta \in \mathscr{A}} \begin{cases} h(s_\alpha(i), s_\beta(i)) + h(s_\alpha(j), s_\beta(j)) & \text{if } (s_\alpha(i), s_\alpha(j)) \in \mathscr{B} \\ & \wedge (s_\beta(i), s_\beta(j)) \in \mathscr{B} \\ 0 & \text{otherwise} \end{cases} \quad (3)$$

This score only considers mutations that support a given base pair. Counterexamples, i.e. sequences where the base pair cannot be realized, are be penalized by another simple ad hoc score, leading to the total score γ:

$$\gamma(i,j) = \gamma'(i,j) + \delta \sum_{s \in \mathscr{A}} \begin{cases} 0 \text{ if } (s(i)s(j)) \in \mathscr{B} \\ 0.25 \text{ if } s(i) \wedge s(j) \text{are gaps} \\ 1 \text{ otherwise} \end{cases} \quad (4)$$

Here, the parameter δ balances between the covariation score supporting a base pair i, j and the penalty for counterexamples that cannot form a base pair.

Since γ is used as a "pseudo" energy during the energy minimization step, we need another parameter, β to balance between the thermodynamic and the phylogenetic score. The default value for both parameters δ and β is 1.

The conservation term γ is also used to decide whether a base pair is considered possible between two alignment columns. Base pairs are only evaluated if γ reaches a certain threshold. This reduces the number of possible base pairs. That is the reason why the speed of `RNAalifold` is usually not much slower than folding a single sequence although in theory the complexity of the algorithm is $\mathscr{O}(n^3 m)$ with n the length of the alignment and m the number of sequences.

2.3 An Improved Covariance Score Based on RIBOSUM Matrices

Many successful applications of `RNAalifold` over the past years demonstrate the power of this relatively simple scoring scheme. However, it suffers from one major drawback: Highly conserved base pairs not necessarily have a high covariance score and therefore only a few counterexamples can destroy a base pair in the consensus structure. In particular, this problem affects alignments with many sequences and high sequence conservation. In an alignment with 500 sequences, for example, three sequences that cannot form a base pair would result in a prohibitive penalty even though all the other sequences have a totally conserved base pair in this position.

A new scoring scheme was introduced in 2008 to overcome this problem [10]. In this model, RIBOSUM like scoring matrices [11] are used as a replacement for the hamming distance-based scores. To obtain these scoring matrices, log-likelihood scores are calculated from alignments of ribosomal RNAs.

$$R(ab, cd) = \log \frac{f(ab, cd)}{f(ac)f(bd)} \tag{5}$$

Here, $f(ab, cd)$ is the frequency that base pair ab is aligned to base pair cd, and $f(ac)$ the frequency that base a is aligned with c. Following the example of the original RIBOSUM and BLOSUM matrices, these log-likelihood scores are computed for different subsets of the rRNA alignments, containing sequences of different sequence conservation. The RIBOSUM matrix to be used to score an alignment is the one that most closely reflects the minimum and maximum sequence distances in the alignment. `RNAalifold` compares the sequences of an input alignment and chooses the appropriate scoring matrix.

The new covariance score γ' can then be written as:

$$\gamma'(i,j) = \frac{1}{2} \sum_{\substack{s_\alpha, s_\beta \in \mathcal{A} \\ s_\alpha \neq s_\beta}} R\left(s_\alpha(i)s_\alpha(j), s_\beta(i)s_\beta(j)\right). \tag{6}$$

The log-likelihood scores are scaled to approximately achieve the same absolute values as the hamming distance scores in Eq. 3. In spite of that, the relative impact of the covariation scoring increased from 5% to 50%, making it necessary to also adapt the factors β and δ. The new default values are $\beta = 0.6$ and $\delta = 0.5$. Overall, RIBOSUM scoring performs about 10% better than hamming distance scoring [10].

2.4 Extending Other Folding Algorithms to Multiple Alignments

It is important to note that all of the extension to the single sequence folding algorithm in `RNAalifold` only apply to the actual energy computations. This means that all other variants of folding algorithms that are available for single sequences can also be adapted to multiple alignments. In particular, partition function folding [12], stochastic backtracking [13], centroid structure [14], and local folding [15] are all implemented for the multiple alignment case in the Vienna RNA package.

3 Using RNAalifold

In the following sections, we briefly demonstrate how to use the RNAalifold program. We assume the reader has access to a UNIX like environment such as GNU/Linux or Mac OS X with the latest Vienna RNA package installed. Refer to **Notes 1** and **2** for details how to install and use the Vienna RNA package.

3.1 A Simple Example

To use RNAalifold, we need a multiple alignment of several homologous sequences. In principle, a pairwise alignment is sufficient although having more sequences clearly improves the results.

RNAalifold expects the alignment to be formatted in CLUSTAL W format (*see* **Note 3** on how to obtain and format alignments). As a short example, we use an alignment of 20 different isolates of Peach latent mosaic viroid. The short regions covers the so-called hammerhead ribozyme, a self-cleaving structural RNA. To predict the consensus secondary structure of this motif we run RNAalifold on the alignment file:

```
$ RNAalifold hammerhead.aln
```

which gives the following result:

```
20 sequences; length of alignment 55.
CAAAAGUCUGGGCUAA_GCCCACUGAUGAGUCGCUGAGAUGCGACGAAACUUUUG
((((((..((((....))))........((((......)))))...))))))
 minimum free energy = -18.58 kcal/mol (-15.36 + -3.21)
```

The output shows the consensus sequence of the alignment and the predicted consensus structure in dot bracket notation. Every base pair is denoted by a pair of brackets "(" and ")" while unpaired positions are shown as dots ".". RNAalifold also reports a "minimum free energy" of the consensus structure of $-18.58 = -15.36 + (-3.21)$ kcal/mol. The value consists of two terms. The first term is the average free energy of the structures and the second part is the covariance term. A negative covariance term indicates the presence of many compensatory or consistent mutations supporting the structure. In contrast to the average free energy, the covariance term has no biophysical meaning and thus the RNAalifold score is more precisely referred to as a "pseudo-energy."

As mentioned earlier, RNAalifold not only implements Zuker's algorithm to find the minimum free energy, but it also implements McCaskill's algorithm [12] to calculate base pair probabilities. To run both variants of the algorithm we use the command line switch "-p":

```
$ RNAalifold -p hammerhead.aln
```

The base pair probabilities can be visualized in a so-called dotplot. This and other ways to visualize the structure predictions are described in the next section.

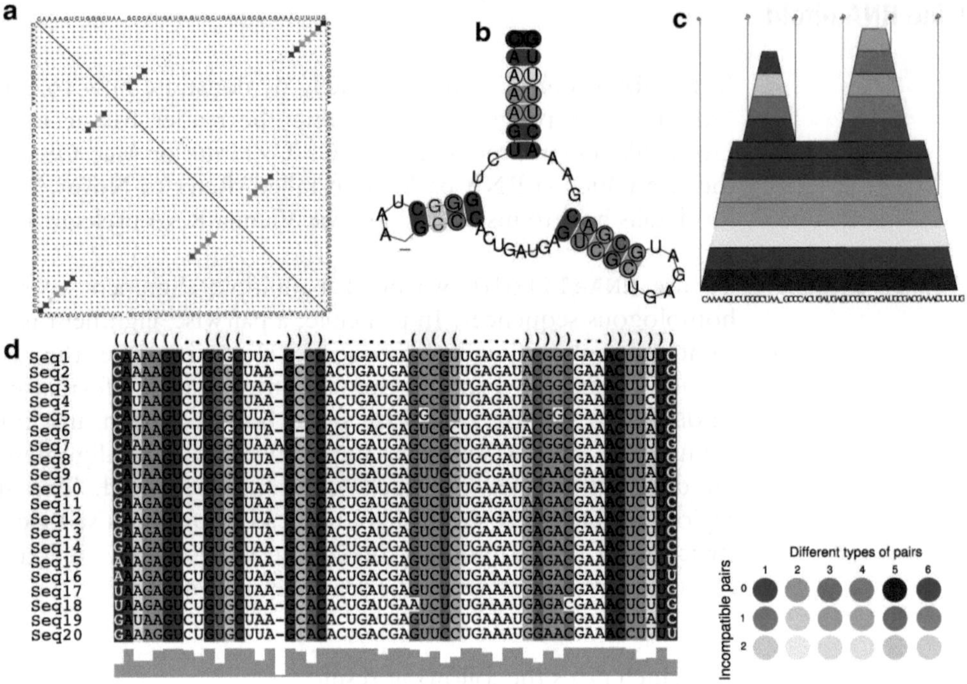

Fig. 1 Visualizations of consensus secondary structures. A color version of this figure can be found online at http://www.tbi.univie.ac.at/papers/SUPPLEMENTS/MiMB/ (**a**) Dotplot of base pair probabilities (*upper right*) and pairing matrix of the optimal consensus structure (*lower left*). (**b**) Traditional RNA representation (**c**) Mountain plot (**d**) Sequence alignment with dot bracket notation of consensus structure

3.2 Visualizing the Results

By default `RNAalifold` generates two representations of the consensus structure (`alirna.ps`) and if run with "`-p`" a dotplot of pair probabilities (`alidot.ps`). However, with additional tools it is possible to create a variety of other visualizations (Fig. 1). In all of these representations the mutational patterns are highlighted using a color code. That way mutations that support a predicted structure (consistent/compensatory mutations) or disrupt a predicted structure (incompatible mutations) can be easily identified. Before we explain this code in more detail, we demonstrate how to generate these pictures on the hammerhead example (*see* **Note 2** for how to view EPS formatted pictures on Linux and OS X).

The dotplot representation (Fig. 1a) is created by default. Every element (i, j) in the matrix represents a base pair formed between i and j. The upper right half of the matrix represents the pair probabilities; the size of every dot is drawn proportional to the probability of this base pair. The lower left half of the matrix represents the optimal structure, i.e. the structure with minimum free (pseudo-)energy. Comparing both parts of the dotplot can be useful to find out whether the optimal predicted structure is well defined or whether there are many alternative structures possible. In the hammerhead example both dotplots are nearly identical because the prediction is supported by many compensatory mutations which are not compatible with alternative structures.

To obtain a color coded representation of the RNA structure as shown in Fig. 1b, we use the Perl script `colorrna.pl`. This script takes the (black and white) structure plot in `alirna.ps` and the dotplot in `alidot.ps` and writes a colored structure plot which is saved to the file `colored_rna.ps`:

```
$ colorrna.pl alirna.ps alidot.ps > colored_rna.ps
```

To produce a mountain plot (Fig. 1c), we simply run the script `cmount.pl` on the dotplot file `alidot.ps` and save the results in a new postscript file `mountain.ps`:

```
$ cmount.pl alidot.ps > mountain.ps
```

In a mountain plot each base pair (i, j) is represented by a trapeze with a baseline from i to j and a height proportional to the probability of base pair (i, j). Unpaired bases form plateaus in this representation.

To show the consensus structure as a dot bracket string on the sequence alignment itself (Fig. 1d) and to color the alignment columns, we use the script `coloraln.pl`. This script reads the original alignment file and the structure file `alirna.ps` and generates a postscript file which is saved in `color_aln.ps`:

```
coloraln.pl -s alirna.ps hammerhead.aln
    > colored_aln.ps
```

All the different representations use the same color code (a colored version of Fig. 1 is available at: http://www.tbi.univie.ac.at/papers/SUPPLEMENTS/MiMB/). If a paired position (i, j) is formed by the same base pair in every sequence of the alignment, the base pair is shown in red. If i, j is supported by 2, 3, 4, 5, 6 different base pairs, they are colored in brown, green, turquoise, blue, and violet, respectively. In this particular example we see positions that have 2, 3, and 5 different base pairs. For example, the outermost base pair is colored in blue because we observe C·G, G·C, A·U,G·U, U·G providing compelling evidence for selective pressure on this base pair. On the other hand, if one or more sequences cannot form a base pair present in the consensus structure, these positions are shown in pale versions of the respective color. If there are more than two inconsistent base pairs, the position is not colored and appears white.

The colored alignment (Fig. 1d) is particularly useful to quickly assess the structure conservation. Many columns in saturated colors other than red indicate a conserved structure well supported by compensatory mutations. This representation also allows to identify sequences that do not fold into the consensus structure.

In practice, some sequences might be misaligned, have changed their structure, form nonstandard base pairs, or contain sequence errors. All of which leads to incompatible mutations that appear without coloring. Also the hammerhead example contains a few positions in sequences 5, 6, and 18 that are incompatible with the consensus although the general structure is very well conserved in all sequences.

4 Advanced Usage

4.1 Refining Structure Prediction of Individual Sequences

`RNAalifold` predicts the consensus structure of a set of aligned RNA sequences. However, there may be parts of the molecules that are not conserved, e.g. because they are not essential for the function or because they have been the subject to recent structural changes. Furthermore, it is not uncommon that the length of conserved stems and loops varies somewhat between the sequences.

To get a potentially more accurate secondary structure of a single RNA sequence in an alignment, one can use a two-step process. First the consensus structure is predicted using `RNAalifold`. As a second step, the consensus structure is used as a *structural constraint* for single structure prediction using `RNAfold`. The ViennaRNA package provides a Perl script `refold.pl` that uses the output of `RNAalifold` and the input alignment to generate the constraints for `RNAfold` for every sequence in the alignment.

The `refold.pl` approach is especially useful for large alignments with extensive unconserved regions. A good example is the very diverse RNase P RNA class A. We used `RNAalifold` to predict a structure for *Vibrio cholerae* from an alignment of 305 bacterial species and compared it to an experimentally well-established reference structure. The consensus prediction is relatively poor and only achieves a sensitivity and specificity of 38.2% and 68.1%, respectively, of correctly predicted base pairs. Refolding the sequence with constraints from the consensus structure leads to a much improved sensitivity of 59.3% at the cost of a slightly lower specificity of 62.4%.

```
$ RNAalifold < rnasep.aln > rnasep.alifold
$ refold.pl rnasep.aln rnasep.alifold | RNAfold -C
```

Note that consensus base pairs that lead to a positive energy contribution in a particular sequence will not be part of the new prediction for this sequence. Also base pairs that are mutually exclusive to those of the consensus structure will not be included. This can lead to a largely unfolded structure if many consensus base pairs are affected. Consequently, sequences that are largely unstructured where the consensus structure contains base pairs are most likely wrongly aligned or do not share the same consensus structure as the rest of the alignment.

4.1.1 Local Structure Prediction in Long Alignments

The prediction of secondary structures for long (>500nt) RNA molecules has two major drawbacks: As the algorithm complexity scales cubic in time, the computation time becomes prohibitive for very long sequences. Furthermore, the quality of the predictions decreases rapidly with the length of the molecules. In practice this is not necessarily a problem. For many applications there is no biological reason to assume that a long RNA molecule (e.g., an mRNA transcript of several kilobases in length) has a defined global structure. Often local structural features such as regulatory elements in an mRNA are of interest. In such cases it is useful to predict *local structural components* only.

The program `RNALalifold` is included in the ViennaRNA package since version 2.0. It predicts local structures with a maximum base pair span L (usually L is between 100 and 200 nt). It uses the algorithm by Hofacker et al. [15] to calculate "locally stable closed structures." "Closed" means that there is always an outer base pair i, j if a substructure for the sequence interval i, j is reported (Exceptions are unpaired bases that contribute favorable "dangling energies" for the structure). "Local stability" means that a substructure starting at base i is only reported if there are no substructures with better or equal energy starting at bases $i-1$ or $i+1$.

While `RNALalifold` is not guaranteed to get all important substructures, it usually helps to get a reasonable list of candidates. As an example we calculate local structures for a 1.1-kb human region on chromosome X that contains a cluster of miRNAs (miR-106a, miR-20b, miR-19b2, miR-92a, and miR-363). We use a 12-way alignment of this region with human aligned to 11 other mammalian species.

For a maximum base pair span of 120, the example can be computed like this:

```
$ RNALalifold -L 120 mirsequences.aln
```

The output consists of a list of locally stable structures with the pseudo free energy of the local structure and the start and stop columns in the alignment:

```
12 sequences; length of alignment 1171.

...

(((.(((.....))).))). ( -2.10)  995 - 1014
(((........))). ( -0.28)  976 -  990
((..((((..................))))..)). ( -1.87)  946 -  977

...
```

Because of space constraints in this book, we have only shown three of the predicted local structures. In total RNALalifold predicts 33 structures in this region including all 5 known miRNA precursors. As a comparison, the single sequence variant RNALfold predicts 125 local structures. This clearly shows that using consensus structure prediction increases the specificity in structure prediction. A single sequence can fold easily into some random structure even though it has no biological meaning. In contrast, it is difficult to find a consensus structure by chance in an alignment of sequences that have accumulated many random mutations. However, the fact that RNAalifold or RNALalifold predicts a consensus structure does not necessarily mean that it is a biologically relevant structure. To answer this particular question other algorithms and metrics have been developed [16–18]. As a ready-to-use implementation to detect biologically relevant structures the program RNAz [19, 20] can be used.

5 The RNAalifold Web-Server

An alternative way to use RNAalifold is the public web-server at http://rna.tbi.unvie.ac.at presenting and explaining all available options and allows to predict consensus structures and all the various visualization without installing software locally (Fig. 2).

6 Alternative Methods and New Approaches

An alternative to thermodynamic folding algorithms are pure probabilistic methods. Stochastic context free grammars (SCFG) are widely used to model RNA structures (*see* Chapter 5). They have also been used to address the consensus folding problem. Pfold [21] explicitly models the evolutionary relationship of the sequences in the alignment using a phylogenetic tree and the RNA structures using a simple grammar. Since both components are probabilistic in nature, Pfold can easily combine them in one framework to calculate the probability distribution of all secondary structures for a given alignment and phylogenetic tree.

Seemann et al. extended Pfold by incorporating pair probabilities from thermodynamic folding [22]. Their program PETFold thus unifies evolutionary and thermodynamic information in this attractive framework.

Despite the differences between these algorithms in how RNA structures are modeled and how evolutionary information is incorporated, all need to solve the same problem of predicting an optimal structure from their model. In the original implementation

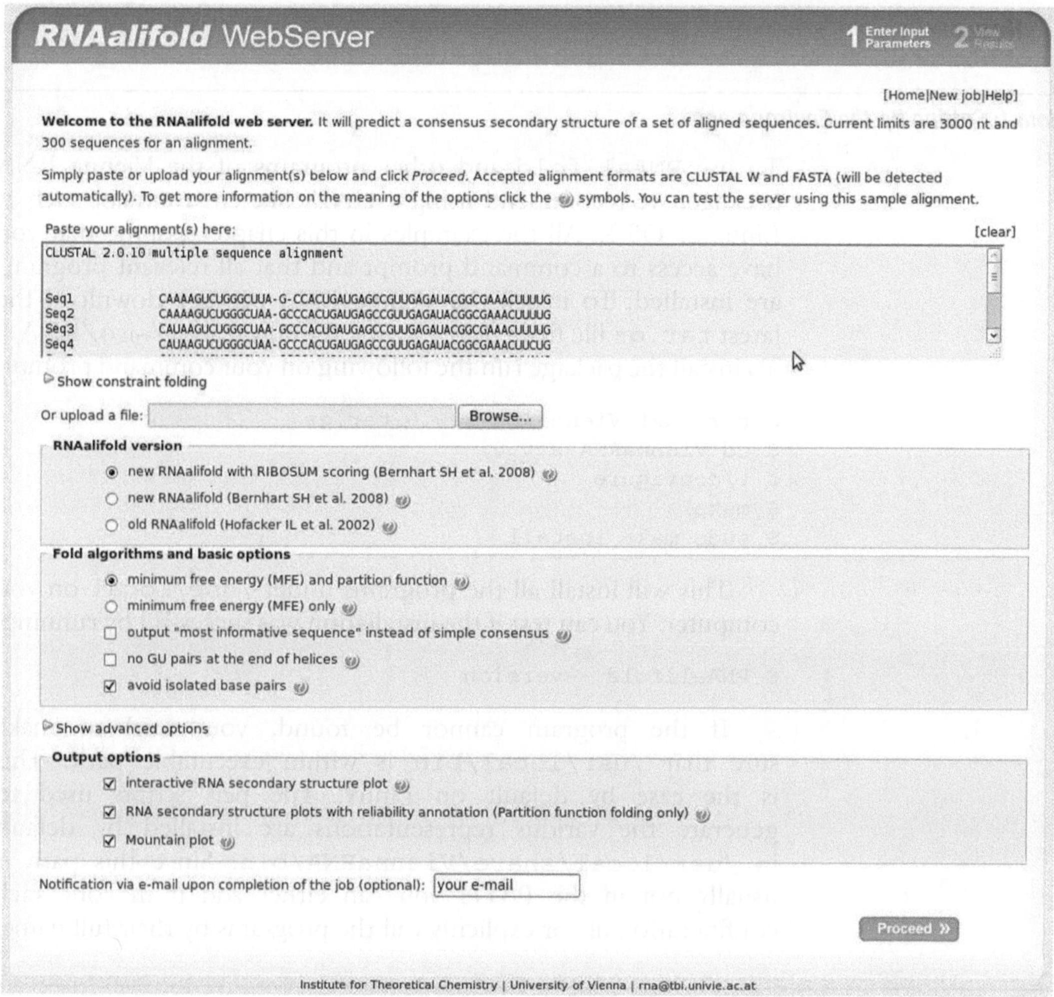

Fig. 2 Screenshot of the `RNAalifold` web-server available at: rna.tbi.univie.ac.at

of `Pfold` this was solved by finding the maximum likelihood solution which corresponds to the minimum free energy solution of `RNAalifold`. A later version of `Pfold` introduced a maximum expected accuracy (MEA) approach. The idea of MEA is to find a structure that when compared to the pair probabilities of all possible structures maximizes the expected base pair accuracy [23, 24]. Also `PETFold` uses this approach and a few other variants of this idea in the context of consensus structure prediction have been proposed [25–27]. In its latest version, also `RNAalifold` predicts MEA structures from the pair probabilities.

7 Notes

Note 1: Setting Up the Environment

To use RNAalifold and other programs of the Vienna RNA package, we recommend using a UNIX like environment such as Linux or OS X. All the examples in this chapter assume that you have access to a command prompt and that all relevant programs are installed. To install the Vienna RNA package download the latest tar.gz file from http://www.tbi.univie.ac.at/~ivo/RNA/. To install the package run the following on your command prompt:

```
$ tar -xzf ViennaRNA-2.0.0.tar.gz
$ cd ViennaRNA-2.0.0
$ ./configure
$ make
$ sudo make install
```

This will install all the programs under /usr/local on you computer. You can test if the installation was successful by running:

```
$ RNAalifold --version
```

If the program cannot be found, you need to make sure that /usr/local/bin is within executable path. This is the case by default on Linux. The perl scripts used to generate the various representations are installed by default in /usr/local/share/ViennaRNA/bin. Since this path is usually not in the PATH you can either add it in your shell configuration file or explicitly call the programs by their full name.

Note 2: Practical Tips to Use Programs from the Vienna RNA Package

All programs presented in this chapter come with detailed online information. To get help for any programs of the Vienna RNA package you can run:

```
$ RNAalifold --help
$ man RNAalifold
```

The perl scripts contain embedded documentation which can be viewed using perldoc:

```
$ perldoc -F 'which refold.pl'
```

All the graphical output is formatted as EPS. There are various programs you can use to view an EPS image. Typically the following command will display an eps file on Linux:

```
$ gv image.ps
```

On OS X you can use the open command:

```
$ open image.ps
```

EPS images usually can be imported and edited by common graphics programs. Moreover, all EPS images produced by the Vienna RNA package are pure vector graphics that can be scaled and converted to other formats without loss of quality.

Note 3: Creating and Formatting Sequence Alignments

`RNAalifold` expects a multiple sequence alignment in `CLUSTAL W` format. You can use the program `CLUSTAL W` itself which natively produces that output. But also more modern alignment programs such as MUSCLE [28] can create `CLUSTAL W` formatted output suitable for `RNAalifold`. If your alignment program does not support the `CLUSTAL W` format or you are using prebuilt alignments provided in a different format, you might need to convert the format. A typical `CLUSTAL W` alignment looks like this example:

```
CLUSTAL 2.0.10 multiple sequence alignment

Seq1  CAAAAGUCUGGGCUUA-G-CCACUGAUGAGCCGUUGAGAUACGGCGAAA
      CUUUUG
Seq2  CAAAAGUCUGGGCUAA-GCCCACUGAUGAGCCGUUGAGAUACGGCGAAA
      CUUUUG
Seq3  CAUAAGUCUGGGCUAA-GCCCACUGAUGAGCCGUUGAGAUACGGCGAAA
      CUUUUG
Seq4  CAUAAGUCUGGGCUAA-GCCCACUGAUGAGCCGUUGAGAUACGGCGAAA
      CUUCUG
```

There is no formal specification for the program but the first line should start with the word "CLUSTAL." After two empty lines the actual alignment data starts. Every entry has a unique name (no white spaces) and separated by a variable number of spaces a sequence string. For `RNAalifold` it can contain the letters A,U,T,G,C, and a dash ("-") for gaps. For long alignments, the data can be broken up into blocks. This is optional but if the data is given in multiple blocks, each block must contain the same list of names in the same order.

Acknowledgements

Stefan Washietl was supported by an Erwin Schrödinger Fellowship of the Austrian Fonds zur Förderung der Wissenschaftlichen Forschung. Stephan H. Bernhart was funded by the Austrian GEN-AU project "Noncoding RNA." We thank Ivo Hofacker and Benjamin Holmes for comments on the manuscript.

References

1. Washietl S, Pedersen JS, Korbel JO, Stocsits C, Gruber AR, Hackermüller J, Hertel J, Lindemeyer M, Reiche K, Tanzer A, Ucla C, Wyss C, Antonarakis SE, Denoeud F, Lagarde J, Drenkow J, Kapranov P, Gingeras TR, Guigó R, Snyder M, Gerstein MB, Reymond A, Hofacker IL, Stadler PF (2007) Structured RNAs in the ENCODE selected regions of the human genome. Genome Res 17:852–864. doi:10.1101/gr.5650707
2. Gardner PP, Giegerich R (2004) A comprehensive comparison of comparative RNA structure prediction approaches. BMC Bioinformatics 5: 140. doi:10.1186/1471-2105-5-140
3. Gardner PP, Wilm A, Washietl S (2005) A benchmark of multiple sequence alignment programs upon structural RNAs. Nucleic Acids Res 33:2433–2439. doi:10.1093/nar/gki541
4. Wilm A, Mainz I, Steger G (2006) An enhanced RNA alignment benchmark for sequence alignment programs. Algorithms Mol Biol 1:19. doi:10.1186/1748-7188-1-19
5. Höchsmann M, Voss B, Giegerich R (2004) Pure multiple RNA secondary structure alignments: a progressive profile approach. IEEE/ACM Trans Comput Biol Bioinform 1: 53–62. doi:10.1109/TCBB.2004.11
6. Sankoff D (1985) Simultaneous solution of the RNA folding, alignment and protosequence problems. SIAM J Appl Math 45:810–825
7. Noller HF, Kop J, Wheaton V, Brosius J, Gutell RR, Kopylov AM, Dohme F, Herr W, Stahl DA, Gupta R, Woese CR (1981) Secondary structure model for 23s ribosomal RNA. Nucleic Acids Res 9(22):6167–6189. doi:10.1093/nar/9.22.6167
8. Hofacker IL, Fekete M, Stadler PF (2002) Secondary structure prediction for aligned RNA sequences. J Mol Biol 319:1059–1066. doi:10.1016/S0022-2836(02)00308-X
9. Zuker M, Stiegler P (1981) Optimal computer folding of large RNA sequences using thermodynamics and auxiliary information. Nucleic Acids Res 9:133–148.
10. Bernhart SH, Hofacker IL, Will S, Gruber AR, Stadler PF (2008) RNAalifold: improved consensus structure prediction for RNA alignments. BMC Bioinformatics 9:474. doi:10.1186/1471-2105-9-474
11. Klein RJ, Eddy SR (2003) RSEARCH: finding homologs of single structured RNA sequences. BMC Bioinformatics 4: 44. doi:10.1186/1471-2105-4-44
12. McCaskill JS (1990) The equilibrium partition function and base pair binding probabilities for RNA secondary structure. Biopolymers 29: 1105–1119. doi:10.1002/bip.360290621
13. Ding Y, Lawrence CE (1999) A bayesian statistical algorithm for RNA secondary structure prediction. Comput Chem 23(3–4): 387–400.
14. Ding Y, Chan CY, Lawrence CE (2005) RNA secondary structure prediction by centroids in a boltzmann weighted ensemble. RNA 11(8): 1157–1166. doi:10.1261/rna.2500605
15. Hofacker IL, Priwitzer B, Stadler PF (2004) Prediction of locally stable RNA secondary structures for genome-wide surveys. Bioinformatics 20:186–190. doi:10.1093/bioinformatics/btg388
16. Washietl S, Hofacker IL (2004) Consensus folding of aligned sequences as a new measure for the detection of functional RNAs by comparative genomics. J Mol Biol 342: 19–30. doi:10.1016/j.jmb.2004.07.018
17. Gruber AR, Bernhart SH, Hofacker IL, Washietl S (2008) Strategies for measuring evolutionary conservation of RNA secondary structures. BMC Bioinformatics 9:122. doi:10.1186/1471-2105-9-122
18. Gesell T, Washietl S (2008) Dinucleotide controlled null models for comparative RNA gene prediction. BMC Bioinformatics 9:248. doi:10.1186/1471-2105-9-248
19. Gruber AR, Findeiß S, Washietl S, Hofacker IL, Stadler PF (2010) RNAz 2.0: improved noncoding RNA detection. Pac Symp Biocomput 15:69–79
20. Washietl S (2007) Prediction of structural noncoding RNAs with RNAz. Methods Mol Biol 395:503–526
21. Knudsen B, Hein J (2003) Pfold: RNA secondary structure prediction using stochastic context-free grammars. Nucleic Acids Res 31: 3423–3428
22. Seemann SE, Gorodkin J, Backofen R (2008) Unifying evolutionary and thermodynamic information for RNA folding of multiple alignments. Nucleic Acids Res 36:6355–6362. doi:10.1093/nar/gkn544
23. Do CB, Woods DA, Batzoglou S (2006) CONTRAfold: RNA secondary structure prediction without physics-based models. Bioinformatics 22:e90–e98. doi:10.1093/ bioinformatics/btl246
24. Lu ZJ, Gloor JW, Mathews DH (2009) Improved RNA secondary structure prediction by maximizing expected pair accuracy. RNA 15:1805–1813. doi:10.1261/rna.1643609
25. Kiryu H, Kin T, Asai K (2007) Robust prediction of consensus secondary structures using averaged base pairing probability matrices. Bioinformatics 23:434–441. doi:10.1093/bioinformatics/btl636

26. Hamada M, Kiryu H, Sato K, Mituyama T, Asai K (2009) Prediction of RNA secondary structure using generalized centroid estimators. Bioinformatics 25:465–473. doi:10.1093/bioinformatics/btn601
27. Hamada M, Sato K, Asai K (2011) Improving the accuracy of predicting secondarystructure for aligned RNA sequences. Nucleic Acids Res 39:393–402. doi:10.1093/ nar/ gkq792
28. Edgar RC (2004) MUSCLE: multiple sequence alignment with high accuracy and high throughput. Nucleic Acids Res 32: 1792–1797. doi:10.1093/nar/gkh340

Chapter 8

SCFGs in RNA Secondary Structure Prediction: A Hands-on Approach

Zsuzsanna Sükösd, Ebbe S. Andersen, and Rune Lyngsø

Abstract

Stochastic context-free grammars (SCFGs) were first established in the context of natural language modelling, and only later found their applications in RNA secondary structure prediction. In this chapter, we discuss the basic SCFG algorithms (CYK and inside–outside algorithms) in an application-centered manner and use the pfold grammar as a case study to show how the algorithms can be adapted to a grammar in a nonstandard form. We extend our discussion to the use of grammars with additional information (such as evolutionary information) to improve the quality of predictions. Finally, we provide a brief survey of programs that use stochastic context-free grammars for RNA secondary structure prediction and modelling.

Key words SCFGs, CYK algorithm, Inside–outside algorithm, Pfold

1 Introduction

Stochastic context-free grammars (SCFGs), also known as probabilistic context-free grammars (PCFGs), were first invented in the purely computational context of natural language modelling, and to this day are widely used in computational linguistics [1–3]. It is perhaps surprising that SCFGs found applications in RNA secondary structure modelling too – but RNA secondary structure and natural languages have many things in common.

In both cases, our concern is producing a string of symbols from a particular alphabet. In natural languages, the alphabet is made up of the vocabulary of the language; in the case of RNA secondary structure, we have nucleotides in particular structural configurations. In both cases it is also a requirement that the symbols are chosen and connected to each other according to particular rules, which can formally be described by a grammar. The rules can describe connections between distant parts of the string (sentence). For example, in the English language a verb

Jan Gorodkin and Walter L. Ruzzo (eds.), *RNA Sequence, Structure, and Function: Computational and Bioinformatic Methods*, Methods in Molecular Biology, vol. 1097, DOI 10.1007/978-1-62703-709-9_8,

must always be connected to a subject, but they need not be immediately adjacent to each other. In RNA secondary structure, any pairing nucleotide must have its pairing partner somewhere in the structure.

Having a grammar to describe a language enables us to formally ask and answer questions about the language itself: for example, we can determine whether a particular string can be produced in the language or not. We can also generate arbitrary strings that fulfill the criteria of the language. Extending the grammar with probabilities, we can ask how likely a sentence is to be produced in the language, or the probability with which two particular words occur in a particular configuration in a sentence.

In this chapter, we will introduce the basic formalism of these ideas and describe how they can be used for the prediction and modelling of RNA secondary structure. Our goal here is an introductory, "hands-on" approach, with focus on providing an intuitive understanding of the subject and practical examples of its use, rather than formal proofs. For a more rigorous treatment of the subject, the reader is advised to consult the references listed at the end of this chapter.

2 What Is an SCFG?

A grammar (also known as formal grammar or transformational grammar) can be thought of as a framework for producing particular types of strings using "placeholders," which turn into symbols from the alphabet according to predefined rules. A grammar thus consists of: an alphabet (terminal symbols, which make up the words of the language), variables (nonterminal symbols, which function as the "placeholders"), and rewriting rules (productions). A string is a concatenation of terminal and nonterminal symbols. A production is of the form $\alpha \rightarrow \beta$, where in general, α and β can be any strings, with the restriction that α must contain at least one nonterminal symbol. The function of a grammar is to convert, in a stepwise fashion, a single nonterminal symbol (the designated start symbol) into a string of terminals from the alphabet of the grammar.

Definition 1: A sequence generating grammar is a tuple (V, Σ, P, S) where

- V is a finite set of nonterminal symbols (variables), conventionally denoted by uppercase letters
- Σ is a finite set of terminal symbols (alphabet), conventionally denoted by lowercase letters

- P is a finite set of productions of the type $\alpha \to \beta$, where $\alpha \in (V \cup \Sigma)^* V (V \cup \Sigma)^*$ and $\beta \in (V \cup \Sigma)^*$.[1]
- $S \in V$ is the start symbol

An example of a grammar is the Knudsen–Hein grammar, which is used in the pfold program for RNA secondary structure prediction.

Example 1: The Knudsen–Hein grammar contains three nonterminal symbols: $\{S, L, F\}$ (of which S is the start symbol), the alphabet: $\{s, d\}$, and six production rules. Alternative productions from the same left-hand side are separated by | ("or").

$$\begin{aligned} S &\to L \mid LS \\ L &\to s \mid dFd \\ F &\to dFd \mid LS \end{aligned}$$

Generating a single s (representing a single unpaired nucleotide) can be done as:

$$S \Rightarrow L \Rightarrow s$$

A small stem-loop could be generated as:

$$S \Rightarrow L \Rightarrow dFd \Rightarrow dLSd \Rightarrow dsSd \Rightarrow dsLd \Rightarrow dssd$$

Note that the derivation connects the two d's to each other grammatically. The first d is like an opening paranthesis with the second d as its closing pair; this is indeed the basis of the dot-bracket representation of RNA secondary structures. In the case of the KH grammar, the semantic interpretation of a derivation is the secondary structure itself.

We will use the grammar from Example 1 throughout this chapter, and will in the following refer to it as the KH grammar for brevity.

Producing a string using a grammar is also known as "deriving" or "parsing" the string. A useful way of representing the parse of a string is a parse tree (*see* Fig. 1). Two derivations of a string differ only if they have different parse trees. If there exists a string that can be produced in more than one way by the grammar, the grammar is said to be *syntactically ambiguous*. As the same RNA sequence can usually fold into many different secondary structures, RNA secondary structure grammars are often syntactically ambiguous by design. A one-to-one correspondence between parse trees and RNA secondary structures is also desirable; in contrast, if the same secondary structure can be represented by more than one parse tree, then the grammar is said to be *semantically ambiguous*.

[1] The symbol * is the Kleene star operator; it is a widely used regular expression and denotes any (possibly empty) string that is produced by concatenating elements drawn from the set. The elements can occur any number of times and in any order in this string.

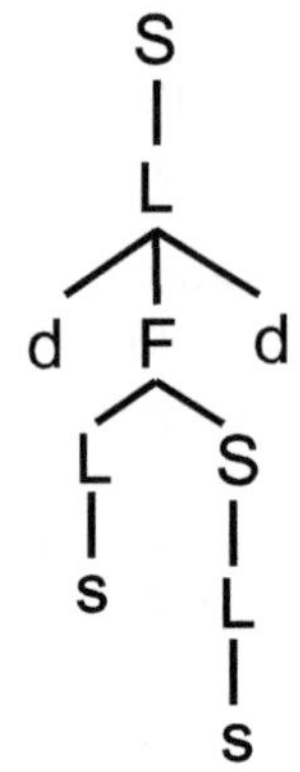

Fig. 1 The parse tree of the derivation of the string *dssd* using the KH grammar. This string can only be derived in one way

Example 2: The KH grammar is syntactically ambiguous. For example, the string *dssdssdssd* can be parsed in two different ways; the details are left to the reader as an exercise. However, the KH grammar is semantically unambiguous, because there is a one-to-one correspondence between RNA secondary structures and parse trees. Chapter 5 in this book discusses the issue of ambiguity in greater detail.

Grammars can be classified on the basis of their production rules into what is known as the Chomsky-hierarchy. Context-free grammars form a class in this hierarchy and are defined by only having a single nonterminal symbol on the left-hand side of all of the production rules. The context-free designation is because one can always rewrite the nonterminal symbol, irrespective of the context in which it occurs. This is a convenient model for RNA secondary structure, where insertions of stem-loops can happen anywhere without affecting the rest of the structure; this is known as "nesting" structures. However, context-free grammars are not naturally suited for modelling pseudoknotted structures. The KH grammar is context-free.

Note that there can be several equivalent grammars that describe the same language and can be converted to each other by rule transformations. This is illustrated for one rule of the KH grammar in Example 3 below.

Example 3: The rule $L \rightarrow dFd$ can be "expanded" to a set of equivalent rules:

$$L \rightarrow NM, \quad N \rightarrow d, \quad M \rightarrow FK, \quad K \rightarrow d.$$

Thus: $L \Rightarrow NM \Rightarrow dM \Rightarrow dFK \Rightarrow dFd$.

It is desirable to use a standard for writing equivalent grammars, such that any general grammar has an equivalent grammar in the standard. These standards are called normal forms, and the most common normal form for context-free grammars is the Chomsky normal form.

Definition 2: A context-free grammar (V, Σ, P, S) is in Chomsky normal form if all of its production rules are of the form:

- $N \rightarrow AB$ (rules of this type are called bifurcations) or
- $N \rightarrow \sigma$ or
- $S \rightarrow \epsilon$

where $N, A, B \in V$, $\sigma \in \Sigma$, ϵ is the empty string and $A, B \neq S$.

It can be proven that every context-free grammar can be written in Chomsky normal form. A grammar in Chomsky normal form is also clearly always context-free. The transformation of a given context-free grammar to Chomsky normal form, however, is not always trivial.

As noted above, the KH grammar provides alternative production rules from the same nonterminals, giving us a choice every time we need to parse a nonterminal. If we associate each of these choices with a probability, such that the total probability of the choices from the same nonterminal is 1, we get a stochastic context-free grammar.

Definition 3: A stochastic grammar is a grammar (V, Σ, P, S) extended with a weight function $w : P \mapsto \mathbb{R}$, where w for each left-hand side is a probability distribution over the possible right-hand sides it can be replaced with. The probability of a derivation is the product of the probabilities of the productions used in that derivation. A stochastic context-free grammar is a context-free grammar that is stochastic.

Derivations with higher probabilities are said to be more consistent with the grammar than derivations with lower probabilities. The probability of a string under the grammar is the sum of the possible derivations of that string under the grammar.

3 SCFG Algorithms

We are now in the position to ask some useful questions about SCFGs:

1. Can a particular string be produced by the grammar?
2. What derivation has the highest probability under the grammar?
3. How can a particular string be produced by the grammar?
4. What is the total probability of all derivations of a particular length?
5. What is the total probability under a grammar describing RNA secondary structure that two particular positions form a pair?

Efficient algorithms have been developed to answer these questions computationally.

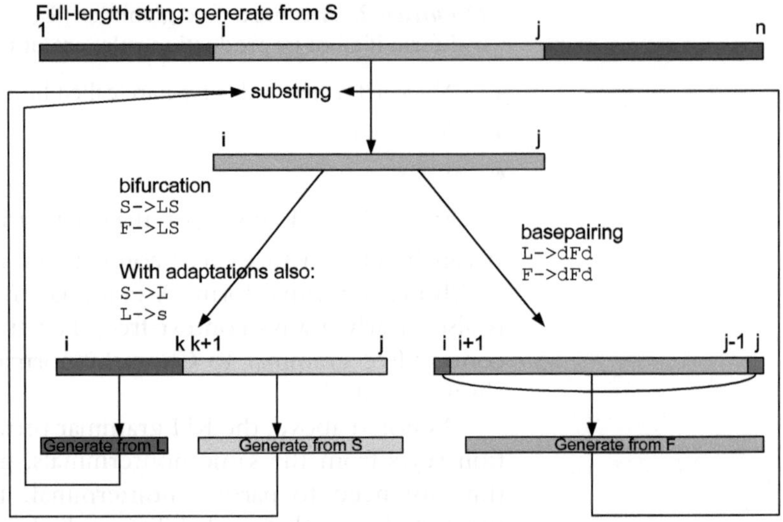

Fig. 2 To determine whether a string can be derived under the KH grammar, one must consider each subsequence recursively. Some subsequences, however, would be calculated several times in a recursive approach; for this reason, dynamic programming is used

3.1 The Basic CYK Algorithm

The Cocke-Younger-Kasami algorithm (CYK algorithm) is a parsing algorithm designed to answer the question: Can a string be derived under the grammar? The algorithm is based on the observation that a string can be derived from a nonterminal $N \in V$ if and only if the string can be derived from the right-hand side of at least one of the productions that have N on their left-hand side.

Example 4: To determine whether a string (secondary structure) $c_1 \dots c_n$ can be derived under the KH grammar, we can begin to reason intuitively:

- $c_1 \dots c_n$ can be derived if and only if one can write a sequence of productions under the grammar such that $S \to \cdots \to c_1 \dots c_n$.
- Consequently, $c_1 \dots c_n$ can be derived if either a derivation $L \to \cdots \to c_1 \dots c_n$ or a derivation $LS \to \cdots \to c_1 \dots c_n$ exists.
- For each of those cases, we must examine the possibilities:
  - If $L \to \cdots \to c_1 \dots c_n$ exists, then either the rule $L \to s$, or the rule $L \to dFd$ must be used. $L \to s$ requires $n = 1$ and $c_1 = s$. $L \to dFd$ requires that $F \to \cdots \to c_2 \dots c_{n-1}$ exists, $c_1 = d$ and $c_n = d$. We need to check which one (if either) is the case.
  - For $LS \to \cdots \to c_1 \dots c_n$, we need to check for every $1 \leq k < n$ whether the pair of derivations $L \to \cdots \to c_1 \dots c_k$ and $S \to \cdots \to c_{k+1} \dots c_n$ exists.

This argument naturally leads to a recursion (*see* Fig. 2). Let the boolean (true/false) array `P[i,j,N]` denote whether $c_i \dots c_j$ can be derived starting from nonterminal $N \in V$ of the context-free grammar (V, Σ, P, S) (assuming it is converted to Chomsky normal form). The string $c_1 \ldots c_n$ can be produced if and only if `P[1,n,S]` is true, and the value of this element can be determined using the following recursion relation:

$$P[i,j,N] = \begin{cases} \text{TRUE if } (S \rightarrow \epsilon) \in P & i = j + 1 \text{ and } N = S \\ \text{TRUE if } (N \rightarrow c_i) \in P & i = j \\ \text{TRUE if } \exists k : i \leq k < j \text{ and } A, B \in V: & \\ \quad (N \rightarrow AB) \in P \text{ AND } P[i,k,A] & \\ \quad \text{AND } P[k+1,j,B] & 1 \leq i < j \leq n \\ \text{FALSE} & \text{otherwise} \end{cases} \quad (1)$$

A direct implementation of this recursion, however, would be very slow, as most cells would have to be visited several times. A computational trick to speed up the evaluation of `P[1,n,S]` is to do the calculations bottom-up (i.e., starting from shorter substrings and progressing to the full-length string), and store the partial results; this is known as dynamic programming. For grammars in Chomsky normal form, the CYK algorithm can be written up directly from the recursion relations:

```
initialize all P[i,j,N] to false

for i = 1 to n do:
  for each production N -> c_i
    set P[i,i,N] = true
for each j = 1 to n do:
  for each i = 1 to j-1 do:
    for each k = i to j-1 do:
      for each production N -> AB do:
        if P[i,k,A] and P[k+1,j,B] then
           set P[i,j,N] = true

 if P[1,n,S] then
   string can be derived under grammar
 else
   string can not be derived under grammar
```

It can be shown that the CYK algorithm has a time complexity of $O(n^3|P|)$ and a space complexity of $O(n^2|V|)$, where $|P|$ is the number of rules and $|V|$ is the number of nonterminals in the grammar.

The CYK algorithm can also be applied to context-free grammars that are not in Chomsky normal form (such as the KH grammar as presented above), but this requires more thought. One approach is to convert the grammar to Chomsky normal form;

however, this often requires the creation of more nonterminal symbols and rules, increasing both the time and space complexity of the algorithm. For this reason, the CYK algorithm is typically adapted to match the format of the grammar.

Example 5: An adaptation of the basic CYK algorithm to the KH grammar would be, given an input structure $c_1 \ldots c_n$ (a string of *s*'s and *d*'s):

```
1  initialize all P[i,j,S], P[i,j,L], P[i,j,F] to
   false

2  for i = 1 to n do:
3     if c_i = 's' then set P[i,i,L] = true

4  for each j = 1 to n do:
5     for each i = 1 to j-1 do:
6       if j-i > 2 then
7          if P[i+1,j-1,F] and c_i = 'd'
             and c_j = 'd' then
8               set P[i,j,L] = true
9               set P[i,j,F] = true

10      for each k = i to j-1 do:
11          if P[i,k,L] and P[k+1,j,S] then
12              set P[i,j,F] = true
13              set P[i,j,S] = true

14      if P[i,j,L]
15         set P[i,j,S] = true

16 if P[1,n,S] = true then
17    structure can be produced by the KH grammar
18 else
19    structure cannot be produced by the KH grammar
```

It is important to note that with grammars not in Chomsky normal form, one must pay great attention to the order of calculations. For example, line 14 here should not be placed before line 8; `P[i,j,L]` must be fully evaluated before its value is needed.

The adapted CYK algorithm for the KH grammar is independent of nucleotide sequence, just like the grammar itself. To make the algorithm (and the grammar) more useful, one must introduce nucleotide-dependence. One way of doing this would be to simply introduce an extended alphabet and more production rules into the grammar (e.g., $L \rightarrow s$ would become $L \rightarrow a|c|g|u$, and $N \rightarrow dFd$ would become $N \rightarrow aFu|uFa|cFg|gFc$, assuming that only A-U and G-C pairs are allowed) and adapt the CYK algorithm to this new grammar. Equivalently, one can start the modifications at the algorithm level, as illustrated below in Example 6.

Example 6: To find out if the grammar can produce the input string $c_1 \ldots c_n$ (which now represents a sequence of nucleotides), only two lines in the previous algorithm need to be modified. Firstly, any nucleotide can be single-stranded so we can write:

```
    set P[i,i,L] = true
```

Secondly, we can force the algorithm not to allow basepairing between nucleotides c_i and c_j unless they are complementary (e.g., A-U, G-C). We will denote complementarity between nucleotides at positions i and j as $c_i \sim c_j$. We can then write:

```
            if P[i+1,j-1,F] and c_i ~ c_j then
```

The basic algorithm presented here only allows us to answer whether a string can be derived from a grammar at all. In our particular example above, the nucleotide-dependent, KH-grammar adapted CYK algorithm answers the question: Can the KH grammar fold this sequence? It turns out the answer will always be `true`, as any sequence will fold into some structure, in the worst case the single-stranded (unfolded) structure.

The basic CYK algorithm presented in this section might not appear to be very useful for RNA secondary structure prediction, but it does provide a versatile framework where only small modifications are required to produce other useful algorithms.

Example 7: The KH grammar (or alternatively the algorithm) can, for example, be adapted to couple sequence to structure, and answer the question: Can the KH grammar fold this particular sequence into this particular structure? To do this, one can introduce more terminal symbols into the grammar to distinguish between pairing and unpairing nucleotides – so instead of s, we would have s_a, s_c, etc., and instead of d, we would have d_a, d_c etc., and the number of rules would again increase. Alternatively, the algorithm can be modified to produce the same result without an increase in complexity – the details of this are left to the reader as an exercise.

3.2 Highest Probability Parse

In RNA secondary structure prediction, we are not as interested in whether a sequence can fold into a structure as in *what structure* it is most likely to fold into. As discussed previously, the probabilities in the grammar make some structures more likely (more consistent with the grammar) than other structures. The CYK algorithm presented above does not take these probabilities into account – it only gives a binary true/false answer. A natural question is: what is the probability of the highest probability parse of a string under the grammar? In biological terms, this is the probability of the best structure the grammar can produce for that sequence (if the grammar is semantically unambiguous).

The basic CYK algorithm can be easily adapted to reveal this for a stochastic context-free grammar by simply considering the `P[i,j,N]` array to be an array of numbers, replacing `and` operations with products of probabilities, and making decisions on the basis of whether the potential new value is higher than the previous one. The modified algorithm for a stochastic context-free grammar in Chomsky normal form is:

```
initialize all P[i,j,N] to 0

for i = 1 to n do:
  for each production N -> c_i
    set P[i,i,N] = P(N->c_i)

for each j = 1 to n do:
  for each i = 1 to j-1 do:
    for each k = i to j-1 do:
      for each production N -> AB do:
        if P[i,k,A]*P[k+1,j,B]*P(N->AB)
           > P[i,j,N] then
           set P[i,j,N] =
              P[i,k,A]*P[k+1,j,B]*P(N->AB)

P[1,n,S] is the probability of the best parse
```

For grammars not in Chomsky normal form (such as the KH grammar) the algorithm can be adapted similarly to the examples in Subheading 3.1.

This algorithm still does not return that final structure. However, it is possible to further modify it to provide the actual parse tree of the highest probability parse. One way of doing this is storing which choice was made in an array of backpointers, and backtracking through that array after the value of `P[1,n,S]` is known.

Example 8: Suppose that the following choice is made in the CYK algorithm:

```
        if P[i,k,A]*P[k+1,j,B]*P(N->AB)
           > P[i,j,N] then
           set P[i,j,N] =
              P[i,k,A]*P[k+1,j,B]*P(N->AB)
```

In this case, one could set the cell `T[i,j,N]` in the backpointer array to point to the cells `T[i,k,A]` and `T[k+1,j,B]` indicating the values that were needed to produce the final value in `T[i,j,N]`.

After the highest probability for the full parse, `P[1,n,S]`, is known, one can start from `T[1,n,S]` and trace back through `T`, reading off the parse tree branch-by-branch, until the full parse tree is produced. This parse tree is the most probable way of parsing the string $c_1 \ldots c_n$ under the grammar.[2]

The unmodified (boolean) CYK algorithm is rarely used in RNA secondary structure prediction, for reasons outlined above. The modified CYK algorithm (complete with backtracking), however, is very common and is typically referred to as "the CYK algorithm" without further elaboration.

3.3 The Inside–Outside Algorithm

The CYK algorithm can be used to determine the best parse of the sequence of using the grammar. However, one might also be interested in the *total* probability that the grammar can generate the string $c_1 \ldots c_n$ at all. In a similar vein as before, we can talk about the total probability of generating any substring $c_i \ldots c_j$, as well as the probability of generating everything outside of $c_i \ldots c_j$ in the string $c_1 \ldots c_n$.

Let us therefore define inside variables e, which represent the sum of the probabilities of all the possible ways of generating $c_i \ldots c_j$ from nonterminal N:

$$e(i,j,N) = P(N \rightarrow c_i \ldots c_j) \quad (2)$$

Similarly, we can define outside variables f, which represent the sum of the probabilities of all the possible ways of generating everything outside of $c_i \ldots c_j$, given that N generates $c_i \ldots c_j$:

$$f(i,j,N) = P(S \rightarrow c_1 \ldots c_{i-1} N c_{j+1} \ldots c_n) \quad (3)$$

Before we delve into the algorithmic details of how to calculate them, it is worth spending a moment reflecting on the power of the inside and outside variables. They can, in fact, be used to calculate many useful probabilities under the model of the grammar – some examples of this are shown below.

Example 9: In the KH grammar, there are two rules that produce basepairs: $L \rightarrow dFd$, and $F \rightarrow dFd$. The total probability that either one of these rules is used to parse $c_i \ldots c_j$ (in the middle of the full string, $c_1 \ldots c_n$) is actually the probability that positions i and j form a basepair under the SCFG model:

$$P_d(i,j) = \frac{f(i,j,L)e(i+1,j-1,F)P(L \rightarrow c_i F c_j)}{P(S \rightarrow c_1 \ldots c_n)} + \frac{f(i,j,F)e(i+1,j-1,F)P(F \rightarrow c_i F c_j)}{P(S \rightarrow c_1 \ldots c_n)} \quad (4)$$

Example 10: In a more general case, the probability of $c_i \ldots c_j$ being derived from N is given by the product $e(i,j,N)f(i,j,N)$.

[2] In practice, we need not even store the array of backpointers at all. Starting from $P[1,n,S]$, one can simply do the calculations "backwards" and determine which choice must have been made in each step.

The probability that $c_i \ldots c_j$ is derived from an initial application of the particular rule $N \to AB$ is

$$P(N \to AB \to c_i \ldots c_j) = \frac{f(i,j,N) \sum_{k=1}^{j-1} P(N \to AB) e(i,k,A) e(k+1,j,B)}{P(S \to c_1 \ldots c_n)} \quad (5)$$

As we shall see later, the quantities in this example are used for "training" the grammar (setting its probabilities).

3.3.1 Inside Algorithm

The inside algorithm is identical to the CYK algorithm for finding the highest probability parse, except that all maxima are replaced with sums. Consequently, its time complexity is also $O(n^3|P|)$ and the space complexity is $O(n^2|V|)$.

3.3.2 Outside Algorithm

The outside algorithm can be thought of as employing "reverse logic" compared to the CYK/inside algorithms. For every rule in the grammar $\alpha \to \beta$, in the CYK/inside algorithms we examine the right-hand side matching the nonterminal we want the value of the CYK/inside variable for. In the outside algorithm, we examine the left-hand side instead. Therefore we also start from the longest subsequences ($f(1, n, S) = 1$) and progress to shorter ones.

For a grammar in Chomsky normal forms, the outside algorithm is:

```
1  initialize f[1,n,S] to 1
   and all f[1,n,N] to 0

2  for each i = 1 to n do:
3    for each j = n to i do:
4      for each k = j to n do:
5        for each production N -> AB do:
6              increment f[i,j,A] with
               f[i,k,N]*e[j+1,k,B]*P(N->AB)
7      for each k = 1 to i do:
8        for each production N -> BA do:
9              increment f[i,j,A] with
               f[k,j,N]*e[k,i-1,B]*P(N->BA)
```

The outside algorithm is illustrated in Fig. 3.

In this algorithm, lines 4–6 represent the situation where we embed $c_i \ldots c_j$ in a longer subsequence: $c_i \ldots c_j c_{j+1} \ldots c_k$ (expressed from N), which we then split into two parts: $c_i \ldots c_j$ (expressed from A) and $c_{j+1} \ldots c_k$ (expressed from B). $f(i,j,A)$ requires that $c_i \ldots c_j$ is *excluded* from the complete sequence – so a contribution to $f(i,j,A)$ from this particular splitting involves a combination of the probability of excluding all of $c_i \ldots c_k$ ($f(i,k,N)$) but at the same time also expressing the extra needed part $c_{j+1} \ldots c_k$ ($e(j+1,k,B)$).

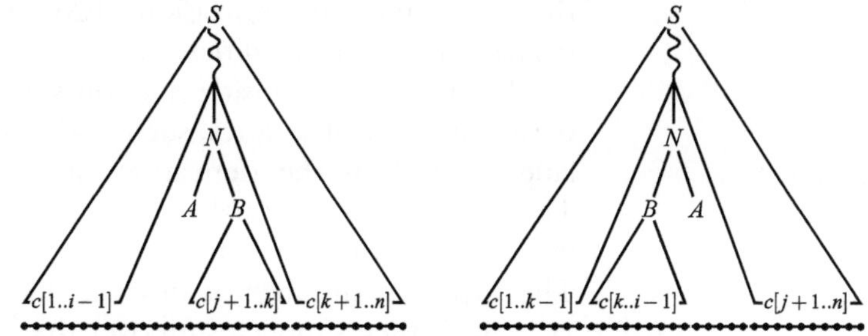

Fig. 3 An illustration of the two parts of the outside algorithm: the algorithm progressively fills out the subsequence not yet derived

Similarly, in lines 7–9 we consider the reverse situation, where $c_i \ldots c_j$ is embedded in $c_k \ldots c_{i-1} c_i \ldots c_j$, and $B \to c_k \ldots c_{i-1}$ and $A \to c_i \ldots c_j$. Again, for $f(i, j, A)$ we add the contribution from excluding the longer string $c_k \ldots c_j$ $(f(k, j, N))$ combined with including $c_k \ldots c_{i-1}$ $(e(k, i-1, B))$.

As is the case with the CYK/inside algorithms, the outside algorithm can also be adapted to grammars in non-Chomsky normal forms. However, the order in which operations are carried out requires attention, just as we have seen in the case of the CYK algorithm.

The computational complexity of the outside algorithm can be shown to be the same as the complexity CYK/inside algorithm.

3.3.3 Training a Grammar

So far we have assumed that the probabilities in the grammar are given – i.e., that our model is fully parametrized. But at some point the parameters need to be set so that the grammar best models what it is meant to model. In RNA secondary structure prediction, grammars are usually designed to model structures that are found in nature. Setting the probabilities of the grammar to the optimal values is termed "training" it.

If one has a dataset where the correct parses of some strings are known, the SCFG is trained by simply setting probability of each rule on the basis of the number times it was used in the derivations. In RNA secondary structure prediction, one will typically first obtain the parse trees from an input dataset of structures[3]: this can be done by using the CYK algorithm to parse all the input data (this will reveal the parse tree for every input structure).

If one has a dataset where the parses are not known (for example, if no RNA structures are known at all, or if one assumes that the structures we are trying to predict aren't like the ones already known), it is still possible to train the grammar so it produces the highest probability predictions. This is done through

[3]They can also be sequence-structure pairs, depending on the model.

the expectation maximization (EM) algorithm, which uses the inside–outside probabilities.

Recall that the inside and outside variables can be used to derive useful probabilities, such as the probability that a particular rule is used in the derivation of a subsequence (*see* Eq. 5). This probability is actually the *expectation* of that rule (i.e., the expected fraction of times that rule is used to parse $c_i \dots c_j$). The expected number of times a rule is used in the parsing of $c_1 \dots c_n$ is then simply a sum over the expectations for all possible subsequences. The probabilities in the SCFG can then be set from these expectations.

The training process in the EM algorithm is done in an iterative fashion. The details of this are beyond the scope of this chapter, but in short, the steps are as follows:

1. The probabilities are initialized to some (possibly random) values.
2. The expectation step: compute the expectation of each rule using the inside–outside algorithm with the current probabilities.
3. The maximization step: update the probabilities to the expectations determined in step 2.
4. Repeat 2–3 until the probabilities converge (or until some arbitrary cutoff value).

In this way, the grammar learns what probabilities will best describe the input data.

4 Discussion

4.1 SCFGs vs. Thermodynamic Models

The thermodynamic prediction of RNA secondary structure is discussed in Chapters 3 and 4 of this book. While it might seem that the two approaches have little in common, in some ways they are more alike than different. The KH grammar presented in this chapter can actually be thought of as a very simple thermodynamic model in itself: it gives scores between 0 and 1 to every basepair (distinguishing stacking basepairs from opening ones), single-stranded nucleotide or loop. Conversely, Rivas and Eddy have shown that Zuker's thermodynamic model can be converted to an SCFG by obtaining the probabilities of productions from the appropriate thermodynamic constants [4]. Both SCFG-based and thermodynamic models are formulated in terms of the optimization of an objective function: thermodynamic methods minimize free energy, SCFG methods maximize probability. In both cases, the optimization is expressed through recursion relations and implemented through dynamic programming algorithms, with the same $O(n^3)$ computational complexity.

However, the two models are also fundamentally different in their underlying scientific concepts and assumptions. Thermodynamic methods are based on a physical, energy-driven model for RNA folding, and obtain their parameters from calorimetric experiments on short oligonucleotides. SCFG-based methods, in contrast, are a form of machine learning, where the focus is on modelling the complete structures observed in nature in the best possible way, and then producing structures "like them," without assumptions about how the folding actually happens in reality.

SCFG-based methods are also inherently probabilistic. The advantage of this is that the theory of probability and statistics is very well developed in pure mathematics, and all that "machinery" can be applied to SCFGs in a rigorous way. SCFG models can also be extended with other models within the same probabilistic framework, in order to improve the quality of predictions, as we shall see in Subheading 4.2.

On the other hand, thermodynamic methods have the advantage that they work with information that is much easier to relate to. The prediction of the free energy contribution of a particular basepair in an RNA structure, for example, is a clear statement about the physical world that (at least in theory) can be tested experimentally. The probability of the same basepair under an abstract SCFG model is much harder to interpret.

It is also interesting to observe that an SCFG will always produce geometrically distributed loop lengths; this is not true for thermodynamic models.

4.2 Pfold: Extending the SCFG Model with Phylogenies

If we were to fold an arbitrary RNA molecule with the basic KH grammar presented in this chapter, we would quickly find that the predictions are highly inaccurate – significantly worse than thermodynamic predictions of the same sequences. The reason for this is that the KH SCFG, by itself, is a much less sophisticated model for RNA folding than thermodynamic models with hundreds of experimentally measured parameters, including a large number of special cases, such as extremely stable tetraloops. However, the KH grammar was not really meant to be used on its own, but in combination with covariance information derived from an alignment of the sequence with homologous sequences.

One way to think about this is that the SCFG proposes a "first guess" probability distribution over secondary structures – in Bayesian terms, this is the *prior probability distribution* over the structures. If additional information is known (such as mutual information between columns of an alignment, derived using an evolutionary model), the prior distribution of secondary structures can be adjusted ("evaluated") using conditional probabilities, so the final, *posterior probability distribution* assigns a probability to each secondary structure on the basis of how well it matches both the SCFG and the phylogenetic model. This has been done in pfold by slight modifications to the inside–outside algorithm: in effect, the KH grammar coupled to the evolutionary model is an SCFG that has the columns of the alignment as its terminals. Grammars like this are known as phylo-grammars, and pfold became the first program to use one. The accuracy obtained by pfold in the prediction of the consensus structure of RNA structural alignments is significantly higher than that of purely thermodynamic programs for individual structure predictions of the same sequences. The caveat is that obtaining a good structural alignment is itself an unsolved problem (*see* Chapter 17): the alignment depends on the structure, but the structure is what we are trying to find using the alignment.

Some programs have been written to align and fold RNA sequences at the same time (*see* Subheading 6.3). However, the accurate prediction of RNA secondary structure remains an open problem. In practice, researchers interested in identifying RNA secondary structures bioinformatically will often iterate between different methods and manually adjust the results of each program on the basis of their background knowledge of the particular sequence.

4.3 Problems and Solutions

The main limitation of the SCFG algorithms discussed above is their computational complexity (this complexity, however, is the same as for algorithms based on thermodynamic models). This was prohibitive in the early days of SCFG-based methods; with the advance of more powerful computer hardware, recent efforts have been made to parallelize the algorithms [5], and even building specialized hardware [6].

Another well-known problem that occurs with the use of SCFGs is numerical underflow. The reason for this is the multiplication of a large number of probabilities (numbers between 0 and 1) with each other. Computer architectures can only represent a finite number of digits, resulting in the rounding of numbers with large negative exponents to zero. A common way of solving this problem is to use log-probabilities instead: products become sums, and the addition operation is typically implemented with the use of a lookup table. Another approach has been to use an extended exponent datatype to represent numbers that otherwise would give underflow [5].

Lastly, SCFG-based RNA secondary structure prediction, just like thermodynamic methods, suffers from the issue of optimizing the objective function over a very complex RNA secondary structure space. Structures with very similar probabilities are not necessarily similar, and very similar structures can have very different probabilities. Therefore the definition of the "best" structure being the one with the highest probability (computed by the CYK algorithm) has been called into question. In pfold, the solution of choice has been to optimize the expectation of predicted positions instead of simply finding the highest probability structure. This optimization is done through an additional specially designed recursion.[4]

5 Recommended Reading

For more details on the formal theory of grammars, we recommend classic computer science textbooks, such as Languages and Machines by Thomas Subkamp [8]. For more on grammars in the context of RNA secondary structure and general sequence analysis, Biological Sequence Analysis by Durbin et al. [9] offers an excellent introduction.

6 Notes

Since their first application in RNA secondary structure modelling, SCFGs have proved to be extremely versatile and have appeared in a large number of popular computational tools. In this section, we provide an overview of the most well-known programs that use SCFGs in RNA secondary structure prediction and modelling.

6.1 Comparative Secondary Structure Prediction

6.1.1 Pfold/PPfold

Pfold [10, 11] uses the KH grammar described in this chapter and couples it to a phylogenetic model. The original pfold webserver is located at http://www.daimi.au.dk/~compbio/pfold/. Pfold takes an RNA alignment as input and returns a consensus secondary structure of the alignment. Each sequence of the alignment is annotated with a "stem-extended" secondary structure. Other outputs of the program include a dotplot over the likelihood of all possible basepairs, the phylogenetic tree calculated for the sequences on the basis of the evolutionary model, and reliability scores for every prediction in the final structure.

Recently, a multithreaded version of pfold has been developed [5]: this program is called PPfold and can be downloaded from http://birc.au.dk/software/ppfold/. PPfold is capable of solving the structure of longer alignments and can be run on any operating system with Java support. Command-line options are available for the advanced user; otherwise, the program has a standalone version with minimal graphical user interface for the selection of the input alignment, and it can also be downloaded as a full-featured plugin to the CLC Workbenches.

[4] A similar approach has recently been applied to a thermodynamic model, see [7].

6.1.2 RNA-Decoder

RNA-Decoder [12], like pfold, not only predicts the secondary structure of alignments, but it also takes into account the known protein-coding context of RNAs. It employs an SCFG together with a set of phylogenetic models designed to describe the overlapping evolutionary constraints.

6.1.3 Xrate

Xrate [13] is an interpreter for phylo-grammars. Its capabilities include maximum likelihood phylogeny, ancestral sequence reconstruction, alignment annotation, and model estimation. It can be downloaded as part of the DART package at http://biowiki.org/DART. Detailed instructions for its installation and use are found on the same site.

6.2 Similarity Search and Gene Finding

6.2.1 Infernal

Infernal [14] is a program to search DNA sequence databases for RNA structure and sequence similarities. It is based on covariance models, which are a special case of SCFGs. Infernal is also discussed in detail in Chapter 9. The popular Rfam database [15, 16] is based on Infernal. The Rfam database can be accessed at http://rfam.sanger.ac.uk/.

6.2.2 tRNAscan-SE

TRNAnscan-SE [17] is one of the several programs developed for detecting noncoding RNA genes in genomes; it detects tRNAs at very high sensitivities using SCFGs. The tRNAscan-SE webserver can be accessed at http://selab.janelia.org/tRNAscan-SE/.

6.2.3 Evofold

Evfold [18] identifies functional RNA structures in multiple-sequence alignments. It is based on a phylo-SCFG and exploits the differences of the substitution process in stem-pairing and unpaired regions to make its predictions. Evofold can be accessed at http://users.soe.ucsc.edu/~jsp/EvoFold/.

6.3 Simultaneous Folding and Aligning of Homologous RNAs

6.3.1 Consan

Consan [19] develops the original Sankoff algorithm [20] for simultaneous folding and alignment; it uses pair stochastic context-free grammars as a unifying framework for scoring pairwise alignment and folding at the same time. There is also a constrained version of the algorithm, which assumes knowledge of a few confidently aligned positions (pins). The pins are selected based on the posterior probabilities of a probabilistic pairwise sequence alignment. The program can be downloaded from http://selab.janelia.org/software/consan/.

6.3.2 Contrafold

Contrafold [21] is based on conditional log-linear models; this class of probabilistic models generalize on SCFGs by using discriminative training. Discriminative models are trained by maximizing conditional likelihood rather than joint likelihood. The Contrafold webserver can be accessed at http://contra.stanford.edu/contrafold/server.html. The program can also be downloaded from the same website.

6.3.3 Stemloc

Stemloc [22, 23] generates pairwise RNA structural alignments based on pair stochastic context-free grammars. Stemloc can be accessed through the DART library at http://biowiki.org/DART. Detailed instructions for its use can be found on the same site.

Acknowledgments

ZS would like to thank Robert Giegerich and Paula Tataru for their comments on the manuscript, and Christine Heitsch and her group at Georgia Tech for useful discussions.

References

1. Chomsky N (1956) Three models for the description of language. IRE Trans Inf Theory 2(3):113–124
2. Younger DH (1967) Recognition and parsing of context-free languages in time n3. Inf Control 10(2):189–208
3. Baker JK (1979) Trainable grammars for speech recognition. Speech communication papers for the 97th meeting of the acoustical society of America, pp 547–550, Boston, MA, 1979
4. Rivas E, Eddy SR (2000) The language of RNA: A formal grammar that includes pseudoknots. Bioinformatics 16(4): 334–340
5. Sükösd Z, Knudsen B, Værum M, Kjems J, Andersen ES (2011) Mulithreaded comparative RNA secondary structure prediction using stochastic context-free grammars. BMC Bioinformatics 12:103
6. Xia F, Dou Y, Zhou D, Li X (2010) Fine-grained parallel RNA secondary structure prediction using SCFGs on FGA. Parallel Comput 36:516–530
7. Lu ZJ, Gloor JW, Mathews DH (2009) Improved RNA secondary structure prediction by maximizing expected pair accuracy. RNA 10:1805–1813
8. Sudkamp TA (2005) Languages and machines: An introduction to the theory of computer science, 3rd edn. Addison Wesley, Reading, MA
9. Durbin R, Eddy SR, Krogh A, Mitchison G (1998) Biological sequence analysis: Probalistic models of proteins and nucleic acids. Cambridge University Press, Cambridge
10. Knudsen B, Hein J (1999) RNA secondary structure prediction using stochastic context-free grammars and evolutionary history. Bioinformatics 15(6):446–454
11. Knudsen B, Hein J (2003) Pfold: RNA secondary structure prediction using stochastic context-free grammars. Nucleic Acids Res. 31(13):3423–3428
12. Pedersen JS, MeyerI, Forsberg R, Simmonds P, Hein J (2004) A comparative method for finding and folding RNA secondary structures within protein-coding regions. Nucleic Acids Res. 32:4925–4936
13. Klosterman P, Uzilov A, Bendana Y, Bradley R, Chao S, Kosiol C, Goldman N, Holmes I (2006) Xrate: a fast prototyping, training and annotation tool for phylo-grammars. BMC Bioinformatics 7(1):428
14. Nawrocki EP, Kolbe DL, Eddy SR (2009) Infernal 1.0: Inference of RNA alignments. Bioinformatics 25:1335–1337
15. Griffiths-Jones S, Moxon S, Marshall M, Khanna A, Eddy SR, Bateman A (2005) Rfam: annotating non-coding RNAs in complete genomes. Nucleic Acids Res. 33: D121–D124
16. Gardner PP, Daub J, Tate JG, Nawrocki EP, Kolbe DL, Lindgreen S, Wilkinson AC, Finn RD, Griffiths-Jones S, Eddy SR, Bateman A (2009) Rfam: updates to the RNA families database. Nucleic Acids Res. 37: D136–D140
17. Lowe TM, Eddy SR (1997) tRNAscan-SE: a program for improved detection of transfer RNA genes in genomic sequence. Nucleic Acids Res. 25:955–964
18. Pedersen JS, Bejerano G, Siepel A, Rosenbloom K, Lindblad-Toh K, Lander ES, Kent J, Miller W, Haussler D (2006) Identification and classification of conserved RNA secondary structures in the human genome. PLoS Comput Biol 2(4):e33
19. Dowell RD, Eddy SR (2006) Efficient pairwise RNA structure prediction and alignment using sequence alignment constraints. BMC Bioinformatics 7:400
20. Sankoff D (1985) Simultaneous solution of the RNA folding, alignment and protosequence problems. SIAM J Appl Math 45(5): 810–825

21. Do CB, Woods DA, Batzoglou S (2006) Contrafold: RNA secondary structure prediction without physics-based models. Bioinformatics 22(14):90–98
22. Bradley RK, Pachter L, Holmes I (2008) Specific alignment of structured RNA: Stochastic grammars and sequence annealing. Bioinformatics 24(23): 2677–2683
23. Holmes I (2005) Accelerated probabilistic inference of RNA structure evolution. BMC Bioinformatics 6:73

Chapter 9

Annotating Functional RNAs in Genomes Using Infernal

Eric P. Nawrocki

Abstract

Many different types of functional non-coding RNAs participate in a wide range of important cellular functions but the large majority of these RNAs are not routinely annotated in published genomes. Several programs have been developed for identifying RNAs, including specific tools tailored to a particular RNA family as well as more general ones designed to work for any family. Many of these tools utilize covariance models (CMs), statistical models of the conserved sequence, and structure of an RNA family. In this chapter, as an illustrative example, the Infernal software package and CMs from the Rfam database are used to identify RNAs in the genome of the archaeon *Methanobrevibacter ruminantium*, uncovering some additional RNAs not present in the genome's initial annotation. Analysis of the results and comparison with family-specific methods demonstrate some important strengths and weaknesses of this general approach.

Key words Covariance models, Infernal, Rfam, Stochastic context-free grammars, Homology search, Genome annotation, tRNA, rRNA, SRP RNA, RNase P RNA, CRISPR, Riboswitch

1 Introduction

Genome annotation is the identification of functional sequence elements in an organism's genome. Knowledge of the presence and location of these sequence elements coupled with understanding of their functional roles helps reveal the types of biological processes that take place in the organism as well as the evolutionary history of that organism. Classes of functional sequence elements include protein-coding genes, non-coding RNA elements, promoter elements, enhancers, as well as others. Functional RNA elements are RNAs that are not translated into proteins, but rather carry out their biological function directly as RNAs. Much like proteins, many of these RNAs fold into a specific three-dimensional structure that is integral to their function. For convenience, I will refer to functional RNAs as simply RNAs in this chapter.

RNAs play vital roles in many cellular processes. For example, transfer RNAs (tRNAs) and ribosomal RNAs (rRNAs) play central roles in the translation of messenger RNAs into proteins.

Jan Gorodkin and Walter L. Ruzzo (eds.), *RNA Sequence, Structure, and Function: Computational and Bioinformatic Methods*, Methods in Molecular Biology, vol. 1097, DOI 10.1007/978-1-62703-709-9_9,

Spliceosomal RNAs (such as U1, U2, U4, U5, and U6) interact with proteins as part of a ribonucleoprotein complex (RNP) responsible for splicing introns from many eukaryotic pre-mRNAs [1]. Small nucleolar RNAs (snoRNAs) are members of RNPs that guide post-transcriptional modification during the maturation of rRNAs and other RNA genes in archaea and eukarya [2]. The SRP (signal recognition particle) RNA is part of an RNP involved in transporting proteins within cells [3]. Ribonuclease P (RNase P) RNA is a vital part of an RNP that processes precursor tRNAs through cleavage of a 5′ leader sequence [4].

Many other RNA elements play key roles in gene regulation, such as microRNAs (miRNAs) that act by binding to specific target mRNAs in eukaryotes via basepairing, affecting the expression of the target [5]. Riboswitches are structured RNA elements typically occurring in the 5′ untranslated region (UTR) of protein-coding genes over which they exert translational or transcriptional control through binding of small metabolites, which cause a structural change in the riboswitch. They often control genes involved in the transport or biosynthesis of their target metabolite [6]. The bacterial 6S RNA promotes more general gene regulation by binding directly to RNA polymerase and repressing its activity during stationary phase of bacterial growth [7].

Other RNA elements are important for defending cells against viruses and transposons. Small-interfering RNAs (siRNAs) are 21–25 nucleotide long RNAs in eukaryotes often derived from exogenous RNAs that are recognized by the protein complex RISC, ultimately leading to degradation of the exogenous RNA [8]. In archaea and bacteria, a similar defense system is encoded in CRISPR (Clustered Regularly Interspaced Short Palindromic Repeats) elements, which are short 24–48 nucleotide repeats which have been predicted to form hairpin structures separated by similar length spacers of foreign DNA from past exposures to parasites such as viruses (phage). RNAs from CRISPR elements are constitutively expressed and guide silencing of complementary foreign DNA or RNA [9].

The phylogenetic distribution of different types of RNAs varies. Some, such as those involved in ubiquitous cellular processes like tRNAs, rRNAs, RNAse P RNA, and SRP RNA, exist in all three domains of life. Others are widespread in one or two of the domains but not the other(s), such as snoRNAs in archaea and eukarya, 6S RNA in bacteria, and microRNAs in eukarya. Finally, some exist within only a specific clade of a domain, ranging in size anywhere from a phylum (e.g., SmY RNAs, found in RNPs predicted to be involved in trans-splicing in nematodes [10]) down to only a few species (e.g., OxyS, a global regulator induced in response to oxidative stress in certain members of one family of gammaproteobacteria, including *E. coli* [11]).

Despite the widespread importance of functional RNAs, the large majority of them are typically not annotated in published genomes, whereas it is generally accepted that most protein-coding genes are. This discrepancy is at least partially due to some unique challenges in identifying RNAs. In this chapter, I will discuss this issue in the context of genome annotation using freely available software programs. I will first compare and contrast protein and RNA annotation in genomes, highlighting challenges specific to RNAs, and then some current methods for RNA annotation will be introduced. I will then focus on one specific method: the use of covariance models (CMs) and detail the strengths and weaknesses of the CM approach through a practical example of RNA annotation of an archaeal genome. Finally, I will compare and contrast the general CM-based approach with the use of family-specific tools designed to identify homologs of particular RNA family.

1.1 Genome Annotation of Proteins and RNAs

The annotation of protein-coding genes in a genome typically consists of two major steps:

Step 1. Predict protein-coding gene sequences.

Step 2. Assign putative functional annotation to the predicted proteins using homology-based search tools.

In the first step, only subsequences of the genomes that correspond to open-reading frames (ORFs) need to be evaluated as possible protein-coding genes. This greatly reduces the possible search space relative to all possible subsequences of the genome and represents an important distinction between protein annotation and functional RNA annotations of genomes, for which there is no analog of the open-reading frame signal. For archaea and bacteria, accurate programs exist for performing step 1, such as the popular Glimmer program, which can correctly identify about 99% of protein-coding genes with known functions [12]. Similar programs have been designed for eukaryotic genomes including Genscan [13], GeneID [14], and Genemark [15], among others, but these are generally less accurate due to the higher complexity of eukaryotic genomes versus archaeal and bacterial genomes.

Given a set of predicted protein-coding genes, step two aims to functionally annotate these genes based on previous functional annotation of genes with similar sequences, which are predicted to be homologous. This step is carried out using homology search tools like HMMER [16] or BLASTP [17] to search various target databases such as Pfam [18], COG [19], the NCBI NR database [20], and others.

As mentioned, unlike proteins, functional RNAs are not contained within open reading frames and so a different strategy is required for their identification. Many RNAs do, of course, contain signals in their promoter regions that can be, and have been, exploited when searching for RNAs [21], but these are often clade- or even species-specific, hampering any general approach by requiring specific foreknowledge of the genome being studied. However, there are several RNA genefinder programs that attempt to address the RNA analog of step 1 of the protein-coding gene scheme by identifying regions that conserve a statistically significant secondary structure, indicating a structural RNA gene. However, these programs are much less reliable than protein genefinders and in particular suffer from high false positive rates [22–24] limiting their utility for RNA annotation.

Consequently, RNA annotation is typically performed using known RNAs as queries for homology searches against the entire genome being studied. This is similar to step 2 of protein annotation from above, but homology searches for RNAs are less powerful than for proteins for several reasons: RNAs tend to be shorter than proteins (often about 100 nucleotides, as opposed to 200–300 amino acids [25]), and the search must be carried out at the RNA/DNA level instead of at the protein level, which reduces statistical power due to the smaller alphabet size and the degeneracy of the genetic code [26]. To cope with the reduced statistical signal, the most successful RNA homology search programs take advantage of the structural conservation of many functional RNAs by scoring a combination of the conserved sequence and the secondary structure of an RNA family [27]. Many basepaired nucleotides in a conserved RNA structure tend to covary over evolutionary time-scales to maintain complementarity, often by changing from one Watson–Crick basepair to another (A:U or C:G) or to a G:U wobble basepair. This covariation offers a useful statistical signal that can be used in addition to sequence conservation when searching for homologous structural RNAs. Sequence- and structure-based tools can be divided into two classes: family-specific methods that are designed for a particular RNA family, and general tools that can work for any family.

1.1.1 Family-Specific RNA Search Methods

Table 1 summarizes some popular family-specific programs for identifying RNAs in genomes. The most widely used RNA homology search tool targets the single largest gene family, tRNAs. The tRNAscan-SE program [28] uses a powerful statistical model called a covariance model (CM) that scores candidates based on both their sequence and predicted secondary structures. CMs are more sensitive (able to find more true homologs) than sequence-only-based searches but are much slower, due to the higher complexity of their scoring algorithms (as discussed in Subheading 1.2) to the point of being impractical when searching large sequence databases. CMs outperform sequence-based methods particularly well for families like tRNA that are short (about 70 nucleotides) and exhibit low levels of sequence similarity while maintaining a highly conserved secondary structure. To deal with the slow search speed of CMs, tRNAscan-SE uses

Table 1
Some popular family-specific tools for identifying RNAs in genomes

Program	Type of RNA	Prefilter stage	Final stage	Reference
tRNAscan	tRNA	Sequence-based; tRNA-specific	CMs	[28]
Aragorn	tRNA, tmRNA	*None*	tRNA/tmRNA-specific heuristic	[29, 30]
Arwen	Mitochondrial tRNA	*None*	Mito tRNA-specific heuristic	[31]
RNAmmer	rRNA (5S,5.8S,SSU, LSU)	Small "spotter" Profile HMMs	Profile HMMs (full-length)	[32]
SRPscan	SRP RNA	Sequence/structure Pattern (RNABOB)	CMs	[33]
Bcheck	RNase P RNA	Sequence/structure Pattern (RNABOB)	CMs	[34]

fast tRNA-specific prefilters that remove a large fraction of the database, leaving only promising subsequences to be evaluated by the slow CM methods. The result is a tool fast enough to search large mammalian genomes on a desktop computer in a few hours.

The strategy of using fast family-specific filters prior to a CM-based search is employed by other family-specific RNA search tools. For example, the Bcheck program [34] uses the sequence and structure-based pattern matching program RNABOB [35] as a fast prefilter for CMs to identify RNase P genes. RNABOB, like other pattern matching programs, identifies subsequences that can fold into a particular structure based on user-defined constraints. SRPscan [33] uses the same strategy with signal recognition peptide (SRP) RNA patterns and CMs to identify SRP RNAs.

Other sequence- and structure-based tools do not use CMs. Aragorn is a tRNA and tmRNA finder [29] that uses a tRNA-specific search algorithm that searches for part of the highly conserved B-box consensus sequence as an alignment seed and expands a structure-aware alignment around that seed. Aragorn's sensitivity is similar to tRNAscan-SEs but it is about an order of magnitude faster. The Arwen program [31] from the developers of Aragorn detects tRNA sequences in mitochondria in a similar manner.

Structure-based methods are not necessary for all RNAs. For example, the small and large subunit ribosomal RNAs (SSU and LSU rRNA) differ markedly from tRNA both in their size (about 1,500 nt and 3,000 nt, respectively) and high level of sequence conservation, and sequence-based homology search methods perform well for these RNAs. They are sometimes annotated using the pairwise sequence similarity search tool BLASTN [17] with homologous query sequences from closely related species, or with the RNAmmer tool [32] based on sequence-based profile hidden Markov models (discussed in more detail in Subheading 1.2).

Some family-specific methods cannot directly be used to scan genomes. For example, the SnoReport program [36] uses pattern descriptors as filters for an SVM-based classification, but requires the target sequences be short candidate snoRNAs, not genome-length sequences.

With the exception for tRNAscan-SE and RNAmmer, none of these tools are commonly used for annotating a genome prior to its publication in a database, as demonstrated by a sampling of 14 published genomes (five archaea, five bacteria, and four eukarya) in NCBI's GenBank database shown in Table 2. Notably, for one of the bacterial genomes listed, *Citrobacter rodentium*, 56 RNAs other than tRNAs and rRNAs were annotated using the Infernal software package and the Rfam database. Infernal implements general CM search methods that can be used for any RNA family, and Rfam contains CMs for about 1,500 RNA families. In the remainder of the chapter I will discuss CMs, Infernal and Rfam, and their potential for large-scale annotation of RNAs in genomes.

1.2 Covariance Models

Covariance models (CMs) are probabilistic models of the sequence and secondary structure of an RNA family [37, 38]. They are constructed from multiple sequence alignments of known homologs of the family that are annotated with a consensus secondary structure. A CM is useful for searching databases for homologs of the family it models and for creating sequence- and structure-based multiple sequence alignments of those homologs.

CMs are stochastic context-free grammars (SCFGs), introduced in Chapters 5 and 8 of this book. More specifically, they are profile SCFGs, analogous to profile hidden Markov models, commonly used for linear sequence analysis of protein domain families, but with added complexity for modeling a conserved secondary structure. Like CMs, a profile HMM is constructed from an alignment of homologous sequences (but without structure annotation).

The key feature of a profile model is its *position-specificity* [39]: each position of the input alignment is modeled independently. This allows profile methods to take into account the level of conservation at each position when scoring/aligning candidate family members, by defining a scoring system that weighs matches and mismatches at highly conserved positions more than at highly variable positions.

For example, take the toy RNA family depicted in Fig. 1, represented by the ungapped alignment of eight RNA homologs of length 11. Imagine a simple sequence-profile model that scans a target database shifting a length 11 window one nucleotide at a time looking for putative homologs of this family. If the target subsequence contains a nucleotide observed in at least one of the

Table 2
Summary of RNA annotations in published genomes. Counts were taken from "NCBI Genome" RefSeq annotation for the listed genomes (www.ncbi.nlm.nih.gov/sites/genome). Archaeal and bacterial genomes were selected as the first five published in 2010 according to NCBI (www.ncbi.nlm.nih.gov/genomes/lproks.cgi) for which a Refseq entry and a referenced publication were available, as of March 30, 2011. Eukaryotes were selected from to be representative, from "complete" genomes according to NCBI (www.ncbi.nlm.nih.gov/genomes/leuks.cgi) as of April 29, 2011. One genome from each "group" (fungi, protists, plant, animal) was chosen. "rRNA" includes 5S, SSU, LSU, and 5.8S for eukaryotes only. Abbreviations: TIGR: The Institute for Genomic Research, RAS: Russian Academy of Sciences, JGI: Joint Genome Institute, NML: National Microbiology Laboratory. GenBank accessions for archaeal and bacterial genomes: *M. rum.*: CP001719.1, *H. vol.*: CP001956.1, *H. jeo.*: CP002062.1, *A. sac.*: CP001742.1, *M. mar.*: CP001710.1, *C. rod.*: FN543502.1, *B. den.*: CP001750.1, *P. sta.*: CP001848.1, *L. mon.*: CP002062.1, *C. ucy.*: CP001602.1. NCBI Genome project RefSeq ID for eukaryotic genomes: *C. dub.*: 38659, *L. bra.*: 19185, *A. tha.*: 116, *M. mus.*: 169

Organism [reference]	Sequencing center, Country	tRNAs Method	#	rRNAs Method	#	Other RNAs Method	#
Archaea							
Methanobrevibacter ruminantium [40]	AgResearch, New Zealand	tRNAscan-SE	58	BLASTN	8		0[a]
Haloferax volcanii [41]	TIGR, USA	tRNAscan-SE	52	BLASTN	6		0[a]
Halalkalicoccus jeotgali [42]	Kyung Hee Univ., Korea	tRNAscan-SE	49	RNAmmer	3		0
Acidilobus saccharovorans [43]	RAS, Russia	tRNAscan-SE	45	RNAmmer	3		0
Methanothermobacter marburgensis [44]	G. August Univ. Germany	tRNAscan-SE	40	RNAmmer	2	*Unknown*	2
Bacteria							
Citrobacter rodentium [45]	Sanger Institute, UK	tRNAscan-SE	86	*Unknown*	22	Infernal & Rfam	56
Bifidobacterium dentium [46]	Univ. of Parma, Italy	tRNAscan-SE	55	BLASTN	13		0
Pirellula staleyi [47]	JGI, USA	*Unknown*	46	*Unknown*	3	*Unknown*	3
Listeria monocytogenes [48]	NML, Canada	tRNAscan-SE	58	RNAmmer	15		0
Cyanobacterium UCYN-A [49]	UC Santa Cruz, USA	tRNAscan-SE	36	search_for_rnas	6		0

(continued)

Table 2
continued

Eukaryotes							
Candida dubliniensis [50]	Sanger Institute UK	*Unknown*	101		0	*Unknown*	11
Leishmania braziliensis [51]	Sanger Institute UK		0		0	*Unknown*	6
Arabidopsis thaliana [52]	multiple centers	tRNAscan-SE, tRNAscan	688	BLASTN	14	*Unknown*	689
Mus musculus [53]	multiple centers	tRNAscan-SE	509[a]	*Unknown*	5	*Unknown*	4059[a]

[a] Numbers from NCBI (shown here) are inconsistent with explicitly mentioned counts given in the referenced publication for this genome

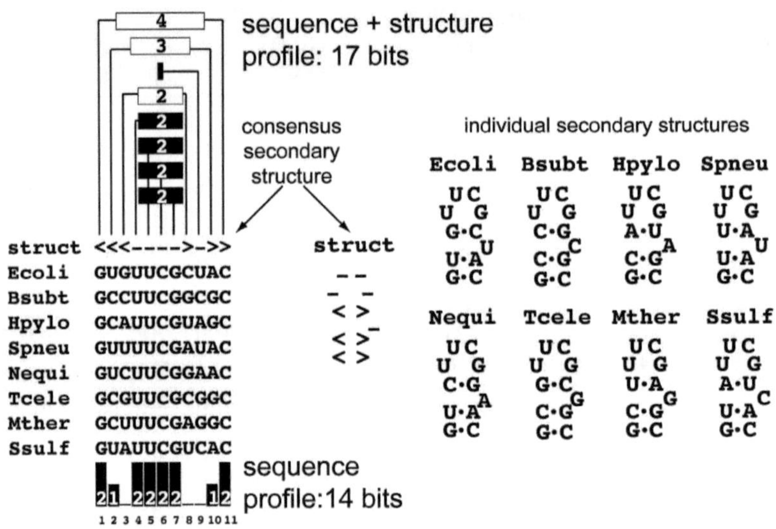

Fig. 1 Information in a sequence-only versus a sequence and structure profile. The eight sequence alignment for a fabricated RNA family used to build both of the profiles is on the *left*. The `struct` line denotes the consensus secondary structure of the family, with basepaired columns indicated by matching nested < and > characters and connected by lines at *top* of the figure. The structure is ignored by the sequence-only profile but used in the sequence and structure profile to define dependencies between basepaired columns. The eight individual secondary structures, defined by imposing the consensus structure on each sequence, are shown on the *right*. Boxes with internal numbers at *top* and *bottom* of the alignment indicate the number of bits per position from the sequence (*black*), or per basepair from the structure (*white*). This figure is similar to one from [54]

eight known homologs at every position, then it is considered a match, otherwise it is a mismatch. In this scenario, the specificity of positions is defined by their conservation in the known homologs. For example, alignment positions 4–7 are completely conserved, containing UUCG for all eight homologs, while positions 3 and 8 are completely variable (25% A, C, G and U), meaning that a putative homolog must contain UUCG at positions 4–7 but can be any nucleotide at positions 3 and 8.

In general, conserved positions are more informative than variable ones as to whether a sequence matches a profile or not. It is possible to quantify the amount of information in a sequence profile based on the alignment the profile was built from. Completely conserved positions contain two *bits* of information [38], because they specify a single choice out of four possible choices, corresponding to answering two yes/no questions to narrow four possibilities down to one. A position that contains two nucleotides, with each at half of the positions, contains one bit of information (e.g., positions 2 and 10 in Fig. 1). A completely variable position contains zero bits of information, because any of the four nucleotides will match. The total amount of information in a profile indicates the likelihood of a match to the profile in a random sequence database. For the 14 bit sequence profile corresponding to the alignment in Fig. 1, we expect a match once in every $2^{14} = 4{,}096$ nucleotides in a random, so-called *iid* (independent, identically distributed) sequence database in which each nucleotide has an equiprobable chance of being observed at each position (25% chance of being A, C, G, or U).

In the case of structural RNAs, we can increase the information of a profile by considering the conserved consensus structure of the family as well as the conserved sequence, allowing us to better discriminate good matches to the model, which represent putative homologs, from background, nonhomologous sequence when searching sequence databases. This is achieved by considering both halves of basepaired positions simultaneously when scoring a sequence against a profile. For example, in Fig. 1, positions 3 and 8 form a basepair in the consensus structure. These two positions are completely variable at the sequence level and so contribute zero bits of information to a sequence profile. However, of the $4 \times 4 = 16$ possible basepairs at these positions, only the four possible Watson–Crick basepairs (A:U, U:A, C:G, G:C) exist in the homologs. By specifying that a match to the profile must contain a Watson–Crick basepair at these positions we've gained two additional bits of information (by reducing 16 possibilities to 4, corresponding again to answering two yes/no questions). Importantly, modeling structure will only add information in cases where the sequence varies and paired positions covary to maintain a basepair in the structure. For example, in Fig. 1 positions 1 and 11 pair in the consensus structure, but are completely conserved in sequence, each contributing two bits to a sequence profile, and collectively contributing four bits to a sequence and structure

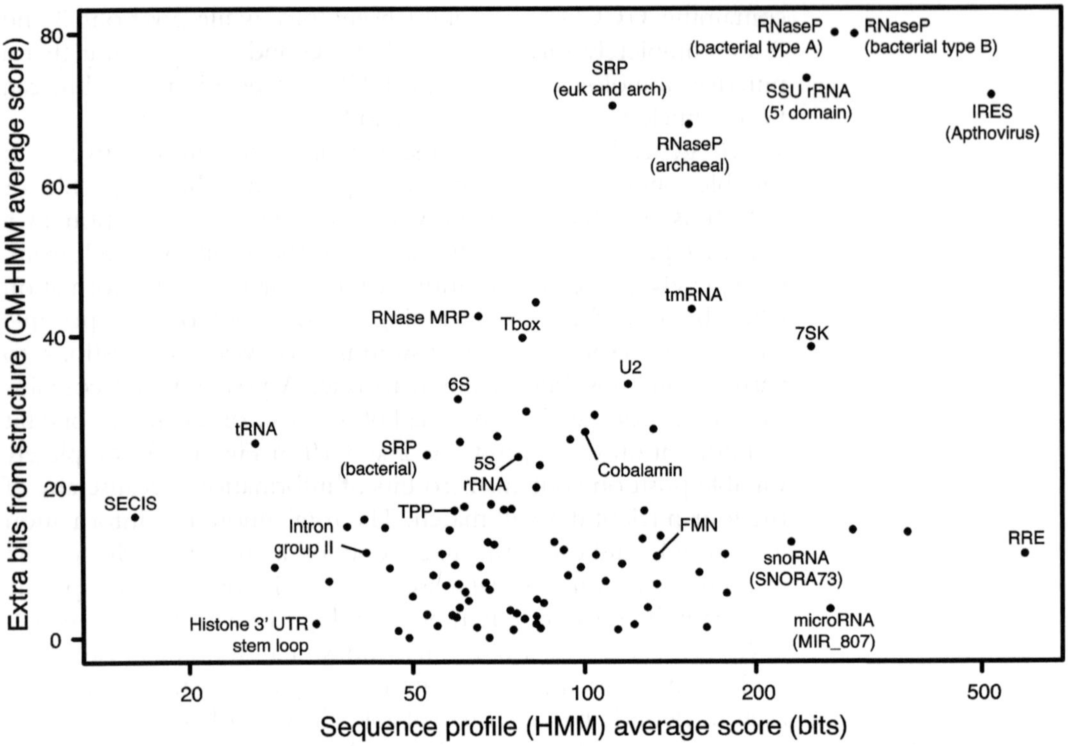

Fig. 2 Additional information (in bits) gained by sequence and structure profiles (CMs) versus sequence-only profiles (HMMs) for various RNA families. Sequence and structure profiles are most advantageous for families with less primary sequence information (*towards left*) and more secondary structure information (*towards top*), so Rfam families that gain the most from including secondary structure terms in a homology search are those *toward the upper left quadrant*. Data shown for the 95 Rfam release 9.1 [55] families with 50 or more sequences in the *seed* alignment. For each family, the seed alignment was used to build two profile models, a CM and a profile HMM. From each model, 10,000 sequences were generated and scored, and the average score per sampled sequence was calculated. Several of the outlying points are labeled by the name of RNA family as given by Rfam. Note that the x-axis is drawn on a log scale. Models were built and sequences were generated and scored using Infernal version 1.0 programs cmbuild, cmemit, and cmalign

profile (reducing 16 possibilities to 1), thus contributing the same amount of information to either a sequence-only or a sequence and structure profile.

For this example, we gain three additional bits of information from considering structure, decreasing our chances of finding a match in a random database by a factor of $2^3 = 8$, from once every $2^{14} = 4{,}096$ nucleotides to once every $2^{17} = 32{,}764$ nucleotides. For real functional RNAs the additional amount of information gained from modeling structure varies widely. Figure 2 shows the information in a sequence and structure profile (CM) versus a sequence-only profile (HMM) for about 100 RNA families. Some RNAs, like tRNA, include about as much information in their structure as in their sequence, while for others, the increase is relatively modest. Note that for most families, modeling structure contributes at least 10 additional bits of information, which corresponds to lowering the expected chance of a false positive in a random database (i.e., the E-value of a database hit) by three orders of magnitude ($2^{10} = 1{,}024$).

1.2.1 CMs Are Probabilistic Models

In the previous example, sequences were either matches or mismatches to a profile, a simple yes/no scheme that offers no information on how *good* a match is. As SCFGs, CMs are importantly different from this simple match/mismatch paradigm in that they assign probabilities to the alignment of each nucleotide of a target sequence to each position of the profile, instead of a binary yes/no decision for a match/mismatch. Additionally, in a CM, nucleotides can be inserted and deleted relative to the consensus model, corresponding to an alignment of a gap in the model to a nucleotide in the target, and a consensus nucleotide in the model to a gap in the target, respectively. To facilitate the handling of insertions and deletions, CMs are organized as a binary tree of states, with each single-stranded position or basepair of the consensus sequence and structure modeled by separate match, insert, and delete states, corresponding to a consensus match, insertion after, or deletion of the relevant position/pair in the model. The topology of the tree mirrors the branching pattern of the consensus structure. States are connected to a subset of other states by ***transitions***, each associated with a probability, and nucleotides are emitted by (aligned to) match and insert states according to state-specific emission probabilities. CM states and transitions are equivalent to SCFG nonterminals and production rules. Given a particular alignment and secondary structure, the CM grammar formalism unambiguously dictates the construction of a particular tree topology of states and possible transitions between those states. The emission and transition probabilities for each state are then defined as mean posterior estimates based on the observed counts in the input alignment position(s) modeled by the state and a mixture Dirichlet prior (for emissions) or single component Dirichlet prior (for transitions). The details of this construction and parameterization procedure are not particularly relevant here and so are omitted; for more information, *see* [37, 38, 56–58].

1.2.2 Scoring Sequences with the CM Inside and CYK Algorithms

Given a fully parameterized model M and a target sequence s, CM implementations, such as Infernal, calculate a log-odds score that the sequence was generated by the CM versus by a background null model R. The null model typically used is a simple generative model of 25% A, C, G, and U, from which the probability of generating any sequence of length L is simply

0.25^L. This log-odds score is calculated by the CM Inside dynamic programming algorithm as:

$$S_{\text{Inside}} = \log_2 \frac{P(s|M)}{P(s|R)} = \sum_{\pi} \frac{P(s,\pi|M)}{P(s|R)}, \tag{1}$$

where π is a particular state path (i.e., alignment to the model, equivalent to a SCFG parse tree) through M that could have generated sequence s. The CM CYK dynamic programming [38] algorithm calculates a similar score: the log-odds score that s was generated by the maximum likelihood state path $\hat{\pi}$ that could have generated s, versus the same null model R. Specifically CYK calculates:

$$S_{\text{CYK}} = \log_2 \frac{P(s,\hat{\pi}|M)}{P(s|R)}. \tag{2}$$

The Inside score is more appropriate for determining if a sequence is homologous based on the model because it effectively integrates out the nuisance variable of the state path of the sequence in question. However, the CYK algorithm is also useful in practice because, due to details of their implementations in Infernal, CYK is about three times faster than Inside, and the CYK score approximates the Inside score well for most high scoring sequences of interest (because a single path accounts for a large fraction of the total probability mass of all paths). To accelerate searches, Infernal uses CYK as a filter for Inside, as explained later.

As presented above, the Inside and CYK algorithms compute log-odds scores for a complete target sequence s, but in practice RNA homologs are relatively short regions within long sequences. Infernal implements variants of Inside and CYK that scan along a target sequence scoring all possible subsequences as potential homologs. Because the log-odds scores are of base 2, the scores are in units of bits. An Inside score of x bits for a target sequence means that the sequence was $y = 2^x$ times more likely to have been generated by the CM than by the background model; for $x = 10$, y is 1024, and for $x = 20$, y is 1,048,576.

CM search algorithms are computationally expensive. Empirically, CYK scales $O(LN^{2.4})$ for a model of N consensus positions and a database of length L [57]. Inside has the same asymptotic time complexity as CYK, but is roughly three times slower in practice. Search times with the standard CYK and Inside algorithms are often impractically slow. For example, to search a typical sized archaeal genome (about 6 million bases (Mb), two strands of a 3 Mb genome) with a tRNA model of 71 consensus positions and an SRP RNA model of 302 consensus positions using the Inside algorithm requires about 2.5 and 21 CPU-hours, respectively. To repeat the same searches on the 3 Gb chimpanzee genome requires 0.3 and 2.5 CPU-years.

In contrast, the profile HMM Viterbi and Forward algorithms, which are analogous to CYK and Inside, scale $O(LN)$ [38]. Consequently, HMM searches take far less time than CM searches, especially for large models. For example, searching the chimpanzee genome with profile HMM models of tRNA and SRP using Infernal version 1.0's implementation of the Forward algorithm require 8 and 30 CPU-hours, respectively, making them about 300 times and 700 times faster than the CM Inside algorithm. More recent implementations of profile HMM algorithms are even faster. The HMMER3 software package uses heuristic filters to rapidly remove the majority of the database quickly and only applies the Forward algorithm to the surviving fraction, resulting in 100- to 1,000-fold acceleration for profile HMM searches for protein families at a negligible cost to sensitivity [59, 60].

1.3 Infernal

Infernal is a software package that implements CM methods. It includes programs to build a CM from an alignment (cmbuild), search a target sequence database with a CM (cmsearch), and create multiple sequence alignments of putative homologs with a CM (cmalign). Additionally, models are "calibrated" with the cmcalibrate program prior to using cmsearch. Calibration enables the reporting of expectation values (E-values) for putative homologs found in database searches. Infernal is an updated version of the Cove software package [61] which is used by tRNAscan-SE and SRPscan.

To alleviate slow search speeds, the latest version of Infernal (v1.1) executes multiple rounds of filtering of the target database prior to using Inside, the slowest but most sensitive CM search algorithm. The earliest rounds of the filter pipeline use a profile HMM to rapidly scan each target sequence and identify subsequences that may contain high-scoring hits to the CM based on sequence conservation alone. These filters are very similar to those employed in the HMMER3 pipeline [60], albeit with different survival thresholds such that a larger fraction of the database is expected to survive. The relaxed thresholds are important to ensure that hits with low sequence similarity but high structural similarity to the model will survive to the downstream CM stages of the pipeline. Subsequences that survive the profile HMM filters are then scored with a constrained version of the CM CYK algorithm. The CYK constraints are derived from a profile HMM alignment of the target subsequence, and limit the range of positions of the subsequence that are permitted to align to each state of the CM. These constraints are enforced as bands on the CYK dynamic programming matrix and result in a significant acceleration versus standard, non-banded CYK, especially for large RNAs (often up to or exceeding 100-fold acceleration) [62, 63]. Subsequences surviving the HMM banded CYK filter are evaluated with the Inside algorithm, again using HMM-derived bands, to assign their final scores. For more details on Infernal's filter pipeline, *see* [58]. The pipeline accelerates typical CM searches by three to four orders of magnitude versus non-filtered, non-banded Inside-only searches at a small cost to sensitivity and enables CM searches of large genomes in a reasonable amount of time.

Parallelization is another strategy Infernal uses for decreasing running times when a compute cluster is available. The cmalign, cmsearch, and cmcalibrate programs are implemented in coarse-grained parallel MPI versions allowing, for example, a search of a large vertebrate genome to finish faster by spreading the search across multiple nodes of a cluster.

1.4 Rfam

Rfam is a database of RNA families, each represented by a CM and two different multiple sequence alignments called a *seed* and a *full* alignment [64]. The seed alignment is a manually curated alignment of representative members of the family that is used to construct a CM using Infernal's cmbuild program. The CM is then searched against a large sequence database called RFAMSEQ based on a particular release of EMBL [65]. For each family, an Rfam curator chooses a bit score threshold, called the gathering threshold (GA), that separates the first clear false positive from trusted true homologs. All hits with bit scores above this threshold are extracted and aligned to the model to create the full alignment, which is not refined further. The most current release of Rfam (10.1) includes 1,973 RNA families and annotates 2,756,313 regions in the 170 Gb RFAMSEQ database, each of which was scored by a model above its GA threshold and is included in a full alignment. Notably, the CMs provided by Rfam come pre-calibrated and so will report E-values when used by cmsearch.

2 Using Infernal to Annotate Structural RNAs in an Archaeal Genome

In this section, I'll guide the reader through an exercise of using Infernal and Rfam to annotate functional RNAs in the genome of *Methanobrevibacter ruminantium* (GenBank accession CP001719.1), a methanogenic archaeon that lives in the stomachs of ruminant mammals such as cows [40]. This particular archaeon was chosen because the analysis of the search results illustrates some important considerations regarding the Infernal/Rfam strategy for genome annotation of RNAs, as discussed later in Subheading 2.1. For this exercise, it is assumed that the reader is familiar with a command-line Unix environment and has some experience writing simple scripts. The specific instructions here correspond to release 10.1 of the Rfam database and version 1.1 of Infernal. If you are using a more recent version of Rfam than 11.0, you should follow slightly different instructions; see the "Notes" section at the end of this chapter.

Step 1. Download and install Infernal 1.1.
Step 2. Download the 102 CMs from Rfam 10.1 that match at least one archaeal sequence from RFAMSEQ.
Step 3. Convert the Infernal 1.0 Rfam CM file to Infernal 1.1 format.
Step 4. Download the *M. ruminantium* genome sequence from NCBI.
Step 5. Run CM searches against the genome.
Step 6. Analyze the results.

Step 1. Download and install Infernal 1.1. Go to http://infernal.janelia.org/ and download version 1.1 of Infernal, then unzip and untar it. The user's guide will be in *infernal-1.1/Userguide.pdf*, which contains installation instructions. For a basic installation, simply execute *./configure; make* from the infernal-1.1 directory. This will create Infernal executable files in the *infernal-1.1/src/* directory, for example the programs *cmsearch* and *cmconvert*, which we'll use here. For steps 3 and 5 to work, you'll need to make sure that these programs are in your path (so that when you type *cmsearch* it executes the *cmsearch* program you just built). To install these programs in system-wide directories, execute *make install.* See the user's guide for more information on installation. (At the time of writing, the most current available version of Infernal is actually 1.1rc1, the first *release candidate* for version 1.1. As you read this it is likely that the final version 1.1 will be available, or perhaps even a newer version. Note that the results here may only be exactly reproducible using version Infernal version 1.1rc1 and Rfam release 10.1.)

Step 2. Download the 102 CMs from Rfam 10.1 that match at least one archaeal sequence from RFAMSEQ. Go to http://rfam.sanger.ac.uk/ and click on *Taxonomy Search* on the left-hand side of the page and search for *archaea*. The next page should report that 102 families were found.

Next, download the 102 CMs from Rfam. At the time of writing, users can either download all 1,973 CMs in the database in a single file, or one at a time each in a separate file. The easiest option is probably to download all the models and then write a script to select the desired 102. To do this, create a text file called *arc.102.list* and copy the names of the 102 families into it. Then, from http://rfam.sanger.ac.uk/ click on *FTP* at the top of the page, and click on *CURRENT* and download the file *Rfam.cm.gz.*

Create a new directory and place the files *arc.102.list* and *Rfam.cm.gz* in it and decompress *Rfam.cm.gz* with *gunzip*. Next, you'll have to write a script to extract the 102 CMs that are listed in *arc.102.list* from the *Rfam.cm* file. There are many ways to do this. There are two important aspects of the CM file format that you'll need to know about. The first is that the *Rfam.cm* file is a concatenation of 1,973 individual CM files, each beginning with a line that reads *INFERNAL-1 [1.0]*, and ending with the line *//*. Secondly, the name of the model appears immediately after the *INFERNAL-1 [1.0]* line. The extraction script will need to read through the file, printing out only those lines from the models listed in *arc.102.list*. Save the 102 models as the file *arc.102.old.cm*.

Step 3. Convert the Infernal 1.0 Rfam CM file to Infernal 1.1 format. Perform the conversion using Infernal 1.1's *cmconvert* program, with the command:

```
cmconvert arc.102.old.cm > arc.102.cm
```

The conversion should take about 3 min.

Step 4. Download the Methanobrevibacter ruminantium *genome sequence from NCBI.* Go to the ENTREZ search page: http://www.ncbi.nlm.nih.gov/sites/gquery, search for *CP001719.1*, and follow the link for the *Nucleotide* database. A page should load reading: *Methanobrevibacter ruminantium M1 chromosome, complete genome*. To download the genome in FASTA format, click on the *Send* link on the right-hand side of the page and select *Complete Record*, *File* and FASTA format; then click *Create file*. A file called *sequences.fasta* should download. Rename this file *mrum.fa* and move it to the directory where you've stored the converted CM files from step 3.

Step 5. Run CM searches against the genome. Now you are ready to search the *M. ruminantium* genome with the Rfam CMs using Infernal. CM searches are computationally expensive, but not impractical. All 102 searches should take less than 10 min and can be executed with a single command:

```
cmsearch --cut_ga --tblout mrum.tbl arc.102.cm mrum.fa
  > mrum.cmsearch
```

This command includes two options. The `--cut_ga` option tells the program to set the score threshold for reporting a hit to each model as that model's manually curated Rfam gathering threshold. The `--tblout mrum.tbl` option specifies that a tabular version of the search results be printed to a file called *mrum.tbl* as explained below in step 6.

Step 6. Analyze the results When the search finishes running you should have two new files in your directory called *mrum.cmsearch* and *mrum.tbl* containing the search results. The former file includes information on the high-scoring regions or *hits* in the genome to each model, including the bit scores and E-values of those hits as well as alignments of each of the hits to its respective model. The latter file contains much of the same information but in a simplified one-line-per-hit format which can be easily parsed by scripts one might use to analyze the results. Next, we'll take a closer look at example results from each of these files.

Open the *mrum.tbl* file and look at the first few lines (these have been split in half below to fit on the page):

```
#target name                 accession query name            accession mdl mdl from   mdl to seq from ...
#--------------------------- --------- -------------------- --------- --- -------- -------- -------- ...
gi|288541968|gb|CP001719.1| -         5S_rRNA              -          cm        1      119   766016 ...

 ...  seq to strand trunc pass   gc  bias  score   E-value inc description of target
 ... ------- ------ ----- ---- ---- ----- ------ --------- --- --------------------
 ...  766137      +    no    1 0.48   4.0   58.9  1.4e-11 !   Methanobrevibacter ruminantium M1, comp...
```

As indicated by the column names, this line reports a hit from position 766016 to 766137 of the genome (with target sequence name gi|288541968|gb|CP001719.1| in the *mrum.fa* file) to the 5S_rRNA Rfam CM, with a bit score of 58.9 bits and an E-value of $1.4e{-}11$. This E-value indicates that the probability of finding a hit with this bit score or higher is about 10^{-11} in a database the size of this genome. Because this E-value is so low, we can be confident that this region is indeed a homolog of 5S rRNA. As discussed later, annotators can be fairly confident of hits with E-values up to about $1e-5$ in archael and bacterial genomes. Additionally, we know that this hit has a bit score that is at least as high as the Rfam gathering (GA) threshold for the 5S_rRNA model. In fact, *all* the hits in our search results will meet or exceed the GA threshold for their respective models because we chose the `-cut_ga` option when running cmsearch. In addition to considering the E-value and bit score of a hit, inspection of the alignment of the hit to the query CM is often useful when determining whether a hit is a real homolog or not. Alignments are not contained in the tabular output files, but are included in the standard cmsearch output. To find the alignment of this hit to the tRNA model in the *mrum.cmsearch* file, it is useful to know that the file is organized into 102 sections, one for each query CM. Each CM's section begins with "Query:" at the beginning of a line followed by the CM name, and then a list of hits ranked by E-value (lowest to highest), and then the hit alignments for all reported hits in the same ranked order.

The first tRNA hit alignment in *mrum.cmsearch* is for the 69.8 bit hit from positions 735136 to 735208 of the *M. ruminantium* genome, target sequence gi|288541968|gb|CP001719.1|, organized into a single block of six lines, shown in Fig. 3. (Longer alignments, such as the second tRNA alignment in this file, will be split into multiple blocks of six lines each.) The second line of each block ends with `CS` and shows the secondary structure of the consensus tRNA molecule modeled by the CM. This structure is shown as the leftmost secondary structure at the bottom of Fig. 3. In the alignment, positions of the model that are paired are indicated by either parentheses or brackets (i.e. `()`, `<>`) and the left and right half of pairs are identified by matching left and right parentheses or brackets from outside to inside as in a mathematical formula. For example, the first (leftmost) '`(`' matches the last

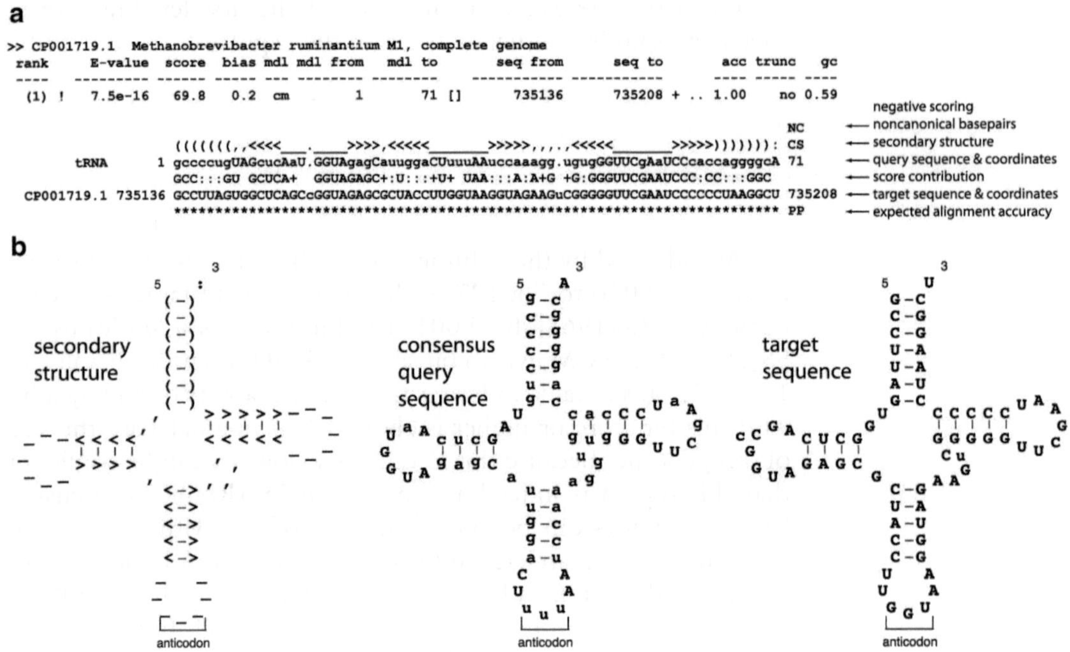

Fig. 3 A sequence- and structure-based alignment of a predicted tRNA in *M. ruminantium* to the Rfam 10.1 tRNA CM. (**a**) Raw output from cmsearch showing the alignment of positions 735136 to 735208 of the target sequence CP001719.1 (renamed from gi|288541968|gb|CP001719.1| to save space) to the model. Scores and alignment annotation are explained in the text. (**b**) Secondary structure diagrams of the consensus tRNA structure annotation from (**a**), the consensus query tRNA sequence from the CM, and the predicted tRNA homolog from the target genome

(rightmost) ')' indicating that these two positions are basepaired with each other. Similarly the second '(' matches the second to last ')', and so on. The difference between parentheses and brackets indicates levels of nesting. For example, the parentheses depict the acceptor stem between the 5′ and 3′ ends of the tRNA (the top stem in the structures in Fig. 3), while the three other stems are indicated by brackets because they are independent stems and are fully contained between the two halves of the acceptor stem. Other RNAs in Rfam, such as SSU_rRNA_archaea and RNase_P_arch, have more than two nesting levels of stems in their consensus secondary structures, and to handle these cases additional characters are used (e.g., {}, []).

The first line of the alignment ends with NC. This line indicates negative scoring noncanonical basepairs, these are basepairs in the target sequence which receive a negative score to the model and are not either Watson–Crick (A:U, U:A, C:G, G:C) or wobble (G:U, U:G) basepairs. A negative score is assigned to basepairs that are less probable in their particular position of the CM than in the random background model, i.e. less than 1/16 = 0.0625. There are zero such basepairs in this target sequence, so this line is entirely blank. If any such basepairs did exist (there are some examples in other alignments in this file), they would be highlighted with a `v` character in this line.

The third line of the alignment shows the query model consensus sequence. This is defined as the highest scoring nucleotide or basepair at each position, with capital letters being highly conserved and lowercase letters being less well conserved. Dots in this line indicate a position where the target sequence has inserted one or more residues. The fifth line shows the target sequence, in this case ranging from position `735136` to `735208`. Lowercase nucleotides here, such as the single lowercase `c` in this line, indicate inserted nucleotides relative to the consensus model. The fourth line in each block indicates how well the query and target align to each other and is meant to help the user quickly judge the quality of the alignment when examining a putative homolog. If a nucleotide *N* is present, then the target has the most probable nucleotide *N* aligned at that position. If a blank space or non-alphabetic character is present, then the target contains either a gap or a nucleotide other than the most probable one. Of these cases, a "`+`" or "`:`" occurs when the target nucleotide receives a positive score for the model, either for single stranded positions (`+`) or basepaired positions (`:`). A blank space occurs if either the target nucleotide is a gap or if the target nucleotide receives a negative score to the model. As explained earlier, a positive score indicates the nucleotide or basepair is more probable than the random background model (i.e., has higher probability than 0.25 or 0.0625, respectively).

Finally, the sixth line ends in `PP` and indicates the confidence level, or expected accuracy, of each position of the alignment. Each position receives a single character summarizing its posterior probability. A `0` means 0–5%, a `1` means 5–15%, and so on; a `9` means 85–95%, and a `*` means 95–100% posterior probability. In this alignment all positions are `*` indicating they are all very confidently aligned correctly, but there are examples of more ambiguous alignments elsewhere in this file. As you might expect, alignment positions with low confidence are often nearby insertions and deletions.

2.1 Important Considerations Regarding Infernal Predictions

Table 3 reports the number of hits found in the *M. ruminantium* genome with scores that exceed the Rfam gathering threshold for each of the 102 families with at least one such hit. There are 128 total predicted RNAs from eight different Rfam families (after removing overlaps and hits with marginal E-values, as explained more below), as opposed to only 66 from four different families in the NCBI GenBank and Refseq annotation (accessions CP001719.1 and NC_013790.1). Closer scrutiny of these results offers insights into some important strengths, weaknesses, and caveats of using Infernal and Rfam for genome annotation. Below, these issues are listed and explained using specific examples from the results.

Table 3
Infernal predicted RNAs in the archaeon *Methanobrevibacter ruminantium* (GenBank accession CP001719.1). All Rfam 10.1 families [64] for which Infernal finds at least one hit above the Rfam bit-score gathering threshold (GA) are shown. Also shown is LSU rRNA, which is not in Rfam as discussed in the text. Non-obvious column heading descriptions: "# in GenBank": number of RNAs in GenBank annotation; "believed": hits believed to be real homologs, these are all nonoverlapping hits with E-values below *0.01* except for the single CRISPR-DR42 hit (*E=0.0045*) which is likely a false positive (the choice of *0.01* here is discussed in Subheading 2.1); "unique": number of nonoverlapping Infernal hits, overlaps of more than 50% the length of the shorter sequence were removed by keeping the hit with the lowest E-value amongst the overlapping hits; "total": total Infernal hits, including overlaps; "best hit": bit scores and E-values of the best scoring hits out of total hits for each family. The following sets of families shared overlapping hits, the family with the lowest E-value for all overlaps is listed first: SSU_rRNA_archaea and SSU_rRNA_bacteria (2 hits), CRISPR-DR2 and CRISPR-DR39 (60 hits), and sR2, sR1, and snoPyro_CD (1 hit)

Rfam family ID	Rfam type	Rfam GA bit thresh	# in Gen-bank	Infernal hits above Rfam GA thr # hits Believed	Unique	Total	Best hit Bit	E-value
tRNA	Gene; tRNA;	24.0	58	59	59	59	69.8	7.5*e*–16
5S_rRNA	Gene; rRNA;	16.0	4	3	3	3	58.9	1.4*e*–11
LSU rRNA			2					
SSU_rRNA_archaea	Gene; rRNA;	658.0	2	2	2	2	1483.0	0
SSU_rRNA_bacteria	Gene; rRNA;	600.0	0	0	0	2	1090.7	0
Archaea_SRP	Gene;	87.0	0	1	1	1	183.7	6.2*e*–52
RNaseP_arch	Gene; ribozyme;	53.0	0	1	1	1	193.9	3.5*e*–63
FMN	Cis-reg; riboswitch;	40.0	0	1	1	1	110.8	8.6*e*–28

Table 3
(continued)

CRISPR-DR2	Gene; CRISPR;	22.0	0	60	61	61	28.2	0.0043
CRISPR-DR39	Gene; CRISPR;	20.0	0	0	2	62	26.5	0.019
CRISPR-DR42	Gene; CRISPR;	19.2	0	0	1	1	20.2	0.0045
sR2	Gene; snRNA;	20.0	0	1	1	1	27.5	9*e*–06
	snoRNA; CD-box;							
sR1	Gene; snRNA;	21.0	0	0	0	1	23.7	0.00017
	snoRNA; CD-box;							
snoPyro_CD	Gene; snRNA;	20.0	0	0	0	1	27.1	0.0018
	snoRNA; CD-box;							
sR11	Gene; snRNA;	16.0	0	0	1	1	16.8	0.021
	snoRNA; CD-box;							
Total			66	128	136	200		

1. Hits from different models can overlap.
It is not uncommon for a single region of a genome or target database to be hit by multiple models from Rfam. There are several reasons why this may occur. First, some families are evolutionarily related to each other. An example of this in the *M. ruminantium* results are the SSU_rRNA_archaea and SSU_rRNA_bacteria models, which model archaeal SSU ribosomal RNA and bacterial SSU ribosomal RNA, respectively. The SSU rRNA is ancient and predates the split of the three domains, so a homology search method that identifies cross-matches between these families is correct. Note that, as expected, the score for the hit to the archaeal model is much higher than the bacterial model (1483.0 versus 1090.7). Another example are the overlaps between nearly all of the roughly 60 CRISPR-DR2 and CRISPR-DR39 hits. Both of these CRISPR CMs have 30 consensus positions, and they share some sequence and structural similarity. When overlapping hits are encountered, it is recommended practice to keep the hit with the better E-value. In most cases this will be the hit with the higher bit score, but not always. If both hits have the same E-value (as with the two hits of E-value 0 to the SSU models), keep the one with the higher bit score.
2. Hits with marginally significant E-values should be carefully examined or thrown out.
To avoid misannotating RNAs in a genome, the E-values of predicted hits should be considered, even for hits above the Rfam GA bit score threshold. In these searches, use of the cmsearch `--cut_ga` option dictated that only hits exceeding the GA threshold be reported. Because we searched with 102 models, the highest-scoring false positive hit to any family we expect is about 1/102, which is roughly 0.01. However, because we also only consider hits above the GA bit score thresholds and for some families those thresholds correspond to E-values below 0.01, this calculation is only roughly accurate. In general, when performing N searches, the highest-scoring false positive hit should have an E-value of roughly $1/N$ because that E-value literally means that we expect $1/N$ such hits from a particular search, so $N*1/N = 1$ such hits are expected in N searches. As the predicted E-value of the highest-scoring false positive, 0.01 is a reasonable E-value threshold to use during annotation. As shown by the differences between the "believed" and "unique" columns of Table 3, doing so in this analysis would lead to the removal of the following unique (nonoverlapping) hits: the single hit to the sR11 model and the two hits to the CRISPR-DR39 model because these three hits have E-values above 0.01. One additional "unique" hit is not counted in the "believed"

column: the CRISPR-DR42 model with an E-value of 0.0045. This E-value is slightly below the 0.01 threshold but is ruled out because CRISPR sequences almost always occur as tandem repeats as explained in the next section.

In fact, because Infernal E-values are not perfectly accurate, even more conservative E-value thresholds are often used in practice. For example, only hits with E-values below $1e-5$ were considered in a recent survey of SmY RNAs in nematodes [10]. In our results, dropping the E-value threshold from 0.01 to $1e-5$ would additionally exclude only the 60 remaining CRISPR-DR2 hits. However, as discussed next, closer scrutiny of these CRISPR-DR2 predictions suggests they are in fact real homologs. Even stricter E-values may be necessary for searches of complex genomes which strongly invalidate the assumptions made by the Infernal E-value machinery. For example, large vertebrate genomes that include high numbers of tandem repeats can pose particular problems for Infernal, as discussed in point 5 below.

3. Expert knowledge of a family can help verify an Infernal prediction.

Often it is possible to gain corroborating evidence that an Infernal prediction is either a true homolog or not based on additional knowledge of the family that cannot be modeled by a CM. For example, CRISPR genes are Clustered Regularly Interspaced Short Palindromic Repeats that are separated by spacers of similar length. The Infernal CRISPR-DR2 predictions follow this pattern: 60 out of 61 are identical 30 nucleotide subsequences, and 59 of those 60 are separated by between 91 and 98 nucleotides (the remaining spacer is 160 nucleotides). In contrast, the 61st hit to this model occurs about 500 Kb away from the cluster of 60 and has a marginal E-value of 0.11 suggesting it is probably not a real CRISPR element. By the same reasoning, the single hit to the CRISPR-DR42 model is likely a false positive hit even though it has a more significant E-value of 0.0045.

Another example involves the single highly significant ($E = 8.8e-28$) hit to the FMN riboswitch model. Because riboswitches tend to occur in the 5′ untranslated region of genes involved in metabolism of a particular ligand, flavin mononucleotide (FMN) for this particular switch, additional evidence that a predicted riboswitch is real is often obtained by examining the function of downstream protein-coding genes. In this case, the nearest gene is annotated as a *ribB* gene, a 3,4-dihydroxy-2-butanone 4-phosphate synthase, which is involved in riboflavin metabolism, and importantly, the predicted riboswitch is on the same strand and is 5′ of the coding sequence of *ribB*. These data suggest the Infernal prediction is in fact an FMN riboswitch.

4. Some RNA families are not included in Rfam, others are represented by models that are not full-length.
There are two common reasons for a family's absence from Rfam. Firstly, it may have just been discovered. Novel families continue to be discovered at a rapid pace [10, 66–68] making it difficult for the limited number of Rfam curators to incorporate all of them into the database. Secondly, some RNAs are so large that running the Rfam search pipeline for them would take an impractical amount of compute time due to the high complexity of the CM Inside and CYK algorithms. A glaring omission from Rfam that falls into the second category is LSU rRNA models. However, as mentioned earlier, LSU is highly conserved at the sequence level and using CM methods to identify them is unnecessary because more efficient sequence-based methods can do the job well, such as the profile HMM approach taken by the RNAmmer program [32]. In the future, Rfam could take a similar approach and use profile HMMs for LSU searches. However, currently an Infernal user aiming for a complete annotation of RNAs in a genome would need to run an additional program such as RNAmmer or BLASTN to annotate full length LSU sequences.
Some families in Rfam are represented by alignments and models that do not cover the full length of the sequence family. Two examples are the group I and group II self-splicing intron models. These models are not full length because the high variability in the sequence and structure of homologs makes construction of a reasonable global structural alignment difficult. In such cases, only the well-conserved core regions are modeled, and the variable parts are omitted. While largely incomplete models like these are rare, nearly complete models that may not include the complete $5'$ and $3'$ ends of the RNA are more common. One reason for this is that the structures and alignments for some Rfam families are based mainly on predictions from comparative sequence analysis and experiments to determine the precise start and end points of the non-coding transcripts have not yet been performed.
5. Eukaryotic sequences offer additional challenges.
For Infernal searches of eukaryotic genomes, new issues arise and some of the problems discussed above can become more severe. Eukaryotic genomes contain certain types of sequence elements largely absent from archaea and bacteria that can lead to high-scoring false positives in CM searches, namely pseudogenes and repeats. For example, some short interspersed nuclear elements (SINEs) are derived from pol-III transcribed RNAs like tRNA or SRP RNA. Examples include Alu sequences, which are common in primates, numbering greater than one million in the human genome. Pseudogenes

of U6, 7SK, and Y RNAs are also common [24]. These elements will often score high to a CM due to their homology with the original RNA family from which they were derived. An example of this problem is shown in Table 4, which contains results of Infernal searches and family-specific searches for selected families in four eukaryotic genomes (described more in the next section). The table shows that thousands of tRNA-derived SINEs are identified by tRNAscan-SE in the mouse genome (*Mus musculus*). These elements were noted in the original publication for that genome [53]. tRNAscan-SE identifies 26,201 regions in the genome, 22,918 of which are reported as pseudogenes, and 3,283 of which are predicted as tRNA genes. All but about 500 of these were discounted after closer inspection as likely non-functional SINE repeats in [53].

Additionally, less complex inverted tandem repeats often score high against any CM with a single stem loop (such as miRNAs) or otherwise simple secondary structure because opportunities for stretches of Watson–Crick basepairing between nearby regions in these elements are abundant. Large numbers of high-scoring false positives greatly complicate the analysis of Infernal results because while it is desirable to set a single E-value threshold for all families, in reality, certain families will require special treatment.

2.2 Comparison of Infernal to Family-Specific Methods

In the *M. ruminantium* searches, Infernal was able to improve upon the existing RNA annotation by finding two probable tRNAs missed by tRNAscan-SE (Fig. 4), suggesting that Infernal can be more sensitive than family-specific methods in some cases. For further comparison, I used some popular family-specific methods and the corresponding Rfam CMs with Infernal to search the fourteen genomes listed in Table 2. A comparison of the results is shown in Tables 4 (eukaryotic genomes), 5 (archaeal genomes), and 6 (bacterial genomes).

The Infernal results largely agree with the tRNAscan-SE, SRPscan, and Bcheck results. This is not surprising considering that all of these programs, including Infernal, are using CMs with sequence-based filters. The main difference is in the design of those filters. For Infernal, profile HMMs built from the CM are used, whereas for the others, sequence and structural characteristics of the specific families being modeled have been exploited to enable stricter filtering in some cases. The stricter filtering not only can enable faster searches in some cases (e.g., Aragorn searches for tmRNAs in bacteria) but can also cause high-scoring hits to be missed, such as Bcheck's failure to identify any RNaseP RNAs in *M. ruminantium* (Table 5). Additional examples are the SRP RNA prediction by Infernal in *M. ruminantium* which is missed by SRPscan, and the three archaeal tRNAs predicted by

```
>> CP001719.1  Methanobrevibacter ruminantium M1, complete genome
 rank     E-value  score  bias mdl mdl from   mdl to       seq from      seq to       acc trunc   gc
 ----   --------- ------ ----- --- -------- --------    ----------- -----------      ---- ----- ----
 (54) !     1e-08   44.7   0.2  cm        1       71 []      361145      361075 - .. 0.99    no 0.44

                                                                                            NC
                    (((((((,,<<<<_______>>>>,<<<<<_______>>>>>,,,,<<<<<_______>>>>>))))))): CS
       tRNA       1 gccccugUAGcucAaUGGUAgagCauuggaCUuuuAAuccaaaggugugGGUUCgAaUCCcaccaggggcA 71
                    G :::UGU :C::A U G A::G:+ ::G +U+ +AA C:: + +U: :GGU C+A+UCC: :CA::: C
 CP001719.1 361145 GGUAAUGUGACCGACUAGAAAGGUGGCUGUUUGCAAAGCAGUUUAUUUCGGUGCAAUUCCGGACAUUAUCU 361075
                    ***************888****************************************************** PP
```

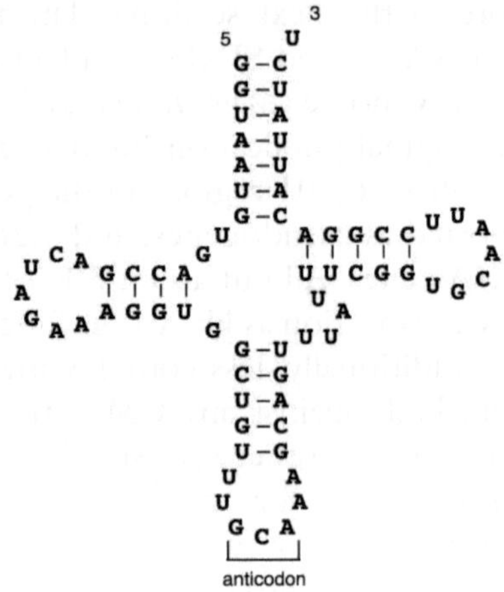


```
>> CP001719.1  Methanobrevibacter ruminantium M1, complete genome
 rank     E-value  score  bias mdl mdl from   mdl to       seq from      seq to       acc trunc   gc
 ----   --------- ------ ----- --- -------- --------    ----------- -----------      ---- ----- ----
 (57) !   4.7e-08   42.3   0.1  cm        1       71 []      363803      363716 - .. 0.98    no 0.50

                                                                                                        NC
                    (((((((,,..<<<<___..___.._>>>>,<<<<<_______>>>>>,,.............,.,<<<<<_______>>>>>))))))): CS
       tRNA       1 gccccugUA..GcucAaU..GGU..AgagCauuggaCUuuuAAuccaaag............g.ugugGGUUCgAaUCCcaccaggggcA 71
                    ::C:::GUA  GC::A+U  GGU  A::GCA :G:ACUU+  AU:C:                +:UGGGUUC+AAUCCCA:C:::G::A
 CP001719.1 363803 AGCUGUGUAguGCCAAGUcaGGUcuAAGGCAGCGGACUUAGGAUUCGUG-gaguuagucuca-uCAUGGGUUCAAAUCCCAUCGCAGCUA 363716
                    ********9889*************************************.8**********5.5************************** PP
```

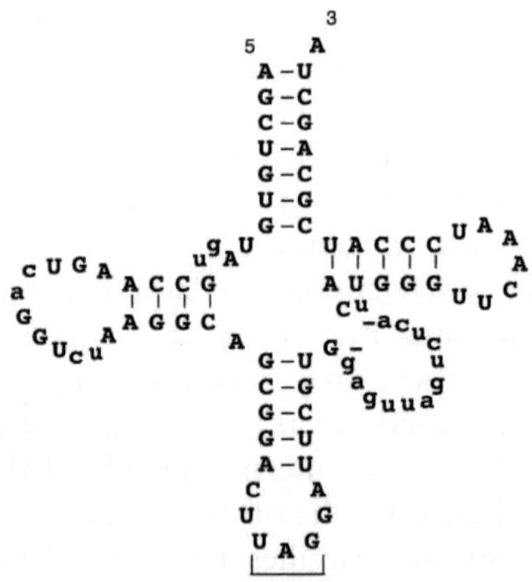

Fig. 4 Two Infernal-predicted tRNAs in the *Methanobrevibacter ruminantium* genome that are not predicted by tRNAscan-SE. The target sequence CP001719.1 has been renamed from gi|288541968|gb|CP001719.1| to save space. The cmsearch output for each alignment to the Rfam 10.1 tRNA model is shown above the corresponding predicted secondary structure. Nucleotides inserted relative to the Rfam consensus model are in *lowercase*

Table 4
Comparison of predictions by Infernal and family-specific methods for various RNAs in four eukaryotic genomes. Abbreviated genome names are fully listed in Table 2. RNAmmer does not do 5.8S rRNA searches, so the corresponding cells are left blank in the table. Column headings and program versions, options and cutoffs are the same as described in caption of Table 5, except that for Infernal searches only hits with bit scores above the Rfam GA cutoff *and* E-values below *1e-5* were considered. Rfam 10.1 models used for each Infernal search: "tRNA": RF00005, "RNase P RNA": RF00009, "5S rRNA": RF00001, "SSU rRNA": RF01960,"5.8S rRNA": RF00002

		Organism (eukarya)								
		C. dub. 14.6 Mb		*L. bra.* 31.4 Mb		*A. tha.* (119.7 Mb)		*M. mus.* (2654.9 Mb)		Avg time
Family	Software	hits	unq	hits	unq	hits	unq	hits	unq	(seconds)
tRNA	Infernal	123	0	82	0	676	5	442	11	359.8
	tRNAscan-SE	130	7	82	0	699	28	26,201[a]	25,770	3481.5
tRNA	Infernal	123	36	82	1	676	44	442	14	359.8
	Aragorn	89	2	85	4	666	34	1,656	1,228	130.0
RNase P RNA	Infernal	1	1	0	0	0	0	17	11	651.1
	Bcheck	0	0	0	0	0	0	6	0	218.4
5S rRNA	Infernal	2	0	9	1	497	3	115	115	134.3
	RNAmmer	2	0	8	0	498	4	0	0	3741.2
SSU rRNA	Infernal	1	0	0	0	4	0	2	0	3203.0
	RNAmmer	1	0	0	0	5	1	2	0	3450.2
LSU rRNA	Infernal									
	RNAmmer	1	1	0	0	4	4	3	3	3816.3
5.8S rRNA	Infernal									
	RNAmmer	1	1	0	0	2	2	2	2	173.3

[a]22,918 of these are annotated as pseudogenes by tRNAscan-SE

Table 5

Comparison of predictions by Infernal and family-specific methods for various RNAs in five archaeal genomes. Abbreviated genome names are fully listed in Table 2. Genome sizes in millions of bases (Mb) are shown in parentheses underneath genome names. Columns labeled "hits" include total number of predictions, and those labeled "unq" include unique hits that are not found with the method on the adjacent line. Average timings are reported in seconds ("(secs)"). The following Rfam 10.1 models were used for each Infernal search: "tRNA": RF00005, "RNase P RNA": RF00373, "SRP RNA": RF01857, "5S rRNA": RF00001, "SSU rRNA": RF01959. LSU rRNA Infernal searches were not performed because Rfam 10.1 has no LSU model. All programs were run in default mode, except when options were necessary to restrict searches to the specific family and/or domain being tested. SRPscan was run in fast mode with non-Alu models. Infernal's cmsearch was run with the `--cut_ga` option which sets the reporting bit score threshold as the family-specific Rfam GA cutoff discussed in the text. Program versions used: Infernal v1.1rc1; tRNAscan-SE v1.23; Aragorn v1.2; Bcheck v0.6; web version of SRPscan available at http://bio.lundberg.gu.se/srpscan/; RNAmmer v1.2. Because no downloadable version of SRPscan was available, times were measured manually via stopwatch on their web site and so are approximate. All other times were measured as single execution threads on 2.66 GHz Intel Xeon Gainestown (X5550) processors

		Organism (archaea)										
		M. rum. (2.9 Mb)		*H. vol.* (4.0 Mb)		*H. jeo.* (3.7 Mb)		*A. sac.* (1.5 Mb)		*M. mar.* (1.6 Mb)		Avg time
Family	Software	hits	unq	hits	unq	hits	unq	hits	unq	hits	unq	(seconds)
tRNA	Infernal	59	2	49	1	46	0	43	0	38	0	2.2
	tRNAscan-SE	58	1	51	3	49	3	45	2	40	2	27.6
tRNA	Infernal	59	3	49	2	46	3	43	5	38	1	2.2
	Aragorn	56	0	54	7	48	5	39	1	38	1	0.9
RNase P RNA	Infernal	1	1	1	0	1	0	1	0	1	0	28.8
	Bcheck	0	0	1	0	1	0	1	0	1	0	13.7
SRP RNA	Infernal	1	1	1	0	1	0	1	0	1	0	10.0
	SRPscan	0	0	1	0	1	0	1	0	1	0	8.0
5S rRNA	Infernal	3	3	2	0	1	0	1	1	3	3	2.0
	RNAmmer	0	0	2	0	1	0	0	0	0	0	16.1
SSU rRNA	Infernal	2	0	2	0	1	0	1	0	2	0	31.7
	RNAmmer	2	0	2	0	1	0	1	0	2	0	16.4
LSU rRNA	Infernal											
	RNAmmer	2	2	2	2	1	1	1	1	2	2	18.8

Infernal but not by tRNAscan-SE, two of which are shown in Fig. 4. Conversely, because Infernal's HMM filters do not consider structure they could miss some high-scoring hits that the other methods find. In these searches, this is exemplified by putative tRNAs predicted by Aragorn and tRNAscan-SE that are not found by Infernal. However, with the exception of tRNA for several genomes, and 5S and SSU rRNA in *A. thaliana*, Infernal finds all of the hits that the family-specific methods report.

Besides being faster in some cases, family-specific tools offer some other important advantages over Infernal, such as offering

Table 6
Comparison of predictions by Infernal and family-specific methods for various RNAs in five bacterial genomes. Abbreviated genome names are fully listed in Table 2. Column headings and program versions, options and cutoffs are the same as described in caption of Table 5, except that for SRPscan, "rare TRRC tetraloop" was used for *P. staleyi* and "common GRRA tetraloop" was used for all others. Rfam 10.1 models used for each Infernal search: "tRNA": RF00005, "tmRNA": RF00023, "RNase P RNA": RF00010 and RF00011, "SRP RNA": RF00169 and RF01854, "5S rRNA": RF00001, "SSU rRNA": RF00177

		Organism (bacteria)										
		C. rod. (5.4 Mb)		*B. den.* (2.6 Mb)		*P. sta.* (6.2 Mb)		*L. mon.* (3.1 Mb)		*C. ucy.* (1.4 Mb)		Avg time
Family	Software	hits	unq	hits	unq	hits	unq	hits	unq	hits	unq	(seconds)
tRNA	Infernal	85	1	57	1	46	2	57	0	37	0	2.0
	tRNAscan-SE	84	0	56	0	46	2	58	1	37	0	34.5
tRNA	Infernal	85	1	57	1	46	1	57	0	37	0	2.0
	Aragorn	87	3	56	0	49	4	59	2	37	0	1.1
tmRNA	Infernal	1	0	1	0	1	0	2	1	1	0	40.9
	Aragorn	1	0	1	0	1	0	1	0	1	0	2.3
RNase P RNA	Infernal	1	0	1	0	1	0	2	0	1	0	28.0
	Bcheck	1	0	1	0	1	0	1	0	1	0	12.8
SRP RNA	Infernal	1	0	1	0	1	0	2	0	1	0	29.2
	SRPscan	1	0	1	0	1	0	1	0	1	0	4.0
5S rRNA	Infernal	8	0	6	1	1	0	5	0	2	0	2.9
	RNAmmer	8	0	5	0	1	0	5	0	2	0	20.2
SSU rRNA	Infernal	7	0	4	0	1	0	5	0	2	0	34.0
	RNAmmer	7	0	4	0	1	0	5	0	2	0	20.4
LSU rRNA	Infernal											
	RNAmmer	7	7	4	4	1	1	5	5	2	2	27.9

additional information relevant to the annotations. For example, tRNAscan-SE reports on the tRNA type in its predictions based on the anticodon sequence, as well as whether the tRNA contains an intron or is a predicted pseudogene. Also, some of the family-specific CM-based tools include more CMs than are present in Rfam, and the models are built from more carefully curated input alignments than those in Rfam in some cases. For example, Bcheck includes two archaeal RNase P CMs, while Rfam includes only one. Using more and/or better models can lead to more accurate or more complete annotations.

The comparison of RNAmmer with Infernal highlights an important difference between profile HMMs (as implemented in RNAmmer) and CMs. While most of the predictions agree, RNAmmer failed to recognize some 5S rRNA candidates in archaea that Infernal finds. This suggests that the additional statistical power gained by modeling the conserved 5S rRNA secondary structure is critical for the CM in these cases.

There are other RNA search tools that have not been tested here for various reasons. SnoReport[36] requires candidate RNA sequences as input and cannot scan along genome length sequences. Snoscan [69] and snoGPS [70] which identify C/D box snoRNAs and H/ACA box snoRNAs, respectively, and take advantage of user-specified ribosomal RNA sequences that include potential methylation/pseudouridylation sites for the predicted snoRNAs, complicating a potential comparison with a general CM approach. The Riboswitch finder [71] and RibEx [72] tools are only available via webservers that do not accept genome length sequences. The microRNA detection program RNAmicro [73] was not tested because it requires an alignment of orthologous sequences as input. Pattern search tools, such as RNA-motif [74], and RNABOB [35] were not tested because libraries of patterns analogous to Rfam CMs that would enable analogous searches are not readily available. Other tools, such as RNA-PATTERN [75], are not freely available for download.

3 Conclusion

As demonstrated here by the example analysis of the *M. ruminantium* genome, using Infernal and Rfam instead of more commonly used tools can lead to a more complete annotation of RNAs in genomes. The Infernal results contain important information on the biology of *M. ruminantium* that is absent from its initial GenBank annotation. For example, the existence of 60 CRISPR hits indicates that *M. ruminantium* can likely acquire resistance against viruses through the CRISPR system [76]. Additionally, the presence of a high-scoring hit (110 bits, E-value of $8.8e-28$) to the FMN riboswitch model strongly indicates that this archaeon encodes a true riboswitch which may control expression of at least some genes involved in riboflavin biosynthesis through binding of FMN to this structural element. This is especially interesting because riboswitches primarily exist in bacteria. Further, a single RNase P RNA and a single SRP RNA have been predicted, which is expected but still relevant because these RNAs were not annotated in the initial publication of the genome.

For annotation of functional RNAs in genomes, the general Infernal/Rfam approach is comparable in both speed and sensitivity to the use of family-specific tools that utilize specialized filters or search algorithms. The total time required for the 102 *M. ruminantium* searches was about 6 min on a single CPU. For families for which specific search tools exist, their results largely agree with Infernal (Tables 4–6). Importantly though, the Rfam database includes a growing number of CMs of families for which specific search tools capable of scanning genomes do not exist, which can be used by Infernal to search for as-of-yet undiscovered homologs. Infernal has the added advantage of convenience for annotation pipeline developers: it is a single program that works for most RNA families, making it easier to incorporate and maintain in a pipeline than multiple family-specific programs.

4 Notes

If you are working through the *M. ruminantium* genome annotation example in Subheading 2 and have access to an Rfam release more recent than 11.0 that was based on Infernal 1.1, you should be able to simplify the six-step annotation process. The first important change is that you do not need to write a script to extract the archaeal CMs from the *Rfam.cm* file. Instead, use Infernal's *cmfetch* program (see the Infernal user's guide for details). Secondly, you can skip step 3, the CM conversion step. The other steps should be followed as written in Subheading 2 but your results will likely be slightly different. For example, you will probably be working with more than 102 CMs.

Acknowledgements

I thank Sean Eddy, Tom Jones and Travis Wheeler for useful discussions and critical comments on the manuscript.

References

1. Burge CB, Tuschl T, Sharp PA (1999) Splicing of precursors to mRNAs by the spliceosomes. In Gesteland RF, Cech TR, Atkins JF (eds) The RNA World, 2nd edn. Cold Spring Harbor Laboratory Press, Cold Spring Harbor, New York, pp 525–560
2. Eliceiri GL (1999) Small nucleolar RNAs. Cell Mol Life Sci 56:22–31
3. Lewin R (1982) Surprising discovery with a small RNA. Science 218:777–778
4. Frank DN, Pace NR (1998) Ribonuclease P: Unity and diversity in a tRNA processing ribozyme. Annu Rev Biochem 67:153–180
5. Bushati N, Cohen S (2007) microRNA functions. Annu Rev Cell Dev Biol 23: 175–205
6. Henkin TM (2008) Riboswitch RNAs: using RNA to sense cellular metabolism. Genes Dev 22:3383–3390
7. Wassarman KM, Storz G (2000) 6S RNA regulates *E. coli* RNA polymerase activity. Cell 101:613–623
8. Meister G, Tuschl T (2004) Mechanisms of gene silencing by double-stranded RNA. Nature 431:343–349
9. Horvath P, Barrangou R (2010) CRISPR/Cas, the immune system of bacteria and archaea. Science 327:167–170
10. Jones TA, Otto W, Marz M, Eddy SR, Stadler PF (2009) A survey of nematode SmY RNAs. RNA Biol 6:5–8
11. Altuvia S, Zhang A, Argaman L, Tiwari A, Storz G (1998) The *Escherichia coli* OxyS regulatory RNA represses FhlA translation by blocking ribosome binding. EMBO J 17:6069–6075
12. Delcher AL, Bratke KA, Powers EC, Salzberg SL (2007) Identifying bacterial genes and endosymbiont DNA with Glimmer. Bioinformatics 23:673–679
13. Burge C, Karlin S (1997) Prediction of complete gene structures in human genomic DNA. J Mol Biol 268:78–94
14. R. Guig (1998) Assembling genes from predicted exons in linear time with dynamic programming. J Comput Biol 5:681–702
15. Lomsadze A, Ter-Hovhannisyan V, Chernoff YO, Borodovsky M (2005) Gene identification in novel eukaryotic genomes by self-training algorithm. Nucleic Acids Res 33:6494–6506

16. Eddy SR (2011) HMMER—biosequence analysis using profile hidden Markov models. Accessed date April 29, 2011. [http://hmmer.janelia.org/]
17. Altschul SF, Madden TL, Schaffer AA, Zhang J, Zhang Z, Miller W, Lipman DJ (1997) Gapped BLAST and PSI-BLAST: A new generation of protein database search programs. Nucleic Acids Res 25:3389–3402
18. Finn RD, Mistry J, Tate J, Coggill P, Heger A, Pollington JE, Gavin OL, Ceric G, Forslund K, Holm L, Sonnhammer ELL, Eddy SR, Bateman A (2010) The Pfam protein families database. Nucleic Acids Res 38:D211–D222
19. Tatusov RL, Galperin MY, Natale DA, Koonin EV (2000) The COG database: a tool for genome-scale analysis of protein functions and evolution. Nucleic Acids Res 28: 33–36
20. Sayers EW, Barrett T, Benson DA, Bolton E, Bryant SH, Canese K, Chetvernin V, Church DM, DiCuccio M, Federhen S, Feolo M, Fingerman IM, Geer LY, Helmberg W, Kapustin Y, Landsman D, Lipman DJ, Lu Z, Madden TL, Madej T, Maglott DR, Marchler-Bauer A, Miller V, Mizrachi I, Ostell J, Panchenko A, Phan L, Pruitt KD, Schuler GD, Sequeira E, Sherry ST, Shumway M, Sirotkin K, Slotta D, Souvorov A, Starchenko G, Tatusova TA, Wagner L, Wang Y, Wilbur WJ, Yaschenko E, Ye J (2011) Database resources of the National Center for Biotechnology Information. Nucleic Acids Res 39:D38–D51
21. Argaman L, Hershberg R, Vogel J, Bejerano G, Wagner EG, Margalit H, Altuvia S (2001) Novel small RNA-encoding genes in the intergenic regions of *Escherichia coli*. Curr Biol 11:941–950
22. Babak T, Blencowe BJ, Hughes TR (2007) Considerations in the identification of functional RNA structural elements in genomic alignments. BMC Bioinformatics 8:33
23. Meyer IM (2007) A practical guide to the art of RNA gene prediction. Brief Bioinform 8:396–414
24. Griffiths-Jones S (2007) Annotating noncoding RNA genes. Annu Rev Genomics Hum Genet 8:279–298
25. Brocchieri L, Karlin S (2005) Protein length in eukaryotic and prokaryotic proteomes. Nucleic Acids Res 33:3390–3400
26. Pearson WR (1996) Effective protein sequence comparison. Methods Enzymol 266:227–258
27. Freyhult EK, Bollback JP, Gardner PP (2007) Exploring genomic dark matter: A critical assessment of the performance of homology search methods on noncoding RNA. Genome Res 17:117–125
28. Lowe TM, Eddy SR (1997) tRNAscan-SE: A program for improved detection of transfer RNA genes in genomic sequence. Nucleic Acids Res 25:955–964
29. Laslett D, Cänback B (2004) ARAGORN, a program to detect tRNA genes and tmRNA genes in nucleotide sequences. Nucleic Acids Res 32:11–16
30. Laslett D, Canback B, Andersson S (2002) BRUCE: a program for the detection of transfer-messenger RNA genes in nucleotide sequences. Nucleic Acids Res 30: 3449–3453
31. Laslett D, Cänback B (2008) ARWEN: a program to detect tRNA genes in metazoan mitochondrial nucleotide sequences. Bioinformatics 24:172–175
32. Lagesen K, Hallin P, Rødland EA, Staerfeldt HH, Rognes T, Ussery DW (2007) RNAmmer: consistent and rapid annotation of ribosomal RNA genes. Nucleic Acids Res 35: 3100–3108
33. Regalia M, Rosenblad MA, Samuelsson T (2002) Prediction of signal recognition particle RNA genes. Nucleic Acids Res 30: 3368–3377
34. Yusuf D, Marz M, Stadler PF, Hofacker IL (2010) Bcheck: a wrapper tool for detecting RNase P RNA genes. BMC Genomics 11:432
35. Eddy SR (2005) RNABOB—fast pattern searching for RNA secondary structures. [ftp://selab.janelia.org/pub/software/rnabob/]
36. Hertel J, Hofacker IL, Stadler PF (2008) SnoReport: computational identification of snoRNAs with unknown targets. Bioinformatics 24:158–164
37. Eddy SR, Durbin R (1994) RNA sequence analysis using covariance models. Nucleic Acids Res 22:2079–2088
38. Durbin R, Eddy SR, Krogh A, Mitchison GJ (1998) Biological Sequence Analysis: Probabilistic Models of Proteins and Nucleic Acids Cambridge University Press, Cambridge ISBN 0521629713
39. Gribskov M, McLachlan AD, Eisenberg D (1987) Profile analysis: Detection of distantly related proteins. Proc Natl Acad Sci USA 84:4355–4358
40. Leahy SC, Kelly WJ, Altermann E, Ronimus RS, Yeoman CJ, Pacheco DM, Li D, Kong Z, McTavish S, Sang C, Lambie SC, Janssen PH, Dey D, Attwood GT (2010) The genome sequence of the rumen methanogen *Methanobrevibacter ruminantium* reveals new possibilities for controlling ruminant methane emissions. PLoS One 5:e8926
41. Hartman AL, Norais C, Badger JH, Delmas S, Haldenby S, Madupu R, Robinson J, Khouri H, Ren Q, Lowe TM, Maupin-Furlow J, Pohlschroder M, Daniels C, Pfeiffer F, Allers T, Eisen JA (2010) The complete genome

sequence of *Haloferax volcanii* DS2, a model archaeon. PLoS One 5:e9605
42. Roh SW, Nam YD, Nam SH, Choi SH, Park HS, Bae JW (2010) Complete genome sequence of *Halalkalicoccus jeotgali* B3(T), an extremely halophilic archaeon. J Bacteriol 192:4528–4529
43. Mardanov AV, Svetlitchnyi VA, Beletsky AV, Prokofeva MI, Bonch-Osmolovskaya EA, Ravin NV, Skryabin KG (2010) The genome sequence of the crenarchaeon *Acidilobus saccharovorans* supports a new order, Acidilobales, and suggests an important ecological role in terrestrial acidic hot springs. Appl Environ Microbiol 76: 5652–5657
44. Liesegang H, Kaster AK, Wiezer A, Goenrich M, Wollherr A, Seedorf H, Gottschalk G, Thauer RK (2010) Complete genome sequence of *Methanothermobacter marburgensis*, a methanoarchaeon model organism. J Bacteriol 192:5850–5851
45. Petty NK, Bulgin R, Crepin VF, Cerdeo-Trraga AM, Schroeder GN, Quail MA, Lennard N, Corton C, Barron A, Clark L, Toribio AL, Parkhill J, Dougan G, Frankel G, Thomson NR (2010) The *Citrobacter rodentium* genome sequence reveals convergent evolution with human pathogenic *Escherichia coli*. J Bacteriol 192:525–538
46. Ventura M, Turroni F, Zomer A, Foroni E, Giubellini V, Bottacini F, Canchaya C, Claesson MJ, He F, Mantzourani M, Mulas L, Ferrarini A, Gao B, Delledonne M, Henrissat B, Coutinho P, Oggioni M, Gupta RS, Zhang Z, Beighton D, Fitzgerald GF, O'Toole PW, van Sinderen D (2009) The *Bifidobacterium dentium Bd1* genome sequence reflects its genetic adaptation to the human oral cavity. PLoS Genet 5:e1000785
47. Clum A, Tindall BJ, Sikorski J, Ivanova N, Mavrommatis K, Lucas S, Glavina T, Nolan M, Chen F, Tice H, Pitluck S, Cheng JF, Chertkov O, Brettin T, Han C, Detter JC, Kuske C, Bruce D, Goodwin L, Ovchinikova G, Pati A, Mikhailova N, Chen A, Palaniappan K, Land M, Hauser L, Chang YJ, Jeffries CD, Chain P, Rohde M, Gker M, Bristow J, Eisen JA, Markowitz V, Hugenholtz P, Kyrpides NC, Klenk HP, Lapidus A (2009) Complete genome sequence of *Pirellula staleyi* type strain (ATCC 27377). Stand Genomic Sci 1: 308–316
48. Gilmour MW, Graham M, Van Domselaar G, Tyler S, Kent H, Trout-Yakel KM, Larios O, Allen V, Lee B, Nadon C (2010) High-throughput genome sequencing of two *Listeria monocytogenes* clinical isolates during a large foodborne outbreak. BMC Genomics 11:120
49. Tripp HJ, Bench SR, Turk KA, Foster RA, Desany BA, Niazi F, Affourtit JP, Zehr JP (2010) Metabolic streamlining in an open-ocean nitrogen-fixing cyanobacterium. Nature 464:90–94
50. Jackson AP, Gamble JA, Yeomans T, Moran GP, Saunders D, Harris D, Aslett M, Barrell JF, Butler G, Citiulo F, Coleman DC, de Groot PW, Goodwin TJ, Quail MA, McQuillan J, Munro CA, Pain A, Poulter RT, Rajandream MA, Renauld H, Spiering MJ, Tivey A, Gow NA, Barrell B, Sullivan DJ, Berriman M (2009) Comparative genomics of the fungal pathogens *Candida dubliniensis* and *Candida albicans*. Genome Res 19(12):2231–2244. doi:10.1101/gr.097501.109
51. Peacock CS, Seeger K, Harris Dn, Murphy L, Ruiz JC, Quail MA, Peters N, Adlem E, Tivey A, Aslett M, Kerhornou A, Ivens A, Fraser A, Rajandream MA, Carver T, Norbertczak H, Chillingworth T, Hance Z, Jagels K, Moule S, Ormond D, Rutter S, Squares R, Whitehead S, Rabbinowitsch E, Arrowsmith C, White B, Thurston S, Bringaud F, Baldauf SL, Faulconbridge A, Jeffares D, Depledge DP, Oyola SO, Hilley JD, Brito LO, Tosi LR, Barrell B, Cruz AK, Mottram JC, Smith DF, Berriman M (2007) Comparative genomic analysis of three *Leishmania* species that cause diverse human disease. Nat Genet 39(7):839–847. doi:10.1038/ng2053
52. Theologis A, Ecker JR, Palm CJ, Federspiel NA, Kaul S, White O, Alonso J, Altafi H, Araujo R, Bowman CL, Brooks SY, Buehler E, Chan A, Chao Q, Chen H, Cheuk RF, Chin CW, Chung MMK, Conn L, Conway AB, Conway AR, Creasy TH, Dewar K, Dunn P, Etgu P, Feldblyum TV, Feng J, Fong B, Fujii CY, Gill JE, Goldsmith AD, Haas B, Hansen NF, Hughes B, Huizar L, Hunter JL, Jenkins J, Johnson-Hopson C, Khan S, Khaykin E, Kim CJ, Koo HL, Kremenetskaia I, Kurtz DB, Kwan A, Lam B, Langin-Hooper S, Lee A, Lee JM, Lenz CA, Li JH, Li Y, Lin X, Liu SX, Liu ZA, Luros JS, Maiti R, Marziali A, Militscher J, Miranda M, Nguyen M, Nierman WC, Osborne BI, Pai G, Peterson J, Pham PK, Rizzo M, Rooney T, Rowley D, Sakano H, Salzberg SL, Schwartz JR, Shinn P, Southwick AM, Sun H, Tallon LJ, Tambunga G, Toriumi MJ, Town CD, Utterback T, Van Aken S, Vaysberg M, Vysotskaia VS, Walker M, Wu D, Yu G, Fraser CM, Venter JC, Davis RW (2000) Sequence and analysis of chromosome 1 of the plant *Arabidopsis thaliana*. Nature 408: 816–820
53. Waterston RH, Lindblad-Toh K, Birney E, Rogers J, Abril JF, Agarwal P, Agarwala R, Ainscough R, Alexandersson M, An P, Antonarakis SE, Attwood J, Baertsch R, Bailey J, Barlow K,

Beck S, Berry E, Birren B, Bloom T, Bork P, Botcherby M, Bray N, Brent MR, Brown DG, Brown SD, Bult C, Burton J, Butler J, Campbell RD, Carninci P, Cawley S, Chiaromonte F, Chinwalla AT, Church DM, Clamp M, Clee C, Collins FS, Cook LL, Copley RR, Coulson A, Couronne O, Cuff J, Curwen V, Cutts T, Daly M, David R, Davies J, Delehaunty KD, Deri J, Dermitzakis ET, Dewey C, Dickens NJ, Diekhans M, Dodge S, Dubchak I, Dunn DM, Eddy SR, Elnitski L, Emes RD, Eswara P, Eyras E, Felsenfeld A, Fewell GA, Flicek P, Foley K, Frankel WN, Fulton LA, Fulton RS, Furey TS, Gage D, Gibbs RA, Glusman G, Gnerre S, Goldman N, Goodstadt L, Grafham D, Graves TA, Green ED, Gregory S, Guig R, Guyer M, Hardison RC, Haussler D, Hayashizaki Y, Hillier LW, Hinrichs A, Hlavina W, Holzer T, Hsu F, Hua A, Hubbard T, Hunt A, Jackson I, Jaffe DB, Johnson LS, Jones M, Jones TA, Joy A, Kamal M, Karlsson EK, Karolchik D, Kasprzyk A, Kawai J, Keibler E, Kells C, Kent WJ, Kirby A, Kolbe DL, Korf I, Kucherlapati RS, Kulbokas EJ, Kulp D, Landers T, Leger JP, Leonard S, Letunic I, Levine R, Li J, Li M, Lloyd C, Lucas S, Ma B, Maglott DR, Mardis ER, Matthews L, Mauceli E, Mayer JH, McCarthy M, McCombie WR, McLaren S, McLay K, McPherson JD, Meldrim J, Meredith B, Mesirov JP, Miller W, Miner TL, Mongin E, Montgomery KT, Morgan M, Mott R, Mullikin JC, Muzny DM, Nash WE, Nelson JO, Nhan MN, Nicol R, Ning Z, Nusbaum C, O'Connor MJ, Okazaki Y, Oliver K, Overton-Larty E, Pachter L, Parra G, Pepin KH, Peterson J, Pevzner P, Plumb R, Pohl CS, Poliakov A, Ponce TC, Ponting CP, Potter S, Quail M, Reymond A, Roe BA, Roskin KM, Rubin EM, Rust AG, Santos R, Sapojnikov V, Schultz B, Schultz J, Schwartz MS, Schwartz S, Scott C, Seaman S, Searle S, Sharpe T, Sheridan A, Shownkeen R, Sims S, Singer JB, Slater G, Smit A, Smith DR, Spencer B, Stabenau A, Stange-Thomann N, Sugnet C, Suyama M, Tesler G, Thompson J, Torrents D, Trevaskis E, Tromp J, Ucla C, Ureta-Vidal A, Vinson JP, Von Niederhausern AC, Wade CM, Wall M, Weber RJ, Weiss RB, Wendl MC, West AP, Wetterstrand K, Wheeler R, Whelan S, Wierzbowski J, Willey D, Williams S, Wilson RK, Winter E, Worley KC, Wyman D, Yang S, Yang SP, Zdobnov EM, Zody MC, Lander ES (2002) Initial sequencing and comparative analysis of the mouse genome. Nature 420:520–562

54. Eddy SR (2006) Computational analysis of RNAs. Cold Spring Harb Symp Quant Biol 71:117–128
55. Gardner PP, Daub J, Tate JG, Nawrocki EP, Kolbe DL, Lindgreen S, Wilkinson AC, Finn RD, Griffiths-Jones S, Eddy SR, Bateman A (2009) Rfam: Updates to the RNA families database. Nucleic Acids Res 37:D136–D140
56. Eddy SR (2002) A memory-efficient dynamic programming algorithm for optimal alignment of a sequence to an RNA secondary structure. BMC Bioinformatics 3:18
57. Nawrocki EP, Eddy SR (2007) Query-dependent banding (QDB) for faster RNA similarity searches. PLoS Comput Biol 3:e56
58. Nawrocki EP, Eddy SR (2012) The Infernal 1.1 user's guide. Accessed date July 1, 2012. [http://infernal.janelia.org/]
59. Eddy SR (2008) A probabilistic model of local sequence alignment that simplifies statistical significance estimation. PLoS Comput Biol 4:e1000069
60. Eddy SR (2011) Accelerated profile HMM searches. PLoS Comp Biol 7:e1002195
61. Eddy SR (1996) COVE—covariance models of RNA secondary structure. [ftp://selab.janelia.org/pub/software/cove/]
62. Brown MP (2000) Small subunit ribosomal RNA modeling using stochastic context-free grammars. Proc Int Conf Intell Syst Mol Biol 8:57–66
63. Nawrocki EP (2009) Structural RNA Homology Search and Alignment Using Covariance Models. PhD thesis, Washington University School of Medicine
64. Gardner PP, Daub J, Tate J, Moore BL, Osuch IH, Griffiths-Jones S, Finn RD, Nawrocki EP, Kolbe DL, Eddy SR, Bateman A (2011) Rfam: Wikipedia, clans and the "decimal" release. Nucleic Acids Res 39:D141–D145
65. Leinonen R, Akhtar R, Birney E, Bonfield J, Bower L, Corbett M, Cheng Y, Demiralp F, Faruque N, Goodgame N, Gibson R, Hoad G, Hunter C, Jang M, Leonard S, Lin Q, Lopez R, Maguire M, McWilliam H, Plaister S, Radhakrishnan R, Sobhany S, Slater G, Ten Hoopen P, Valentin F, Vaughan R, Zalunin V, Zerbino D, Cochrane G (2010) Improvements to services at the *E*uropean *N*ucleotide *A*rchive. Nucleic Acids Res 38:D39–D45
66. Weinberg Z, Barrick JE, Yao Z, Roth A, Kim JN, Gore J, Wang JX, Lee ER, Block KF, Sudarsan N, Neph S, Tompa M, Ruzzo WL, Breaker RR (2007) Identification of 22 candidate structured RNAs in bacteria using the CMfinder comparative genomics pipeline. Nucleic Acids Res 35: 4809–4819
67. Weinberg Z, Perreault J, Meyer MM, Breaker RR (2009) Exceptional structured noncoding RNAs revealed by bacterial metagenome analysis. Nature 462:656–659
68. Weinberg Z, Wang JX, Bogue J, Yang J, Corbino K, Moy RH, Breaker RR (2010) Comparative genomics reveals 104 candidate structured RNAs from bacteria, archaea,

and their metagenomes. Genome Biol 11:R31
69. Lowe TM, Eddy SR (1999) A computational screen for methylation guide snoRNAs in yeast. Science 283:1168–1171
70. Schattner P, Decatur WA, Davis CA, Fournier MJ, Lowe TM (2004) Genome-wide searching for pseudouridylation guide snoRNAs: analysis of the *Saccharomyces cerevisiae* genome. Nucleic Acids Res 32: 4281–4296
71. Bengert P, Dandekar T (2004) Riboswitch finder–a tool for identification of riboswitch RNAs. Nucleic Acids Res 32: W154–W159
72. Abreu-Goodger C, Merino E (2005) RibEx: a web server for locating riboswitches and other conserved bacterial regulatory elements. Nucleic Acids Res 33: W690–W692
73. Hertel J, Stadler PF (2006) Hairpins in a haystack: Recognizing microRNA precursors in comparative genomics data. Bioinformatics 22:e197–e202
74. Macke TJ, Ecker DJ, Gutell RR, Gautheret D, Case DA, Sampath R (2001) RNAMotif, an RNA secondary structure definition and search algorithm. NAR 29:4724–4735
75. Kazanov MD, Vitreschak AG, Gelfand MS (2007) Abundance and functional diversity of riboswitches in microbial communities. BMC Genomics 8:347
76. Barrangou R, Fremaux C, Deveau H, Richards M, Boyaval P, Moineau S, Romero DA, Horvath P (2007) CRISPR provides acquired resistance against viruses in prokaryotes. Science 315:1709–1712

Chapter 10

Class-Specific Prediction of ncRNAs

Peter F. Stadler

Abstract

Many RNA families, i.e., groups of homologous RNA genes, belong to RNA classes, such as tRNAs, snoRNAs, or microRNAs, that are characterized by common sequence motifs and/or common secondary structure features. The detection of new members of RNA classes, as well as the comprehensive annotation of genomes with members of RNA classes is a challenging task that goes beyond simple homology search. Computational methods addressing this problem typically use a three-tiered approach: In the first step an efficient and sensitive filter is employed. In the second step the candidate set is narrowed down using computationally expensive methods geared towards specificity. In the final step the hits are annotated with class-specific features and scored. Here we review the tools that are currently available for a diverse set of RNA classes.

Key words RNA gene finding, RNA secondary structure, Motifs, Descriptors, Filtering, Gene annotation

1 Introduction

Class-specific tools for RNA-annotation can be viewed as an intermediate between homology-based annotation and *de novo* prediction. While homology search is restricted, by definition, to the finding novel relatives of already known members of RNA families, we focus here on tools that recognize combinations of specific features that define functional classes of RNAs which, typically, comprise many different families that are unrelated by descent and hence very dissimilar at sequence level.

Class-specific RNA annotation thus can be seen as a generalization of the classical problem of finding protein-coding genes. Capitalizing on common features, such as open reading frames, splice site patterns, polyadenylation signals, start and stop codons, and specific distributions of codon usage, gene finding tools—recently reviewed, e.g., in [1]—are designed to work independently of any features of the protein that is encoded.

Jan Gorodkin and Walter L. Ruzzo (eds.), *RNA Sequence, Structure, and Function: Computational and Bioinformatic Methods*, Methods in Molecular Biology, vol. 1097, DOI 10.1007/978-1-62703-709-9_10,

Some, but not all, of the class-specific ncRNA finders make use of general-purpose descriptor-based search tools such as `RNAmotif` [2], `RNAbob` [3], `ERPIN` [4], or `HyPa` [5], as part of their work flow. A recent addition to the descriptor-based tools is `Structator` that implements extremely fast searches for sequence-structure patterns [6]. A few studies used generic descriptions for particular RNA classes, such as tRNAs [7] or the U5 snRNA [8] and searched genomic DNA with these descriptors only. In general, however, a more elaborate post-processing is required to increase specificity.

The typical architecture of a class-specific RNA gene finder is three-tiered, Table 1: The first step is the fast and efficient detection of candidate loci in the target genome. Typically, characteristic sequence motifs, specified, e.g., as a regular expression or a position-specific scoring matrix, or a local secondary structure motif is determined. In practice, this step is often implemented as a stepwise filtering procedure. At this point, the goal is sensitivity, aiming at an ideally loss-less filtering of the genomic input sequence. In the second step, the initial hit is reevaluated using a computationally much more expensive approach, i.e., a co-variance model. Now the focus is on specificity. The third and final step consists of a detailed annotation of the hit, identifying the exact location, in particular, of characteristic sequence and structure motifs and, if applicable, a sub-classification.

2 Methods

Despite their common general outline, there are many important differences between the available class-specific tools. The most commonly used ones are reviewed in some detail in this section. Since microRNAs are covered separately in Chapter 20, we do not discuss microRNA gene finders here.

2.1 tRNAs

Transfer RNAs (tRNAs) are among the best-conserved and evolutionary oldest sequences [9]. Present in all three domains of life they share (with few exceptions discussed below) a common, cloverleaf-shaped secondary structure. Eukaryotic tRNA genes contain furthermore internal promoter elements for polymerase III, the box A and box B elements, which account for a strong sequence signal common to all tRNAs. On the other hand, they have been a prolific source of repetitive elements (SINEs) with which they still share these promoter elements.

2.1.1 General Tools

`tRNAscan-SE` [10] is the most commonly used and most accurate software tool among the class-specific RNA gene predictors. It is, in fact, a composition of several algorithms: In the initial scanning phase it uses both (a variant of) `tRNAscan` [11] and Pavesi's

Table 1
Summary of the most frequently used tools for class-specific ncRNA detection

Tool	Initial search	Refinement	Annotation	Ref.
tRNAs				
`tRNAscan`	Secondary structure (`tRNAscan` [11]) and sequence motifs (`EufindtRNA` [12])	Covariance model (`cove` [13])	Anticodon, introns in anticodon loop	[10]
ARWEN	Hairpin structures of C-, D-, and T-arm	Complete structure of three types of mt-tRNAs	Anticodon	[15]
`MitFi`	`Infernal` for individual mt-tRNAs	Comparison of conflicting predictions	Secondary structure, anti-codons	[16]
`SPLITX`	BHB 5' and 3' sequences, secondary structure of combinations	`tRNAscan-SE`	BHB, splice-site, anticodon	[24]
rRNAs				
`RNAmmer`	`Spotter`: core HMM	Full HMM	rRNA type	[30]
tmRNA				
`ARAGORN`	T- and A-arm, resume motif	Secondary structure	tRNA motif, resume site, peptide tag	[15, 29]
SRP RNA				
`SRPscan`	Helix 8 with `rnabob`	Covariance models (`coves`)	Secondary structure	[35]
RNAse P RNA				
`bcheck`	Sequence-structure motifs with `rnabob`	Covariance models (`infernal`)	Secondary structure, pseudoknot	[41]
snoRNAs				
`fisher`	H and ACA boxes, interaction with possible target	Secondary structure folding	Sequence and structure patterns	[57, 58]
`snoGPS`	Sequence motifs, including target interaction	Stem loops, folding energy	Sequence and structure patterns	[59]
`snoSeeker`	Sequence and structure motifs	Conservation or target	Sequence, structure, conservation	[66]
`snoReport`	Sequences boxes	Constrained folding	Support vector machine	[67]

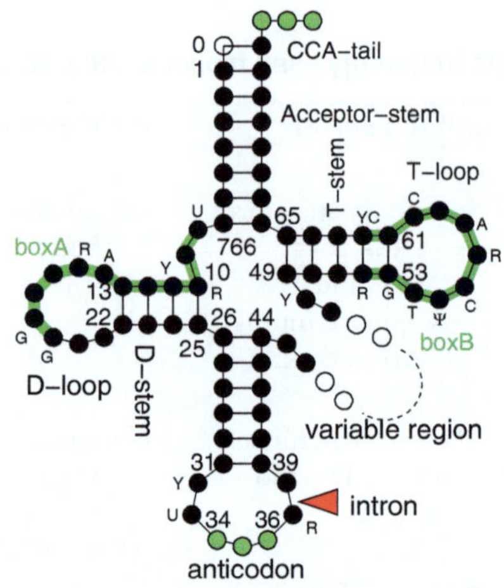

Fig. 1 Organization of a typical tRNA. Conserved nucleotides, the internal promoter elements, and the canonical intron position is indicated. The CCA tail is added post-transcriptionally in most organism but genomically encoded in some cases

algorithm `EufindtRNA` [12] to detect candidate tRNAs. While `tRNAscan` uses the base-pairing rules to identify cloverleaf-like structures, `EufindtRNA` is sequence-based, using position-specific weight matrices (PSSMs) to detect box A and box B and looks for the poly-T stretch of the terminator element, Fig. 1. In stage 2, the combined candidates, with short stretches of flanking sequence, are scanned with the local alignment algorithm `covels` [13] for the exact location of tRNA. A heuristics scoring is used to distinguish tRNAs from likely pseudogenes. In the final state, the detected tRNAs are annotated. Their secondary structure is determined with a global structure alignment to a covariance model (`coves` [13]), the anti-codon is detected, and tRNA-introns are identified as stretches of five or more consecutive non-consensus nucleotides within anticodon loop. Throughout the procedure, special rules are applied to account for the aberrant structure of the selenocysteine tRNA.

A sensitivity of well above 99% and a false discovery rate better than 1 in 10^{10} nt has been reported for `tRNAscan-SE` [10]. `ARAGORN` [14] achieves an increased sensitivity on prokaryotic sequences. Its heuristic first attempts to find a potential T-arm of tRNA, then it searches the upstream region for a box A motif and a D-arm structure. If successful, the sequences flanking the intermediate candidate are searched for connecting base pairs forming the A-stem, and the C-arm is constructed between the D- and A-arms.

2.1.2 Metazoan mt-tRNAs

Animal mitochondria have extremely reduced genomes typically encoding 22 tRNAs (one each for 18 of the 20 canonical amino acid, and two tRNAs with distinct anticodons for both serine and leucine). Many of these tRNAs have unusual structures, e.g. lacking the T- or the D-loop. Since animal mitogenomes are usually shorter than 20 kb, specificity and speed is of less concern than sensitivity.

`ARWEN` [15] employs a search heuristic based on that of the earlier `ARAGORN` tool [14]. It first searches for a possible C-arm characterized as hairpin with a 5–6 bp stem and a 6–8 bp loop and then attempts to extend this partial structure with possible D-arm (upstream) and a T-arm (downstream) and then detects possible base pairs of the A-stem. In the second step, `ARWEN` combines these candidate elements in a complete tRNA with a least three of the four stems and scores the result for each of the three template types (D-replacement loop, TV-replacement loop, and standard cloverleaf). Secondary structures and anti-codons are reported. `ARWEN` is optimized for sensitivity and has a substantial false discovery rate. By construction, it also misses mt-tRNAs with even more deviant structures.

`mitfi` [16] uses separate covariance models for all 22 animals tRNAs and, because of the small size of the mitogenomes, uses `infernal` already in the first pass. The tool then compares the scoring over overlapping hits and also annotates both multiple copies including degraded pseudogenes. This makes it particularly suitable as a component for a comprehensive annotation pipeline for animal mitogenomes [17].

2.1.3 Split tRNAs

Introns are a common phenomenon in tRNAs. In eubacteria, tRNA introns are self-splicing group I introns [18]. Under most circumstances, these tRNAs are not recognizable by the presently available software tools. In eukarya, tRNA introns are small and (almost) invariably interrupt the anticodon loop 1 base downstream of the anticodon, i.e., between the canonical nucleotide positions 37 and 38. The tRNA introns of archaea often reside in the same position. However, there are frequent exceptions. While introns in the canonical position are efficiently dealt with by both `tRNAscan-SE` and `ARAGORN`, specialized tools are required to recognize other cases.

The parasite *Nanoarchaeum equitans*, furthermore, produces functional tRNA from separate genes, one encoding the 5'-half and the other the 3'-half [19]. Such split tRNAs, and even more complex systems of tRNA fragments that are individually transcribed and further trans-spliced to generate multiple tRNAs [20] have been described in several archaean genomes.

In Eukarya, tRNAs are found at unusual positions in the genomes of nucleomorphs, i.e., the remnant nuclei of eukaryotic, secondary endosymbionts of cryptophytes and chlorarachniophytes. Permuted tRNAs are also found in nucleomorphs as well as closely related free-living algae. Here, the 5' and 3' halves of the tRNA-gene are positioned in reverse on the genome [21].

`Split-tRNA-Search` [22] assumes that tRNAs are only split at the canonical position in the anti-codon loop. It searches for the sequence patterns characteristic of the 3' and 5' halves of the tRNAs and then checks for the presence of possible base pairs that could form the A- and C-stems. It also uses `tRNAscan-SE` to ensure specificity.

In contrast to the situation in eukarya, the majority of archaeal exon–intron boundaries form a folded RNA structure, termed the bulge-helix-bulge (BHB) motif, consisting of 2- or 3-nucleotide bulges separated by a 4-bp helix, see, e.g., [23]. Both `SPLITX` [24] and its predecessor `SPLITS` [25] start by predicting possible BHB sites using PWMs modeling the 5' and 3' sequences of the BHB sites [26]. Combinations of individual hits are then tested for compliance with the minimal BHB secondary structure model and a sufficient folding energy (using `RNAeval`) to determine candidates for complete tRNA genes. These are then passed to `tRNAscan-SE` for evaluation and annotation.

2.2 tmRNA

In bacteria, trans-translation recycles ribosomes entrapped at the 3' ends of mRNAs that lack a natural stop codon. The main player in this reaction is tmRNA (SsrA-RNA), a bi-functional RNA that acts as both a tRNA and an mRNA: The alanine-charged tmRNA enters at the ribosomal A-site, translation shifts to the resume codon in tmRNA, and continues to a stop codon at the end of the small reading frame. This adds a short peptide tag to the incomplete nacsent polypeptide, which is recognized as a degradation signal, and releases the ribosome [27]. A recent review [28] summarizes the variation of tmRNA genes. Besides the circular permutation, which produces two-piece tmRNAs, functional tmRNA genes may also be interrupted by mobile elements such as group I introns, genomic islands, and palindromic elements. Endosymbiont tmRNAs, furthermore, tend to lose secondary structure and length in the mRNA-like region.

`BRUCE` [29] first searches for the T-arm of the tRNA-Ala-like domain. Then sequence intervals of about 500 nt up- and downstream of the initial hit are searched for a possible A-stem. If a 5' A-stem is found downstream of the T-stem, this is taken as an indication of a permuted tmRNA gene. Next a consensus motif is searched for in a location that depends on whether a canonical or a permutated tmRNA is assumed. For each tmRNA candidate, the secondary structure of the tRNA-like domain, the position of the resume consensus motif, and the amino acid sequence of the proteolysis tag are predicted. `ARAGORN` [15], which also predicts tRNAs, adds two additional structural requirements: a hairpin at the 3'end of the tag peptide sequence and a hairpin structure upstream of the tag peptide, which may be part of a pseudo-knot. tmRNA genes with large interruptions are not predicted.

2.3 Ribosomal RNAs

RNAmmer [30] uses hidden Markov models trained on data from the 5S ribosomal RNA database and 16S/18S and 23S/28S rRNA alignments from the European ribosomal RNA database project. The 75 most conserved consecutive columns in these alignment were used to train a small model called spotter that is used a fast, nearly loss-less, preprocessor. HMMer [31] was used to train the models and is used as search engine. A similar tool, Meta_RNA, is geared in particular towards metagenomics datasets [32].

2.4 RNAse P, RNAse MRP, and SRP RNAs

2.4.1 Signal Recognition Particle RNA

The signal recognition particle (SRP) is involved in the targeting of protein to cellular membranes. Its RNA component is present in all domains of life. Substantial variations in size and number of RNA secondary structure elements are observed between different phylogenetic groups. A recent review [33] classified SRP RNAs into seven structural groups: Archaea, Bacteria with a small (4.5S) SRP RNA, Bacteria with a large (6S) SRP RNA, Fungi (Ascomycota), Metazoa group, Protozoa group, and Plants, see also [34].

SRPscan [35] starts by searching for the ubiquitously conserved helix 8 motif using a slightly modified version of RNAbob as search engine. Optionally, more specific descriptor can be used for fungal, plant, and other Eukaryotic structures. Then covariance models (with or without the Alu domain) mapped to the candidate sequence using coves [13].

2.4.2 RNAse P and MRP RNA

RNase P RNA is a ribozyme involved in the processing of pre-tRNAs. It can be found in almost all organisms and is present in most organellar genomes, although it is absent in, e.g., animal mitochondria [36]. So far, there is compelling evidence for the loss of RNase P RNA only in a single organism, the archaeon *Nanoarchaeum equitans* [37]. It is not unlikely, however, that plants, red algae, and heterokonts [38] also do not have an RNAse P RNA. The recent discovery of RNAse P RNA in *Pyrobaculum* [39], on the other hand, suggests that the structural diversity of this class of ncRNAs is larger than anticipated.

Most RNAse P RNAs have a length between 250 and 550 nt, comprising two structural domains with up to 19 conserved stems. There are five regions with strong sequence conservation, designated CR-I to CR-V, including the P4 pseudoknot composed by CR-I and CR-V [40]. A simple pipeline combining a search for these motifs and the subsequent evaluation of the candidates with a dedicated RNAmotif pattern is described in [40]. It is not available as a public tool, however.

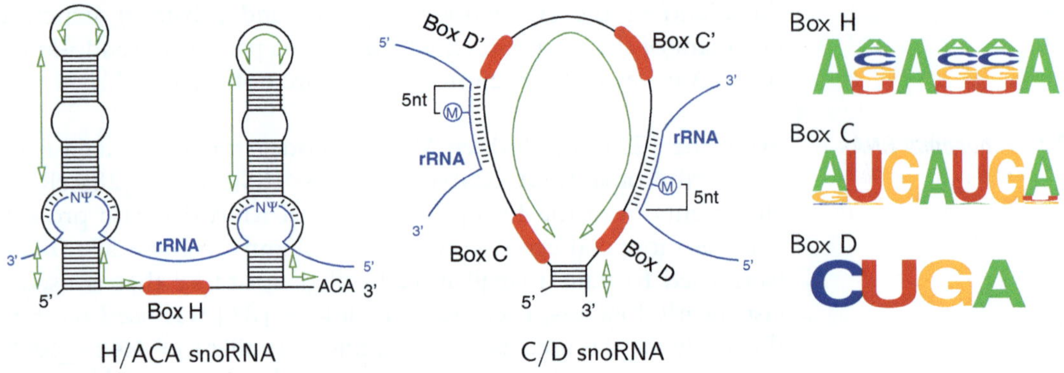

Fig. 2 Structural features and conserved sequence boxes of the two snoRNA classes

The RNAse P RNA gene finding tool `bcheck` [41] utilizes specialized descriptors for the best-conserved parts of the seven subfamilies identified in the literature: archaeaA, archaeaM, bacteriaA, bacteriaB, eukaryotic nuclear, and two fungal subtypes [42, 43]. The initial descriptor-based search is performed with `rnabob`. For the subfamilies arcA, bacA, and nuc two variants with different sensitivity are used. If the more selective pattern fails, a more promiscuous version is used. Initial candidates are then evaluated with `infernal` [44] against covariance models for the seven subtypes. First the (short) candidate sequences are matched in local alignments. Positive results are then extended and aligned globally.

RNase MRP closely resembles RNase P but recognizes distinct substrates including pre-rRNA and mRNA [45]. RNase MRP RNA and telomerase RNA, furthermore, are part of a mammalian RNA-dependent RNA polymerase [46]. Despite the structural similarities of RNAse P and RNAse MRP [34, 47, 48] there is at present no dedicated search tool for RNAse MRP RNAs.

2.5 snoRNAs

Small nucleolar RNAs (snoRNAs) are an abundant class of non-coding RNAs with a wide variety of cellular functions including chemical modification of RNA, telomere maintenance, pre-rRNA processing, and regulatory activities in alternative splicing. There are two classes of snoRNAs distinguished by both the secondary structure and the presence of characteristic sequence motifs. The common structural features of each class, Fig. 2, can be attributed to the incorporation of snoRNAs into ribonucleoparticles (snoRNP) that share class-specific protein components [49]. The main role of box C/D snoRNAs is to determine the targets for 2'-O-ribose methylation. Box H/ACA snoRNAs, on the other hand, facilitate the conversion of Uracil to pseudouracil (Ψ) in a specific sequence context [50–52]. Most snoRNAs target specific nucleotides in ribosomal RNAs. A subgroup characterized by an additional conserved sequence box localizes to the Cajal body [53].

The Cajal body-specific small nuclear RNAs (scaRNAs) function as guide RNAs just like ordinary snoRNAs. They target mostly pol-II transcribed spliceosomal RNAs [54]. In archaea, homologous classes of snoRNA-like small RNAs are involved in the biogenesis of rRNAs and tRNAs [55].

The interaction of the snoRNA with its target(s) is mediated by base pairing of a specific region of the snoRNAs with almost perfectly complementary sequence motifs in the target RNA. For box C/D snoRNAs this binding region has a length of 7–20 nts and ends exactly five nucleotides upstream of the 5′-end of the D- and/or D′-box. The duplex may contain a few mismatches, while bulges are forbidden. In contrast, H/ACA snoRNAs position the target U by means of two specific interactions of the flanking target RNA sequence with the complementary sequence of the recognition loop of the snoRNA [56], *see* Fig. 2.

2.5.1 SnoRNA Gene Finders

The first generation of approaches to snoRNA gene finding combines secondary structure prediction with the recognition of the characteristic sequence boxes and putative targets. Software tools of this type, therefore, require the rRNA and possibly snRNA sequences as additional input. The tools `fisher` and `snoGPS` detect box H/ACA snoRNAs. `SnoScan` and `SNO.pl` focus on box C/D snoRNAs.

The program `fisher` [57, 58] first searches for the H-box pattern and then identifies the interaction regions based on the sequence of a candidate target sequences. Then an ACA box (with the pattern `AHA`) is identified. The secondary structure of the region between the H and ACA boxes computed and only acceptable folds are retained. A hairpin structure to the left of the H-box may also be required. The retained candidates are then scored based on primary and secondary structure.

The `snoGPS` pipeline [59, 60] separately tries to detect the two stem-loop substructures using a series of tests including a search for the sequence boxes and the evaluation of distances between various elements and an assessment of the folding energy. Scores are determined for individual features. These scores were trained from a set of known box H/ACA snoRNAs.

`SnoScan` [61] searches for the terminal stem, the conserved C and D boxes, the optional C' and D' boxed and complementarity to a putative target. A modified version of `SnoScan` that allows more user-defined modification of patterns was applied to the genome of *Giardia intestinalis* [62]. `SNO.pl` [63] searches for conserved box C/D snoRNA motifs in a database of orthologous introns.

An increasing number of orphan snoRNAs [64, 65] appear not to be involved in modification of rRNAs and snRNAs but interact with mRNAs. The complete transcriptome, however, is too large to provide informative constraints on the interaction region. This has prompted the development of a second generation of snoRNA search tools that avoid the use of target candidates.

`snoSeeker` [66] consists of two separate programs `CDseeker` and `HACAseeker`. Both are conceptually similar to `snoScan` and `snoGPS`, respectively. As an alternative to complementarity to the target sequence conservation can be used as additional source of information. The tool is in particular geared towards the analysis of next-generation sequencing data.

`snoReport` [67] uses pairs of the conserved sequence boxes with a maximum distance (200 nt for the C and D boxes and 120 nt for H and ACA boxes) to define the initial candidates. A constrained folding algorithm (`RNAfold -C` [68]) is then used to check for compatible secondary structures. A feature vector comprising about a dozen entries is extracted from all candidates that can form the prototypical structures. A support vector machine trained on known snoRNAs is used to determine a call probability for box C/D and box H/ACA snoRNAs.

All of the tools described in this subsection use some form of filtering by structure and/or size. They will miss, therefore, atypical snoRNA and scaRNAs such as the U3 RNA [69] and chimeric snoRNAs such as U87 and U88 [54].

2.5.2 Target Prediction for snoRNAs

The increasing number of orphan snoRNAs has prompted the development of dedicated target search algorithms. For box C/D snoRNAs the task is at least conceptually rather straightforward since the nucleotide targeted for methylation is always located at 5 nt upstream of the box D motif in a paired region. `snoTarget` [70] thus derives the binding region from the snoRNA and then employs pattern matching to find candidates in the input sequence. These are then ranked by the co-folding energy of snoRNA and target as computed by `RNAcofold` [71] . `PLEXY` [72] uses the much faster alignment-like `RNAplex` algorithm [73], which uses a somewhat simplified energy model. Then a set of rules characterizing the box C/D snoRNA–target interaction structure [74] is used as a filter. The duplex energies are used as a ranking criterion. `RNAsnoop`, finally, is an efficient and reliable tool for predicting the much more complex interactions of box H/ACA snoRNAs with their targets [75].

2.6 Miscellaneous Tools

2.6.1 SECIS Elements

SECIS elements are structured RNA elements located in the 3' UTR in both eukaryotic and archaeal mRNAs that direct the recoding of the UGA codon to selenocystein. `SECISearch` [76] uses both primary sequence patterns, using `PatScan`, and free energy criteria for predicted RNA secondary structures, using `RNAfold`, to detect SECIS elements.

2.6.2 UTRscan

`UTRscan` is a general purpose pattern matcher for UTR motifs defined in the UTRSite Database, a manually curated database of search patterns collected from the published literature and revised by experts for particular motifs [77].

2.6.3 Bacterial Terminator Hairpins

`TransTermHP` [78] searches for Rho-independent transcription terminators which consist of a short, low-energy hairpin, flanked by a downstream oligo-T stretch and an upstream oligo-A stretch. The algorithm uses a window of length 6 containing three Ts as anchor for the 5'end of the T-tail and employs a fast dynamic programming algorithm to find a hairpin structure upstream of this position, as well as heuristic scoring functions for the A- and T-tails. Candidates are then re-scored depending on the genomic GC content.

2.6.4 Group I Introns

The class-specific search tool `CITRON` [79] is not in common use any more. The Group I Intron Sequence and Structure Database GISSD [80], like `Rfam`, uses the general search tool `infernal` instead. An alternative approach is taken by `G1fold` [81], which is based on the so-called thermodynamic matchers. TDMs are restricted folding algorithms that force the input sequence to conform with a prescribed fold. The difference between constrained and unstrained folding energy is then used to evaluate candidates [82].

3 Notes

Many of the tools discussed in the previous section have become available also as web services. These are listed in Table 2.

4 Perspectives

Despite the large collection of class-specific tools for RNA bioinformatics, there are several important classes of RNAs for which no dedicated tools exist. Maybe the two most important cases in eukaryotes are the RNase MRP RNA, which is a distant relative of RNAse P [38, 48], and telomerase RNA. The latter is known only in vertebrates, in a few fungi, in ciliates, in *Arabidopsis*, and possibly in plasmodium [83]. Its extreme size variation, the absence of highly conserved regions, and a very fast evolution at sequence level place it among the hardest cases for homology search [84].

An interesting alternative to the tools described above was explored in a series of homology search studies for 7SK RNA,

Table 2
Web resources for class-specific RNA gene finding

Tool	URL	Ref.
ARAGORN	http://130.235.46.10/ARAGORN/	[14]
ARWEN	http://130.235.46.10/ARWEN/	[15]
bcheck	http://rna.tbi.univie.ac.at/cgi-bin/bcheck/	[41]
BRUCE	http://130.235.46.10/BRUCE/	[29]
MitFi	http://mitos.bioinf.uni-leipzig.de	[16]
RNAmmer	http://www.cbs.dtu.dk/services/RNAmmer/	[30]
SECISearch	http://genome.unl.edu/SECISearch.html	[76]
snoscan	http://lowelab.ucsc.edu/snoscan/	[61, 89]
snoGPS	http://lowelab.ucsc.edu/snoGPS/	[59, 89]
snoseeker	http://genelab.zsu.edu.cn/snoseeker/	[66]
SRPscan	http://bio.lundberg.gu.se/srpscan/	[35]
TransTermHP	http://transterm.cbcb.umd.edu/	[78]
tRNAscan-SE	http://lowelab.ucsc.edu/tRNAscan-SE/	[10, 89]
UTRscan	http://itbtools.ba.itb.cnr.it/	[77]

Y RNAs, and vault RNAs [85–87]. These less well-studied classes of vertebrate RNAs are transcribed by RNA polymerase III and share rather conserved promoter elements. These promoter-based sequence patterns were used as a filtering step. Conceivably, it is viable to construct specific search tools of sufficient specificity for the various types of pol-III promoters [88].

Outside the eukarya, on the other hand, class-specific RNA finders are almost nonexistent, with the notable exception of `bcheck` for RNAse P RNA and `tRNAscan-SE` for tRNAs. The cause for this imbalance is probably that vast majority of bacterial RNAs has a phylogenetically rather limited distribution, and a least in part a lack of systematic descriptions of many of these RNA families. We suspect that significant progress is possible in particular for the more widespread families. This would be of particular interest, e.g., for the 6S RNA, which is thought to be ubiquitous among eubacteria, but has remained undetected in many phyla.

References

1. Solovyev V, Kosarev P, Seledsov I, Vorobyev D (2006) Automatic annotation of eukaryotic genes, pseudogenes and promoters. Genome Biol 7 Suppl 1:S10.1–12
2. Macke TJ, Ecker DJ, Gutell RR, Gautheret D, Case DA, Sampath R (2001) RNAMotif, an RNA secondary structure definition and search algorithm. Nucleic Acids Res 29(22):4724–4735
3. Eddy S (2005) `rnabob`. ftp://selab.janelia.org/pub/software/rnabob/. Accessed 9 Nov 2013
4. Gautheret D, Lambert A (2001) Direct RNA motif definition and identification from multiple sequence alignments using secondary structure profiles. J Mol Biol 313: 1003–1011
5. Gräf S, Strothmann D, Kurtz S, Steger G (2001) HyPaLib: a database of RNAs and RNA structural elements defined by hybrid patterns. Nucleic Acids Res 29:196–198
6. Meyer F, Kurtz S, Backofen R, Will S, Beckstette M (2011) `Structator`: fast

index-based search for RNA sequence-structure patterns. BMC Bioinformatics 12:214

7. Tsui V, Macke T, Case DA (2003) A novel method for finding tRNA genes. RNA 9:507–517
8. Collins LJ, Macke TJ, Penny D (2004) Searching for ncRNAs in eukaryotic genomes: maximizing biological input with `RNAmotif`. J Integr Bioinform 1:6
9. Eigen M, Lindemann BF, Tietze M, Winkler-Oswatitsch R, Dress AWM, von Haeseler A (1989) How old is the genetic code? Statistical geometry of tRNA provides an answer. Science 244:673–679
10. Lowe TM, Eddy SR (1997) tRNAscan-SE: a program for improved detection of transfer RNA genes in genomic sequence. Nucl. Acids Res. 25:955–964
11. Fichant GA, Burks C (1991) Identifying potential tRNA genes in genomic DNA sequences. J Mol Biol 220:659–671
12. Pavesi A, Conterio F, Bolchi A, Dieci G, Ottonello S (1994) Identification of new eukaryotic tRNA genes in genomic DNA databases by a multistep weight matrix analysis of transcriptional control regions. Nucleic Acids Res 22:1247–1256
13. Eddy SR, Durbin R (1994) RNA sequence analysis using covariance models. Nucleic Acids Res 22:2079–2088
14. Laslett D, Canbäck B (2004) ARAGORN, a program to detect tRNA genes and tmRNA genes in nucleotide sequences. Nucleic Acids Res 32:11–16
15. Laslett D, Canbäck B (2008) ARWEN: a program to detect tRNA genes in metazoan mitochondrial nucleotide sequences. Bioinformatics 24:172–175
16. Jühling F, Pütz J, Bernt M, Donath A, Middendorf M, Florentz C, Stadler PF (2012) Improved systematic tRNA gene annotation allows new insights into the evolution of mitochondrial tRNA structures and into the mechanisms of mitochondrial genome rearrangements. Nucleic Acids Res 40:2833–2845
17. Donath A, Bernt M, Jühling F, Externbrink F, Florentz C, Fritzsch G, Pütz J, Middendorf M, Stadler PF (2013) MITOS: Standardizing and improving metazoan mitochondrial genome annotation. Mol Phylog Evol 69:313–319
18. Haugen P, Simon DM, Bhattacharya D (2005) The natural history of group I introns. Trends Genet 21:111–119
19. Randau L, Münch R, Hohn MJ, Jahn D, Söll D (2005) Nanoarchaeum equitans creates functional tRNAs from separate genes for their 5'- and 3'-halves. Nature 433:537–541
20. Fujishima K, Sugahara J, Kikuta K, Hirano R, Sato A, Tomita M, Kanai A (2009) Tri-split tRNA is a transfer RNA made from 3 transcripts that provides insight into the evolution of fragmented tRNAs in archaea. Proc Natl Acad Sci USA 106:2683–2687
21. Maruyama S, Sugahara J, Kanai A, Nozaki H (2010) Permuted tRNA genes in the nuclear and nucleomorph genomes of photosynthetic eukaryotes. Mol Biol Evol 27:1070–1076
22. Muench R, Randau L (2003) `Split-tRNA-Search`. http://www.prodoric.de/sts/. Accessed 9 Nov 2013
23. Kim YK, Mizutani K, Rhee KH, Nam KH, Lee WH, Lee EH, Kim EE, Park SY, Hwang KY (2007) Structural and mutational analysis of tRNA intron-splicing endonuclease from *Thermoplasma acidophilum* DSM 1728: catalytic mechanism of tRNA intron-splicing endonucleases. J Bacteriol 189:8339–8346
24. Sugahara J, Yachie N, Arakawa K, Tomita M (2007) *In silico* screening of archaeal tRNA-encoding genes having multiple introns with bulge-helix-bulge splicing motifs. RNA 13:671–681
25. Sugahara J, Yachie N, Sekine Y, Soma A, Matsui M, Tomita M, Kanai A (2006) SPLITS: a new program for predicting split and intron-containing tRNA genes at the genome level. In Silico Biol 6:411–418
26. Marck C, Grosjean H (2003) tRNomics: analysis of tRNA genes from 50 genomes of Eukarya, Archaea, and Bacteria reveals anticodon-sparing strategies and domain-specific features. RNA 8:1189–1232
27. Keiler KC (2008) Biology of trans-translation. Annu Rev Microbiol 62:133–151
28. Mao C, Bhardwaj K, Sharkady SM, Fish RI, Driscoll T, Wower J, Zwieb C, Sobral BW, Williams KP (2009) Variations on the tmRNA gene. RNA Biol 6:355–361
29. Laslett D, Canbäck B, Andersson S (2002) BRUCE: a program for the detection of transfer-messenger RNA genes in nucleotide sequences. Nucleic Acids Res 30:3449–3453
30. Lagesen K, Hallin P, Rødland EA, Stærfeldt H-H, Rognes T, Ussery DW (2007) RNAmmer: consistent and rapid annotation of ribosomal RNA genes. Nucleic Acids Res 35:3100–3108
31. Eddy S (1998) Profile hidden markov models. Bioinformatics 14:755–763
32. Huang Y, Gilna P, Li W (2009) Identification of ribosomal RNA genes in metagenomic fragments. Bioinformatics 25:1338–1340
33. Rosenblad MA, Larsen N, Samuelsson T, Zwieb C (2009) Kinship in the SRP RNA family. RNA Biol 6:508–516
34. Donath A, Findeiß S, Hertel J, Marz M, Otto W, Schulz C, Stadler PF, Wirth S (2010) Non-coding RNAs. InCaetano-Anolles G (ed)

Evolutionary genomics and systems biology. Wiley-Blackwell, Hoboken, NJ, pp 251–293
35. Regalia M, Rosenblad MA, Samuelsson T (2002) Prediction of signal recognition particle RNA genes. Nucleic Acids Res 30:3368–3377
36. Walker SC, Engelke DR (2008) A protein-only RNase P in human mitochondria. Cell 135:412–414
37. Randau L, Schröder I, Söll D (2008) Life without RNase P. Nature 453:120–123
38. Piccinelli P, Rosenblad MA, Samuelsson T (2005) Identification and analysis of ribonuclease P and MRP RNA in a broad range of eukaryotes. Nucleic Acids Res 33:4485–4495
39. Lai LB, Chan PP, Cozen AE, Bernick DL, Brown JW, Gopalan V, Lowe TM (2010) Discovery of a minimal form of RNase P in *Pyrobaculum*. Proc Natl Acad Sci USA 107:22493–22498
40. Li Y, Altman S (2004) In search of RNase P RNA from microbial genomes. RNA 10:1533–1540
41. Yusuf D, Marz M, Stadler PF, Hofacker IL (2010) `Bcheck`: a wrapper tool for detecting RNase P RNA genes. BMC Bioinformatics 11:432
42. Brown JW (1999) The Ribonuclease P Database. Nucleic Acids Res 27:314
43. Frank DN, Adamidi C, Ehringer MA, Pitulle C, Pace NR (2000) Phylogenetic-comparative analysis of the eukaryal ribonuclease P RNA. RNA 6:1895–1904
44. `Infernal 1.0`: Inference of RNA Alignments (2009) Nawrocki, e. p. and kolbe, d. l. and eddy, s. r. Bioinformatics 25:1335–1337
45. Esakova O, Krasilnikov AS (2010) Of proteins and RNA: the RNase P/MRP family. RNA 16:1725–1747
46. Maida Y, Yasukawa M, Furuuchi M, Lassmann T, Possemato R, Okamoto N, Kasim V, Hayashizaki Y, Hahn WC, Masutomi K (2009) An RNA-dependent RNA polymerase formed by TERT and the RMRP RNA. Nature 461:230–235
47. Dávila López M, Rosenblad MA, Samuelsson T (2009) Conserved and variable domains of RNase MRP RNA. RNA Biol 6:208–220
48. Woodhams MD, Stadler PF, Penny D, Collins LJ (2007) RNAse MRP and the RNA processing cascade in the eukaryotic ancestor. BMC Evol Biol 7:S13
49. Reichow SL, Hamma T, Ferré-D'Amaré AR, Varani G (2007) The structure and function of small nucleolar ribonucleoproteins. Nucleic Acids Res 35:1452–1464
50. Samarsky DA, Fournier MJ, Singer RH, Bertrand E (1998) The snoRNA box C/D motif directs nucleolar targeting and also couples snoRNA synthesis and localization. EMBO 17:3747–3757
51. Bachellerie JP, Cavaillé J, Hüttenhofer A (2002) The expanding snoRNA world. Biochimie 84:775–790
52. Terns MP, Terns RM (2002) Small nucleolar RNAs: Versatile *trans*-acting molecules of ancient evolutionary origin. Gene Expr 10:17–39
53. Gall JG (2003) The centennial of the Cajal body. Nat Rev Mol Cell Biol 4:975–980
54. Darzacq X, Jády BE, Verheggen C, Kiss AM, Bertrand E, Kiss T (2002) Cajal body-specific small nuclear RNAs: a novel class of 2'-O-methylation and pseudouridylation guide RNAs. EMBO J 21:2746–2756
55. Ziesche SM, Omer AD, Dennis PP (2004) RNA-guided nucleotide modification of ribosomal and non-ribosomal RNAs in Archaea. Mol Microbiol 54:980–993
56. Ni J, Tien AL, Fournier MJ (1997) Small nucleolar RNAs direct site-specific synthesis of pseudouridine in ribosomal RNA. Cell 89:565–573
57. Edvardsson S, Gardner PP, Poole AM, Hendy MD, Penny D, Moulton V (2002) A search for H/ACA snoRNAs in yeast using MFE secondary structure prediction. Bioinformatics 19:865–873
58. Freyhult E, Edvardsson S, Tamas I, Moulton V, Poole AM (2008) `Fisher`: a program for the detection of H/ACA snoRNAs using MFE secondary structure prediction and comparative genomics—assessment and update. BMC Res Notes 1:49
59. Schattner P, Decatur WA, Davis CA, Ares M, Fournier MJ, Lowe TM (2004) Genome-wide searching for pseudouridylation guide snoRNAs: analysis of the *Saccharomyces cerevisiae* genome. Nucleic Acids Res 32:4281–4296
60. Schattner P, Barberan-Soler S, Lowe TM (2006) A computational screen for mammalian pseudouridylation guide H/ACA RNAs. Bioinformatics 12:15–25
61. Lowe TM, Eddy SR (1999) A computational screen for methylation guide snoRNAs in yeast. Science 19:1168–1171
62. Chen XS, Rozhdestvensky TS, Collins LJ, Schmitz J, Penny D (2007) Combined experimental and computational approach to identify non-protein-coding RNAs in the deep-branching eukaryote *Giardia intestinalis.* Nucleic Acids Res 35:4619–4628
63. Fedorov A, Stombaugh J, Harr MW, Yu S, Nasalean L, Shepelev V (2005) Computer identification of snoRNA genes using a Mammalian Orthologous Intron Database. Nucleic Acids Res 33:4578–4583
64. Hüttenhofer A, Kiefmann M, Meier-Ewert S, O'Brien J, Lehrach H, Bachellerie JP, Brosius J (2001) RNomics: an experimental approach that identifies 201 candidates for novel, small,

non-messenger RNAs in mouse. EMBO J 20:2943–2953
65. Rogelj B (2006) Brain-specific small nucleolar RNAs. J Mol Neurosci 28:103–109
66. Yang JH, Zhang XC, Huang ZP, Zhou H, Huang MB, Zhang S, Chen YQ, Qu LH (2006) snoSeeker: an advanced computational package for screening of guide and orphan snoRNA genes in the human genome. Nucleic Acids Res 34:5112–5123
67. Hertel J, Hofacker IL, Stadler PF (2008) `snoReport`: Computational identification of snoRNAs with unknown targets. Bioinformatics 24:158–164
68. Hofacker IL, Fontana W, Stadler PF, Bonhoeffer LS, Tacker M, Schuster P (1994) Fast folding and comparison of RNA secondary structures. Monatsh. Chem 125:167–188
69. Marz M, Stadler PF (2009) Comparative analysis of eukaryotic U3 snoRNAs. RNA Biol 6:503–507
70. Bazeley PS, Shepelev V, Talebizadeh Z, Butler MG, Fedorova L, Filatov V, Fedorov A (2008) snoTARGET shows that human orphan snoRNA targets locate close to alternative splice junctions. Gene 408:172–179
71. Bernhart SH, Tafer H, Mückstein U, Flamm C, Stadler PF, Hofacker IL (2006) Partition function and base pairing probabilities of RNA heterodimers. Algorithms Mol Biol 1:3
72. Kehr S, Bartschat S, Stadler PF, Tafer H (2011) `PLEXY`: Efficient target prediction for box C/D snoRNAs. Bioinformatics 27:279–280
73. Tafer H, Hofacker IL (2008) `RNAplex`: a fast tool for RNA-RNA interaction search. Bioinformatics 24:2657–2663
74. Chen CL, Perasso R, Qu LH, Amar L (2007) Exploration of pairing constraints identifies a 9 base-pair core within box C/D snoRNA-rRNA duplexes. J Mol Biol 369:771–783
75. Tafer H, Kehr S, Hertel J, Stadler PF (2010) `RNAsnoop`: Efficient target prediction for box H/ACA snoRNAs. Bioinformatics 26:610–616
76. Kryukov GV, Castellano S, Novoselov SV, Lobanov AV, Zehtab O, Guigó R, Gladyshev VN (2003) Characterization of mammalian selenoproteomes. Science 300:1439–1443
77. Grillo G, Turi A, Licciulli F, Mignone F, Liuni S, Banfi VA, Gennarino S, Horner DS, Pavesi G, Picardi E, Pesole G (2010) UTRdb and UTRsite (release 2010): a collection of sequences and regulatory motifs of the untranslated regions of eukaryotic mRNAs. Nucleic Acids Res 38:D75–D80
78. Kingsford CL, Ayanbule K, Salzberg SL (2007) Rapid, accurate, computational discovery of Rho-independent transcription terminators illuminates their relationship to DNA uptake. Genome Biol 8:R22
79. Lisacek F, Diaz Y, Michel F (1994) Automatic identification of group I intron cores in genomic DNA sequences. J Mol Biol 235:1206–1217
80. Zhou Y, Lu C, Wu QJ, Wang Y, Sun ZT, Deng JC, Zhang Y (2008) GISSD: Group I intron sequence and structure database. Nucleic Acids Res 36:D31–D17
81. Töpfer A (2011) Prediction of group I introns under structure variation. Master's Thesis, University of Bielefeld
82. Höchsmann T, Höchsmann M, Giegerich R (2006) Thermodynamic matchers: strengthening the significance of RNA folding energies. Comput Syst Bioinformatics Conf, pp 111–121
83. Podlevsky JD, Bley CJ, Omana RV, Qi X, Chen JJ (2008) The telomerase database. Nucleic Acids Res 36:D339–D343
84. Menzel P, Gorodkin J, Stadler PF (2009) The tedious task of finding homologous noncoding RNA genes. RNA 15:2075–2082
85. Gruber A, Kilgus C, Mosig A, Hofacker IL, Hennig W, Stadler PF (2008) Arthropod 7SK RNA. Mol Biol Evol 25:1923–1930
86. Mosig A, Guofeng M, Stadler BMR, Stadler PF (2007) Evolution of the vertebrate Y RNA cluster. Theory Biosci 126:9–14
87. Stadler PF, Chen JJ, Hackermüller J, Hoffmann S, Horn F, Khaitovich P, Kretzschmar AK, Mosig A, Prohaska SJ, Qi X, Schutt K, Ullmann K (2009) Evolution of vault RNAs. Mol. Biol. Evol. 26:1975–1991
88. Dieci G, Fiorino G, Castelnuovo M, Teichmann M, Pagano A (2007) The expanding RNA polymerase III transcriptome. Trends Genet 23:614–622
89. Schattner P, Brooks AN, Lowe TM (2005) The tRNAscan-SE, SnoScan and snoGPS web servers for the detection of tRNAs and snoRNAs. Nucleic Acids Res 33:W686–689

Chapter 11

Abstract Shape Analysis of RNA

Stefan Janssen and Robert Giegerich

Abstract

Abstract shape analysis is a method to learn more about the complete Boltzmann ensemble of the secondary structures of a single RNA molecule. Abstract shapes classify competing secondary structures into classes that are defined by their arrangement of helices. It allows us to compute, in addition to the structure of minimal free energy, a set of structures that represents relevant and interesting structural alternatives. Furthermore, it allows to compute probabilities of all structures within a shape class. This allows to ensure that our representative subset covers the complete Boltzmann ensemble, except for a portion of negligible probability.

This chapter explains the main functions of abstract shape analysis, as implemented in the tool RNA shapes. It reports on some other types of analysis that are based on the abstract shapes idea and shows how you can solve novel problems by creating your own shape abstractions.

Key words Boltzmann ensemble, Partition function, RNA shapes, Shape abstraction, Representative structures, Pseudoknot solver

1 Introduction to Abstract Shape Analysis

Abstract shapes classify RNA secondary structures by their arrangement of helices, allowing for different level of detail regarding stretches of unpaired bases. We start this chapter explaining how shape abstraction works and what you can do with its implementation as provided by the tool RNA shapes. But as we go along, keep in mind that the idea of shape abstraction is more general. There are many features in RNA structure which you can capture by designing your own shape abstraction function. In doing so, you can make your RNA folding algorithm aware of this feature and report results accordingly. This will be explicated by example near the end of the chapter.

1.1 Diving Deeper into the RNA Folding Space

According to the rules of base pairing, an RNA molecule can fold into a very large number of secondary structures. Structure prediction based on a well-established thermodynamic model [1]

Jan Gorodkin and Walter L. Ruzzo (eds.), *RNA Sequence, Structure, and Function: Computational and Bioinformatic Methods*, Methods in Molecular Biology, vol. 1097, DOI 10.1007/978-1-62703-709-9_11,

(see also Chapter 3 in this book) is performed by tools such as MFOLD [2] (and its successor UNAFOLD [3]), RNAFOLD [4], and RNASTRUCTURE [5] (see also Chapter 4 in this book). They predict the structure with minimum free energy (MFE). The MFE structure is also the most likely individual structure for the given molecule, but tells us little about all its competing, alternative foldings.

Often, there is good reason not to rely on a single, predicted structure, albeit of minimal free energy. Inaccuracy of energy parameters, varied physiological conditions or temperature, unaccounted base modifications, interaction with ligands or other RNA molecules, and many other circumstances may cause the functional structure to be different from the MFE structure. But still—in any case, the functional structure lies hidden among the near-optimal foldings of our molecule. Hence, we seek means to analyze these foldings more completely.

Abstract shape analysis is a systematic way to extract comprehensive information about the folding alternatives of an RNA molecule. Shape abstraction is a mathematical version of what humans do when they communicate about RNA structure: We speak of "a hairpin structure with some bulges or internal loops" for miRNA precursors, of "a cloverleaf structure" for tRNAs, "two adjacent hairpins" for *oxyS* RNA, and, for lack of a compact name, "the arrangement of helices P3–P9 typical of a group I intron." The number and arrangement of helices characterizes a particular structure quite comprehensively, while abstracting from details such as loop sizes or length of helical regions. Shape abstraction captures the arrangement of helices in a secondary structure.

Postponing formal definitions to Section 2.1, we see that the idea of shape abstraction partitions the folding space into different classes of structures, each class characterized by an abstract shape. If our structure prediction program can compute with abstract shapes, we may solve more specific problems, such as "Compute the best folding into a cloverleaf shape, whether or not it is the MFE folding."

Shape abstraction can, in principle, be incorporated into any of the structure prediction programs mentioned before, as it integrates smoothly with the dynamic programing algorithms used in structure prediction. But currently, only the programs RNA shapes [6–9] and RNALISHAPES [10] compute with abstract shapes.

Let us compare this idea to a popular alternative method. Several current software packages (including RNA shapes) support the sampling of suboptimal structures from their Boltzmann distribution. Structures then are clustered according to a structure metric, often a base pair distance. Clusters can be considered as structural classes, and cluster size can be used to estimate class probabilities. This achieves a similar effect as abstract shape analysis, yielding representative structures which are different enough to be of interest, and also provides some assertion of their relevance. The difference with abstract shape analysis is that abstract shapes are defined a priori, no metric is required, and probabilities are not estimated but exact.

1.2 Functional Overview of Abstract Shape Analysis

Computing with abstract shapes allows us to perform two main types of analysis, which we will discuss in detail in this chapter.

1. *Simple shape analysis* computes a small set of *representative structures*: We can ask for near-optimal structures $s_1, s_2, s_3, \ldots$ of different shape, such that each one is the optimal structure of its shape class. This gives a representative, but compact overview of the alternatives in the folding space.
2. *Probabilistic shape analysis* accumulates Boltzmann probabilities for all structures per shape, to find the probability that a molecule will fold into a structure of that shape. Or in other words, in a population of many copies of the same RNA molecule in a cell, this tells us the proportion of them which attains that shape.

Probabilistic shape analysis is more costly in times of computational effort, but provides stronger information. It assigns a weight to the representative structures—for example, if the shape classes of s_1 and s_2 together cover (say) 95% probability, we may rule out that there are relevant structures of any other shape. There are many variants and other tasks that rely on shape analysis in some way.

This chapter is organized as follows: In the remainder of this section, we give some background, relating abstract shapes to other ideas that have been used to conquer the folding space. Section 2 covers the formal definitions of shape abstraction and related concepts. You may skip this on first reading, as at the start of Section 3, we give a short, informal résumé. Sections 3 and 4 describe the two main applications, computing shape representative structures and computing shape probabilities. In Section 5, we show how to use shape abstraction to solve some custom problems. Finally, in Section 6 we list additional functions provided by the RNA shapes package and point to other uses of the abstract shapes idea, e.g. in RNA motif search.

1.3 From Minimum Free Energy Folding to Abstract Shape Analysis

RNA secondary structure prediction based on thermodynamics has become an indispensable tool in RNA research, despite its well-known shortcomings [1, 11]. There are limitations in the underlying energy model [12] (with respect to ion concentration, temperature, and entropic effects), influences of co-transcriptional folding [13], and mechanisms like RNA thermometers and riboswitches [14, 15], where the prediction of an "optimal" structure, even when correct, tells only half the story. Hence, much past and ongoing work has been devoted to improving this state of affairs.

When studying a single sequence at hand, the only way forward is to give a more complete account of the folding space of the molecule. Is the MFE structure *dominant* in the sense that all other, near-optimal structures, are very similar to it? Are there two or more succinctly different structures near the MFE value? Or is the folding essentially undefined, as there is a large number of (dis-similar) folding alternatives near the energetic minimum? Such an analysis requires criteria about what "similar", "distinct", and "dominate" should mean.

The state of an RNA molecule must be seen as a Boltzmann ensemble of structures, with low energy structures more likely than the others. The relation between energy and probability of a certain structure is captured by Boltzmann's law, as given later. The challenge of folding space analysis is to determine whether there is some family of structures in this ensemble that is internally similar, distinct from the rest, and together dominates the probabilities of all other families. If any, then the dominating family should be the functionally relevant one.

Individual structures can be assigned probabilities by means of Boltzmann statistics. This approach requires to compute the partition function. In words, the partition function is the sum of exponentially weighted energies of all structures. The probability of an individual structure is its relative contribution to this sum.

Intrinsic to this approach is that the probability is proportional to the (Boltzmann-weighted) energy of a structure. There is no possibility that an individual structure has a higher probability than a structure with lower free energy, and the MFE-structure is always the most probable one; albeit with an individual probability that is often very close to zero. This fact has already been stated by McCaskill in [16], where he proceeds to show how to extract *new* information from partition function calculations. Instead of computing the probability of a complete structure, the probabilities of the smallest structural units, i.e. individual base pairs, are determined by what has become known and widely used as the McCaskill algorithm. Probability dotplots are a popular visualization of base pair probabilities.

Another route is followed by the RNAsubopt program, released by Peter Schuster's group in Vienna in 1998 [17], and part of their Vienna RNA package [18]. It can give a non-heuristic enumeration of near-optimal structures. However, there is an "embarrassingly large" (McCaskill) number of such structures in the vicinity of the energy minimum, and the problem remains how to derive significant observations from such exhaustive information.

In [19, 20], Ding and Lawrence introduced a statistical sampling algorithm which is implemented in the tool Sfold. In each step of the recursive backtracing procedure, base pairs and the structural element they belong to are sampled according to their probability, obtained from the partition function. Features of the sampling procedure are that each run is likely to produce a different sample and that the same structure can be sampled multiple times, where the MFE structure has the highest probability. Still, the MFE structure is not guaranteed to be present in the sample, especially for long sequences. Sampled structures are clustered, and cluster centroids can be considered as representing the relevant structural alternatives in the molecule's folding space. The authors could show that sampling the folding space of the Spliced Leader of *Leptomonas collosoma* yields structures from two families. These two families, which were defined by manual alignment of the sampled structures, correspond to the alternating structures of this conformational switch. This improves over a non-probabilistic sampling procedure yielding similar results [21, 22]. At the time of this writing, SFOLD was no longer available, but other tools such as RNAFOLD and RNA shapes have incorporated the sampling technique.

Relating shape abstraction to the aforementioned approaches is easy: Rather than sampling or exhaustively enumerating structures and clustering them thereafter, structures are classified a priori into classes defined by their shapes. The virtues of this approach rest with four facts:

1. No heuristics and no nondeterminism are involved; results are reproducible.
2. In contrast to clusters based on some distance measure, abstract shapes are easy to interpret intuitively.
3. Shapes are meaningful across sequences, independent of sequence composition and length, making it possible to ask for conserved shapes in related RNAs.
4. Shape abstraction is a flexible concept, which can work with different levels of abstraction. It can even be redefined to accomplish quite unrelated tasks, as we will exemplify in Section 5.

However, there is also a *caveat*: Of course, there is diversity *within* a shape class, and it is growing for longer sequences. At some point, representing the whole shape class by a single representative structure may obscure relevant information.

Let us now proceed to introduce shape abstraction formally.

2 Mathematical Framework

We assume the reader has a good background on RNA folding and the underlying thermodynamic model, for example from studying the earlier chapters of this book, such as Chapters 2, 3, and 4.

2.1 Definitions: Folding Space and Shape Abstraction

In this section we give the basic definitions for abstract shape analysis, without introducing concrete representations of structures and shapes yet. We shall use letters x for sequences, s for structures, and p for shapes.

2.1.1 Thermodynamics

Let x be an RNA sequence and $F(x)$ its folding space, i.e. the set of all feasible secondary structures of x. $E(s)$ denotes the free energy of structure s, and we write $s^*(x)$ for a structure of x which has minimal free energy. Computing $s^*(x)$ is the classical RNA folding problem under the MFE criterion.

Actually, one should not speak of *the* MFE structure, but this is commonly done. There may be several structures which achieve the energetic minimum—this is but one of the reasons why a single MFE prediction may be misleading. When there are in fact two MFE structures, there is no guarantee that they share any common features.

The Boltzmann weight of a structure s is defined as

$$B(s) = e^{\frac{-E(s)}{RT}} \tag{1}$$

where $E(s)$ is the energy of structure s in kcal/mol, R is the universal gas constant (0.00198717 kcal/K), and T is the temperature in Kelvin.

Boltzmann weights are summed up to compute the partition function

$$Q(x) = \sum_{s \in F(x)} B(s) \tag{2}$$

and the probability $Prob(s)$ of an individual secondary structure $s \in F(x)$ is defined as:

$$Prob(s) = B(s)/Q(x) \tag{3}$$

The minimum free energy structure $s^*(x)$ and the partition function $Q(x)$ are computed by classical dynamic programming algorithms. Given both, the probability of an individual structure is easily computed. It is typically very small. For an tRNA x[1] of length 77, for example, we obtain $E(s^*(x)) = -30.7$ kcal/mol, and $Prob(s^*(x)) = 2.1137 \cdot 10^{-11}$.

[1]To be concrete, this is gb:X02584.1/1-77, GCCAAGGUGGCAGAGUUCGGCCCAACGCAUCCGCCUGCAGAGCGGAACCCCCGCCGGUUCAAAUCCGGCCCUUGGCU.

2.1.2 Shape Abstraction

Let us write F (without an x) for the set of *all* RNA secondary structures, and let P denote the set of abstract shapes. How shapes are actually represented will be discussed later, when we will define several shape abstraction functions and different types of shapes. Generally, *shape abstraction* is a function $\pi : F \rightarrow P$. The shape space of RNA sequence x under abstraction function π is

$$P_\pi(x) = \pi(F(x)) = \{\pi(s) \mid s \in F(x)\}. \tag{4}$$

We omit the π in $P_\pi(x)$ when it is clear from the context. Defining $s_1 \equiv_\pi s_2$ if and only if $\pi(s_1) = \pi(s_2)$, it is clear that $\equiv_\pi$ partitions $F(x)$ into equivalence classes

$$F_p(x) = \{s \mid s \in F(x), \pi(s) = p\}, \tag{5}$$

which are called *shape classes.*

2.1.3 Properties of Shape Classes

Shape classes can be assigned representatives in a natural way: The *shape representative structure* $\overset{*}{s_p}(x)$ of shape p is the minimum free energy structure of the given shape, i.e.

$$\overset{*}{s_p}(x) = argmin_s\{E(s)|s \in F_p(x)\} \tag{6}$$

Shape representative structure are often nicknamed "shreps."

Shape representative structures allow to order their shapes by the increasing shrep energy, $E(\overset{*}{s_{p_1}}(x)) \leq E(\overset{*}{s_{p_2}}(x)), \ldots$. This is called the *energy ranking of shapes* . Naturally, for p_1, the top shape under this ranking, $\overset{*}{s_{p_1}}(x) = s^*(x)$.

As with MFE structures, the shreps so defined are not necessarily unique, and neither is their ranking. Due to the sophistication of the energy model, two different structures rarely have the same energy, but it may occur—even for the optimal $s^*(x)$.[2] In such a case, we assume that a decision is made in some consistent way, e.g. based on a lexicographic ordering of a concrete representation of structures. Hence, we shall ignore this situation in the sequel and continue to speak of *the* MFE structure and *the* shrep, as is common in the literature.

The accumulated Boltzmann weight and the probability of shape class p is given by

$$B(p,x) = \sum_{s \in F_p(x)} B(s) \tag{7}$$

$$Prob(p,x) = \sum_{s \in F_p(x)} Prob(s) = B(p,x)/Q(x) \tag{8}$$

[2]This was observed in 23 cases out of 306 in [23]—so the situation is rare, but real.

In analogy with the energy ranking, we can also define a *probability ranking of shapes*, in decreasing order of $Prob(p, x)$. Although both rankings often agree (and especially so when there is a dominant shape), it is not uncommon that shapes switch positions between the two rankings.

Given our earlier observation that shape classes are a partitioning of $F(x)$, no structure is accounted for twice in the sum below, and we note the

Theorem

$$Q(x) = \sum_{p \in P(x)} B(p, x) \tag{9}$$

This fact will play a central role for the efficient computation of shape probabilities.

While individual structure probabilities are typically very small, a shape class can hold a large number of members, and hence, the shape probabilities often reach meaningful values. For miRNA precursors, for example, we typically find a probability larger than 0.9 for the hairpin shape. The nice thing about shape probabilities (in contrast to folding energies) is that such values are independent of the base composition, and hence are meaningful across sequences from different organisms. Still, the significance of a large shape probability is dependent on sequence length, in the sense that a highly dominating shape is more easily achieved in the smaller folding space of a shorter sequence.

2.2 Computational Tasks

At this point, we can precisely state the computational problems to be discussed in this chapter.

Problem 1: Shape Representative Structures. Given sequence x, shape abstraction function π, and energy threshold η, compute the set of shape representative structures within that range:

$$\begin{aligned} shreps(x, \eta) = \{(p, s, E(s)) \mid\ & s \in F(x), p = \pi(s), \\ & s = s_p^*(x), E(s) \leq E(s^*(x)) + \eta\} \end{aligned}$$

A solution to this problem will provide a good overview of the folding space of x. Only a small number of shapes should meet a reasonable threshold, and their shreps, having different shapes, should be different enough to be interesting.

Problem 2: Shape Probabilities. Given sequence x and shape abstraction function π, compute probabilities of all shapes in the folding space of x:

$$shapeprobs(x) = \{(p, Prob(p, x)) \mid p \in P(x)\}$$

Grammar *OverDangle*.
$\mathcal{A} = \{A,C,G,U\}$,
$V = \{struct,...,region\}$,axiomis *struct*.

production rule		rule name
struct	→ *x struct*	*sadd(x,s)*
	| *closed struct*	$cadd(s_1,s_2)$
	| ε	*nil()*
closed	→ *stack*	*init(s)*
	| *hairpin*	*init(s)*
	| *leftB*	*init(s)*
	| *rightB*	*init(s)*
	| *iloop*	*init(s)*
	| *multiloop*	*init(s)*
stack	→ *x closed y*	*sr(x,s,y)*
hairpin	→ $x_1\ x_2$ *region* $y_2\ y_1$	$hl(x_1,x_2,s,y_2,y_1)$
leftB	→ $x_1\ x_2$ *region closed* $y_2\ y_1$	$bl(x_1,x_2,lr,s,y_2,y_1)$
rightB	→ $x_1\ x_2$ *closed region* $y_2\ y_1$	$br(x_1,x_2,s,rr,y_2,y_1)$
iloop	→ $x_1\ x_2$ *region closed region* $y_2\ y_1$	$il(x_1,x_2,lr,s,rr,y_2,y_1)$
multiloop	→ $x_1\ x_2$ *ml_comps* $y_2\ y_1$	$ml(x_1,x_2,s,y_2,y_1)$
ml_comps	→ *x ml_comps*	*sadd(x,s)*
	| *closed ml_comps*1	$cadd(s_1,s_2)$
*ml_comps*1	→ *xml_comps*1	*sadd(x,s)*
	| *closed ml_comps*1	$cadd(s_1,s_2)$
	| *closed*	*id(s)*
	| *closed region*	$addss(s_1,s_2)$
region	→ *x region*	*sadd(x,s)*
	| *x*	*base(x)*

Fig. 1 Grammar *OverDangle*. The rules are schematic. For any joint occurrence of *x* and *y* or x_i and y_i in a rule, each legal base pair must be substituted

2.3 Representations of RNA Secondary Structures and Their Abstract Shapes

To precisely specify the mapping from structures to abstract shapes, we need concrete representations for both. We describe folding spaces by context free grammars, and individual structures as trees. We use grammars and rule names in the same way as it is done in Chapter 5 in this book. *See* Fig. 1.

The building blocks of RNA secondary structure are unpaired regions, base pairs, helices composed from successive, "stacking" base pairs, from bulges and internal loops inside helices, and multiloops. RNA shapes uses two grammars (for different subtasks) to define how structures are composed from substructures. Here we show the simpler one, grammar *OverDangle*, omitting only minor details. Note the similarity to the grammar *RNAfeatures* in Chapter 5. The grammar *OverDangle* prevents the occurrence of lonely base pairs—each structural component is closed by a double base pair, at least.

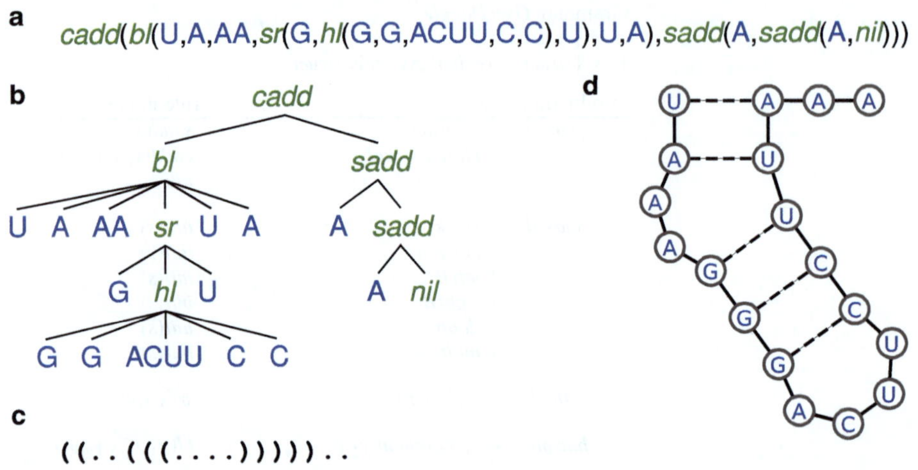

Fig. 2 An example structure s_1 shown (**a**) as a formula, (**b**) as a tree according to grammar *OverDangle*, (**c**) as a dot-bracket string, and (**d**) in graphical form

2.3.1 Representation of Structures

The structural components are also reflected in the rule names, which we use to represent individual structures as trees. *See* Fig. 2b.

2.3.2 Evaluating Structures

When it comes to evaluating structures, we view the rules names as scoring functions and the trees as formulas. Figure 2a gives an example of a structure as a formula composed from these function symbols. You may see the rule names as the set of functions from a generic Java interface, which may be implemented in different ways and over different concrete data types. We shall use the term *evaluation algebra* for a particular implementation.

Therefore, these functions are typed. Let A stands for the "alphabet" of bases in RNA, $\{A, C, G, U\}$, A^+ for nonempty sequences over A, and V for the generic data type of whatever values we want to compute from structures.

$sadd : A, V \rightarrow V$	A single, unpaired base left of a structure
$cadd : V, V \rightarrow V$	A structure composed from two adjacent subwords of the sequence
$nil : \varepsilon \rightarrow V$	The empty word
$init : V \rightarrow V$	The begin of a helix
$sr : A, V, A \rightarrow V$	A base pair stacked upon another base pair
$hl : A, A, V, A, A \rightarrow V$	Two base pairs, enclosing an unpaired region
$bl : A, A, V, V, A, A \rightarrow V$	Two base pairs, enclosing a bulge on the 5'-side of a helix

$br : A, A, V, V, A, A \to V$	Two base pairs, enclosing a bulge on the 3'-side of a helix
$il : A, A, V, V, V, A, A \to V$	Two base pairs, enclosing an internal loop (two-sided bulge)
$ml : A, A, V, A, A \to V$	Two base pairs, enclosing several substructures, also called a multiloop
$id : V \to V$	A structure
$addss : V, V \to V$	A stretch of unpaired bases right of a structure
$base : A \to V$	A single, unpaired base

Since our functions are typed, one cannot put them together quite arbitrarily (such as in $hl(ACA, base, CU)$). But even so, there are still some formulas, technically type-correct, which we do not accept as structure representations. For example, $ml(A, a, hl(C, G, s, C, G), U, U)$ is really not what we call a multiloop and should rather be represented by $sr(A, sr(A, hl(C, G, s, C, G), U), U)$. In a proper multiloop $ml(A, A, s, U, U)$, the substructure s must branch into at least two helices, otherwise it should be represented as a bulge or an internal loop. Our grammar takes care that no illegal structures will be considered for evaluation by the folding algorithm.

2.3.3 Deriving Properties from Structures

Properties of a structure can be derived by a suitable evaluation algebra. When s is a structure and $\mathcal{E}$ is such an algebra, we write $\mathcal{E}(s)$ for this interpretation.

Let us specify two algebras: Algebra *BasePair* counts the number of base pairs in a structure. Algebra *DotBracket* determines the dot-bracket representation of a structure. On the right-hand side of equations defining *DotBracket*, we omit string quotes and concatenate strings by juxtaposition.

Algebra *DotBracket*		Algebra *BasePair*	
$sadd(x, s)$	$= .s$	$sadd(x, s)$	$= s$
$cadd(s_1, s_2)$	$= s_1\ s_2$	$cadd(s_1, s_2)$	$= s_1 + s_2$
$nil()$	$= \varepsilon$	$nil()$	$= 0$
$init(s)$	$= s$	$init(s)$	$= s$
$sr(x, s, y)$	$= (s)$	$sr(x, s, y)$	$= 1 + s$
$hl(x_1, x_2, s, y_2, y_1)$	$= ((s))$	$hl(x_1, x_2, s, y_2, y_1)$	$= 2 + s$
$bl(x_1, x_2, lr, s, y_2, y_1)$	$= ((lr\ s))$	$bl(x_1, x_2, lr, s, y_2, y_1)$	$= 2 + s$
$br(x_1, x_2, s, rr, y_2, y_1)$	$= ((s\ rr))$	$br(x_1, x_2, s, rr, y_2, y_1)$	$= 2 + s$
$il(x_1, x_2, lr, s, rr, y_2, y_1)$	$= ((lr\ s\ rr))$	$il(x_1, x_2, lr, s, rr, y_2, y_1)$	$= 2 + s$
$ml(x_1, x_2, s, y_2, y_1)$	$= ((s))$	$ml(x_1, x_2, s, y_2, y_1)$	$= 2 + s$
$id(s)$	$= s$	$id(s)$	$= s$
$addss(s_1, s_2)$	$= s_1\ s_2$	$addss(s_1, s_2)$	$= s_1$
$base(x)$	$= .$	$base(x)$	$= 0$

```
CGUCUUAAACUCAUCACCGUGUGGAGCUGCGACCCUUCCCUAGAUUCGAAGACGAG
((((((...(((..(((...))))))...(((..((.....))..)))))))))..

Shape Level 5:    [[][]]
Shape Level 4:    [[][[]]]
Shape Level 3:    [[[]][[]]]
Shape Level 2:    [[_[]][_[]_]]
Shape Level 1:    [_[_[]]_[_[]_]]_
```

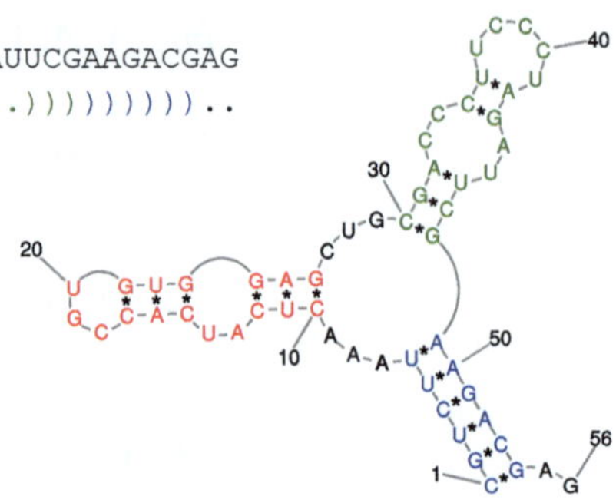

Fig. 3 Five shape abstraction levels implemented in RNA shapes

Considering our example structure s_1 of Fig. 2, we obtain $DotBracket(s_1)$ = ((..((((....)))))).., and $BasePair(s_1)$ = 5.

2.3.4 Free Energies and Partition Function Algebras

$MFE(s)$ computes the energy of s, and $BWE(s)$ the Boltzmann-weighted energy. To compute energies or Boltzmann weights, summed up for the partition function, these algebra functions call upon the energy parameters of the thermodynamic model. This is not different from the use of the energy model in Chapter 4. We do not reproduce these algebras here, but the interested reader can inspect them in the RNA shapes package.

2.3.5 Representations of Abstract Shapes

Abstract shapes represent the arrangement of helices in an RNA secondary structures, abstracting from helix lengths, and–to a varying extent–also from the presence of unpaired bases within and between helices. They are all defined by suitable shape abstraction algebras.

We represent shapes as strings. A helix, i.e. a series of stacked base pairs, is represented by a single pair of square brackets, [ ], and a sequence of unpaired bases is represented by an underscore _. We do never represent the bases in a hairpin loop in shapes, as the hairpin loop *must* contain at least three unpaired bases; the _ would always be there and hence not designate a characteristic structural feature. RNA shapes provides shape abstraction algebras π_1 through π_5 for five different levels of abstractness, which differ in the amount of detail about unpaired regions they retain. Figure 3 gives an overview.

Here are the algebras π_1 and π_5.

Shape abstraction algebra π_1		Shape abstraction algebra π_5	
$sadd(x, s)$	$= \_s$	$sadd(x, s)$	$= s$
$cadd(s_1, s_2)$	$= s_1\ s_2$	$cadd(s_1, s_2)$	$= s_1\ s_2$
$nil()$	$= \_$	$nil()$	$= \_$
$init(s)$	$= s$	$init(s)$	$= s$
$sr(x, s, y)$	$= s$	$sr(x, s, y)$	$= s$
$hl(x_1, x_2, s, y_2, y_1)$	$= []$	$hl(x_1, x_2, s, y_2, y_1)$	$= []$

Table 1
Runs of unpaired bases are represented by a single underscore where indicated on levels 1 and 2. On higher levels, internal loops (levels 3 and 4) and bulges (level 3 only) are implicitly indicated by splitting the enclosing helix, but no _ is shown. Note, for example, that a level 3 shape [[ ]] tells us that there is a stem interrupted by a single bulge or internal loop (but not which), while a level 5 shape [[ ]] does not exist. Helix interruptions are ignored on level 5 and the above level 3' shape becomes [ ]

Unpaired bases represented for	External loop	Multiloop	Internal loop	Bulge
Level 1	YES	YES	YES	YES
Level 2	NO	NO	YES	YES
Level 3	NO	NO	Implicit	Implicit
Level 4	NO	NO	Implicit	NO
Level 5	NO	NO	NO	NO

$bl(x_1, x_2, lr, s, y_2, y_1)$	$= [\_\ s]$	$bl(x_1, x_2, lr, s, y_2, y_1)$	$= s$
$br(x_1, x_2, s, rr, y_2, y_1)$	$= [s\ \_]$	$br(x_1, x_2, s, rr, y_2, y_1)$	$= s$
$il(x_1, x_2, lr, s, rr, y_2, y_1)$	$= [\_\ s\ \_]$	$il(x_1, x_2, lr, s, rr, y_2, y_1)$	$= s$
$ml(x_1, x_2, s, y_2, y_1)$	$= [s]$	$ml(x_1, x_2, s, y_2, y_1)$	$= [s]$
$id(s)$	$= s$	$id(s)$	$= s$
$addss(s_1, s_2)$	$= s_1\ s_2$	$addss(s_1, s_2)$	$= s_1\ s_2$
$base(x)$	$= \_$	$base(x)$	$= \varepsilon$

Again, we write string concatenation as juxtaposition, under the rule that concatenating two _ always results in a single _. Considering our example structure s_1 of Fig. 2, we obtain $\pi_1(s_1) =$ [_[]]_ and $\pi_5(s_1) =$ [].

On level 1, which is the most concrete level, π_1 creates an underscore _ for each stretch of unpaired bases, except inside hairpin loops. Each helix part, i.e. a contiguous series of stacked base pairs, is represented by a single pair of square brackets, [...]. Helix interruptions by bulges or internal loops lead to several nested brackets, interspersed with _ as in [_ [ ] _ [ ]].

On Level 5, which is the most abstract shape level, π_5 ignores all unpaired bases. ε denotes the empty string which is produced as their shape representation. Helices interrupted by bulges and internal loops will be represented by a single pair of brackets (rather than one for each helix part). Level 5 has no "_" in its shapes, except for one: the completely open structure has shape "_".

Shape abstraction levels π_2, π_3, and π_4 are defined by similar algebras which are not reproduced here. Consult Fig. 3 or Table 1, which summarizes which structural features are reflected at each shape level.

Table 2
Asymptotic number of shapes of length *n*. We include the asymptotics of concrete structures as level 0 shapes

Shape level	Number of shapes
0	$1.84892^n \cdot 1.48483 \cdot n^{-3/2}$
1	$1.47667^n \cdot 3.04214 \cdot n^{-3/2}$
2	$1.37736^n \cdot 3.61323 \cdot n^{-3/2}$
3	$1.27614^n \cdot 4.19348 \cdot n^{-3/2}$
4	$1.26197^n \cdot 4.42176 \cdot n^{-3/2}$
5	$1.20259^n \cdot 5.12777 \cdot n^{-3/2}$

2.3.6 Reduction of Complexity by Shape Abstraction

The main motivation for the use of abstract shapes is to focus on a small number of shreps rather than investigating a large space of near-optimal foldings. For a short sequence of (say) length 80, it is typically feasible to look at all level 5 shreps, while considering those at level 2 or 1 would already be quite demanding. Just as $|F(x)|$ grows exponentially with $|x|$, so does $|\pi(F(x))|$, albeit much slower. The asymptotics of shapes have been analyzed extensively in [24] and [25]. We report results from the latter article, as the shape abstractions considered there coincide with π_1–π_5 as defined here and actually implemented in the tool RNA shapes.

The number of shapes for *all* sequences of length n is shown in Table 2, assuming a minimal hairpin loop length of 3 and no lonely base pairs. Generally, the asymptotic formulas take the form

$$a^n \cdot b \cdot n^{-3/2}.$$

Here we can see that the base of the exponential growth function is reduced by shape abstraction, for example from 1.84892 for secondary structures (level 0 shapes, if you wish) to 1.20259 for level 5. This gives an impression about the relative reduction achieved by different levels, but does not tell us how many shapes we might have to deal with when analyzing a concrete sequence. Of greater interest would be the expected number of shapes for a sequence of length n. This combinatorial problem is an open challenge. Not even general form of the asymptotic formula is known. Empirical measurements by S. Abdoulhak (personal communication) and our own, performed on a sample of 10.000 sequences of varying length, suggest an expected number of shapes for π_5 also following a law of $f(n) = a^n \cdot b \cdot n^{-\frac{3}{2}}$ with $a \approx 1.138$ and $b \approx 15.415$. Naturally, the variance must be high, as a sequence like CCCCC...CCCC has only a single shape (the shape _), while CGCGCG...CGCGCG has a very large number of shapes.

2.4 Deriving Structures from Sequences via Dynamic Programming

Any dynamic programming algorithm for RNA structure analysis is based on a grammar which describes the folding space. RNA shapes uses two different grammars. *MicroState*, which is very similar to *OverDangle* shown above, is used for simple shape analysis. A more sophisticated grammar, *MacroState*, is used for probabilistic shape analysis. A *parser*, which can be constructed automatically from the grammar, reads an RNA sequence x and constructs the folding space $F(x)$. In this chapter, we do not worry about how grammars are designed and how the parser works. The important aspect is that the parser does not explicitly construct structures represented as trees. Instead, it builds formulas, which, as they emerge, are passed on to an evaluation algebra, such as *DotBracket* (returning structures in dot-bracket notation) or *MFE* (returning the free energy value).

These algebras are augmented with an objective function. The objective function may be used for simply reporting values derived from all candidates, but typically, it performs minimization or maximization. It may also report all near-optimal candidates up to a certain energy threshold, and so on.

Writing G for the grammar as well as for the corresponding parser, E for the evaluation scheme, and x for the input sequence, a call to

$$G(E, x)$$

performs the analysis described by E on the structures described by G. This shorthand notation will be used for the computational tasks discussed in the sequel.

3 Computing Shape Representative Structures

3.1 Résumé of Shape Abstraction

We recall what was introduced more formally in the previous section. Alternative secondary structures are mapped to abstract shapes, which capture their arrangement of helices, but not their position and size. Different levels of abstraction provide different detail about the presence of single stranded regions (in bulges) or connecting branches in a multiloop. For example, for structure

“(((((...(((...)))......))...(((...))))))”,

its level-5 shape is

“[[][]]”,

whereas its level-2 shape is

“[[_[]_]_[]]”.

All the foldings of a sequence having the same shape constitute a shape class, and within it, the structure with least free energy is called the shape representative structure, also called shrep.

The folding space of an RNA sequence is described by a grammar G_S, for example *OverDangle* as shown in Fig. 1.

For extracting information about the structures in the folding space, we use algebras such as *MFE* for free energy minimization, *DotBracket* for printing a structure, and π_i for computing shape abstraction, level i. These algebras are used in combination (operator *). On the command line of the tool RNA shapes, these algebras and combinations are options to the program call. On the interactive web site, you can choose them via buttons.

In the sequel, we will work with the following evaluation algebras, defined either herein or in the RNA shapes package.

Name	Value Computed	Objective	Defined In	Computes
DotBracket	Dotbracket string	Enum.	Section 2.3	Printable structure representation
MFE	Free energy	Min	Package	Structure of minimum free energy
MFE(η)	Free energy	Min	Package	Structure upto threshold η above MFE
BWE	Boltzmann weights	Sum	Package	Boltzmann-weights and partition function
Count	Number	Sum	Section 2.3	Number of structures
$\pi_1 \ldots \pi_5$	Shape string	Enum.	Section 2.3	Abstract shape
STEM	Classifier	Enum.	Section 5	Resolved pseudoknots

3.2 Minimum Free Energy Folding

If you want to use RNA shapes simply as an MFE-based structure prediction program, you can do it: Just call

$$G_S(MFE, x)$$

or

$$G_S(MFE{*}\,DotBracket, x)$$

The first call will produce the MFE value only, while the second call will produce the MFE value and the optimal structure as a dot-bracket string. In fact, should there be several co-optimal structures, all will be reported. This is an example for the use of algebra products: The algebra product operator "*" allows to obtain additional information, here the *DotBracket* string, about the candidate(s) which are reported via the first algebra, here the MFE value.

3.3 Adding Shape Information

Consider calling

$$G_S(\pi_5, x)$$

or

$$G_S(\pi_5 * DotBracket, x).$$

The first call reports all different shapes existing in $F(x)$. No energy computation is involved. The second call, for each shape, also reports the structures that belong to it as *DotBracket* strings. This is instructive to do, but feasible only for very short sequences. In particular, the output of the second call can be voluminous.

3.4 Computing k-Best Shape Representatives

What we really want to do is combine shape abstraction and energy minimization. There we go:

$$G_S(\pi_5 * MFE, x)$$

or

$$G_S(\pi_5 * MFE * DotBracket, x)$$

This gives all shapes and the minimal MFE value within each shape, and in the second case, also each shape representative structure–where it is not unique, several per shape.

Reporting this information about *all* shapes may be uninteresting. We are interested in the near-optimal shreps only. To this end, the *MFE* algebra may be used with a threshold parameter, η, causing it to report all energy values within η percent of the optimum. Used without shape analysis, a call like

$$G_S(MFE(\eta) * DotBracket, x)$$

would be equivalent to using the program RNAsubopt [17]. Again, voluminous output will result, as the vicinity of the MFE structure holds very many structures which only differ in a base pair or two. To combine this with shape analysis, we use another product operation "⊗", which restricts the near-optimals reported to one per shape, for as many shapes as pass the threshold. The call

$$G_S(\pi_5 \otimes MFE(C) * DotBracket, x)$$

computes only those shape representative structures below the threshold. This is the first, main function of RNA shapes.

3.5 Examples

Our first example is the analysis of a microRNA precursor. We first show the two calls to RNA shapes that are most commonly made. To compute simple shape analysis of the sequence x, we call $G_S(\pi_5 * MFE * DotBracket, x)$. In this case, it yields three shreps. (In the output, results are arranged (MFE, DotBracket, Shape)).

```
Sequence from MirBase: Caenorhabditis elegans lin-4 stem-loop
Description: lin-4 is found on chromosome II in Caenorhabditis elegans [1] and is complementary to
sequences in the 3' untranslated region (UTR) of lin-14 mRNA. lin-4 acts to developmentally
repress the  accumulation of lin-14 protein. This repression is essential for the proper timing of
numerous events of  Caenorhabditis elegans larval development [2].

>cel-lin-4 MI0000002
AUGCUUCCGGCCUGUUCCCUGAGACCUCAAGUGUGAGUGUACUAUUGAUGCUUCACACCUGGGCUCUCCGGGUACCAGGACGGUUUGAGCAGAU

RNA shapes -e 10 AUGCUUCCGGCCUGUUCCCUGAGACCUCAAGUGUGAGUGUACUAUUGAUGCUUCACACCUGGGCUCUCCGGGUACCAGGACG
  GUUUGAGCAGAU
-41.20  .((((((((.((((..(((.((((.((((.((((((((((......))))).))))).)))).)))).)))...)))).)))
        ...)))))...  []
-33.10  ......(((.((((..(((.((((.((((.((((((((((......))))).))))).)))).)))).)))...)))).)))
        (((....))).  [][]
-32.60  .((((((((.((((..(((.((((.((((.(((..((....))..(((....)))))).)))).)))).)))...)))).)))
        ...)))))...  [[][]]
```

MFE values already suggest that the single stem-loop shape dominates the folding space, but to know for sure, we next compute shape probabilities according to $G_S(\pi_5 * BWE, x)$.

```
RNA shapes -p AUGCUUCCGGCCUGUUCCCUGAGACCUCAAGUGUGAGUGUACUAUUGAUGCUUCACACCUGGGCUCUCCGGGUACCAGGACGGUU
              UGAGCAGAU
0.9999968  []
0.0000032  [][]
```

A probability of 0.98 or higher for the stem-loop shape is typical for a microRNA precursor. It tells us that all other foldings are irrelevant. The third shape from the previous call has a probability so low that it has been suppressed from the output.

While most users are happy with this type of analysis, let us explicate how we can analyze the folding space (of any sequence, not just this one) in depth by calling RNA shapes with different evaluation algebras. Let us start with two simple questions: How large is $F(x)$, or in other terms:

$G_S(Count, x)$:

```
13,879,950,756,355
```

This call shows us how large the folding space of this molecule is–about 10^{13} structures are possible according to the rules of base pairing–not bad for 106 nucleotides. Many of these will have positive folding energies (up to +36.10 kcal/Mol), so they will not fold at all in reality. What is the MFE, after all? If this is all we want to know, we call

$G_S(MFE, x)$:

```
-41.20
```

This call computes the MFE value—but not the MFE structure. To obtain the MFE structure, we include algebra *DotBracket* and call

$G_S(MFE^{*}DotBracket, x)$:

```
-41.20 , .((((((((.((((..(((.((((.((((.(((((((.(((.......)))))))))).)))).)))).)))...)))).)))...)))
         ))...
-41.20 , .((((((((.((((..(((.((((.((((.((((((((((.......)))).)))))).)))).)))).)))...)))).)))...)))
         ))...
-41.20 , .((((((((.((((..(((.((((.((((.((((((((((.......))))).))))).)))).)))).)))...)))).)))...)))
         ))...
```

Surprise! There are actually three co-optimal structures. But then, surprise revoked, they only differ in the location of the innermost bulges. So overall, the optimum is very well defined. Next, we wonder about the number of shapes our 10^{13} structures map to. It is an interesting, open algorithmic problem to find their number directly. We have to resort to their enumeration (and with RNA shapes, this is a call you should be cautious to make).

$G_S(\pi_5, x)$:

```
_
[]
[][]
[[][]]
[][][]
[][[][]]
... 6080 more shapes ...
[[[[[][]][][]][[][]]][]]
[[[[[][]][[][]]][[][]]][]]
[[[][][][[][[][[][]]]]]][]]
```

So, the π_5 abstraction maps 10^{13} structures to 6,089 shapes. Do you wonder how the structures are distributed over the shapes? Find out by calling

$G_S(\pi_5{*}Count, x)$:

```
_ 1
[] 289,700,811,231
[][] 204,746,223,564
[[][]] 1,415,885,109,683
[][][] 200,875,411,893
[][[][]] 547,121,144,761
... 6080 more shapes ...
[[[[[][]][][]][[][]]][]] 1,776
[[[[[][]][[][]]][[][]]][]] 72
[[[][][][[][[][[][]]]]]][]] 192
```

It is reassuring to see that the open shape '_' has only a single structure mapped to it (the completely unfolded one). If any structure was accounted for twice, all partition function and probability computations would be incorrect. To compute the MFE value within each shape, we call

$G_S(\pi_5{*}MFE, x)$:

```
_ , 0
[] , -41.20
[][] , -33.10
[[][]] , -32.60
[][][] , -27.50
[][[][]] , -29.20
... 6083 more shapes omitted
```

Normally, we also want to see the shreps, so let us add *DotBracket* back in:

$G_S(\pi_5{*}MFE^{*}DotBracket, x)$:

```
_ ,        0 , ..........................................................................
               ................
[] ,      -41.20 , .((((((((.((((..(((.((((.((((.(((((((.(((.......)))))))))).)))).)))).)))...))))
               .)))...)))))...
[] ,      -41.20 , .((((((((.((((..(((.((((.((((.((((((((((.......)))).)))))).)))).)))).)))...))))
               .)))...)))))...
[] ,      -41.20 , .((((((((.((((..(((.((((.((((.((((((((((.......))))).))))).)))).)))).)))...))))
               .)))...)))))...
[][] ,    -33.10 , ......(((.((((..(((.((((.((((.(((((((.(((.......)))))))))).)))).)))).)))...))))
               .)))(((....))).
[][] ,    -33.10 , ......(((.((((..(((.((((.((((.((((((((((.......)))).)))))).)))).)))).)))...))))
```

```
                      .)))(((....))).
[][] ,      -33.10 , ......(((.((((..(((.((((.((((.((((((((((.......))))).))))).)))).)))).)))...))))
                      .)))(((....))).
[[][]] ,    -32.60 , .((((((((.((((..(((.((((.((((.(((..((....))..((((....)))))).)))).)))).)))...))))
                      .)))...)))))...
[][][] ,    -27.50 , .((....)).((((..(((.((((.((((.(((((((.(((.......)))))))))).)))).)))).)))...))))
                      ...((....))....
[][][] ,    -27.50 , .((....)).((((..(((.((((.((((.((((((((((.......)))).)))))).)))).)))).)))...))))
                      ...((....))....
[][][] ,    -27.50 , .((....)).((((..(((.((((.((((.((((((((((.......))))).))))).)))).)))).)))...))))
                      ...((....))....
[][[][]] , -29.20 , ..((.....))((((((((.((((.((((.(((((((.(((.......)))))))))).)))).)))).))..(((...
                      ..)))..)))))))..
[][[][]] , -29.20 , ..((.....))((((((((.((((.((((.((((((((((.......)))).)))))).)))).)))).))..(((...
                      ..)))..)))))))..
[][[][]] , -29.20 , ..((.....))((((((((.((((.((((.((((((((((.......))))).))))).)))).)))).))..(((...
                      ..)))..)))))))..
... 7925 more answers omitted
```

Here we find that not only the MFE structure but also the shreps of other shapes are not unique. But again, their differences are marginal.

What would we do without abstract shapes? The following corresponds to a call of RNASUBOPT:

$G_S(MFE(\eta)^{*} DotBracket, x)$ for η = 5% of MFE (= –41.20):

```
-41.20 , .((((((((.((((..(((.((((.((((.((((((((((.......))))).))))).)))).)))).)))...)))).)))...)))
         ))...
-41.20 , .((((((((.((((..(((.((((.((((.((((((((((.......)))).)))))).)))).)))).)))...)))).)))...)))
         ))...
-41.20 , .((((((((.((((..(((.((((.((((.(((((((.(((.......)))))))))).)))).)))).)))...)))).)))...)))
         ))...
-40.30 , .((((.(((.((((..(((.((((.((((.((((((((((.......))))).))))).)))).)))).)))...)))).)))....))
         ))...
-40.30 , .((((.(((.((((..(((.((((.((((.((((((((((.......)))).)))))).)))).)))).)))...)))).)))....))
         ))...
-40.30 , .((((.(((.((((..(((.((((.((((.(((((((.(((.......)))))))))).)))).)))).)))...)))).)))....))
         ))...
-40.20 , ..(((((((.((((..(((.((((.((((.((((((((((.......))))).))))).)))).)))).)))...)))).)))...)))
         )....
-40.20 , ..(((((((.((((..(((.((((.((((.((((((((((.......)))).)))))).)))).)))).)))...)))).)))...)))
         )....
-40.20 , ..(((((((.((((..(((.((((.((((.(((((((.(((.......)))))))))).)))).)))).)))...)))).)))...)))
         )....
-40.10 , .((((((((.(((((.(((.((((.((((.((((((((((.......))))).))))).)))).)))).))).)).))).)))...)))
         ))...
-40.10 , .((((((((.(((((.(((.((((.((((.((((((((((.......)))).)))))).)))).)))).))).)).))).)))...)))
         ))...
-40.10 , .((((((((.(((((.(((.((((.((((.(((((((.(((.......)))))))))).)))).)))).))).)).))).)))...)))
         ))...
-40.10 , .((((((((.((((..(((.((((.((((.(((((((.((.........))))))))).)))).)))).)))...)))).)))...)))
         ))...
-39.90 , .((((((((.((((..(((.(((((.((.((((((((((.......))))).))))).)))).)))).)))...)))).)))...)))
         ))...
-39.90 , .((((((((.((((..(((.(((((.((.((((((((((.......)))).)))))).)))).)))).)))...)))).)))...)))
         ))...
-39.90 , .((((((((.((((..(((.(((((.((.(((((((.(((.......)))))))))).)))).)))).)))...)))).)))...)))
         ))...
-39.50 , .((((((((.((((..(((.((((.((((.((((((((..........))).))))).)))).)))).)))...)))).)))...)))
         ))...
-39.50 , .((((((((.((((..(((.((((.((((.((((((((..........)).)))))).)))).)))).)))...)))).)))...)))
         ))...
-39.30 , ..(((.(((.((((..(((.((((.((((.((((((((((.......))))).))))).)))).)))).)))...)))).)))....))
         )....
-39.30 , ..(((.(((.((((..(((.((((.((((.((((((((((.......)))).)))))).)))).)))).)))...)))).)))....))
         )....
-39.30 , ..(((.(((.((((..(((.((((.((((.(((((((.(((.......)))))))))).)))).)))).)))...)))).)))....))
         )....
-39.20 , .((((.(((.(((((.(((.((((.((((.((((((((((.......))))).))))).)))).)))).))).)).))).)))....))
         ))...
-39.20 , .((((.(((.(((((.(((.((((.((((.((((((((((.......)))).)))))).)))).)))).))).)).))).)))....))
         ))...
-39.20 , .((((.(((.(((((.(((.((((.((((.(((((((.(((.......)))))))))).)))).)))).))).)).))).)))....))
         ))...
-39.20 , .((((.(((.((((..(((.((((.((((.(((((((.((.........))))))))).)))).)))).)))...)))).)))....))
         ))...
```

This exemplifies that straight enumeration of near-optimals yields many structures that are too similar to be interesting, unless we study the aspects as refolding trajectories.

What might we get with the following call? $G_S(\pi_5 \otimes MFE(\eta)^{*} DotBracket, x)$:

```
for Eta = 0.05 of MFE (=-41.20)
[] , -41.20 , .((((((((.((((..(((.((((.((((.(((((((.(((.......)))))))))).)))).)))).)))...)))).)))
                ...)))))...
```

```
[] , -41.20 , .((((((((.((((..(((.((((.((((.((((((((((.......)))).)))))).)))).)))).)))...)))).)))
              ...))))))...
[] , -41.20 , .((((((((.((((..(((.((((.((((.((((((((((.......))))).))))).)))).)))).)))...)))).)))
              ...))))))...
```

This gives all shapes which have shreps within 5% of MFE (which is only one shape in this example), and for those shapes, all their co-optimals.

4 Probabilistic Shape Analysis

A shape probability is the sum over the Boltzmann probabilities of all structures within the shape class. For probabilistic shape analysis, RNA shapes uses a grammar *MacroState*, which is much more refined than grammar *OverDangle* used for simple shape analysis. This is necessary for computing correct Boltzmann statistics. The full energy model assigns energies to "dangling" bases, which continue stacking onto an adjacent helix although they are not part of a base pair. These energy contributions are substantial and cannot be ignored. With a simple grammar, either the dangling effect is overestimated, or structures are accounted for in $F(x)$ as many times as they have alternative dangling conformations. This tends to distort probabilities assigned by Boltzmann statistics. To avoid these effects, a new grammar was developed, which first appeared in [7]. A detailed study of the effects of using different grammars has recently been reported [23]. It shows that simpler grammars can approximate Boltzmann probabilities quite well, this is why we do not include the more complex *MacroState* grammar here.

Let us use the name G_B for whatever grammar is used for computing Boltzmann statistics.

4.1 Exhaustive Computation of Shape Probabilities

Let *BWE* be our Boltzmann weight algebra, equipped with an objective function which sums up Boltzmann weights of all candidate structures found.

$$G_B(BWE, x) = Q(x)$$

computes the partition function. To compute shape probabilities, we use

$$G_B(\pi_5 * BWE, x),$$

which yields Boltzmann weights, accumulated per shape. Dividing them by $Q(x)$ yields shape probabilities, as desired. To also report shreps, we use

$$G_B(\pi_5 * (MFE \times BWE), x),$$

where "×" denotes a Cartesian product. This implements the second main function of RNA shapes, probabilistic shape analysis.

Simple as it is, this method has a drawback: Shape probabilities are computed for all shapes, and as indicated earlier, their number grows exponentially with the sequence length, albeit with a base of the exponent close to 1.1. We have no direct method to compute the most likely shape *only*, left alone the most likely shapes up to some threshold.[3] It is not known whether a polynomial time algorithm exists for this problem. Observations on a similar task with Hidden Markov Models make it appear unlikely [26]. However, for the situation when compute time or space is critical, RNA shapes offers two heuristic approaches, which are described next.

4.2 Heuristic Approaches to Rapid Shape Probabilities

There are two types of heuristics which can be employed to obtain shape probabilities more quickly.

A *result heuristics* makes a compromise as to the precision with which shape probabilities are calculated. We can *sample* structures according to their Boltzmann weights, compute their shapes, and the frequency of a shape in the whole sample serves as an estimate of the shape probability. Accuracy can be improved by making the sample larger, with increased likelihood but no guarantee that each "interesting" shape will actually be seen.

A *runtime heuristics* computes shape probabilities one by one and stops when probability has been exhausted up to $1 - T$, where T is a threshold to be supplied. Here, the calculated probabilities are exact, but the speedup in runtime is not guaranteed. The idea is to compute highly probable shapes early, to make the procedure terminate quickly. But which shapes are promising? If we knew the most likely shapes for sure, our problem would already be solved.

Shapes with low energy shreps can be used as candidates, or promising shapes from sampling (cf. above). The method to compute the exact probability of a particular shape p is as follows: Given a "promising" shape p, a grammar G_p is generated automatically, which describes all and only the structures of shape p. The Bellman's GAP system [27] is used to compile the grammar into an executable parser. Now, the call

$$G_p(BWE, x)$$

computes the Boltzmann weight of all p-shaped structures in $F(x)$. According to Eq. 9, dividing by $Q(x)$ yields the shape probability

[3] For the experts in dynamic programming, we remark that this is because there is an exponential number of shapes, and the accumulation of Boltzmann weights does not satisfy Bellman's Principle of Optimality. We cannot focus on the most likely sub-shapes during the construction of larger shapes.

of p. This is done for as many shapes as needed, until the probability threshold is exhausted.

4.3 Examples

One reason to compute shape probabilities is to exclude that there are other structures that compete with the MFE structure. Another reason is that there may be a suboptimal structure which, together with its look-alikes, actually dominates the MFE structure. We choose a UTR sequence with a known protein-binding motif and compute shape probabilities by calling $G_B(\pi_5 * BWE, x)$, with the results of BWE divided by $Q(x)$ to yield probabilities:

```
Sequence from RF00109 "Vimentin 3' UTR protein-binding region"
Description: The vimentin 3' UTR protein-binding region is an RNA element that contains a Y shaped
structure which has been shown to have protein binding activity.[1] The same region has been
implicated in the control of mRNA localisation to the perinuclear region of the cytoplasm,
possibly at sites of intermediate filament assembly.[2] The identity of the proteins involved and
the localisation mechanism are not known.

>Canis lupus familiaris AACN010807753.1/3-67
UCCAUAUCUUAAAGGAACAGCUUUCAAGUGCCUCUCUGCAGUUUUUCAGGAGCGCGAGAUAGAUU

RNA shapes -p UCCAUAUCUUAAAGGAACAGCUUUCAAGUGCCUCUCUGCAGUUUUUCAGGAGCGCGAGAUAGAUU
0.7176863  [[][]]
0.2408858  [][]
0.0388604  []
0.0024303  [][][]
0.0001311  [[][][]]
0.0000060  [][[][]]
```

This shows two shapes with high probabilities (72% and 24%), dominating the rest of the folding space. An interesting phenomenon occurs with their rank. Let us look at the associated structures:

```
RNA shapes UCCAUAUCUUAAAGGAACAGCUUUCAAGUGCCUCUCUGCAGUUUUUCAGGAGCGCGAGAUAGAUU
-11.60     (((..........)))....((.((..(((.(((.(((........)))))))))..)).))...  [][]
-11.40     ....((((((...((((....))))..(((.(((.(((........)))))))))))))))....  [[][]]
-10.60     ....................((.((..(((.(((.(((........)))))))))..)).))...  []
```

Here we have a case where the rank ordering by probability inverts the ranking by shrep energy. The representative structure of the Y-shape [[][]] is in excellent agreement with the protein-binding motif described in [28]. Doing simple MFE analysis, this structure would not have been found.

Shape probabilities are particularly interesting with riboswitches. There, one may expect two competing shapes, which together dominate the Boltzmann ensemble. After all, when a molecule refolds, triggered by a ligand, the structure to fold into should be clearly defined. This has been observed in various examples, but cannot readily be turned into an approach to evaluate switching potential. For example, both conformations may contribute to the same shape, and hence, they do not become visible as alternatives in the shape analysis. This happens, for example, when the two conformations both have (say) shape [][], where the second helix only moves in location, or adds bulges and internal loops. Feel free to try probabilistic shape analysis on riboswitches, but be sure to submit the complete sequence of a switch to RNA shapes. Current Rfam [29] sequences only include the conserved aptamer domain.

5 Do-It-Yourself Shape Abstraction: Creating a Pseudoknot Solver

The idea of shape abstraction is more general than what is provided by the tool RNA shapes. Any attribute, any property of RNA secondary structures that can be computed by an evaluation algebra can be viewed and used as a shape abstraction. If we can classify structures into red, green, and blue ones, we may perform energy minimization in a class-wise fashion, to obtain the red, green, and blue structure of minimal free energy.

You may think that changing an established program such as RNA shapes, written by someone who you don't even know, might be difficult. But RNA shapes is implemented with Bellman's GAP [27], a dynamic programming system that supports abstractness and modularity in dynamic programming. All you have to do is invent your own classification algebra C, and then recompile and call RNA shapes in the form

$$G(C^{*}A, x),$$

where A denotes any kind of analysis to be done class-wise now. We will demonstrate this by developing a new application: resolving pseudoknots in an energetically optimal way.

The majority of RNA tools is not capable to handle pseudoknots, due to their computational complexity. As a compromise to get pseudoknotted structures manageable for these tools, often all crossing stems are resolved.The minimal pseudoknot example contains at least two crossing stems, let us call them α and β. The question is which one should be resolved? Which stem is less important for the remaining structure?

Sandra Smit has implemented five different methods to resolve pseudoknots [30], offered at http://www.ibi.vu.nl/programs/k2nwww/. They are all based on heuristics, such as maximizing numbers of retained base pairs, or favoring longer helices over shorter ones in case of conflict. What is lacking is a method based on free energy, such as the obvious: Given a pseudoknotted structure R, find the un-knotted structure of minimal free energy which use only base pairs from R. Let us call this the $K2N(E)$-Problem for "knotted to nested structure based on energy."

Another concrete tool that deals with this problem is CMBUILD of the INFERNAL package [31]. It gets a multiple RNA sequence alignment and one consensus structure as input and produces a covariance model, which can be used to search for homologous sequences. The architecture of the covariance model follows the consensus structure, which must not contain pseudoknots. INFERNAL is used to create family models for the RFAM database [29], but this database also contains some alignments with annotated pseudoknot consensus structures. In

Table 3
Example of classification algebra *Stem*. The first candidate (3rd row) follows *R* in all base pairs and unpaired bases, except in the crossing stem regions and contains four base pairs of stem α, thus it is classified as *A*. Second candidate is *F*, because it misses one of the "normal" base pairs. Third candidate is also *F*, because none of the "normal" base pairs of *R* are matched. Fourth candidate follows *R* and uses pairs of β, thus it's of type *B*. The last candidate does not use pairs of α or β, but follows *R*, thus it is the only candidate of type *C* in *F(r)*

s	GGCCCCUUGCCGUCGGGCCAGGGGAUACCUGAGCA	
R	. α α α α . . β β β . . . . . α′ α′ α′ α′ . . (((. . . . .))) . . β′ β′ β′	−9.8 kcal/mol
A	. ((((. . . . .)))) . . (((. . . . .)))	−8.0 kcal/mol
F	. ((((. . . . .)))) . . ((. . . . . . .))	−5.4 kcal/mol
F	. ((((. . . . .)))) . . ((((. . .))))	−4.5 kcal/mol
B	. . . . (((. . . . . (((. . . . .))) . .)))	−2.1 kcal/mol
C	. . . . (((. . . . .)))	−2.7 kcal/mol

this situation, INFERNAL seems to destroy the stem with the smaller number of base pairs—and when both are equal, who knows?

Let us now solve the $K2N(E)$ problem by means of a suitable classification algebra. R denotes the (pseudoknotted) structure to be resolved, and r denotes its sequence. Let us reuse the known *OverDangle* grammar to span the folding space $F(s)$. Our new classification algebra *Stem* uses R to keep track of the types of base pairs each candidate $s \in F(r)$ uses. A consistent candidate must agree with R for all "normal" base pairs and unpaired bases except in opening and closing regions for α and β. There, it can leave bases unpaired. This candidate evaluates to (C). Should a consistent candidate also contain a base pair in α, it is classified as (A), or (B) if it uses a base pair of β. All other candidates are *faulty* (F). Note that there cannot be candidates in $F(r)$ which contain base pairs in α and in β, as all structures in $F(r)$ are un-knotted. Consider Table 3 to see how the classification works.

For our new classification algebra *Stem* we need some additional functions to correctly combine results of substructures with respect to their classification.

- $t_{bp}(x, y)$ is a function that returns the type (C, A, B, or F) of the base pair (x, y) by looking it up in R.
- $t_{ss}(x)$ is a function that returns F if the region x of single stranded bases contains a "normal" base pair in R, otherwise it returns C.
- $x \oplus y$ is a binary operation to combine the types of subsolutions x and y. $\oplus$ is defined by the table

$\oplus$		y			
		C	A	B	F
	C	C	A	B	F
x	A	A	A		F
	B	B		B	F
	F	F	F	F	F

Note that there is no result defined for the cases $(x, y) = (A, B)$ and $(x, y) = (B, A)$, because they cannot occur for any candidate $F(s)$.

With these auxiliaries, we can define our algebra *Stem* to classify candidates into types A, B, C, or F.

Algebra *Stem*	
$sadd(x, s)$	$= s \oplus t_{ss}(x)$
$cadd(s_1, s_2)$	$= s_1 \oplus s_2$
$nil()$	$= C$
$init(s)$	$= s$
$sr(x, s, y)$	$= t_{bp}(x, y) \oplus s$
$hl(x_1, x_2, s, y_2, y_1)$	$= t_{bp}(x_1, y_1) \oplus t_{bp}(x_2, y_2) \oplus \quad t_{ss}(s)$
$bl(x_1, x_2, lr, s, y_2, y_1)$	$= t_{bp}(x_1, y_1) \oplus t_{bp}(x_2, y_2) \oplus s \oplus t_{ss}(lr)$
$br(x_1, x_2, s, rr, y_2, y_1)$	$= t_{bp}(x_1, y_1) \oplus t_{bp}(x_2, y_2) \oplus s \oplus t_{ss}(rr)$
$il(x_1, x_2, lr, s, rr, y_2, y_1)$	$= t_{bp}(x_1, y_1) \oplus t_{bp}(x_2, y_2) \oplus s \oplus t_{ss}(lr) \oplus t_{ss}(rr)$
$ml(x_1, x_2, s, y_2, y_1)$	$= t_{bp}(x_1, y_1) \oplus t_{bp}(x_2, y_2) \oplus s$
$id(s)$	$= s$
$addss(s_1, s_2)$	$= s_1$
$base(x)$	$= x$

Using this algebra, the call

$$G(Stem * Count, x)$$

classifies each candidate of $F(s)$ into one of the classes C, A, B, or F and counts how many members each class has.

$$G(Stem * MFE, x)$$

also sorts candidates into their classes, but instead of the size of the class, its energetically most favorable candidate is reported. This solves the $K2N(E)$ problem.

Family RF01072 from the 10.1 Rfam release shall serve as an example. Its alignment contains 27 sequences. The annotated consensus structure is:

$\alpha\alpha\alpha\alpha......\beta\beta\beta\beta\beta\beta\alpha'\alpha'\alpha'\alpha'......\beta'.\beta'\beta'\beta'\beta'\beta'$

Rfam curators decided to break the β stem and retain the α stem. Let us now check if this decision is consistent with our energy-based criterion: The first sequence in the multiple alignment for RF1072 is

x_i = AGUGUUUUUCCCUCCACUUAAAUCGAAGGAU.

The result of $G(Stem^*MFE, x_1)$ is

```
( C ,  0.00 )
( F , -1.90 )
( A ,  0.81 )
( B , -0.40 ).
```

It tells us—ignoring faulty class (F)—that the β stem contributes –0.4 kcal/mol stabilizing energy while α destabilizes the conformation by 0.81 kcal/mol. A clear vote to keep stem β and remove α.

For another sequence x_2 = AGUGUUUCGGCUUCCACUUAAAUCG AAAGCC,
$G(Stem^*MFE, x_2)$ yields

```
( C ,  0.00 )
( F , -2.70 )
( A , -3.12 )
( B , -0.40 )
```

This time stem α is the winner. If we look at all 27 sequences of RF01072, 24 vote for β and just 3 for α. Thus, our pseudoknot solver contradicts the Rfam decision and suggests the following un-knotted consensus structure:

$$..........\beta\beta\beta\beta\beta\beta..........\beta'.\beta'\beta'\beta'\beta'\beta'$$

It is worth to keep an eye on class (C), because there are some rare cases (for example, family RF00390), where it is energetically favorable to break α and β stems.

6 The RNA shapes Package and Related Software

6.1 The RNA *shapes* Package

The RNA shapes tool has been online since 2004 [6] and available as a software package since 2006 [9]. A number of diverse features have been added since its first release. It is available (open source) for download, and also online as an interactive web site and as a webservice. The source code is interesting if you want to experiment with your own ideas about shape abstraction, as we exemplified in the previous section. The program is written in the algebraic dynamic programming style and implemented with Bellman's GAP [27], which means that adding in a new evaluation algebra is indeed as easy as claimed above.

When you visit the RNA shapes web site at bibiserv.cebitec. uni-bielefeld.de, you are first asked to make up your mind on the task you want to use it for. The first two modes of usage are the ones explained in detail in this chapter:

- *Shapes* computes shape representative structures, up to an energy threshold you can specify.

- *Probs* computes shape probabilities, in addition to the shape representative structures.

Other features are provided for convenience:

- *Rapid Shapes* [32] is the runtime heuristic that allows you to compute the probabilities of the most likely shapes (up to a threshold) more quickly than by the exhaustive mode. It is described in Section 4.2.
- *Suboptimal folding* reports for each shape class within an energy threshold not only shape representative structures but also all near-optimal structures below this threshold. Thus, it mimics RNASUBOPT, enriched by shape strings for every structure.
- *Sampling mode* does stochastic sampling from the Boltzmann distribution, as also provided by other RNA folding programs. To realize the result heuristic, mentioned in Section 4.2, sampled structures are translated into shapes strings of the desired abstraction level. The frequencies of those shape strings are used as estimators for their according shape probabilities. A switch allows to also report sampled structures, which are omitted by default.
- *Cast* addresses the situation where you try to find a consensus structure for sequences too diverged to align well, so RNAalifold will not work, and are too long or too many, such that approaches based on the Sankoff algorithm are impractical (cf. Chapters 7 and 13 in this book). This mode, named *RNAcast* in [33], quickly determines *consensus shapes* shared by any number of sequences, gives you their corresponding *shreps* and leaves you alone to decide which set of shreps might be a good choice for the consensus. In particular, when looking for conformational switches, which have several consensus structures, this is a strategy not supported by any other tool.

Finally, there is a service to the theorists. Shape abstraction has raised considerable theoretical interest. The combinatorics of shapes have been analyzed extensively in [24, 25], but some questions are still unresolved.

- *Shape combinatorics* takes you to a web site where you can experiment with shapes and study shape numbers for test sequences empirically.

6.2 Related Problems and Tools

Here we list a few other task and programs which are based on the abstract shapes approach.

RNALISHAPES [10] applies shape analysis to aligned RNA sequences, and hence constitutes a hybrid of RNA shapes and RNAALIFOLD.

Generation of shape matchers is the following problem: Given shape p, derive an algorithm m_p such that for any sequence x,

$$m_p(x) = (s, E(s), Prob(p, x)) \text{ where } s = s_p^*(x)$$

Such a program m_p is called a *thermodynamic matcher* (TDM) for shape p, as it computes how to fold any sequence x into shape p in the thermodynamically most favorable way. It is undefined for a given sequence x when $p \notin \pi(F(x))$. Generation of shape matchers is not available as a stand-alone tool, but used inside the *Rapid Shapes* approach as well as in the following tool.

LOCOMOTIF [34] generates thermodynamic matchers from structure graphics. It can be seen as the thermodynamics- and graphics-based successor to RNAMOTIF [35]. Users define structural motifs via an interactive editor and annotate them with size constraints and sequence motifs. Graphics are then compiled into matchers which can be used to scan genomes for occurrences of the motif.

RNASIFTER [36] is a technique to speed up Rfam searches, when it comes to matching a large number of RNA sequences, say from an RNA-seq experiment, against all models in Rfam. It pre-computes *shape spectra* of all Rfam families and stores them in an index. For a given RNA transcript, the index can be searched quickly to determine those families which have a common shape with the query. Only for the matching families, the expensive covariance model search must be applied.

As indicated by our custom-made shape abstraction in Section 5, this list of approaches building on the abstract shapes idea is likely to extend further in the near future.

Acknowledgment

Many people have contributed to the development of the abstract shapes approach. We gratefully acknowledge the early contributions by Björn Voß, Peter Steffen, Marc Rehmsmeier, and Jens and Janina Reeder.

References

1. Mathews DH, Sabina J, Zuker M, Turner DH (1999) Expanded sequence dependence of thermodynamic parameters improves prediction of RNA secondary structure. J Mol Biol 288(5):911–940. ISSN 00222836. DOI 10.1006/jmbi.1999.2700. URL http://dx.doi.org/10.1006/jmbi.1999.2700
2. Zuker M, Stiegler P (1981) Optimal computer folding of large RNA sequences using thermodynamics and auxiliary information. Nucleic Acids Res 9(1):133–148. ISSN 0305–1048. DOI 10.1093/nar/9.1.133
3. Markham NR, Zuker M (2008) UNAFold: software for nucleic acid folding and hybridization. Meth Mol Biol (Clifton, NJ) 453:3–31. ISSN 1064–3745. DOI 10.1007/978-1-60327-429-6_1
4. Hofacker IL, Fontana W, Stadler PF, Bonhoeffer SL, Tacker M, Schuster P (1994) Fast folding and comparison of

RNA secondary structures. Monatsh Chem 125:167–188. DOI http://dx.doi.org/10.1007/BF00818163. URL http://dx.doi.org/10.1007/BF00818163
5. Reuter J, Mathews D (2010) RNAstructure: software for RNA secondary structure prediction and analysis. BMC Bioinformatics 11(1):129. ISSN 1471–2105. DOI 10.1186/1471-2105-11-129. URL http://www.biomedcentral.com/1471-2105/11/129
6. Giegerich R, Voß B, Rehmsmeier M (2004) Abstract shapes of RNA. Nuclic Acids Res 32(16):4843–4851. DOI 10.1093/nar/gkh779. URL http://nar.oxfordjournals.org/cgi/content/abstract/32/16/4843
7. Voß B, Giegerich R, Rehmsmeier M (2006) Complete probabilistic analysis of RNA shapes. BMC Biol 4(1):5. ISSN 1741–7007. DOI 10.1186/1741-7007-4-5. URL http://www.biomedcentral.com/1741-7007/4/5
8. Reeder J, Giegerich R (2009) RNA secondary structure analysis using the RNA shapes package. Wiley. ISBN 9780471250951. DOI 10.1002/0471250953.bi1208s26. URL http://dx.doi.org/10.1002/0471250953.bi1208s26
9. Steffen P, Voß B, Rehmsmeier M, Reeder J, Giegerich R (2006) RNA shapes: an integrated RNA analysis package based on abstract shapes. Bioinformatics 22(4):500–503. ISSN 1367–4803. URL http://view.ncbi.nlm.nih.gov/pubmed/16357029
10. Voß B (2006) Structural analysis of aligned RNAs. Nucleic Acids Res 34(19):5471–5481. DOI 10.1093/nar/gkl692. URL http://nar.oxfordjournals.org/content/34/19/5471.abstract
11. Doshi KJ, Cannone JJ, Cobaugh CW, Gutell RR (2004) Evaluation of the suitability of free-energy minimization using nearest-neighbor energy parameters for RNA secondary structure prediction. BMC Bioinformatics 5. ISSN 1471–2105. DOI 10.1186/1471-2105-5-105. URL http://dx.doi.org/10.1186/1471-2105-5-105
12. Mathews DH, Turner DH (2006) Prediction of RNA secondary structure by free energy minimization. Curr Opin Struct Biol 16(3):270–278. ISSN 0959–440X. DOI 10.1016/j.sbi.2006.05.010. URL http://dx.doi.org/10.1016/j.sbi.2006.05.010
13. Meyer IM, Miklós I (2004) Co-transcriptional folding is encoded within RNA genes. BMC Mol Biol 5(1). ISSN 1471–2199. DOI 10.1186/1471-2199-5-10. URL http://dx.doi.org/10.1186/1471-2199-5-10
14. Mandal M, Breaker RR (2004) Gene regulation by riboswitches. Nat Rev Mol Cell Biol 5(6):451–463
15. Waldminghaus T, Gaubig LC, Klinkert B, Narberhaus F (2009) The Escherichia coli ibpA thermometer is comprised of stable and unstable structural elements. RNA Biol 6(4):455 – 463. DOI 10.4161/rna.6.4.9014. URL http://www.landesbioscience.com/journals/rnabiology/article/9014/
16. McCaskill JS (1990) The equilibrium partition function and base pair binding probabilities for RNA secondary structure. Biopolymers 29(6–7):1105–1119. ISSN 0006–3525. DOI 10.1002/bip.360290621. URL http://dx.doi.org/10.1002/bip.360290621
17. Wuchty S, Fontana W, Hofacker IL, Schuster P (1999) Complete suboptimal folding of RNA and the stability of secondary structures. Biopolymers 49(2):145–165. ISSN 0006–3525. URL http://view.ncbi.nlm.nih.gov/pubmed/99169417
18. Lorenz R, Bernhart SH, Hoener zu Siederdissen C, Tafer H, Flamm C, Stadler PF, Hofacker IL (2011) ViennaRNA Package 2.0. Algorithms for Molecular Biology 6(1):26. ISSN 1748–7188. DOI 10.1186/1748-7188-6-26 URL http://www.almob.org/content/6/1/26
19. Ding Y, Lawrence CE (2003) A statistical sampling algorithm for RNA secondary structure prediction. Nuclic Acids Res 31(24):7280–7301. ISSN 1362–4962. DOI 10.1093/nar/gkg938. URL http://dx.doi.org/10.1093/nar/gkg938
20. Chan CY, Lawrence CE, Ding Y (2005) Structure clustering features on the Sfold Web server. Bioinformatics 21(20):3926–3928. ISSN 1367–4803. DOI 10.1093/bioinformatics/bti632. URL http://dx.doi.org/10.1093/bioinformatics/bti632
21. Giegerich R, Haase D, Rehmsmeier M (1999) Prediction and visualization of structural switches in RNA. In: Proc. 1999 Pacific Symposium on Biocomputing, pp 126–137. World Scientific, Singapore
22. Voß B, Meyer C, Giegerich R (2004) Evaluating the predictability of conformational switching in RNA. Bioinformatics 20:1573–1582. URL http://bioinformatics.oupjournals.org/cgi/content/abstract/bth129v1?ct
23. Janssen S, Schudoma C, Steger G, Giegerich R (2011) Lost in folding space? comparing four variants of the thermodynamic model for RNA secondary structure prediction. BMC Bioinformatics 12(1):429. ISSN 1471–2105. DOI 10.1186/1471-2105-12-429. URL http://www.biomedcentral.com/1471-2105/12/429

24. Lorenz WA, Ponty Y, Clote P (2008) Asymptotics of RNA shapes. J Comput Biol 15(1):31–63. DOI 10.1089/cmb.2006.0153. URL http://dx.doi.org/10.1089/cmb.2006.0153
25. Nebel ME, Scheid A (2009) On quantitative effects of RNA shape abstraction. Theory Biosci. 128(4):211–225. ISSN 1431-7613. DOI 10.1007/s12064-009-0074-z. URL http://dx.doi.org/10.1007/s12064-009-0074-z
26. Brejová B, Brown DG, Vinař T (2007) The most probable annotation problem in HMMs and its application to bioinformatics. J Comput Syst Sci 73(7):1060–1077. DOI http://dx.doi.org/10.1016/j.jcss.2007.03.011
27. Sauthoff G, Janssen S, Giegerich R (2011) Bellman's GAP - a declarative language for dynamic programming. In: Proceedings of 13th International ACM SIGPLAN Symposium on Principles and Practice of Declarative Programming, PPDP '11. ACM. 12:29–40. ISBN 978-1-4503-0776-5. DOI 10.1145/2003476.2003484. URL http://doi.acm.org/10.1145/2003476.2003484
28. Zehner ZE, Shepherd RK, Gabryszuk J, Fu T-F, Al-Ali M, Holmes WM (1997) RNA-protein interactions within the 3' untranslated region of vimentin mRNA. Nucleic Acids Res 25(16):3362–3370. DOI 10.1093/nar/25.16.3362. URL http://nar.oxfordjournals.org/content/25/16/3362.abstract
29. Gardner PP, Daub J, Tate J, Moore BL, Osuch IH, Griffiths-Jones S, Finn RD, Nawrocki EP, Kolbe DL, Eddy SR, Bateman A (2010) Rfam: Wikipedia, clans and the "decimal" release. Nucleic Acids Res. DOI 10.1093/nar/gkq1129. URL http://nar.oxfordjournals.org/content/early/2010/11/08/nar.gkq1129.abstract
30. Smit S, Rother K, Heringa J, Knight R (2008) From knotted to nested RNA structures: A variety of computational methods for pseudoknot removal. RNA 14(3):410–416. DOI 10.1261/rna.881308. URL http://rnajournal.cshlp.org/content/14/3/410.abstract
31. Nawrocki EP, Kolbe DL, Eddy SR (2009) Infernal 1.0: inference of RNA alignments. Bioinformatics 25(10):1335–1337. DOI 10.1093/bioinformatics/btp157. URL http://bioinformatics.oxfordjournals.org/content/25/10/1335.abstract
32. Janssen S, Giegerich R (2010) Faster computation of exact RNA shape probabilities. Bioinformatics 26(5):632–639. DOI 10.1093/bioinformatics/btq014. URL http://bioinformatics.oxfordjournals.org/cgi/content/abstract/26/5/632
33. Reeder J, Giegerich R (2005) Consensus shapes: an alternative to the Sankoff algorithm for RNA consensus structure prediction. Bioinformatics 21(17):3516–3523. ISSN 1367-4803. URL http://view.ncbi.nlm.nih.gov/pubmed/16020472
34. Reeder J, Reeder J, Giegerich R (2007) Locomotif: from graphical motif description to RNA motif search. Bioinformatics 23(13):i392. DOI doi:10.1093/bioinformatics/btm179
35. Macke TJ, Ecker DJ, Gutell RR, Gautheret D, Case DA, Sampath R (2001) RNAMotif, an RNA secondary structure definition and search algorithm. Nucleic Acids Res 29(22):4724–4735. DOI 10.1093/nar/29.22.4724. URL http://nar.oxfordjournals.org/content/29/22/4724.abstract
36. Janssen S, Reeder J, Giegerich R (2008) Shape based indexing for faster search of RNA family databases. BMC Bioinformatics 9:131. DOI 10.1186/1471-2105-9-131. URL http://dx.doi.org/10.1186/1471-2105-9-131

Chapter 12

Introduction to RNA Secondary Structure Comparison

Stefanie Schirmer, Yann Ponty, and Robert Giegerich

Abstract

Many methods have been proposed for RNA secondary structure comparison, and new ones are still being developed. In this chapter, we first consider structure representations and discuss their suitability for structure comparison. Then, we take a look at the more commonly used methods, restricting ourselves to structures without pseudo-knots. For comparing structures of the same sequence, we study base pair distances. For structures of different sequences (and of different length), we study variants of the tree edit model. We name some of the available tools and give pointers to the literature. We end with a short review on comparing structures with pseudo-knots as an unsolved problem and topic of active research.

Key words RNA structure comparison - Base pair distance - Tree edit distance - Tree alignment distance - Forest alignment

1 Introduction

In many chapters of this book, we study the problem of *assigning* secondary structure to RNA sequences. In the present chapter, we consider this problem solved. We are given a set of RNA sequences, each together with one or more secondary structures. We do not care about their origin—they could be predicted by one of the algorithms in the other chapters, they could be sampled from the folding space, derived from resolved 3D structures, or even generated as a random test set. The question at hand is how to compare these structures. Our motivation might be to cluster them into families, or to evaluate a new prediction tool against others or against a reference database.

In principle, all questions we deal with in sequence comparison re-pose themselves with structures: Pairwise (global) alignment, best local alignment, finding the best fit of a small structure in a larger one. Naturally, algorithms for structure comparison are different from those used in sequence comparison, but in a very systematic way. Sequences have a one-dimensional principle of organization: adjacency. One character follows the other.

Jan Gorodkin and Walter L. Ruzzo (eds.), *RNA Sequence, Structure, and Function: Computational and Bioinformatic Methods*, Methods in Molecular Biology, vol. 1097, DOI 10.1007/978-1-62703-709-9_12,

Structures have two dimensions: adjacency and inclusion. Unless we have pseudo-knots, two RNA helices are either adjacent or one includes the other. Where each alignment step with two sequences recurs on a single subproblem of aligning two suffixes of the sequences, each alignment step with structures creates two subproblems—at least. So, even if you think sequence comparison is a worn-out and dull topic algorithmically, you might find structure comparison quite interesting.

In fact, RNA structure comparison has been a creativity parlor for computer scientists. Fancy representations of RNA structures have been suggested and algorithms working thereon have been proposed. In this chapter, we focus on a small number of approaches which have been widely used and which, in our humble opinion, essentially solve the basic problem to satisfaction—for structures without pseudo-knots.

The structure of this chapter is as follows:

1. We begin with a section on structure representation, as it is the representation on which our algorithms work. Essentially, we argue that representation should be spartanic, and not introduce extra detail that is irrelevant and may confuse the comparison.
2. Then, we focus on the problem of (global) structure comparison. We first discuss the simple scenario, where we compare structures from the folding space of a single sequence. This is the easier problem, as all structures have the same size, and the i-th residue means the same base for each structure.
3. We then study the general problem: comparing arbitrary structures from different sequences of varying length. We present two alternative methods and report on programs that implement them. At the end of this section, we discuss some variations of these problems and algorithms.
4. In the concluding section of this chapter, we report on the difficulties encountered with the comparison of pseudo-knotted structures and give hints to the literature.

2 A Few Words on Secondary Structure Representation

In this chapter, we can safely assume that our reader is familiar with the notions of RNA secondary structure and has already encountered various forms of structure representation. Here, we discuss representations from the viewpoint of their suitability for structure comparison, both by human experts and by computer programs.

2.1 Aah, Squiggle Plots!

The most popular representation of RNA secondary structure, and one of the most intuitive, is the so-called *squiggle plot*. It provides

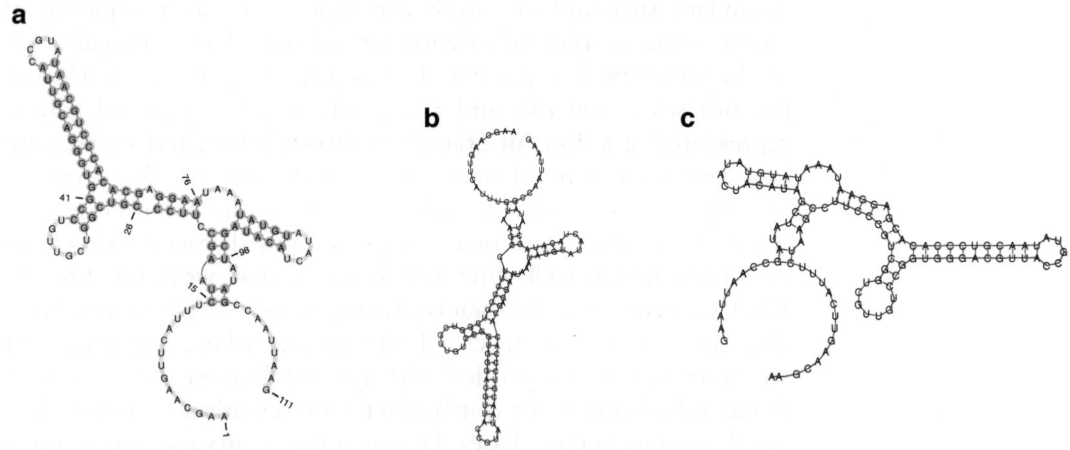

Fig. 1 Three squiggle plots of the same secondary structure, drawn by different programs. (**a**) drawn by pseudoviewer [1], (**b**) drawn by RNAplot [2] with the naview layout option, (**c**) drawn by RNAplot with the simple radial layout option. Note that not only the layout but also the orientation changes between clockwise and counterclockwise

a graphical 2D layout of the helices and loops that make up a secondary structure. Even when structures are quite large, these plots give us a good overview of characteristic structural features.

However, squiggle plots have a shortcoming when it comes to structure comparison. Each structure can be drawn in many different ways. In Fig. 1, you see three plots of the same structure, produced by different algorithms. Why are they different? A graphical visualization leaves much room for choice. The layout of helices asks for an aesthetic design, the angles chosen between branching helices in a multiloop are designed to avoid overlapping of substructures. The drawing of the external loop akin to a hairpin loop in a circular fashion, as you see it in Fig. 1, is a bit of a questionable convention. There is nothing to bring the 5' and the 3' nucleotide into vicinity. So, structure drawings are good for comparison by human inspection only if they are produced by the same algorithm, or if their layout has been engineered by a human expert to exhibit their similarity. After structures have been aligned, visualizing the alignment with a graphics tool (such as *VARNA* [3]) is quite useful for the human inspector. So, while using such plots as computer *input* for structure comparison does not seem a good idea, many programs produce them as output. For embedding such a program in a software pipeline, a second, more computer-oriented representation must be offered.

2.2 Dot-Bracket Representation

A good compromise between human readability and suitability for program input is the *dot-bracket representation*, popularized, for example, by the *ViennaRNA* package [2]. Most commonly, each secondary structure of length n is represented as a sequence of length n that consists of parentheses and dots. Each base pair (i, j) in the structure is represented by a pair of opening and closing parentheses at the i-th and j-th position. Each unpaired base is represented as a dot. An example is shown below, and we are sure you have seen many of those in the other chapters. Some people use angle or curly brackets rather than the round parentheses, and when different types of brackets are combined, pseudo-knots can be represented as well. One nice feature is that, written below the RNA sequence, the dot-bracket string annotates it with structure. This also works for structural alignments, where sequence and structure strings are padded with gap symbols in the same way. A second advantage of dot-bracket representation is that it does *not* show the concrete bases. Hence, it represents a structure *per se* and can be associated with any RNA sequence of the same length—suitable bases for matching brackets provided.

```
       1 2 3 4 5 6 7 8 9 10 11 12 13 14 15 16 17 18 19 20
S1 =   C A G C U C U C A U  G  U  C  C  C  A  C  A  T  A
S2 =   C A G C U C U G U G  U  C  C  C  A  C  A  C  A  A
D1 =   . . . . . . . . ( (  (  (  .  .  .  )  )  )  )  .
D2 =   . . . . . . . ( ( (  (  .  .  .  )  )  )  )  .  .
```

Structures not extending beyond one line can well be read and compared by a human, given some practice. But one or two line breaks already destroy much of the readability. A caveat is the following. While the overall arrangement of helices is fairly easy to see, it is difficult to check if the opening and closing parentheses actually match up well. It requires a little parsing algorithm to check for this property. Do not forget to include this check while writing a program that takes dot-bracket strings as input.

2.3 Base Pair Set Representation

For mathematical purposes, a secondary structure is often defined as a set of base pairs, or more precisely, the set of base paired positions in the primary sequence.

```
       1 2 3 4 5 6 7 8 9 10 11 12 13 14 15 16 17 18 19 20
S1 =   C A G C U C U C A U  G  U  C  C  C  A  C  A  T  A
S2 =   C A G C U C U G U G  U  C  C  C  A  C  A  C  A  A
```

$B_1 = \{\{9,19\},\{10,18\},\{11,17\},\{12,16\}\}$
$B_2 = \{\{8,18\},\{9,17\},\{10,16\},\{11,15\}\}$

This representation is quite impractical for human inspection, and for computer input and output, the mathematical *set* should be represented by a *sorted list* to make comparison easy.

2.4 Tree Representations

The *tree representations* constitute a class of representations that may differ in their level of detail. Generally, a secondary structure is represented by an ordered rooted tree. Trees mathematically

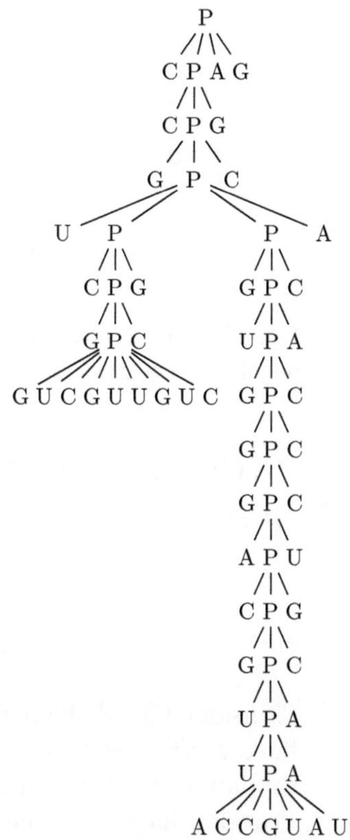

Fig. 2 A tree representation of a y-shaped substructure of the structures plotted in Fig. 1, including residues 25–73

encompass the two organization principles of adjacency and inclusion. A *rooted ordered tree* consists of a root node and a forest of subtrees (inclusion). A *forest* is a possibly empty sequence of trees (adjacency). We normally speak of tree representation, but—given that a tree node has a forest of subtrees—speaking of forest representations would be at least as adequate. Note that there are no empty trees, while forests can be empty.

Figure 2 shows one out of many tree representations. This one combines sequence and structure representation. At the leaves of the tree, we find the primary sequence, in 5' to 3' order. A node labeled *P* denotes a base pair bond between the two bases represented as its left- and rightmost child.

From what we said about squiggle plots, you might expect that we consider trees as unsuitable for communication between programs producing and consuming or comparing structures. But this is not the case, since a tree can be encoded by straightforward formula, allowing for an easy manipulation. For example,

$$P(C, P(C, P(G, P(U, P(C, P(G, P(G, U, C, G, U, U, G,$$

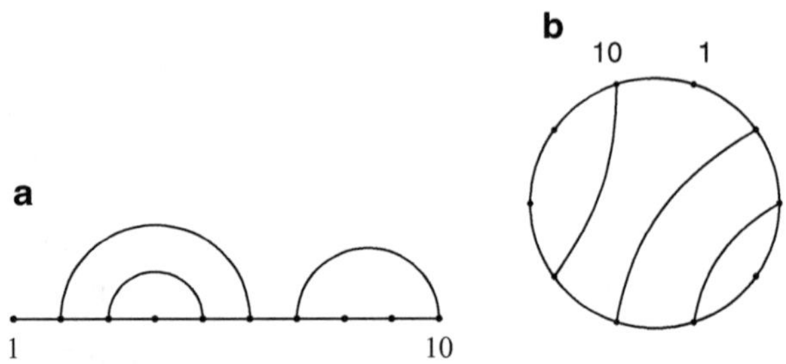

Fig. 3 Two equivalent representations, except for their layout .... (**a**) Arc-annotated sequence. (**b**) Circle representation

$$U, C), C), G), P(G, P(U, P(G, P(G, P(G, P(A, P(C, P(G, P(U,$$
$$P(U, P(A, C, C, G, U, A, U)A), A), C), G), U), C), C), C), A),$$
$$C), A), C), G), A, G)$$

encodes the tree of Fig. 2.

2.5 Further Representations

2.5.1 Arc-Annotated Sequences

Consider Fig. 3. Depending on which lines we draw straight, we have a *circle plot* or an *arc-annotated sequence*. You can also see the latter as a tree, where the leaves are written on a line and the P-nodes have become the arcs. If so, these arcs never cross. But here is the important difference that makes arc-annotated sequences interesting: Once we do allow to connect arbitrary bases by arcs, we can represent more general structures as well. We may allow crossing arcs leading to pseudo-knots, and bases incident to two arcs, modeling triple base interactions. Most of the recent work that considers pseudo-knotted structures uses the arc-annotated sequence representation.

2.5.2 Coarse Graining

Sometimes when we ask for structural similarity, we do not care about resolution on the base pair level. Trees nicely lend themselves also to coarse-grained structure representations. For example, a tree representation can also be reduced to a homeomorphic irreducible tree (HIT) representation, in which all sequences of internal nodes which have only one child are contracted into one node. A weight associated with it might reflect the number of nodes that are combined into this one node [4, 5]. A coarse-grained tree representation, as we show it in Fig. 4, indicates structural elements, such as stems, hairpin loops, internal loops, bulges, multibranch loops (multiloops). A fictitious root node (labeled N in Fig. 4) is added sometimes. It does not correspond to a structural element, but ensures the formation of a tree (preventing a forest).

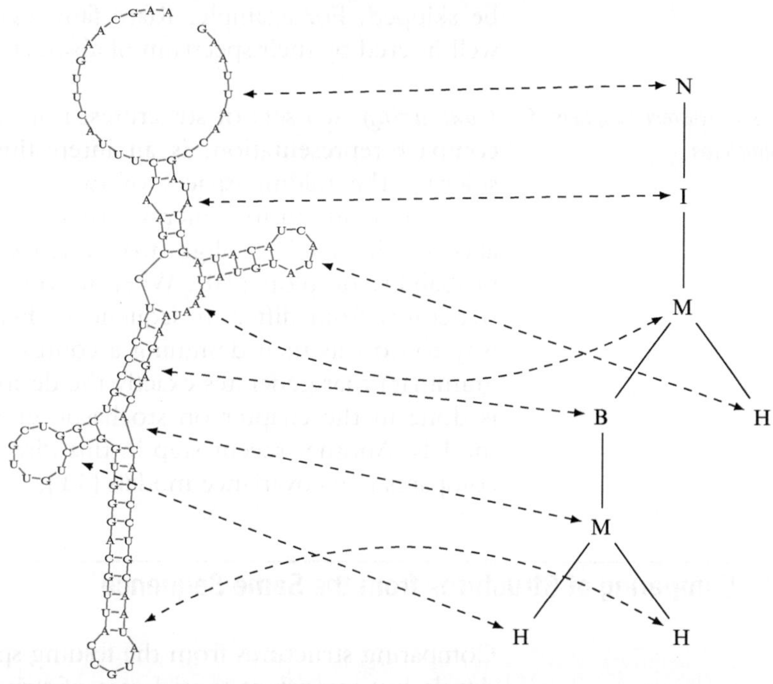

a RNA secondary structure **b** The coarse tree representation

Fig. 4 Coarse-grained tree representation, which represents an RNA secondary structure as a tree of structural building blocks such hairpin loops (H), multiloops (M), bulges (B), internal loops (I). Node N does not represent a structural element, it closes the secondary structure and makes sure the representation forms a tree

Many similar representations are found in the literature. In [6], stems are not included as nodes, but just loops. In [5], the stems may be included and the stem loop size can be used to influence the score function.

Another coarse-grain representation are abstract shapes [7], see also Chapter 11. A Y-shaped structure is denoted "[[][]]," a cloverleaf is "[[][][]]." Shapes can also cover different levels of abstraction. If the presence of bulges is to be indicated, the representation could be "[[_[]_]_]," where the underscores indicate that this simple stem-loop structure has an internal loop and a bulge on the 3' side. It is easy to convert an abstract shape string into a tree, and vice versa, so there is not much essential difference between shapes and other coarse-grain representations.

Coarse graining can be used to organize structure comparison in a hierarchic fashion, proceeding from more abstract to more concrete representations [8]. Coarse graining is also useful when we deal with large numbers of pairwise structure comparisons. Often, when there is no match on the coarse-grain structure, a detailed (and more expensive) comparison of two structures can be skipped. For example, Rfam families have been shown to be well filtered by their spectrum of abstract shapes [9].

2.5.3 Representing Sets of Structures

Comparing two sets of structures, not element-wise but in their compact representation, is an interesting problem. When considering the folding spaces of two sequences, rather than just the two minimum-free-energy structures, the *CONSTRUCT* tool allows to interactively align two sequences based on their base pair probability dotplots [10]. When it comes to representing sets of structures from different sequences, there is probably no better way to do this than designing a context free grammar (or a tree grammar) that generates exactly the desired set. In this book, this is done in the chapter on stochastic grammars and RNA family models. Another recent step in this direction is an approach that compares two covariance models [11].

3 Comparing of Structures from the Same Sequence

Comparing structures from the folding space of one sequence is a simple but important special case of structure comparison, which we cover first. When we deal with a single sequence, there is no need of inserting gaps to exhibit similarity. The methods described here are also applicable for structures of closely related sequences of the *same* length, a situation which may arise, for example, in the analysis of the impact of SNPs on the folding space of an RNA molecule.

For the following explanation, let B_1 and B_2 denote two alternative structures of an RNA sequence.

3.1 Base Pair Distance

Base pair sets are a useful representation in the present case, as "residue i" means the same nucleotide for all compared structures. Methods for comparing these sets can be transferred from set theory. The *symmetric set difference* is a good first approach to evaluate the difference of structures:

Definition 1: The naive *base pair distance* d_{BP} is the cardinality of the symmetric difference between the sets of base pairs B_1 and B_2,

$$d_{BP}(B_1, B_2) = |(B_1 \backslash B_2) \cup (B_2 \backslash B_1)|.$$

where a base pair is given as a pair of positions in the sequence.

Hence, we count the base pairs present in either of the structures, but not in both. This naive base pair distance is a metric. This metric is very strict. All differences have the same weight. Consider as examples the base pair sets B_1 and B_2 we encountered in Subsection 2.3. The two structures are considered as different as can be, $D_{bp}(B_1, B_2) = 8$, as *all* base pairs are different. Intuitively, one would say that the two structures are not that different, as both have the same number of base pairs, in the same arrangement, but shifted one position. These concerns are addressed by variations of the base pair distance, such as our next refinement.

3.2 Hausdorff Distance

The Hausdorff distance measures the distance between sets of points. It captures the "maximum distance of a point in a set to the nearest point in the other set." The Hausdorff distance is a classic maximin function, which fulfills metric properties. Applying the Hausdorff distance to RNA secondary structures means that we interpret their base pair positions as coordinates in a 2D plane. A peculiarity: The sets of base pairs must not be empty, and hence this distance measure is not applicable to the empty (unfolded) structure.

Definition 2: The *Haussdorff distance* d_H is defined in three steps. First, the distance between two base pairs $(i,j) \in B_1$ and $(i',j') \in B_2$ is defined as

$$d\left((i,j),(i',j')\right) = \max\left(|i-i'|,|j-j'|\right).$$

We next formulate the distance between a base pair and a set:

$$d_a\left((i,j),B_2\right) = \min_{(i',j')\in B_2} d\left((i,j),(i',j')\right)$$

Then, the asymmetric distance d_a between two sets of base pairs is defined as:

$$d_a\left(B_1,B_2\right) = \max_{(i,j)\in B_1} \min_{(i',j')\in B_2} d\left((i,j),(i',j')\right).$$

Establishing symmetry gives the final definition of distance d_H:

$$d_H\left(B_1,B_2\right) = \max\left(d_a(B_1,B_2), d_a(B_2,B_1)\right).$$

For our examples used above, we find $d_H(B_1,B_2) = 1$.

The metric d_H can deal reasonably well with shifted base pairs, but also has its weakness. Differences in isolated base pairs can lead to very small or very high distance values depending on the distance to the next base pair. Inspired by the *Hausdorff metric*, a further refined distance d_Z on RNA secondary structures was developed by Zuker et al. [12] to circumvent the problem of isolated base pairs. The distance d_Z was used to filter very similar foldings from the output of near-optimal structures in the early *Mfold* program.

4 Comparing Structures from Different Sequences

In this section, we turn to methods suitable to compare structures of RNA molecules of different primary sequence and, in most cases, of different length. These methods are more general than the base pair metrics, and they are more expensive to compute. We start with a warning about simple ideas that seem plausible, but do not work well. Thereafter, we turn to two models of comparing structures represented as trees: the tree edit and the tree alignment model.

4.1 Ideas That Are Too Simple...

What do we expect from a general method of structure comparison? Here are two examples:

- Assume we are comparing two tRNA structures. One is the classical four-stem cloverleaf structure, the other exhibits a fifth stem that is observed with certain tRNAs. A good comparison method should be able to match up the "standard" stems and show the extra stem as an insertion.
- Assume we are comparing 5' and 3' UTRs of genes. We expect to find there a large number of small hairpins, about ten base pairs and a short hairpin loop each. If among them there are iron responsive elements—characterized as small hairpins with a bulged-out "C" nucleotide, we hope that a good comparison method would allow to match them up automatically.

Hence, a general comparison should allow for sequence insertions and deletions, as well as replacements of bases that preserve the base pairs in the structure. This general problem is tackled by the tree edit models studied in the next two sections. The methods we discuss in the subsequent sections have been shown to master these challenges.

Before we start, let us discuss why we cannot simply use sequence comparison methods. After all, we can represent structures as (dot-bracket) sequences! People familiar with sequence alignment might approach the problem by aligning dot-bracket strings. This might work for *very* similar structures, but in general, strange things will happen.

Consider an example such as

```
((((...))))...((((...))))
((((.................))))
```

Although the shown alignment is certainly an optimal *string* alignment, it is not an alignment that is consistent with structure. Left and right brackets that constitute a base pair in the lower sequence are aligned to left and right brackets in the upper structure—but these do not make up a base pair there! Sequences represent the principle of adjacency, but not that of inclusion.

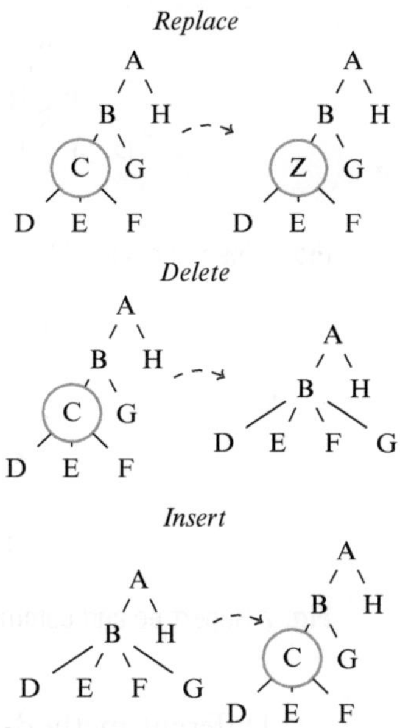

Fig. 5 Tree edit operations: Node replacement, deletion, and insertion

There is no way a sequence alignment algorithm can look back to the right place before aligning two closing brackets. People have tried to overcome this by using a richer sequence alphabet, where `M` denotes a multiloop, `B` a bulge, etc. It does not help. The general message is: No matter how we encode a structure in a string representation, a sequence alignment method will run into cases where it aligns parts of the sequence that do not coincide in structure. We need to make alignment algorithms work on trees.

4.2 The Tree Edit Model

The tree edit model requires finding a series of edit operations that transforms one input tree into the other with minimum overall cost, defined as the accumulated cost of the basic edit operations. We have replacements, insertions, and deletions as edit operations, as they are familiar from sequence comparison models, but here they operate on trees.

4.2.1 Tree Edit Operations

The tree edit operations *Replace*, *Delete*, and *Insert* are shown in Fig. 5.

Each edit operation is associated with a cost, which may depend on the node label that is replaced, deleted, or inserted. When one tree is edited into the other, the accumulated sum of edit operation costs makes up the overall cost of the editing. When we assign costs, we seek to minimize them. Alternatively, we might use similarity scores, which we seek to maximize.

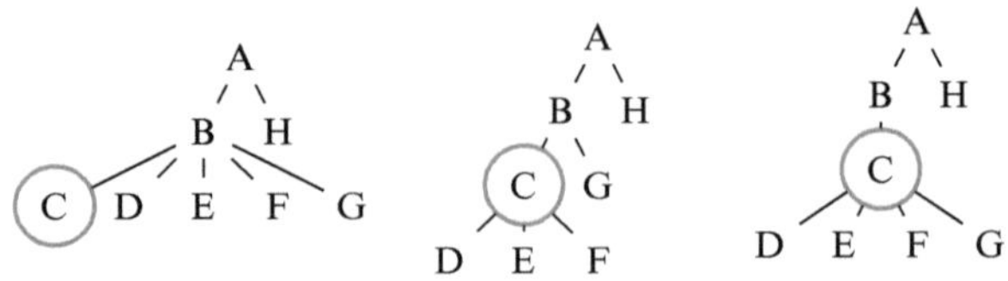

Fig. 6 Inserting node *C* as a child to *B*—three of the possible outcomes

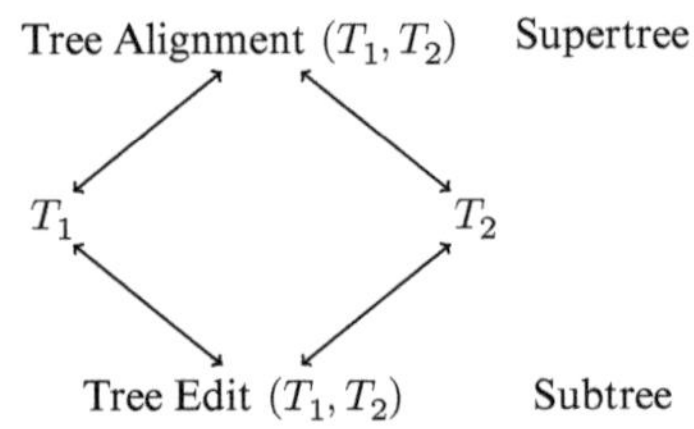

Fig. 7 Supertree and common subtree, via tree alignment and tree edit

Different methods of tree comparison can be based on these edit operations (and there is some confusion about this fact in the literature). It results from some vagueness in the phrase "we edit tree T into tree T' with a sequence of edit operations."

There is an important difference between insertions and deletions—in spite of the fact that they are mutually inverse operations. If we "delete a node" in T, the resulting tree is uniquely defined. If we "insert a node Y below node X" in T', it is not defined what the resulting tree is. The outcome depends on where we insert Y into the siblings of X, and which siblings of X are chosen to become siblings of the new node Y. Figure 6 shows alternative choices for inserting node C below B.

Thus, "a sequence of edit operations transforming tree T into tree T'," a phrase that we find frequently in the literature, cannot really *produce* T' from T, but can relate T and T' when both are given. And it can do so in different ways.

4.2.2 Edit or Align: Subtrees Versus Supertrees

Given our edit operations, different methods of tree comparison can be designed. Two of them have become popular in RNA structure comparison, *tree edit* and *tree alignment*, and we will describe them below. Given trees T_1 and T_2, tree alignment constructs a common supertree, while tree edit constructs a common subtree. Both minimize a cost function associated with the edit operations. Consider Fig. 7. The supertree can be transformed, applying *Delete* operations only, into either T_1 or T_2. Both T_1 and T_2 can be transformed, applying *Delete* operations only, into the common subtree.

It remains to be shown that subtree and supertree can be different. This is seen in Fig. 8. The supertree must have two E-nodes, as they have different sets of children in T and T'.

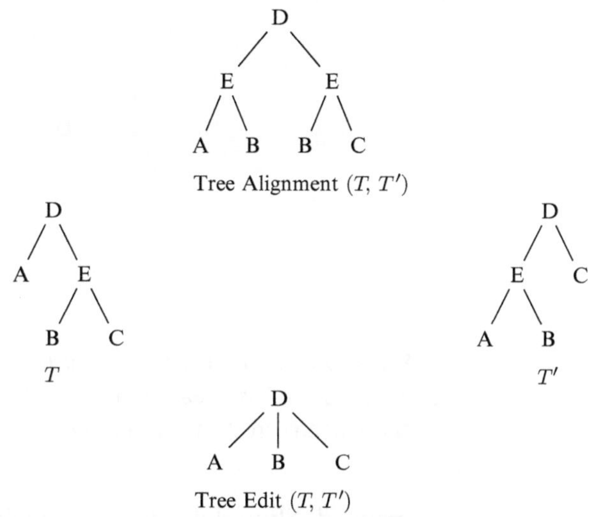

Fig. 8 Example of difference between common supertree and common subtree. Note that the two *B*-nodes in the supertree cannot be joined in the supertree without creating a dag, and the *E*-nodes cannot be joined without making *A* and *C* siblings

We focus on tree edit distance and tree alignment distance here, while further modes of comparing trees are conceivable. Before you venture to design yet another method, be sure to study the article by Rosselló and Valiente [13], where they relate these and other measures under the general notion of tree embeddings.

4.3 Tree Edit Distance

Because node labels generally are not unique, we reference tree nodes by their numbers in preorder. *label*(i) is the label of node number i. Let us write $i_1 \prec i_2$ when i_1 is an ancestor of i_2. Let w be a cost function defined on pairs of node labels.

Definition 3: A *mapping* M is a partial bijection between nodes of T and T' with the following property:
For any two pairs (i_1, j_1) and (i_2, j_2) in M,

1. $i_1 < i_2$ iff $j_1 < j_2$ (order preservation)
2. $i_1 \prec i_2$ iff $j_1 \prec j_2$ (ancestor preservation).

The *cost* of a mapping M is defined as $\sum_{(i,j)\in M} w(i,j)$

Order preservation is the condition we already have in sequence alignment, while ancestor preservation is the tree-specific property. A mapping can be seen as specifying replacements only, where the replaced nodes constitute the common subtree, while all others are deleted. *See* Fig. 9 for an illustration of a mapping between two trees. Note that it is impossible to extend the mapping the two nodes labeled F without violating the constraints. Should we decide to include (F, F), the nodes C and E can no longer be

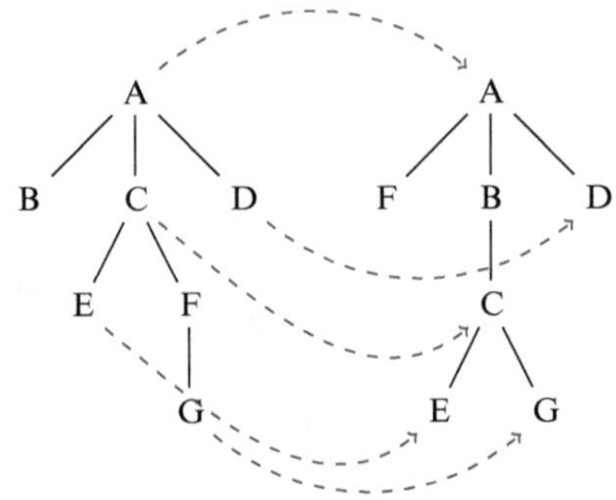

Fig. 9 Example of a legal tree mapping. Given the arcs shown, we could still add an arc B $\rightarrow$ F, whereas an arc F $\rightarrow$ B, B $\rightarrow$ B , or F $\rightarrow$ F would violate the ancestor preservation requirement

mapped. Should we decide to include $(A, E) \in M$, no other node can be mapped, as all nodes are descendents of A in T, while E has no descendent at all in T'.

Definition 4: The *tree edit distance* $d_{\mathrm{TE}}(T, T')$ of two trees T and T' is the minimum cost of a mapping between T and T'.

We can find the mapping of minimal cost by considering all possible edit sequences that touch each node in T and T' exactly once. This can be described by the following recurrences, given in graphical form in Fig. 10.

The figure is to be read as follows: Two adjacent boxes represent the forests to be mapped. The input forests are replaced by (alternatively) a *Replace*, *Delete*, or *Insert* operation plus the related subproblems that arise. Since we only score the replacements (because they constitute the mapping), case R is seen as producing (a^1, b^1) as a contribution to the mapping and $w(a^1, b^1)$ as a contribution to its score, while the D and I cases contribute no local score, but only the (optimal) score of the respective subproblem. All mappings are scored as they are constructed, and the minimal cost is chosen. In an implementation, this requires dynamic programming recurrences that store intermediate results in tables addressed by the subproblem solved. This is technically intricate, and we refer the reader to [5, 14–16] for these details.

4.4 Tree Alignment Distance

While tree edit says nothing about the nodes deleted, the tree alignment method has the advantage that it puts all nodes into a relationship in the common supertree. This gives rise to a representation of tree alignments akin to sequence alignments, either in graphical form as in Fig. 11 or as (aligned) sequences annotated with (aligned) dot-bracket structure.

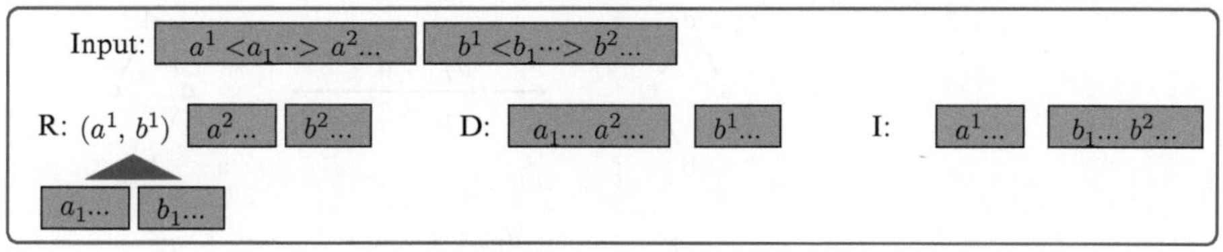

Fig. 10 Case distinction for the tree edit distance algorithm. On top are the two forests that we align, underneath are three cases corresponding to the edit operations.

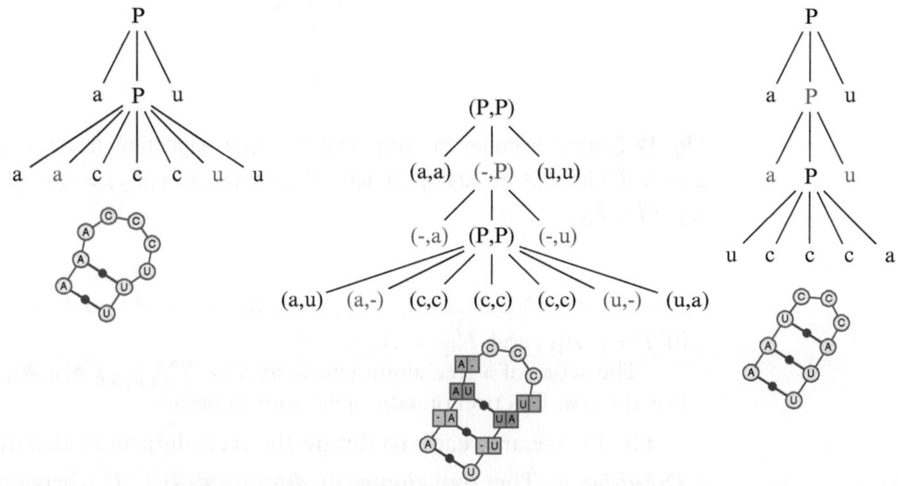

Fig. 11 Graphical representation of a tree alignment. On the left and right are two RNA hairpin structures represented as trees, and in the center is one of their possible tree alignments. Below of each tree is the corresponding squiggle representation (produced with *VARNA* [3])

4.4.1 Tree Alignments

Tree alignments are defined formally akin to sequence alignments. If you see a sequence alignments as a *sequence of columns*, where the columns may hold alphabet characters and a gap symbol, the following definition is obvious to you:

Definition 5: Given ordered labeled trees over some alphabet $\mathcal{A}$, an *alignment tree* is a tree over the pair alphabet $\mathcal{A}_{\text{pair}} = \mathcal{A} \cup \{-\} \times \mathcal{A} \cup \{-\} \setminus \{(-,-)\}$.

In an alignment tree, the node labels represent edit operations, where nodes of type (a, b) are replacements, $(a, -)$ are deletions and $(-, a)$ are insertions.

Given an alignment tree A, it is clear which are the two trees aligned in it. Its projections to the first and second component, $A|_1$ and $A|_2$, are trees on the alphabet $\mathcal{A} \cup \{-\}$ of RNA with gaps. These trees may be contracted by π, where $\pi(T)$ is the tree that results from deleting all nodes with the gap symbol from a tree T. The result of this contraction are the two aligned trees over $\mathcal{A} \cup \{-\}$. This observation gives us an elegant definition for a tree alignment:

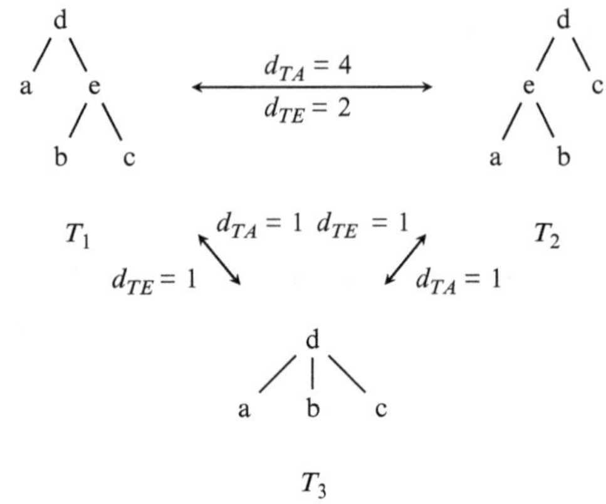

Fig. 12 Counterexample to show that the tree alignment distance does not satisfy the triangle inequality. Under unit cost, $d_{\mathrm{TA}}(T_1, T_2) > d_{\mathrm{TA}}(T_1, T_3) + d_{\mathrm{TA}}(T_2, T_3)$

Definition 6: A tree A over $\mathscr{A}_{\mathrm{pair}}$ is an *alignment of trees* T, T' over $\mathscr{A}$ iff $T = \pi(A|_1)$ and $T' = \pi(A|_2)$.

The score of a tree alignment is $w(A) = \sum_{(a,b) \in A} w(a, b)$, where w is the cost function on edit operations as before.

Finally, we are ready to define the tree alignment distance:

Definition 7: The *tree alignment distance* $d_{\mathrm{TA}}(T, T')$ between two trees is the minimum cost over all possible alignments of the two trees.

$$d_{\mathrm{TA}}(T, T') = \min\{w(A) \mid A \text{ is an alignment of } T \text{ and } T'\}$$

It is interesting to observe that the tree alignment distance is not a metric, as it does not satisfy the triangle inequality. Figure 12 uses unit cost and demonstrates a counterexample. The triangle inequality is violated for d_{TA} (while it generally holds for d_{TE} as long as $w(Replace(a, b))$ is a metric).

4.4.2 Computing Optimal Tree Alignments

To solve the tree alignment problem by finding the alignment with the optimal score, all possible candidate alignments have to be considered in a dynamic programming algorithm. The recurrences are similar to those of tree edit distance, but with important differences that take care that actually a supertree is constructed. Again, we give the recurrences in graphical form, *see* Fig. 13. Again, read these recurrences as the alignment tree spreading out recursively in two dimensions on a sheet of paper. Here, each edit operation chosen leads to a node recorded for the emerging supertree. The *Replace* case leads to two recursive alignment computations for each edit operation. *Delete* and *Insert*, however, lead to splits of certain subforests in all possible ways.

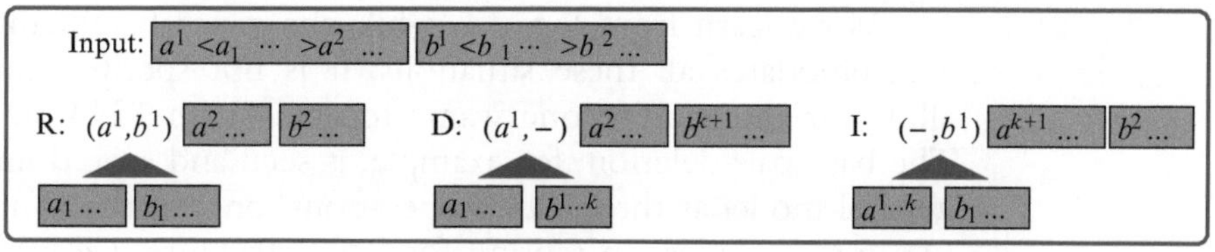

Fig. 13 Case distinction for the forest alignment distance algorithm. Again, $a^1 < a_1 \ldots > a^2$ denotes a forest whose first tree has root label a^1 and subtrees $a_1 \ldots$. Boxes indicate the forests that have to be aligned as (sub)problems. Edit operations are generated as nodes of the emerging alignment forest. On top are the two forests that we align, underneath are three cases corresponding to the edit operations.

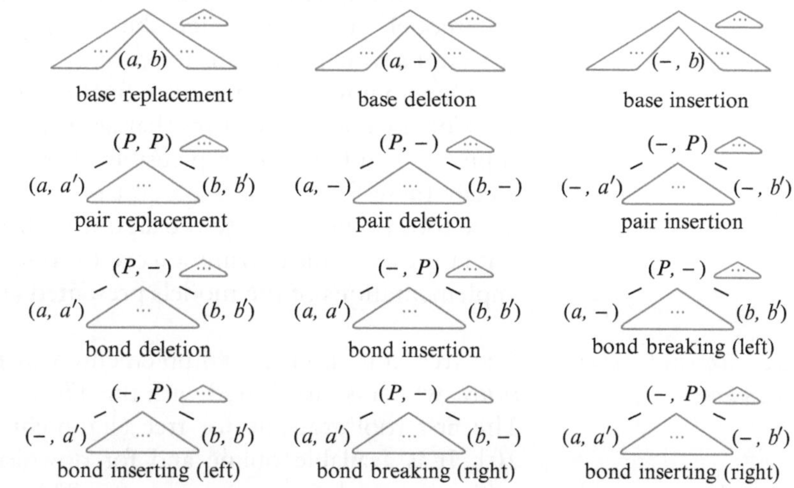

Fig. 14 RNA-specific edit operations for tree alignment

To compute the minimal cost of a tree alignment, the recurrences are executed in reverse, scores added as they arise, and optimal scores of sub-alignments are recorded in the typical dynamic programming fashion as we proceed from smaller to larger alignments. Efficiently storing and accessing these intermediate results requires some care—see [16] for the details.

4.5 Specifics of RNA Structure Comparison Under the Tree Edit Model

4.5.1 Refined Edit Operations for RNA Structure Trees

The two previous sections describe tree comparison models in general. Let us now see what tree editing means when comparing, in particular, RNA secondary structures. Following [17], we distinguish RNA-specific situations of interest, such as losing a single base and a base pair. In Fig. 14 we see how these operations are expressed as tree edits in our particular representation of tree alignments.

In our terminology, *Replace* includes the case where a base or base pair is preserved. The replacement scoring function will, of course, make the important difference between matches and mismatches.

As we learn from Fig. 14, while the general tree edit model accommodates all these situations, it is not specific enough to allow for the scoring one wants to apply with RNA structures. The base pair deletion, for example, is seen and scored under the general model as three *Delete* operations, one for the *P*-node and two for the bases. In scoring for RNA structure, we want to be able to assign a specific score to this situation, different from the sum of three independent deletions. Similarly, bond breaking is a consequence of losing one base of a base pair, so we might not want to score this as two independent events. Therefore, should you plan to implement RNA structure comparison based on a tree edit model, you must refine the general case to precisely those operations which you want to score.

One might take this even further—giving special scores to functionally relevant features that are expected in a certain ncRNA family at hand. It seems plausible that one should just have to specify these features as extra edit operations, supply a score, and some tool would generate the algorithm from such a specification. But at present, this remains a topic for research. Fortunately, good implementations of the models presented above are available.

4.5.2 Tools for Structure Comparison

The tree edit distance is implemented in the program *RNAdistance*, which is distributed with the *ViennaRNA* package [2, 18]. The first tool available for tree alignment was *RNAforester* [19, 20]. It is available online and for download at http://bibiserv.cebitec.uni-bielefeld.de/rnaforester. This program also computes multiple structure alignments by a progressive strategy and has a local similarity mode to find most similar substructures in two larger structures. *RNAforester* has been used, for example, in a large scale study [21] to separate microRNA precursors from other hairpins by structural clustering. Tree alignment is also implemented in *Gardenia* [22] and available online and for download at http://bioinfo.lifl.fr/RNA/gardenia. It is a bit faster than *RNAforester* because it does not explicitly encode the *P*-nodes, thus keeping the trees smaller. The *RNAforester* tool has recently been enhanced to include an affine gap model and to speed up alignments of structures that are already known to be similar by the use of anchorings [16]. At the time of this writing, the beta version is available at http://bibiserv.cebitec.uni-bielefeld.de/rnaforester2.

There are more programs on the market, more to come, and there is even a test site to evaluate such programs, made available by the BRASERO project [23] at http://brasero.labri.fr/.

4.6 Variations on Tree Edit and Tree Alignment Distance

4.6.1 Classical Work on Tree Edit Distance

The tree comparison problem was first studied by Tai in [14] and further developed by Zhang and Shasha [15]. Both articles define the distance between two trees as the weighted number of edit operations that transform one tree into the other. Zhang and Shasha compute the distance with a dynamic programming algorithm in $O(|T_1| \cdot |T_2| \cdot depth(T_1) \cdot depth(T_2))$, which is an improvement to [14].

4.6.2 An Improvement by Path Decomposition

For Klein [24], the edit operations are label modification and edge contraction. The trees are represented as Euler strings (parenthesized strings) and the algorithm deals with a string alignment problem. For comparing substrings of Euler strings, a simplified and less efficient variant of the Zhang–Shasha algorithm is presented. From this starting point, a faster dynamic programming algorithm is developed by using a decomposition of the tree into paths. The algorithm has a worst case bound of $O(n^3 \log n)$ for trees of size n, which is better than the Zhang–Shasha algorithm for rooted ordered trees. For some cases, Zhang–Shasha may be still faster, due to its different path decomposition strategy.

4.6.3 A Linear Tree Edit Distance Algorithm for Similar Ordered Trees

Sometimes, we only want to know the exact distance if the two trees are similar, i.e. their distance lies below a certain threshold. In this situation, a linear time algorithm is possible. Touzet in [25] improves the Zhang–Shasha-Algorithm by pruning the search space, if a bounding number of errors k is given. For the search space, she represents the possible edit operations to transform one tree into the other as an edit graph, where the vertices are the Cartesian product of the two trees node indices in postorder traversal. The incident arcs represent deletion, insertion, and replacement edit operations.

To prune the search space, such that only relevant vertices and arcs remain, she uses three optimization strategies. The first strategy is similar to the k band alignment algorithm for strings, as it computes a band of the edit graph only, constrained by the number of allowed errors k. The other strategies are a boundary for the maximal number of errors, which allows pruning of non-relevant vertices outside that bound, and a strategy for pruning of vertices below a certain depth, based on the number of errors between the compared subtrees. The resulting algorithm constructs an optimal mapping between trees A and B in $O(n \cdot k^3)$ time if the number of errors is bounded by k. For the trivial bound of $k = |A| + |B|$, the algorithm has the same complexity as the Zhang–Shasha algorithm ($O(n^4)$), and the average complexity would be worse compared to Zhang–Shasha's, as some optimizations have been left out.

4.6.4 Alignment Hierarchy and Decomposition Strategies

As a theoretical unifying framework, Blin et al. propose the alignment hierarchy [22]. In [26, 27], Dulucq and Touzet generalize several tree edit distance algorithms by describing them as decomposition strategies. The main algorithms they study are the ones of Klein [24] and Zhang–Shasha [28]. The central idea is the concept of cover strategies. Dulucq and Touzet introduce a general framework of cover strategies, analyze the complexity of cover strategies, and develop a new tree edit distance algorithm, optimal in the cover strategy framework.

4.6.5 Tree Edit Distance with Gaps

Touzet studies edit distance with convex gap weights in [29] and proves that there is no polynomial algorithm for the tree edit distance problem with convex gap weights, unless $P = NP$. She restricts the definition of gaps to complete subtrees (such that all descendants belong to the subtree) and presents a quadratic algorithm for the according tree edit distance.

4.6.6 Seeded Tree Edit Distance

The method of seeded tree alignment [30] is an interesting combination of mapping and alignment. In spite of its name, it constructs mappings rather than alignments.

Seed mappings, a set of node pairs which have to map onto each other, are used to constrain the mappings. The seed mappings preserve the lowest common ancestor relationship, and in this way select a specific common super-tree structure. This makes the mappings compatible with, while still more abstract than, tree alignments. This has been generalized to seed sets determined by exact matching, where the seeds in the set need not be compatible [31]. This leads to the problem of optimally "chaining" compatible seeds, for which an in $O(m^2 \log(m))$ time and $O(m^2)$ space algorithm is presented in [32], where m is the number of seeds.

4.6.7 Classical Tree Alignment

The tree alignment algorithm was introduced as an alternative to the tree edit algorithm by Jiang et al. in [33] for pairwise and multiple alignment. Comparing tree alignment to tree edit distance, Jiang states that the tree alignment corresponds to a restricted tree edit distance, "in which all the insertions precede all deletions" [33]. This is an operational way to express the difference between a common supertree and a common subtree. If all insertions are done first, the common supertree is produced at the point before the first deletion. If deletions come first, their end marks the common subtree.

Jiang's algorithm for ordered trees has time complexity $O(|T_1| \cdot |T_2| \cdot (deg(T_1) + deg(T_2))^2)$, where $deg(T_i)$ is the degree of T_i, so the algorithm is faster than all known ones for the tree edit distance, if the degrees are smaller than the depths of the trees.

4.6.8 Average-Case Complexity of Tree Alignment

Practical compulation of tree alignments showed that the algorithm on average performs better than the asymptotics given above suggest. Note that in the worst case, when $deg(T) \approx |T|$, a runtime of $O(n^4)$ is implied. This is puzzling, as when the degree is almost equal to the tree size, the tree is essentially a sequence (i.e., a root node with $n - 1$ leaves). In fact in [34], Herrbach et al. could show that tree alignment runs in $O(mn)$ *on average*, where m and n are the sizes of the trees. In the proof, they also show that several characteristics of trees, such as the number of closed subforests or prefix/suffix subforests of a tree are in $O(n)$ on average. These results are relevant to the average-case analysis of other tree-based algorithms, too. In the proof, they also show that several characteristics of trees, such as the number of closed subforests or prefix/suffix subforests of a tree are in $O(n)$ on average. These results are also relevant to the average-case analysis of other tree-based algorithms.

4.6.9 Local Similarity in RNA Secondary Structure and Progressive Multiple Alignment

Hoechsmann et al. extended Jiang's tree alignment distance to compute also pairwise local alignments and multiple global alignments [19, 20]. Local alignment maximizes a similarity score (rather than minimizing distance) and is the tree-based analog of the Smith–Waterman algorithm [35] for local sequence similarity. In terms of forests, this means that all pairs of closed subforests (= all substructures) have to be compared to each other as subproblems, which raises the worst-case runtime complexity to $O(|T_1| \cdot |T_2| \cdot deg(T_1) \cdot deg(T_2) \cdot (deg(T_1) + deg(T_2)))$. The multiple structure alignment uses the profile alignment method known from sequence alignment. After computing pairwise all-against-all alignments, profiles are created and successively aligned, starting with nearest neighbors.

4.6.10 Tree Alignment with Affine Gaps and Anchors

Two improvements to the tree alignment algorithm are presented in [16], motivated by its application to RNA structure alignment, where tree alignments tend to appear scattered with small gaps. The first improvement is the introduction of an affine gap cost model. In this model, the costs of opening a gap can be set higher than the costs of extending it, thus favoring few large gaps over many small ones. This appears biologically more plausible, as each a gap indicates an evolutionary event. The algorithm essentially uses seven copies of the recurrence given above, where the subalignments can be in different combinations of no-gap, parent-gap, and sibling-gap mode. The second improvement is a speedup of the alignment when certain nodes in the forest are pre-aligned by a so-called anchoring. While the affine gap model slows down the tree alignment by a constant factor ≈ 7, the anchoring provides a linear speedup depending on the number of anchors.

4.6.11 Alignment Under an Extended Set of Edit Operations

Allali and Sagot [36] extend the set of operations supported by the Zhang–Shasha algorithm [15], adding node fusion and splitting operations. These operations are particularly relevant in the context of a coarse-grain representation (one node per helix), where they allow to match an helix with two consecutive helices, e.g. separated by a bulge. This improved expressivity comes at a cost, and the resulting algorithm has complexity in $O(4^l \cdot d_1^{l+1} \cdot d_2^{l+1} \cdot n_1 \cdot n_2)$, where d_x is the max degree of a node, n_x is the sequence length, and l is the maximal number of consecutive nodes that are fusioned. The increase in complexity is compensated by a hierarchical approach [37], which initially aligns RNAs at a coarse-grain level using the extended set of operations. Then the algorithm *zooms in*, using the higher level alignment as a set of constraints for a refined alignment, and using a classic set of operations.

5 Comparing Structures with Pseudo-Knots: The Next Frontier

In this chapter, we have focused on the methods most commonly used for RNA secondary structure comparison. We have restricted ourselves to plain—non-crossing—secondary structures, leaving aside any elements of 3D structure such as base triplets, or special tertiary motifs such as kink turns and E loops (cf. Chapter 18). In particular, we have excluded pseudo-knots, defined here broadly as any crossing interactions, from our consideration—not because they are unimportant, but because they are the topic of ongoing research. At this point, we see no generally accepted method of comparison for pseudo-knotted structures, and no widely used tool for this purpose.

There are three problems that impede progress in this field: algorithms tend to be sophisticated and computationally expensive, topologically feasible conformations are hard to characterize, and reliable data is relatively scarce—a situation that equally affects the development of pseudo-knot prediction methods.

5.1 An Algorithmic Challenge

From an algorithmic perspective, the problem of comparing pseudo-knots is usually abstracted as the comparison of arc-annotated sequences (cf. Subheading 2.5.1) featuring crossing interactions. This problem has been thoroughly studied through the lens of computational complexity theory, a branch of computer science which aims at characterizing the inherent difficulty of problems. As recently surveyed by Touzet and Blin [38], the results turned out to be quite discouraging. For instance, exactly solving the problem was shown to be NP-hard (i.e., very probably intractable) under any reasonable model for superstructures. It was also proven difficult to approximate accurately and efficiently (MAX-SNP hard) under any realistic sets of operations on RNA (e.g., arc-breaking, arc-altering. . .). Therefore, there is little hope for general algorithms that would align arbitrary arc-annotated sequences both exactly and efficiently.

5.2 Topology to the Rescue

However, while any RNA can be modeled as an arc-annotated sequence, it is noteworthy that not every arc-annotated sequence is a realistic candidate for an RNA conformation. Indeed, some arc-annotated sequences may induce such an intricate structure that reconstructing them in 3D would unavoidably lead to clashes and other highly unstable features!

In a perfect world, the algorithmic difficulty of the problem would arise from such unfeasible instances, and one could devise algorithms whose runtimes would be reasonable for real RNAs. More realistically, the idea that RNA structures may be more constrained than arc-annotated sequences has led to three categories of approaches, each exploiting in some way the restricted topology of real structures. Classes of pseudo-knots have been characterized topologically in [39, 40], where it is shown how to proceed from a chosen pseudo-knot architecture to a folding algorithm.

5.3 From Folding Pseudo-Knots to Their Alignment

A first idea, due to Möhl et al. [41], considers restricted—computationally easy, yet sufficiently expressive—classes of pseudo-knots introduced in the context of RNA folding (*see* Chapter 12). The main rationale is that the scheme underlying a folding algorithm can be transformed, without too much pain, into a dynamic-programming algorithm for the alignment. Moreover, membership to various classes of pseudo-knots can be efficiently tested, as shown by Rastegari and Condon [42]. This suggests a *meta-algorithm*, which starts by determining the class of each structure, and then selects a suitable dynamic-programming algorithm.

The practicality of the resulting method largely depends on the class of each compared structure. For instance, two pseudo-knotted structures of the Rivas and Eddy type [43] lead to a polynomial, yet prohibitive, complexity in $O(n \cdot m^6)$, where n and m are the length of the longer and shorter sequence, respectively. Simpler classes of structures, on the other hand, may be aligned using as little as $O(n \cdot m^4)$ time and, more importantly, only $O(n \cdot m^2)$ memory.

5.4 Parameterized DP-Based Algorithms

Another difficulty related to the previous approach resides in the necessity, for each RNA, to belong to one of the classes studied for the folding problem. While existing classes already achieve a fair level of generality, a single interaction in the wrong place may be sufficient to make the whole approach inapplicable. This is especially problematic for tertiary motifs, which are increasingly considered as central to the notion of structural homology, whereas pseudo-knotted classes of RNAs are usually defined in terms of helices only.

This need for more robust approaches has motivated further algorithmic works, based on the concept of parameterized algorithms. Such an algorithm provides a general solution to an NP-Hard problem, which can be thought as automatically adapting its complexity to the difficulty of the problem. A parameter is introduced, and one aims at finding an algorithm whose complexity is usually exponential on the value of the parameter, but remains polynomial on the length of the instance (here, the cumulated length of both RNAs). Although the parameter may adopt arbitrary large values, its value on real-life instances is hopefully small, and the resulting method may be of practical interest. Another asset of this approach is that the time taken by the algorithm can usually be anticipated, leaving to the user to decide whether to embark in a heavy computation, or to consider alternative options (e.g., simplify the structure, prealign the sequence by adding further constraints. . .).

Möhl et al. [44] describe such an algorithm, based on a parameter k called the crossing number. The final complexity of the algorithm takes $O(n^4 \cdot s^{16k})$, where n is the length of the longest sequence, and s is the number of base pairs involved in a crossing stem. In practice, the value of k seems typically equal to 0 or 1, and s typically grows much slower than the sequence length n, allowing for the alignment of RNAs with pseudo-knots about 400 nucleotide long in a matter of hours.

5.5 Alternative Problem Encodings

A last category of exact methods reformulates the alignment of RNAs with pseudo-knots as a mathematical optimization problem. More precisely, the possible ways to align sequences are encoded as variables (e.g., modeling the position), coupled with a system of (in)equations that describes the constraints weighing on a solution. Generic solvers can then be used to maximize an objective function (i.e., the score of the alignment) under the constraints described by the system of equations. While this allows, in principle, to solve any optimization problem, the practicality of the resulting tool greatly depends on the quality of the encoding, and considerable craftsmanship is usually required to achieve good performance.

The LARA software [45] uses integer programming, a specialized version of this paradigm, to solve the RNA alignment problem in the presence of pseudo-knots. It first encodes the problem as a system of equations and then simplifies it using the Lagrangian Relaxation. This general technique penalizes the violation of certain constraints (here, the symmetry of the matching) instead of enforcing it strictly, making the system easier to solve. A near-optimal alignment is then reconstructed from the values for the variables giving the best score by iterating the solving of the equation system using different penalties. Quite interestingly, this formalism allows to express multiple base pairs per position, or even the probability matrices described in Chapter 4. A standalone implementation is available and can be interfaced with T-Coffee [46] to venture into the realm of multiple sequence alignment with pseudo-knots.

Finally, let us mention a heuristic approach, based on geometric hashing, implemented in the HARP [47] webserver at http://bioinfo3d.cs.tau.ac.il/HARP/. Here, the graph of inter-

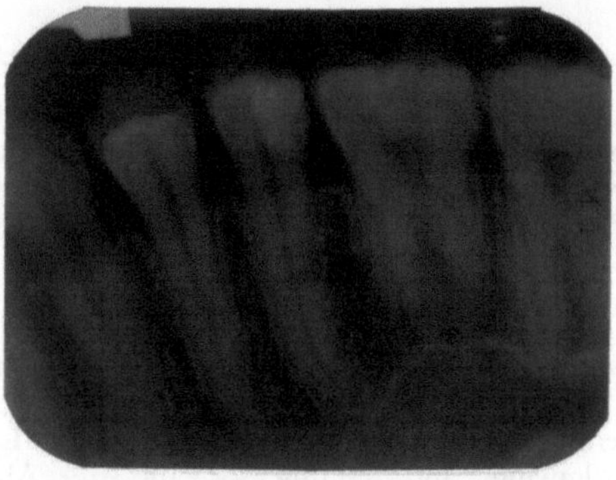

Fig. 15 Before endodontic therapy: filling exhibits pressure on pulp chamber

acting helices is decomposed into elementary triangles, used as building blocks to reconstruct a mapping using a variant of the maximal-weighted matching algorithm. This matching is then extended in a greedy fashion into an *alignment* (for lack of a better word). Despite its non-exact nature and its theoretical $O(n^7)$ worst-case complexity, the approach seems to allow for a decent alignment of large sub-unit ribosomal RNAs.

5.6 Concluding Remark

As can be seen in this short review, few algorithms have been developed for the structural alignment of RNAs featuring pseudo-knots, and even fewer implementations went further than the proof-of-concept stage. Arguably the main open problem is the support of non-canonical base pairs and tertiary motifs. While such structural features are pervasive and conserved throughout evolution, they tend to be poorly handled by exact approaches, at least for two reasons. Firstly, dynamic programming schemes are usually defined at the helix level, and tertiary interactions may be isolated, breaking their scheme. Secondly, considering such interactions leads to multiple partners for a given base, leading to an increase of the theoretical difficulty. Future algorithms may offer more flexibility towards such an inclusion. In the meantime, aligning pseudo-knots requires huge computational resources, and at least equal amount of patience.

Acknowledgment

Thanks go to Dr. Ralph Kandalla for a careful endodontic therapy of the first author, see also Fig. 15. Without his expertise, this chapter would not have been completed in time. Thanks also go to Sonja Klingberg for a careful reading of the manuscript.

References

1. Byun Y, Han K (2009) PseudoViewer3: generating planar drawings of large-scale RNA structures with pseudoknots. Bioinformatics 25(11):1435–1437
2. Hofacker IL, Fontana W, Stadler PF, Sebastian Bonhoeffer L, Tacker M, Schuster P (1994) Fast folding and comparison of RNA secondary structures. Monatshefte für Chemie 125: 167–188
3. Darty K, Denise A, Ponty Y (2009) VARNA: Interactive drawing and editing of the RNA secondary structure. Bioinformatics 25(15):1974–1975
4. Fontana W, Konings DAM, Stadler PF, Schuster P (1993) Statistics of RNA secondary structures. Biopolymers 33(9): 1389–1404
5. Shapiro BA, Zhang KZ (1990) Comparing multiple RNA secondary structures using tree comparisons. Comput Appl Biosci 6(4): 309–318
6. Shapiro BA (1988) An algorithm for comparing multiple RNA secondary structures. Comput Appl Biosci 4(3):387–393
7. Giegerich R, Voß B, Rehmsmeier M (2004) Abstract shapes of RNA. Nucleic Acids Res 32(16):4843–4851
8. Allali J, Sagot M-F (2005) A multiple graph layers model with application to RNA secondary structures comparison. In: String processing and information retrieval. Springer, New York, pp 348–359
9. Janssen S, Reeder J, Giegerich R (2008) Shape based indexing for faster search of RNA family databases. BMC Bioinformatics 9(1):131
10. Wilm A, Linnenbrink K, Steger G (2008) ConStruct: Improved construction of RNA consensus structures. BMC Bioinformatics 9(1):219
11. Höner zu Siederdissen C, Hofacker IL (2010) Discriminatory power of RNA family models. Bioinformatics 26(18):i453–i459
12. Zuker M (1989) The use of dynamic programming algorithms in RNA secondary structure prediction. CRC Press, Boca Raton, RL, pp 159–184
13. Rosselló F, Valiente G (2006) An algebraic view of the relation between largest common subtrees and smallest common supertrees. Theor Comput Sci 362(1):33–53
14. Tai K-C (1979) The tree-to-tree correction problem. J ACM 26(3):422–433
15. Zhang K, Shasha D (1989) Simple fast algorithms for the editing distance between trees and related problems. SIAM J Comput 18(6):1245–1262
16. Schirmer S, Giegerich R (2011) Forest alignment with affine gaps and anchors. In: Combinatorial pattern matching. Springer, New York, pp 104–117
17. Jiang T, Lin G, Ma B, Zhang K (2002) A general edit distance between RNA structures. J Comput Biol 9(2):371–388
18. Lorenz R, Bernhart SH, Höner zu Siederdissen C, Tafer H, Flamm C, Stadler PF, Hofacker IL (2011) ViennaRNA Package 2.0. Algorithm Mol Biol 6(1):26
19. Hoechsmann M, Töller T, Giegerich R, Kurtz S (2003) Local similarity in RNA secondary structures. Proc IEEE Comput Syst Bioinformatics Conference (CSB 2003) 2: 159–168
20. Hoechsmann M, Voß B, Giegerich R (2004) Pure multiple RNA secondary structure alignments: A progressive profile approach. IEEE/ACM Trans Comput Biol Bioinformatics 1:53–62
21. Ritchie W, Legendre M, Gautheret D (2007) RNA stem loops: to be or not to be cleaved by RNAse III. RNA 13(4):457–462
22. Blin G, Denise A, Dulucq S, Herrbach C, Touzet H (2010) Alignments of RNA structures. IEEE/ACM Trans Comput Biol Bioinformatics 7(2):309–322
23. Allali J, Saule C, Chauve C, d'Aubenton Carafa Y, Denise A, Drevet C, Ferraro P, Gautheret D, Herrbach C, Leclerc F, de Monte A, Ouangraoua A, Sagot M-F, Termier M, Thermes C, Touzet H (2012a) Brasero: A resource for benchmarking RNA secondary structure comparison algorithms. Adv Bioinformatics 2012
24. Klein PN (1998) Computing the edit-distance between unrooted ordered trees. In: Proceedings of the 6th annual European Symposium on Algorithms (ESA). Springer, New York, pp 91–102
25. Touzet H (2005) A linear tree edit distance algorithm for similar ordered trees. In: CPM '05: Proceedings of the 16th annual symposium on combinatorial pattern matching, pp 334–345
26. Dulucq S, Touzet H (2003) Analysis of tree edit distance algorithms. In: CPM '03: Proceedings of the 14th annual symposium on combinatorial pattern matching, pp 83–95
27. Dulucq S, Touzet H (2005) Decomposition algorithms for the tree edit distance problem. J Discrete Algorithm 3(2–4):448–471
28. Zhang K, Shasha D (1987) On the editing distance between trees and related problems. Ultra-computer Note 122, NYU C.S TR 310, August 1987

29. Touzet H (2003) Tree edit distance with gaps. Inform Process Lett 85(3):123–129
30. Lozano A, Pinter RY, Rokhlenko O, Valiente G, Ziv-Ukelson M (2008) Seeded tree alignment. IEEE Trans Comput Biol Bioinformatics 503–513
31. Heyne S, Will S, Beckstette M, Backofen R (2009) Lightweight comparison of RNAs based on exact sequence-structure matches. Bioinformatics 25(16):2095–2102
32. Allali J, Chauve C, Ferraro P, Gaillard A-L (2012b) Efficient chaining of seeds in ordered trees. J Discrete Algorithm 14:107–118
33. Jiang T, Wang L, Zhang K (1995) Alignment of trees – an alternative to tree edit. Theor Comput Sci 143(1):137–148
34. Herrbach C, Denise A, Dulucq S (2010) Average complexity of the Jiang-Wang-Zhang pairwise tree alignment algorithm and of a RNA secondary structure alignment algorithm. Theor Comput Sci 411:2423–2432
35. Smith TF, Waterman MS (1981) Identification of common molecular subsequences. J Mol Biol 147(1):195–197
36. Allali J, Sagot M-F (2004) Novel tree edit operations for RNA secondary structure comparison. Algorithms Bioinformatics 412–425
37. Allali J, Sagot M-F (2008) A multiple layer model to compare RNA secondary structures. Software Pract Exp 38(8):775–792
38. Blin G, Touzet H (2006) How to compare arc-annotated sequences: The alignment hierarchy. In: SPIRE, pp 291–303
39. Bon M, Orland H (2011) Tt2ne: a novel algorithm to predict RNA secondary structures with pseudoknots. Nucleic Acids Res 39(14):e93. DOI 10.1093/nar/gkr240. URL http://nar.oxfordjournals.org/content/39/14/e93.abstract
40. Reidys CM, Huang FWD, Andersen JE, Penner R, Stadler PF, Nebel M (2011) Topology and prediction of RNA pseudoknots. Bioinformatics 27(8):1076–1085
41. Moehl M, Will S, Backofen R (2010) Lifting prediction to alignment of RNA pseudoknots. J Comput Biol 17(3):429–442
42. Rastegari B, Condon A (2007) Parsing nucleic acid pseudoknotted secondary structure: Algorithm and applications. J Comput Biol 14: 16–32
43. Rivas E, Eddy SR (1999) A dynamic programming algorithm for RNA structure prediction including pseudoknots. J Mol Biol 285: 2053–2068
44. Möhl M, Will S, Backofen R (2008) Fixed parameter tractable alignment of rna structures including arbitrary pseudoknots. In: Proceedings of the 19th annual symposium on combinatorial pattern matching (CPM 2008)
45. Bauer M, Klau GW (2004) Structural Alignment of Two RNA Sequences with Lagrangian Relaxation. In: Fleischer R, Trippen G (eds) Proceedings of the 15th international symposium ISAAC 2004, vol 3341 of Lecture Notes in Computer Science, pp 113–123. Springer, New York
46. Notredame C, Higgins DG, Heringa J (2000) T-Coffee: A novel method for fast and accurate multiple sequence alignment. J Mol Biol 302:205–217
47. Abraham M, Wolfson HJ (2011) Inexact graph matching by "geodesic hashing" for the alignment of pseudoknoted RNA secondary structures. In: Holub J, Žďárek J (eds) Proceedings of the Prague stringology conference 2011, pp 45–57, Czech Technical University in Prague, Czech Republic. ISBN 978-80-01-04870-2

Chapter 13

RNA Structural Alignments, Part I: Sankoff-Based Approaches for Structural Alignments

Jakob Hull Havgaard and Jan Gorodkin

Abstract

Simultaneous alignment and secondary structure prediction of RNA sequences is often referred to as "RNA structural alignment." A class of the methods for structural alignment is based on the principles proposed by Sankoff more than 25 years ago. The Sankoff algorithm simultaneously folds and aligns two or more sequences. The advantage of this algorithm over those that separate the folding and alignment steps is that it makes better predictions. The disadvantage is that it is slower and requires more computer memory to run. The amount of computational resources needed to run the Sankoff algorithm is so high that it took more than a decade before the first implementation of a Sankoff style algorithm was published. However, with the faster computers available today and the improved heuristics used in the implementations the Sankoff-based methods have become practical. This chapter describes the methods based on the Sankoff algorithm. All the practical implementations of the algorithm use heuristics to make them run in reasonable time and memory. These heuristics are also described in this chapter.

Key words Structural RNA alignment, Simultaneous folding and alignment of RNA sequences, Sankoff algorithm

1 Introduction

RNA function is often examined through the structure formed by the specific RNA molecule. However, predicting an RNA structure using a single sequence is in general not reliable (*see* Chapter 1), and thus using evolutionary conservation when possible is a much better approach. However, an RNA structure is often more conserved than its primary sequence, which gives rise to challenges in comparative analysis of RNA secondary structure. Here the secondary structure is defined as the list of all the Watson–Crick and G-U wobble base pairs in the structure. Alignment tools which can take the secondary structure into account often perform much better than the tools developed for aligning sequences without considering the secondary structures when making alignments of structured RNAs [1]. Sequence similarity only alignment tools like

Jan Gorodkin and Walter L. Ruzzo (eds.), *RNA Sequence, Structure, and Function: Computational and Bioinformatic Methods*, Methods in Molecular Biology, vol. 1097, DOI 10.1007/978-1-62703-709-9_13,

Clustal [2] are thus ill suited for this task, unless there are high sequence similarity between the involved sequences, or they have highly conserved sequence motifs [3, 4].

Thus, a key problem in comparative RNA structure analysis is to obtain a good structural alignment of related RNA sequences. A strategy towards this is to simultaneously predict an alignment and a conserved secondary structure of the RNA sequences.

The RNA sequences can be non-protein-coding RNAs (noncoding RNAs, ncRNAs), or RNA structures within protein-coding RNAs (mRNAs), for example regulatory RNA structures located in the UTRs. Given the structural alignment a large part of the RNA secondary structure can often be worked out and models to search for more related RNAs of the same family can be made as well.

Alignment of structured RNAs differs from the alignment of protein-coding sequences as well as other noncoding sequences since the information embedded in the primary sequences is different. For protein-coding sequences the three nucleotide long amino acid codon gives raise to a signal in the primary sequence. For completely noncoding sequence the primary sequence is evolving neutrally, and there is no extra signal in the primary sequence. For ncRNA molecules there is a signal in the primary sequence, but it differs from the signal in protein-coding regions. The coding signal in protein-coding sequences is located in neighboring nucleotides, the signal in structured RNAs is not. The signal in structured RNAs stems primarily from the base pairs in the structure. In some base pairs the pairing nucleotides are located relatively close to each other, but in others the distance between the pairing nucleotides can be hundreds of nucleotides. To predict the base pairs of an RNA structure the algorithms must consider the long distance base pairs, and this together with the branching of the structures makes the computational complexity of the algorithms high.

In 1985 David Sankoff described an algorithm which can simultaneously fold and align a set of RNA sequences using free energy minimization [5]. This algorithm is referred to as the Sankoff algorithm.

The main problem with the Sankoff algorithm is the time and memory which is required to run it. This algorithm runs in $O(L^{3N})$ time and requires $O(L^{2N})$ memory, where L is the length of the sequences, and N is the number of sequences being folded and aligned. This means that if it takes one time unit to run the algorithm on just two sequences of length L, then it is expected to take $\frac{(2L)^{3\times 2}}{L^{3\times 2}} = 2^6 = 64$ time units to run the algorithm on two sequences of length $2L$. Increasing the number of sequences is even worse since this changes the exponential factor. As described below, today there are several software tools which have found ways to work around these time and memory problems.

This chapter focuses on the methods which stay true to the Sankoff idea of simultaneously folding and aligning sequences. The next chapter describes the methods which separate these two processes. The methods which are discussed in this chapter are all based on either minimizing the free energy of the RNA folding or statistical models based on Stochastic Context Free Grammars (SCFGs). Six methods: Foldalign, Dynalign, LocaRNA, Murlet, Consan, and Stemloc will be described in some detail in this chapter.

2 Principles

The Sankoff algorithm uses structure and sequence similarity to build an alignment and predict a common structure. Thus the assumption is that the sequences can fold into one conserved structure while the nucleotide sequences of the molecules are not necessarily conserved. The basis for this assumption is that compensating base pair changes can allow the primary sequences of RNA molecules to diverge while keeping the structures identical. A compensating base pair change is one where a nucleotide in a base pair is changed to another nucleotide which can also base pair with the partner of the original nucleotide. For example: A `G-C` base pair is changed to a `G-U` base pair, which can then change to an `A-U` base pair. The primary sequences have changed from `G` to `A` and `C` to `U` without breaking the base pair between the two positions.

In Sankoff style algorithms the sequences are folded with the same constraints as for single sequence folding (*see* Chapter 4), but with the added constraint that if a base pair is to be predicted, it must be conserved in all the sequences considered.

The prediction of structure similarity is either based on energy minimization or a probabilistic framework. Energy minimization is based on the nearest neighbor energy model where the main contributions to the energy come from the stacking of base pairs on to each other. Furthermore the length of the unpaired regions also adds to the energy.

Figure 1 shows an example of a grammar which can be used to predict a conserved structure for two RNA sequences. The probabilistic framework used is either base pair probabilities calculated from the free energy minimization and its partition function, *see* Chapter 4, or SCFGs, *see* Chapters 5 and 8. The SCFG methods calculate the most likely structures using a probabilistic model combining structure prediction and alignment.

The sequence similarity is often accounted for using the Ribosum substitution model [6]. The Ribosum substitution matrices are calculated in a fashion similar to the BLOSUM matrices [7],

a

$B \rightarrow abScd$
$B \rightarrow abL$
$S \rightarrow abScd$
$S \rightarrow abL$
$S \rightarrow Lab$
$S \rightarrow LS$
$L \rightarrow abL$
$L \rightarrow Lab$
$L \rightarrow abScd$
$L \rightarrow E$

b

```
gcgaaaacgc
(((....)))
auuc-ccaau
```

c

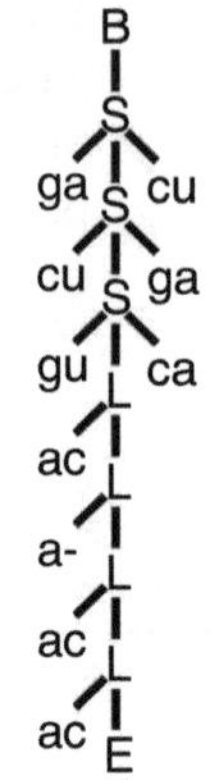

Fig. 1 (**a**) A grammar for a simple pairwise Sankoff alignment. *S* is a stem state, *L* is a loop state, *B* and *E* are the begin and end states. *a*, *c* are the emitted nucleotides or gaps in the first sequence, and *b*, *d* are the emitted nucleotides or gaps in the second sequence. When a *S* stem state is selected then the *a* must base pair with *c* and *b* must base pair with *d*. In the *S* state Eq. 2 is used to score the substitutions. In the *L* state Eq. 1 is used to score the substitution. (**b**) A simple structural alignment. (**c**) The parse tree for the alignment in (**b**) using the grammar from (**a**)

but with two separate matrices. One for unpaired nucleotides, and one for pairs of base paired nucleotides. The base pair substitution matrix contains, for example, a score for substituting an `A-U` base pair with a `G-U` base pair. The Ribosum substitution matrix for single stranded regions is:

$$A^{\text{single}}_{n_i n_k} = \log \frac{f_{n_i n_k}}{f_{n_i} f_{n_k}} \tag{1}$$

where $A^{\text{single}}_{n_i n_k}$ is the cost of substituting the nucleotide n_i with nucleotide n_k at respective positions i and k in the two sequences. Extracted from existing curated structural alignments $f_{n_i n_k}$ is the observed frequencies of substitution of nucleotide n_i with nucleotide n_k. f_{n_i} and f_{n_k} are the observed frequencies of nucleotides n_i and n_k in the structural alignments. Thus the score is a measure of how often a given substitution is observed compared to how often it would have been observed in random sequences. The Ribosum substitution matrix for base pairs is defined in a similar fashion:

$$A^{\text{base pair}}_{p_{ij} p_{kl}} = \log \frac{f_{p_{ij} p_{kl}}}{f_{n_i} f_{n_j} f_{n_k} f_{n_l}} \tag{2}$$

here $f_{p_{ij}p_{kl}}$ is the observed frequency of substitutions between the base pairs p_{ij} and p_{kl}. In Fig. 1 the single strand matrix (Eq. 1) is used to calculate the substitution scores in the loop states L, and the base pairing matrix (Eq. 2) is used in a similar fashion in the stem states S.

3 A Pairwise Structural Alignment Algorithm Example

For a single sequence the minimum free energy can be found by folding the sequence freely, *see* Chapter 4 for details. The folding of a single sequence is done by searching for an optimal structure between positions i and j ($i < j$) and extend (lowering i and increasing j). A simple way to extend this to two sequences is to simultaneously consider the positions i and j in the first sequence and correspondingly the positions k and l ($k < l$) in the second sequence. As depicted in Fig. 2 existing pairwise alignments are extended in a similar fashion as the folding of a single sequence. When aligning positions i to k and j to l then the substitutions

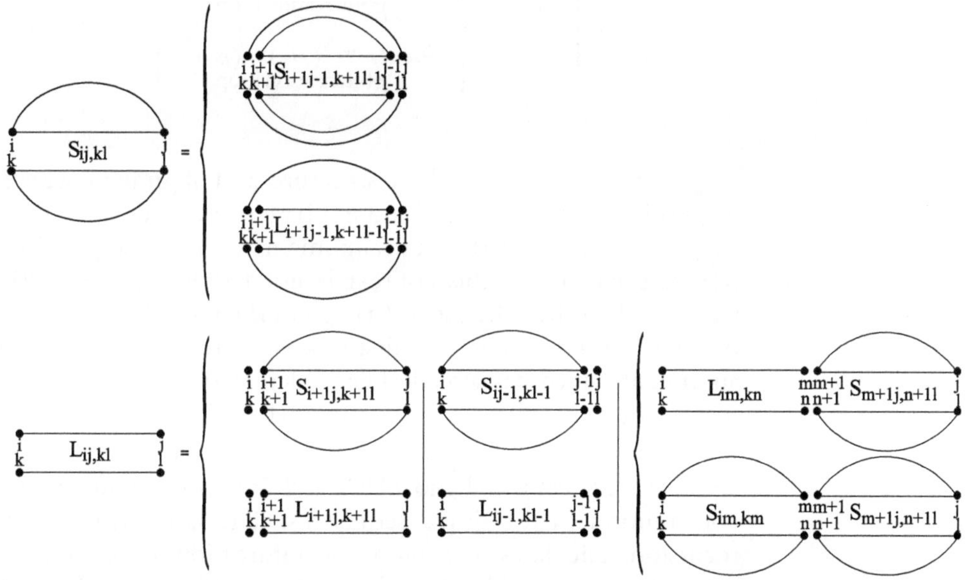

Fig. 2 A graphical representation of Eq. 3. An arch between two dots indicates that the nucleotides at the two positions are base paired. The *top part* of the figure represents the $S_{ij,kl}$ calculation, and the *bottom part* the $L_{ij,kl}$ part of the calculation. For $S_{ij,kl}$ there are two possibilities. Either an existing stem is extended with a base pair (the *upper case*, (a)), or a loop is closed by a base pair (the *lower case*, (b)). For $L_{ij,kl}$ a new loop can be opened on the left side of a stem (*upper left case*, (c)), a loop can be extended to the left (*lower left case*, (d)), a loop can be opened on the right side of a stem (*upper middle case*, (e)), a loop can be extended to the right (*lower middle case*, (f)). The right most cases of $L_{ij,kl}$ (g) join together existing structures. The upper case handles the most common case of a loop state being added to a stem state. The lower case adds together two stem states

are scored using Eqs. 1 and 2. Extending the alignment scheme to multiple sequences follows in a similar fashion adding two new position indexes per sequence.

The recursion for a pairwise version of the Sankoff algorithm using a simple energy model can be exemplified with the following score maximization (*see* Figs. 1 and 2).

$$S_{ij,kl} = \max \begin{cases} S_{(i+1)(j-1),(k+1)(l-1)} + A^{\text{base pair}}_{p_{ij}p_{kl}} \\ \quad + E^{\text{stack}}_{n_i n_j n_k n_l, n_{i+1} n_{j-1} n_{k+1} n_{l-1}} & (a) \\ L_{(i+1)(j-1),(k+1)(l-1)} + A^{\text{base pair}}_{p_{ij}p_{kl}} \\ \quad + E^{\text{loop close}}_{n_i n_j n_k n_l, n_{i+1} n_{j-1} n_{k+1} n_{l-1}} & (b) \end{cases}$$

$$L_{ij,kl} = \max \begin{cases} S_{(i+1)j,(k+1)l} + A^{\text{single}}_{n_i n_k} + E^{\text{loop open}}_{n_i n_k, n_{i+1} n_j n_{k+1} n_l} & (c) \\ L_{(i+1)j,(k+1)l} + A^{\text{single}}_{n_i n_k} + E^{\text{loop extend}} & (d) \\ S_{i(j-1),k(l-1)} + A^{\text{single}}_{n_j n_l} + E^{\text{loop open}}_{n_j n_l, n_i n_{j-1} n_k n_{l-1}} & (e) \\ L_{i(j-1),k(l-1)} + A^{\text{single}}_{n_j n_l} + E^{\text{loop extend}} & (f) \\ \max\limits_{\substack{i<m<j \\ k<n<l}} \begin{Bmatrix} L_{imkn} + S_{(m+1)j,(n+1)l} \\ \quad + E^{\text{bifurcation loop}} \\ S_{im,kn} + S_{(m+1)j,(n+1)l} \\ \quad + E^{\text{bifurcation loop}} \end{Bmatrix} & (g) \end{cases} \tag{3}$$

$S_{ij,kl}$ is the score of the best structure and alignment where the nucleotides at positions i, j and k, l base pair as indicated in Fig. 2. $L_{ij,kl}$ is the score of the best alignment where these positions do not base pair (L in this context is not to be confused with the sequence length). The terms (a)–(f) in the recursion each consists of three types of scores. The first type of score is the score of a sub-alignment which has already been calculated. The second type of score is the cost of substituting the nucleotides from one sequence with those of the other. The last type of scores, E^{stack}, $E^{\text{loop close}}$, $E^{\text{loop open}}$, $E^{\text{loop extend}}$ and $E^{\text{bifurcation loop}}$ corresponds to energy contribution as for single sequences. The last term (g) of the recursion calculates the score of bifurcation-loops. Here two substructures are added together to form one extended structure. The score is simply the sum of the two substructures plus an energy contribution for adding the extra structure. The (g) term has two parts. In the first a loop state is combined with a stem state. In the second two stem states are combined. The recursion is initialized by setting $S_{ii,kk} = 0$ and $L_{ii,kk} = 0$.

In contrast to the more applied energy minimization, the example here maximizes a score to accommodate both energy and substitution in one equation.

The energy model used is simplified compared to the full energy model. For a description of the full energy model, *see* Chapter 4. The model used here includes stacking interactions between neighboring base pairs and stacking between single stranded nucleotides and neighboring base pairs. The energy contribution of loops is taken to be linear in length which differs from the usual logarithmic length dependency. Finally there is an energy contribution for adding a new structure to bifurcation-loops.

Dynamic programming is used to calculate the structural alignment scores in Eq. 3. The dynamic programming technique relies on the fact that the longer alignments are built from the smaller alignments, in a similar way as described in Chapter 1 for single sequences. Dynamic programming is also used to make alignments based on sequence similarity.

4 From Pairwise to Multiple Structural Alignments

When more sequences are structurally aligned, the time and space requirements increase correspondingly. In the naive case where all sequences are folded and aligned simultaneously, the time requirement grows as $O(L^{3N})$ where L is the length of the sequences, and N is the number of sequences. The memory requirement grows as $O(L^{2N})$. This means that for even just three sequences the computational cost becomes too high for relatively short sequences. The usual way to circumvent this problem is to align the sequences in an order directed by a guide tree [8]. The guide tree is built from pairwise alignments of the sequences. The pairwise alignments can not only be full structural alignments but can also just be sequence similarity based. The algorithm for building the multiple alignment then becomes:

1. Make all pairwise alignments
2. From the scores of the pairwise alignments build a tree
3. Use the tree to guide the alignment of sequence pairs, and later the alignment of single sequences or alignments onto existing alignments

To make the process even faster some methods use ordinary fast sequence similarity alignment in step 1 to build the guide tree [9]. The much more expensive Sankoff algorithm is then only used to build the final alignment in step 3.

Another method is used by Stemloc-AMA [10]. This method uses sequence annealing based on the probabilities calculated by pairwise alignments.

Table 1
Method types and implementations

Type/implementation	Dynalign	Foldalign	LocaRNA	Murlet	Consan	Stemloc
Energy	+	+				
Energy probabilistic			+	+		
SCFG					+	+

The six implementations described in this chapter represent three fundamentally different ways of implementing the Sankoff algorithm. The energy implementations make the structural alignment in one process using minimum free energy folding. The energy probabilistic implementations first calculate the base pairing probabilities using the minimum free energy model and its partition function, then align the base pair matrices. The SCFG implementations use Stochastic Context Free Grammars to predict the most likely structural alignments

5 Implementations

There are three main strategies for implementing the Sankoff algorithms: The minimum free energy-based Sankoff method, the probabilistic energy-based Sankoff method, and the SCFG-based method. For examples of implementation using the three different strategies, *see* Table 1.

In the first strategy, the minimum free energy-based Sankoff method, the sequences are aligned while forcing them to conform to the same structure. This is done by only allowing a base pair in the consensus structure when the base pair is conserved between the sequences being aligned. Sequence similarity is either directly integrated into the alignment process by using a scoring scheme which scores the energy of the structure as well as the substitutions between the sequences as in Eq. 3, or sequence similarity is used to refine the final alignment. The minimum free energy strategy is the strategy most similar to the original Sankoff algorithm. The minimum free energy-based strategy is used by implementations like Foldalign [11–14] and Dynalign [15–18].

In the second strategy, the probabilistic energy-based Sankoff method, the sequences are first folded separately using free energy minimization and the partition function to get the base pairing probabilities. The consensus base pairs are then scored using the base pair probability. Strictly, when computing the base pair probabilities independently for each sequence this is a (slight) deviation from the Sankoff approach. In LocaRNA [19, 20] the score of a consensus base pair (during pairwise alignment. LocaRNA can handle multiple sequences) $\tau_{p_{ij}p_{kl}}$ is:

$$\tau_{p_{ij}p_{kl}} = \left(\log \frac{P_{ij}}{P_0^A} \Big/ \log \frac{1}{P_0^A}\right) + \left(\log \frac{P_{kl}}{P_0^B} \Big/ \log \frac{1}{P_0^B}\right) \quad (4)$$

Here P_{ij} is the base pair probability of the base pair between positions i and j in the first sequence, and P_{kl} is the probability of a base pair between positions k and l in the second sequence. P_0^A and P_0^B are the probabilities of random base pairs in the two sequences A and B, respectively. The advantage of the probabilistic energy-based approach is that by only including base pairs with a probability above a fixed cutoff the algorithmic complexity is lowered significantly. A disadvantage is that the base pair probabilities are not being calculated in a truly local fashion, making the resulting alignment at least partially dependent on global features (or a window length). This is, for example, an issue when screening genomic sequence [21]. The Ribosum substitution matrix is used either directly or indirectly to refine the resulting alignment. This strategy is used by LocaRNA [19, 20], Murlet [22], and FoldalignM [23]. The LocaRNA and FoldalignM methods are based on the PMcomp algorithm [9].

In the third strategy, the SCFG-based Sankoff method, the sequences are folded and aligned using SCFGs. These methods are statistical methods which calculate the probabilities of nucleotides base pairing and being aligned. An SCFG is a more general version of the Regular Grammars implemented in Hidden Markov Models (HMMs) and the SCFG can take some long range interactions like base pairing and bifurcation-loops into account. An SCFG starts out with a grammar chosen by the user. Choosing the right grammar is a balance between the level of detail in modeling, the folding of the sequence, and the computational requirement of the algorithm [24, 25]. The grammar consists of a number of hidden states. Each state has a set of probabilities for emitting one or more nucleotides, and a set of probabilities for transitions to the next state. The run time scales linearly with the number of states.

SCFGs are introduced in Chapters 5 and 8. The pair-SCFG described here differs by taking two sequences into account. A pair-SCFG emits nucleotides from two sequences simultaneously. The underlying theory is the same for single sequence SCFGs and the pairwise or multiple sequence SCFGs. In Fig. 1a there are four states: Stem S, Loop L, Begin B, and End E. The Begin and End states are used to initialize and end alignments. The Stem state emits a set of aligned base paired nucleotides, and the loop state emits aligned single stranded loop nucleotides. The Stem state will, for example, have high probabilities for emitting a `G-C` base pair aligned with a `G-C`, `G-U`, or `A-U` base pair, but a low probability for emitting a `G-C` base pairs aligned to an `A-C` base pair, as `A-C` base pairs are much less frequently appearing. The Loop state will typically have a high probability of emitting conserved nucleotides like a `G` aligned to a `G`, and low probabilities for emitting non-conserved nucleotides like a `G` aligned to a `U`. The Ribosum matrices were originally developed to be used in an SCFG [6]. The Stem state (and likewise the Loop state) also have a set of probabilities for transition into the next Stem, Loop, or End state.

Table 2
The heuristics that are used in the implementations

Heuristics/implementation	Dynalign	Foldalign	LocaRNA	Murlet	Consan	Stemloc
Alignment-envelope	+	−	−	+	−	+
Banding	+	+	−	+	−	−
Fold-envelope	−	−	+	+	−	+
Max local alignment length	−	+	−	−	−	−
Pins	−	−	−	−	+	−
Pruning	−	+	−	−	−	−
Skipping	−	−	−	+	−	−

The alignment-envelope dictates which nucleotides can be aligned. Banding limits the length differences and location of subsequences being aligned. The fold-envelope controls which base pairs are allowed. The max local alignment length limits the maximum length of local alignments. Pins are short alignments which the structural alignment must pass through. Pruning stops poor alignments from being included/extended into longer alignments. Skipping limits the number of bifurcation-loop calculations

The probabilities are estimated by training the SCFG on alignments of known ncRNA sequences, see, for example, the discussion of the Ribosum substitution matrices in Subheading 2. This is one of the main differences between the free energy-based methods and the SCFGs. The parameters of the free energy methods are to a large extent experimentally determined (a few of the free energy parameters are also estimated using alignments), where the SCFG parameters are estimated using alignments. When the grammar has be selected, and the SCFG trained, the SCFG model can used to align sequences and make parse trees as seen in Fig. 1b, c. For more elaborate descriptions of SCFGs, *see* Chapters 5 and 8. Methods like Consan [26] and Stemloc [10, 27] are examples of the SCFG type of methods.

6 Heuristics

The main problem with the Sankoff algorithm is that it requires a large amount of computational resources. Equation 3 deals with just two sequences, and still $S_{ij,kl}$ and $L_{ij,kl}$ are four-dimensional matrices, and the last part of the recursion is a six-dimensional calculation. Three sequences require a nine-dimensional calculation and a six-dimensional matrix to store the results. All practical implementations of the Sankoff algorithm therefore use heuristics to bring down the resource requirement. In this section some key ones are described. Table 2 gives an overview over which implementations use which heuristics.

6.1 Alignment-Envelope

The alignment-envelope restricts which nucleotides can be aligned [18, 27]. The probability of two positions being aligned is pre-calculated using an HHM, and the final alignment can only pass through nucleotide pairs which have an alignment probability above a given cutoff. The advantage of the heuristic is that the alignment process is speed up, but the disadvantage is that for highly diverge sequences there may not be much signal left.

6.2 Banding

The idea behind the banding heuristic is that in good alignments the length difference, i.e., the extra number of gaps in one sequence over the other, between the two subsequences being aligned, is small. For local alignment this is used to limit the length difference between two subsequences being aligned [11]. For global alignment this heuristic also places a limit on which positions can be aligned [14, 15]. The advantage of the heuristic is that it greatly reduces time and memory consumption. The disadvantage is that in some alignments it is biologically relevant that the one sequence should contain many more gaps than the other sequence.

6.3 Fold-Envelope

A fold-envelope is a pre-calculation of the allowed base pairs. It is the folding equivalent of the alignment-envelope. For LocaRNA the fold-envelope is calculated by folding a single sequence and calculating the base pairing probability for all possible base pairs. By discarding all base pairs with low probability—$\text{Prob}(p_{ij}) < P_{\text{cut}}$ the computational requirements are significantly reduced [19, 20]. This heuristic is at the core of the energy probability-based Sankoff methods. LocaRNA also reduces the complexity of the bifurcation-loop calculation, step (g) in Eq. 3 by only using a fixed number of structures as the right side structure. For Stemloc the fold-envelope is calculated as the union of the N-best structures predicted using a single sequence folding SCFG [27]. The advantage of the fold-envelope heuristic is that it significantly lowers the run time of the programs. The disadvantage is that true base pairs which are not predicted as probable by single sequence folding are missed.

6.4 Maximum Local Alignment Length

The maximum local alignment length limitation limits the maximum length of the final alignment [11, 13]. This limitation can only be used for local alignment, since a global alignment necessarily covers the full length of the input sequences. The length of the input sequences is not limited by this constraint. The length of the final alignment can be, and most often is, smaller than this maximum length. This is essential for local alignment.

This heuristic makes it possible to locally align sequences of any length, and it reduces time and memory consumption. The cost is that long alignments will be split into several shorter alignments.

6.5 Pins

A pin is a small high similarity alignment which the full alignment must include as a sub-alignment [26]. The pins are pre-calculated by making a local alignment and identifying regions with high similarity. These regions are then used as pins. The difference between pins and an alignment-envelope is that the alignment-envelope defines all pairs of positions which can be aligned, whereas a pin is only a small sub-alignment which the full alignment must pass through. The advantage of the pins is that they limit the alignment space the algorithm must search, but the disadvantages are that it may not be possible to find the pins, or the pins may not be correct.

6.6 Pruning

In the pruning heuristic sub-alignments with a score below a length-dependent cutoff are bared from becoming part of longer alignments [14]. Sub-alignments are discarded if: $S_{ij,kl} < C_S(j - i, l - k)$ or $L_{ij,kl} < C_L(j - i, l - k)$ where C_S and C_L are the length dependent cutoffs for the stem and loop states in Eq. 3, respectively. This greatly reduces the time and memory required to make the alignments while performance is not significantly affected.

6.7 Skipping

The calculation of the bifurcation-loop scores is the most time-consuming part of the Sankoff algorithms (the (g) term of Eq. 3). In the skipping heuristic only a subset of the possible combinations of substructures are joined together [22]. Calculating just every second or third possible combination greatly reduces the running time of the algorithm. The disadvantage is that the optimal solution might be missed.

7 Applications of Sankoff-Based Methods

The Sankoff-based approaches have been applied not only to make structural alignments but also for genomic screens detecting novel structured RNAs. Foldalign was used to search corresponding, but unalignable (in sequence) regions between human and mouse [28]. Dynalign was used to improve genome annotation in *E. coli* and *S. typhi* [29]. In a recent study LocaRNA was used to improve on RNAz screens. The RNAz alignments were originally based on sequence alignments and had a fixed size windows. LocaRNA was used to trim the edges of the predictions [20].

A practical attempt to automate structural alignments of RNA sequences is done in the WAR server, which combines several Sankoff-based as well as non-Sankoff-based approaches for structural multiple alignments of RNA sequences [30]. The non-Sankoff-based approaches are described in the following chapter.

8 Discussion

While the Sankoff algorithm in its nature requires large computational resources, heuristics have over the past decade made the approaches practical [21, 31]. This class of methods remain among the best performing methods for structural alignments of RNAs [1]. The area is still under active development and new faster versions of the algorithms are regularly published, and new heuristics are being developed.

One area of development is the limitation on how often part (g) of Eq. 3 is calculated as used by LocaRNA [20]. This is related to the skipping heuristic of Murlet [22]. These heuristics lower the computational requirements significantly, and further development along this line is likely to be seen in future works. Another example of a heuristic being developed is the integration between databases of sequences and the folding and alignment algorithms. The pruning and Maximum local alignment length heuristics of Foldalign [14] are also being developed to allow longer structural alignments.

The computational methods have shown essential in constructing the core alignment of related RNA sequences starting from unaligned sequences. However, even though the progress has been large and the methods often capture the essential parts of the structure, there are several circumstances where a detailed structure cannot be obtained. These circumstances include (1) the heuristic might reduce the search space too much, (2) the sequences do not contain much covariation (folding a set of identical sequences corresponds to folding a single sequence), (3) the sequences contain vast structural variation (as is the case for e.g., RNAseP and telomerase RNA), (4) even for the full Sankoff model, as for single sequence folding not all folding parameters are known, (5) the tree should also be taken into account [32] to yield better exploitation of the covariance patterns [33].

Still, full or partial structural alignments are useful starting points. RNA editors (*see* Chapter 17) have been implemented to assist in manual refinement of computational derived alignments. Most entries in the Rfam database [34] have been aided by structural alignment programs, and many entries were subsequently manually refined. This strategy was further extended in RNAstar [35] where experimentally determined structures were mapped to the alignments and through additional curation yielded further improvement including taking the isostericity of the base pairs into account.

The future development of the structural features of the algorithms point towards the ability to handle molecules with more than one consensus structure in the context of structural alignments. Examples of such RNA structures are the riboswitches, located in the 5′UTRs of mRNAs and regulate the mRNAs by changing structure in the response to cellular conditions [36]. Work on this has already been made for: Single sequences[37], analysis of already aligned sequences [38], and for alignment and folding multiple sequences using sampling [39]. Algorithms are currently being developed for single sequences taking non-canonical base pairs [40] as well as probing data [41] into account, and these can be generalized into the Sankoff framework. All this point in the direction of even more exciting progress in this area which also hold the long-term potential to integrate with 3D approaches and computational screens for RNA structure in genomic sequence.

Acknowledgements

This work is supported by the Danish Council for Independent Research (Technology and Production Sciences), the Danish Council for Strategic Research (Programme Commission on Strategic Growth Technologies), as well as the Danish Center for Scientific Computing.

References

1. Gardner PP, Wilm A, Washietl S (2005) A benchmark of multiple sequence alignment programs upon structural RNAs. Nucleic Acids Res 33(8):2433–2439
2. Larkin MA, Blackshields G, Brown NP, Chenna R, McGettigan PA, McWilliam H, Valentin F, Wallace IM, Wilm A, Lopez R, Thompson JD, Gibson TJ, Higgins DG (2007) Clustal W and Clustal X version 2.0. Bioinformatics 23(21):2947–2948
3. Washietl S, Hofacker IL (2004) Consensus folding of aligned sequences as a new measure for the detection of functional RNAs by comparative genomics. J Mol Biol 342(1): 19–30
4. Menzel P, Gorodkin J, Stadler PF (2009) The tedious task of finding homologous noncoding RNA genes. RNA 15(12):2075–2082
5. Sankoff D (1985) Simultaneous solution of the RNA folding, alignment and protosequence problems. SIAM J Appl Math 45(5): 810–825
6. Klein RJ, Eddy SR (2003) RSEARCH: finding homologs of single structured RNA sequences. BMC Bioinformatics 4(1):44
7. Henikoff S, Henikoff JG (1992) Amino acid substitution matrices from protein blocks. Proc Natl Acad Sci USA 89(22):10915–10919
8. Higgins DG, Sharp PM (1988) CLUSTAL: a package for performing multiple sequence alignment on a microcomputer. Gene 73(1):237–244
9. Hofacker IL, Bernhart SH, Stadler PF (2004) Alignment of RNA base pairing probability matrices. Bioinformatics 20(14):2222–2227
10. Bradley RK, Pachter L, Holmes I (2008) Specific alignment of structured RNA: stochastic grammars and sequence annealing. Bioinformatics 24(23):2677–2683
11. Gorodkin J, Heyer LJ, Stormo GD (1997) Finding the most significant common sequence and structure motifs in a set of RNA sequences. Nucleic Acids Res 25(18):3724–3732
12. Gorodkin J, Stricklin SL, Stormo GD (2001) Discovering common stem-loop motifs in unaligned RNA sequences. Nucleic Acids Res 29(10):2135–2144
13. Havgaard JH, Lyngso RB, Stormo GD, Gorodkin J (2005) Pairwise local structural alignment of RNA sequences with sequence

similarity less than 40%. Bioinformatics 21(9): 1815–1824
14. Havgaard JH, Torarinsson E, Gorodkin J (2007) Fast pairwise structural RNA alignments by pruning of the dynamical programming matrix. PLoS Comput Biol 3(10):1896–1908
15. Mathews DH, Turner DH (2002) Dynalign: an algorithm for finding the secondary structure common to two RNA sequences. J Mol Biol 317(2):191–203
16. Mathews D (2004) Predicting the secondary structure common to two RNA sequences with Dynalign. Curr Protoc Bioinformatics. Unit 12.4
17. Mathews DH (2005) Predicting a set of minimal free energy RNA secondary structures common to two sequences. Bioinformatics 21(10):2246–2253
18. Harmanci AO, Sharma G, Mathews DH (2007) Efficient pairwise RNA structure prediction using probabilistic alignment constraints in Dynalign. BMC Bioinformatics 8:130
19. Will S, Reiche K, Hofacker IL, Stadler PF, Backofen R (2007) Inferring noncoding RNA families and classes by means of genome-scale structure-based clustering. PLoS Comput Biol 3(4):e65
20. Will S, Joshi T, Hofacker IL, Stadler PF, Backofen R (2012) LocARNA-P: accurate boundary prediction and improved detection of structural RNAs. RNA 18(5):900–914
21. Gorodkin J, Hofacker IL, Torarinsson E, Yao Z, Havgaard JH, Ruzzo WL (2010) *De novo* prediction of structured RNAs from genomic sequences. Trends Biotechnol 28(1):9–19
22. Kiryu H, Tabei Y, Kin T, Asai K (2007) Murlet: a practical multiple alignment tool for structural RNA sequences. Bioinformatics 23(13):1588–1598
23. Torarinsson E, Havgaard JH, Gorodkin J (2007) Multiple structural alignment and clustering of RNA sequences. Bioinformatics 23(8):926–932
24. Dowell RD, Eddy SR (2004) Evaluation of several lightweight stochastic context-free grammars for RNA secondary structure prediction. BMC Bioinformatics 5(1):71
25. Rivas E, Lang R, Eddy SR (2012) A range of complex probabilistic models for RNA secondary structure prediction that includes the nearest-neighbor model and more. RNA 18(2):193–212
26. Dowell RD, Eddy SR (2006) Efficient pairwise RNA structure prediction and alignment using sequence alignment constraints. BMC Bioinformatics 7:400
27. Holmes I (2005) Accelerated probabilistic inference of RNA structure evolution. BMC Bioinformatics 6:73
28. Torarinsson E, Sawera M, Havgaard JH, Fredholm M, Gorodkin J (2006) Thousands of corresponding human and mouse genomic regions unalignable in primary sequence contain common RNA structure. Genome Res 16(7):885–889
29. Uzilov AV, Keegan JM, Mathews DH (2006) Detection of non-coding RNAs on the basis of predicted secondary structure formation free energy change. BMC Bioinformatics 7(1):173
30. Torarinsson E, Lindgreen S (2008) WAR: Webserver for aligning structural RNAs. Nucleic Acids Res 36(Web server issue):W79–W84
31. Gorodkin J, Hofacker IL (2011) From structure prediction to genomic screens for novel non-coding RNAs. PLoS Comput Biol 7(8):e1002100
32. Meyer IM, Miklós I (2007) SimulFold: simultaneously inferring RNA structures including pseudoknots, alignments, and trees using a Bayesian MCMC framework. PLoS Comput Biol 3(8):e149
33. Menzel P, Seemann SE, Gorodkin J (2012) RILogo: visualising RNA-RNA interactions. Bioinformatics 28(19):2523–2526
34. Gardner PP, Daub J, Tate J, Moore BL, Osuch IH, Griffiths-Jones S, Finn RD, Nawrocki EP, Kolbe DL, Eddy SR, Bateman A (2011) Rfam: Wikipedia, clans and the "decimal" release. Nucleic Acids Res 39(Database issue):D141–D145
35. Widmann J, Stombaugh J, McDonald D, Chocholousova J, Gardner P, Iyer MK, Liu Z, Lozupone CA, Quinn J, Smit S, Wikman S, Zaneveld JR, Knight R (2012) RNASTAR: an RNA STructural Alignment Repository that provides insight into the evolution of natural and artificial RNAs. RNA 18(7):1319–1327
36. Breaker RR (2011) Prospects for riboswitch discovery and analysis. Mol Cell 43(6):867–879
37. Ding Y, Lawrence CE (2003) A statistical sampling algorithm for RNA secondary structure prediction Nucleic Acids Res 31(24):7280–7301
38. Voss B (2006) Structural analysis of aligned RNAs. Nucleic Acids Res 34(19):5471–5481
39. Harmanci AO, Sharma G, Mathews DH (2009) Stochastic sampling of the RNA structural alignment space. Nucleic Acids Res 37(12):4063–4075

40. Höner zu Siederdissen C, Bernhart SH, Stadler PF, Hofacker IL (2011) A folding algorithm for extended RNA secondary structures. Bioinformatics 27(13):i129–i136

41. Washietl S, Hofacker IL, Stadler PF, Kellis M (2012) RNA folding with soft constraints: reconciliation of probing data and thermodynamic secondary structure prediction. Nucleic Acids Res 40(10):4261–4272

Chapter 14

RNA Structural Alignments, Part II: Non-Sankoff Approaches for Structural Alignments

Kiyoshi Asai and Michiaki Hamada

Abstract

In structural alignments of RNA sequences, the computational cost of Sankoff algorithm, which simultaneously optimizes the score of the common secondary structure and the score of the alignment, is too high for long sequences ($O(L^6)$ time for two sequences of length L). In this chapter, we introduce the methods that predict the structures and the alignment separately to avoid the heavy computations in Sankoff algorithm. In those methods, neither of those two prediction processes is independent, but each of them utilizes the information of the other process. The first process typically includes prediction of base-pairing probabilities (BPPs) or the candidates of the stems, and the alignment process utilizes those results. At the same time, it is also important to reflect the information of the alignment to the structure prediction. This idea can be implemented as the probabilistic transformation (PCT) of BPPs using the potential alignment. As same as for all the estimation problems, it is important to define the evaluation measure for the structural alignment. The principle of maximum expected accuracy (MEA) is applicable for sum-of-pairs (SPS) score based on the reference alignment.

Key words Structural alignment, Common secondary structure, Base-pairing probability, Maximum expected accuracy

1 Introduction

One of the major purposes of structural alignment of RNA sequences is to find the conserved secondary structures. Therefore, a good structural alignment of RNA sequences is expected to reflect the common secondary structure. The Sankoff algorithm simultaneously optimizes the common secondary structure and the alignment, but suffers from high costs in computation, $O(L^{3n})$ for time and $O(L^{2n})$ for space, where L is the length of the input RNA sequences and n is the number of them. Those costs are too high for practical applications that include long RNA sequences even when we conduct pairwise alignments.

There have been proposed several algorithms that utilize heuristics and approximations for structural alignments in order to

Jan Gorodkin and Walter L. Ruzzo (eds.), *RNA Sequence, Structure, and Function: Computational and Bioinformatic Methods*, Methods in Molecular Biology, vol. 1097, DOI 10.1007/978-1-62703-709-9_14,

avoid the heavy computation of Sankoff algorithm. They have been designed either/both to find the common secondary structures that are consistent to the alignment or/and the alignments that are consistent to the common secondary structures. The key idea to avoid the heavy computation is to predict the common secondary structure and the alignment nonsimultaneously. The two prediction processes are not independent, but they are mutually related. For example, the candidates of the common secondary structures derived from potential non-structural alignments can be used for building the structural alignments It is possible to design an algorithm that align the sequences based on the predicted secondary structure of each sequence. The predicted secondary structures, however, are not usually accurate enough, and the alignment based on the predicted secondary structures is not reliable. It should also be noted that the prediction of secondary structure of each RNA sequence may not reflect the common secondary structure.

Apart from the "2D prediction first" approach, roughly three groups of techniques are used in non-Sankoff approach, though the classification is quite rough and some part of the key concept of each group is also used by the other groups. The first group of algorithms find the candidates of local secondary structures of each sequence, before aligning the remaining parts of the sequences (CARNAC [1, 2], Scarna [3], MXSCARNA [4], CMfinder [5]). This type of algorithms is typically used in pairwise structural alignments. The second group of algorithms maps the structural information, such as base-pairing probabilities (BPPs), to each sequence, and applies the state-of-art techniques of multiple sequence alignments (MAFFT [6], PicXAA-R [7], R-COFFEE [8, 9]). The third group of algorithms maximizes the approximated objective functions that include structural consistency and the alignment score by some optimization techniques (CentroidAlign [10], LARA [11], StrAL [12], MASTR [13], SimulFold [14]).

In this chapter, we introduce the important concepts for the structural alignment in Subheading 2, briefly explain popular software tools in Subheading 3, and discuss the accuracy and the speed in Subheading 4.

2 Methods

Multiple sequence alignment is, even if we don't care about the secondary structures, a computationally expensive problem for exact solution. Therefore, most of the algorithms for multiple (nonstructural) alignments adopt hierarchical progressive process and/or iterative refinement of the initial alignment. In progressive alignments and iterative refinements, pairwise alignments are necessary to align two sequences or two groups of sequences in each iteration. Most of the multiple structural alignment methods adopt same kind of hierarchical processes using pair structural alignments as their subprocess. Therefore, it is necessary to have a method for good pairwise structural alignments to construct multiple structural alignments.

In this section, we introduce key ideas used for non-Sankoff structural alignments: matching the local structures in pairwise structural alignments, consistency of the alignment and the structures, and the score to maximize in structural alignments.

2.1 Extracting the Candidates of Local Structures

The candidates of local structures can be extracted from predicted secondary structures of each sequence. Formation of secondary structures in RNA, however, is not a deterministic process but a thermodynamic process that include probabilistic uncertainty. Therefore, we should not rely on single predictions of secondary structures, but should consider various potential local structures, e.g., candidates of stems, to find the matching of local structures between the sequences. CMfinder [5] extracts the candidates of common structures from the sequences using Covariance Models (CMs). Carnac anchors the positions strongly conserved as the common structures. Scarna extracts the candidate of stems by thresholds in the length of the stems and the BPP, the probability that a specific position i and j in RNA sequence x forms a base pair, defined as:

$$p^{\mathrm{bp}}(i,j|x) = \sum_{\sigma \in S_{ij}(x)} p(\sigma|x). \tag{1}$$

$S_{ij}(x)$ is the set of all the secondary structures that include (x_i, x_j) as one of their base pairs, and $p(\sigma|x)$ is the probability that an RNA sequence x forms a secondary structure σ, given as

$$p(\sigma|x) = \frac{1}{Z_2(x)} \exp\left(-\frac{E(\sigma,x)}{kT}\right), \tag{2}$$

given the free energy of the secondary structure, $E(\sigma, x)$, and constants k and T, where $Z_2(x)$ is the normalization term known as the partition function.

2.2 Aligning Local Structures

If the candidates of local structures are extracted from the initial alignment, the matching of the local structures across the sequences in their pairwise alignment is straightforward. If the candidates of stems are extracted independently from each sequence, the computational costs for finding the corresponding stems is high, because the $5'$ fragments and the $3'$ fragments of the corresponding stems must be consistently matched. The computations were reduced by anchoring the highly conserved regions derived from the initial alignment in CARNAC [1, 2], and by avoiding strict consistencies in the engineered dynamic programming of Scarna [3].

Scarna adopts fixed-length overlapping stem candidates as the unit of pairwise structural alignments. The stem candidates are separated into $5'/3'$ fragments and sorted according to their positions in nucleotides. The stem fragments derived from each RNA sequence are aligned almost independently in $5'$ and $3'$. The overlapping stem candidates that form a single longer stem, however, are carefully treated to match consistently in $5'$ and $3'$. This alignment of stem fragments of course produces in general inconsistent matches, which are removed by a post process. The remaining loop regions are aligned after this post process. The accuracy of pairwise structural alignments by Scarna was slightly inferior to the Sankoff methods, but it was used for the subprocess of multiple alignments in MXSCARNA [4] successfully.

2.3 Consistency of Alignments and Structures

Multiple (nonstructural) sequence alignment is one of the most well-studied subject in sequence analysis and a number of state-of-the-art techniques have been proposed. The problem of multiple structural alignment of RNA sequences has additional difficulties, but still those well-studied techniques are useful. Utilizing the structural information within those techniques is the key to get good structural alignments.

2.3.1 Probabilistic Consistency Transformation

In pairwise alignment of two sequence x and y, Miyazawa model [15] gives the probability of an alignment a as

$$p(a|x,y) = \frac{1}{Z_1(x,y)} \exp\left(-\frac{S(a,x,y)}{T}\right), \tag{3}$$

where $S(a,x,y)$ is the score of alignment a and $Z_1(x,y)$ is the normalization term, the partition function of pairwise alignments. The matching probability, the probability that a position i in x is aligned to a position j in y, is defined as

$$p^m(i,j|x,y) = \sum_{a \in A_{ij}(x,y)} p(a|x,y), \tag{4}$$

where $A_{ij}(x,y)$ is the set of all the alignments that include (x_i, y_j) as one of their aligned pairs.

In progressive multiple sequence alignments, groups of sequences are hierarchically aligned in bottom-up manner. In order to minimize miss-alignments in early stages, the following probabilistic consistency transformation (PCT) [16] is often used to modify the matching probabilities of pairwise alignments by the information of alignments with the other sequences:

$$\tilde{p}^m(i,j|x,y) = 1/N \sum_z \sum_k p^m(i,k|x,z) p^m(k,j|z,y), \quad (5)$$

where z represents the third sequence whose matching probabilities, $p^m(i,k|x,z)$ with x and $p^m(k,j|z,y)$ with y, are used for the transformation. The PCT has an effect of encouraging the pairwise alignment to be consistent to the alignment with other sequences and improves the quality of multiple alignments.

2.3.2 Consistency of Base Pairs

In pairwise structural alignment, it is necessary to consider the consistency not only in the correspondence of bases but also in the correspondence of secondary structures. Before conducting the structural pairwise alignments as one of the subprocesses of a multiple structural alignment, PCT of BPPs is often used for the consistency of the secondary structures and the alignments, without assuming a single optimum alignment as follows:

$$\tilde{p}^{bp}(i,j|x) = 1/N \sum_z \sum_{k,\ell} p^m(i,k|x,z) p^m(j,\ell|x,z) p^{bp}(k,\ell|z). \quad (6)$$

This PCT of BPPs was first used in Murlet [17], which conducts Sankoff algorithm in reduced DP spaces.

The consistency of base pairs is also used for the score of the alignment in MAFFT [6], by adopting the following type of four-way consistency.

$$Q_{x,y}(i,k) = \sum_{j<i,m<k} b^{bp}(j,i|x) p^m(j,m|x,y) b^{bp}(m,k|y) + \sum_{i<j,k<m} b^{bp}(i,j|x) p^{sm}(j,m|x,y) b^{bp}(k,m|y) \quad (7)$$

2.4 The Score to Maximize in Structural Alignments

Whether a solution of a problem is good or not depends on the objective function to optimize. The standard objective function is the score based on substitutions and gaps in sequence alignments to maximize, the free energy in secondary structure prediction to minimize. In structural alignments, the objective function should consider both the alignment and the secondary structure.

2.4.1 Maximum Expected Accuracy in Structural Alignment

We can generally define the maximum gain estimator (MEG) as follows:

$$\hat{\theta}^{MEG} = \arg\max_{\hat{\theta}} \sum_{\theta} G(\theta, \hat{\theta}) p(\theta|D), \quad (8)$$

where $p(\theta|D)$ is the distribution of the parameter θ given the data D, and $G(\theta, \hat{\theta})$ is a gain function of the true parameter θ and its estimator $\hat{\theta}$, whose expected value to maximize.

In many problems in bioinformatics, the maximum likelihood estimator (MLE) has been regarded as the best solution. In secondary structure predictions of RNA, the widely used minimum free energy (MFE) structure is the MLE that maximizes the probability $p(\sigma|x)$ in Eq. 2, and in sequence alignments, the alignment a that maximizes the score $S(a)$ is the MLE that maximizes $p(a|x, y)$ in Eq. 3. Because the MLE maximizes the probability that the estimator is equal to the true parameter, the MLE is formulated as MEG by selecting delta function $\delta(\theta, \hat{\theta})$ as the gain function.

If we have an accuracy measure, we can use in MEG a gain function that is suitable for that measure. This type of estimators is called maximum expected accuracy (MEA) estimators and has been successfully used in bioinformatics, especially in sequence alignments and secondary structure predictions of RNA. The γ-centroid estimator, which was successfully used in CentroidFold for secondary structure prediction of RNA, is a MEG estimator that maximizes the expected value of TN + γTP, where "TN" is the number of true positives and "TP" is the number of true negatives. CentroidAlign [10] applies γ-centroid estimator for the sum-of-pairs score (SPS), which marginalizes all the possibilities of the secondary structure in each RNA sequence for the calculation of the expected value of TN + γTP using an approximation described later. One of the remarkable feature of the γ-centroid estimators is that they are often calculated by simple dynamic programming algorithms using the marginal probabilities. The recursion in the dynamic programming of CentroidAlign is described as follows:

$$M_{u,v} = \max \begin{cases} M_{u-1,v-1} + (\gamma + 1)p_{uv}^{\text{mea}} - 1 \\ M_{u-1,v} \\ M_{u,v-1}, \end{cases} \tag{9}$$

where

$$p_{uv}^{\text{mea}} = \sum_{j:u<j,\ell:v<\ell} p^{pm}(u, j, v, \ell|x, y) + \sum_{i:i<u,k:k<v} p^{pm}(i, u, k, v|x, y) + p^{m}(u, v|x, y). \tag{10}$$

The $p^{pm}(u, j, v, \ell|x, y)$ is the probability that (x_u, x_j) and (y_v, y_ℓ) are matched as base pairs in the alignment.

2.4.2 Approximation of the Probability Distribution by Decomposition

Sankoff model gives the combined distribution of the secondary structures and the alignments, but including this type of distributions requires heavy computational costs. This chapter focuses non-Sankoff methods that avoid this computation. The following decompositions of the probabilities are often used in non-Sankoff approaches.

$$\tilde{p}^{pm}(i,j,k,\ell|x,y) = p^{\mathrm{bp}}(i,j|x)p^{\mathrm{bp}}(k,\ell)p^{m}(i,k|x,y)p^{m}(j,\ell|x,y) \tag{11}$$

$$\tilde{p}^{sm}(i,k|x,y) = p^{s}(i|x)p^{s}(k|y)p^{m}(i,k|x,y) \tag{12}$$

where $\tilde{p}^{pm}(i,j,k,\ell|x,y)$ is an approximation of the marginal probability that (x_i, x_j) and (y_k, y_ℓ) are the base pairs in the secondary structures and that (x_i, y_k) and (x_j, y_ℓ) are aligned in the alignment at the same time, and $\tilde{p}^{sm}(i,k|x,y)$ is an approximation of the marginal probability that neither x_i nor y_k forms any base pair in the secondary structure and that x_i and y_k are aligned in the alignment. $p^s(i|x)$ ($p^s(k|y)$) is the marginal probabilities that $x_i(y_k)$ do not form any base pair.

3 Software Tools

In this section, we introduce non-Sankoff software tools for structural alignments of RNA sequences following the classification into three groups described in previous sections. The summary of those tools are shown in Table 1.

3.1 Group1: Matching Potential Local Structures of the Sequences

In this approach, potential local structures, e.g., candidates of stems, are extracted from each sequence or from the initial alignment, then the corresponding local structure across the sequences are searched.

3.1.1 CARNAC [1, 2] (http://bioinfo.lifl.fr/carnac/)

CARNAC extracts the candidates of stems with a low free energy by using a dynamic programming algorithm (DP), and the anchor regions between the two sequences by sequence similarity. Then, CARNAC predicts the optimal common secondary structure between two RNA sequences based on the extracted stems and the anchor regions. Although CARNAC is designed to predict the common secondary structure, the CARNAC Web Server also produces the multiple alignment (by ClustalW) with the common secondary structure predicted by the above method.

3.1.2 Scarna/MXSCARNA [3, 4] (http://mxscarna.ncrna.org/)

Scarna produces a pairwise structural alignment by aligning the stem candidates of two sequences and the remaining positions. The stem candidates with a fixed length are extracted from each sequence by a threshold in BPPs, the stem candidates are separated into the 5′ components and the 3′ components and

Table 1
Summary of non-Sankoff-based aligners for multiple RNA sequences

Software	Descriptions	Opt.[a]	Time[b]	Ref.
CARNAC	Compute common secondary structure based on stem candidates. The Web server also produces a multiple alignment (by ClustalW) with the common structure	DP	$O(L^3)$[c]	[1, 2]
CentroidAlign	Optimized for SPS, designed based on maximizing expected accuracy (MEA).	DP	$O(L^3 + c^2 dL^2)$	[10]
CMfinder	Based on a covariance model (CM)	DP	$O(L^3 \vert M \vert)$[d]	
LARA	A graph representation of alignment and an integer programming-based approach	IP	N/A	[11]
MAFFT	Conduct a Progressive & iterative alignment with considering base-paring probabilities	DP	$O(L^3)$	[6]
MASTR	A combined cost function that considers sequence conservation and covariation, optimized by MCMC	MCMC	N/A	[13]
MXSCARNA	Heuristic approach for aligning candidates of stems	DP	$O(L^3)$	[4]
PicXAA-R	A modified PCT with base-paring probabilities is employed	DP	$O(L^3)$	[7]
ProbConsRNA	A pair HMM model trained by existing RNA sequences are employed; PCT is also utilized	DP	$O(L^2)$	[16]
R-COFFEE	A variant of T-coffee. The method heuristically incorporates the information of base-paring probabilities	DP	$O(L^3)$	[8, 9]
SimulFold	Simultaneously infers RNA structures including pseudo-knots, alignments, and trees using a Bayesian Markov chain Monte Carlo (MCMC) framework.	MCMC	N/A	[14]
StrAL	A scoring function (for pairwise alignment) that takes into account sequence similarity and up and downstream paring probability is employed	DP	$O(L^3)$	[12]

[a]Methods for obtaining the optimal alignment: *DP* dynamic programming, *IP* integer programming, *MCMC* Markov chain Monte Carlo
[b]Time complexity for pairwise alignment. *L* is denoted as the length of the input RNA sequences
[c]An *empirical* order stated in the authors' paper [1]
[d]*M* is the number of states in the CM

they are aligned separately by engineered dynamic programming, which partially reflects the consistency of the matching structures. MXSCARNA builds multiple structural alignment by progressive process using Scarna as one of its procedures.

3.1.3 CMfinder [5] (http://bio.cs.washington.edu/yzizhen/CMfinder/)

CMfinder is based on a covariance model (CM) for finding local motifs from a set of RNA sequences. CMfinder can produce (local) multiple alignment of input RNA sequences. The time complexity is equal to $O(NL^3|M|)$ where L is the maximum sequence length and $|M|$ is the number of states in the CM.

3.2 Group2: Combining Structural Information into State-of-the-Art Techniques

The objective function in MAFFT contains four-way consistency score in which base-paring probabilities as well as the similarity score of bases are incorporated. In a procedure of multiple alignments in MAFFT, a progressive alignment and iterative refinement to optimize the alignment are employed. MAFFT can use Scarna as its subprocess as an option, which produces reliable structural alignments.

3.2.1 PicXAA-R [7] (http://www.ece.tamu.edu/~bjyoon/picxaa/)

PicXAA-R extends an idea of PicXAA [18], which is based on a modified PCT. PicXAA-R employs the information of common secondary structures by using base-paring probabilities derived from the McCaskill algorithm.

3.2.2 ProbConsRNA [16] (http://probcons.stanford.edu/)

ProbConsRNA is based on the pair HMM model and employs no secondary structure information of each sequence. The software employs a PCT in which the information of the other sequences are incorporated in a pairwise alignment of the specific RNA sequences. A difference between ProbCons and ProbConsRNA is parameters in the model, because the parameters in ProbConsRNA were trained by using RNA sequences. In general, ProbCons is the faster than other aligners that consider secondary structure information, because the time complexity of ProbCons(RNA) is $O(nL^2)$ where n is the number of sequences and L is the length of sequences.

3.2.3 R-COFFEE [8, 9] (http://www.tcoffee.org/Projects_home_page/r_coffee_home_page.html)

The R-COFFEE is an extension of the T-COFFEE, which incorporates the information of secondary structure by using banded base-paring probabilities computed by RNAplfold. A progressive alignment is conducted in the multiple alignment step.

3.3 Group3: Optimizing the Combined Score Function of Structures and Alignments

MASTR employs a combined cost function (for a multiple alignment) that considers sequence conservation and covariation and BPPs. Markov chain Monte Carlo (MCMC) in a simulated annealing (SA) framework is utilized in order to optimize the cost.

3.3.1 MASTR [13] (http://servers.binf.ku.dk/mastr/)

3.3.2 StrAL [12] (http://www.biophys.uni-duesseldorf.de/stral/)

StrAL employs a scoring function (for pairwise alignment) that takes into account sequence similarity and up and downstream paring probability. The multiple alignment is based on a progressive strategy.

3.3.3 CentroidAlign [10] (http://www.ncrna.org/software/centroidalign)

CentroidAlign adopts a novel estimator for multiple alignments of structures RNAs, based on maximizing the expected accuracy (MEA). (The MEA is successfully applied in many problems in bioinformatics [19]). CentroidAlign approximately maximizes the expected SPS of a predicted alignment under a probability distribution of alignments given by marginalizing the Sankoff model. The time complexity of pairwise alignment by CentroidAlign is $O(L^3 + c^2 dL^2)$ where L is the length of input sequences and both c and d are constants independent of L. Finally, the multiple alignment in CentroidAlign is conducted by a fast progressive alignment. In [10], the authors showed that CentroidAlign achieved favorable performance compared to Sankoff-based algorithms such as Murlet and RAF.

3.3.4 LARA [11] (https://www.mi.fu-berlin.de/w/LiSA/Lara)

LARA utilized a graph-based representation for structural alignments and applied an integer linear program (IP) for solving the optimization problem. LARA can predict a common secondary structure with a multiple alignment.

3.3.5 SimulFold [14] (http://www.cs.ubc.ca/~irmtraud/simulfold)

SimulFold simultaneously infers RNA structures including pseudoknots, alignments, and trees using a Bayesian MCMC framework.

4 Notes

4.1 Summary of Non-Sankoff Aligners

The summary of non-Sankoff aligners is shown in Table 1. All the alignment tools that consider secondary structures naturally compute the secondary structures or the BPPs for each sequence. That leads those tools to require at least $O(L^3)$ time for the sequences of length L.

It is important to have the evaluation measure and the "correct" references for estimation problems. In structural alignment of RNA sequences, however, the evaluation measure is not obvious even if we fix the "correct" reference alignment, because the structural alignments include multiple information of the sequences and the structures. The SPS is appropriate if we only care about the number of correctly aligned positions, but the number of correctly predicted common base pairs is important if we want to find the common secondary structure of the sequences.

Acknowledgements

The authors of this chapter thank Hisanori Kiryu, Kengo Sato, Martin Frith, Toutai Mituyama, and the other members in Computational Biology Research Center (CBRC) for discussions and comments for theories and algorithms for RNA sequence analysis.

References

1. Touzet H (2007) Comparative analysis of RNA genes: the caRNAc software. Methods Mol Biol 395:465–474
2. Touzet H, Perriquet O (2004) CARNAC: folding families of related RNAs. Nucleic Acids Res 32:W142–W145
3. Tabei Y, Tsuda K, Kin T, Asai K (2006) SCARNA: fast and accurate structural alignment of RNA sequences by matching fixed-length stem fragments. Bioinformatics 22:1723–1729
4. Tabei Y, Kiryu H, Kin T, Asai K (2008) A fast structural multiple alignment method for long RNA sequences. BMC Bioinformatics 9:33
5. Yao Z, Weinberg Z, Ruzzo WL (2006) CMfinder–a covariance model based RNA motif finding algorithm. Bioinformatics 22:445–452
6. Katoh K, Toh H (2008) Improved accuracy of multiple ncRNA alignment by incorporating structural information into a MAFFT-based framework. BMC Bioinformatics 9:212
7. Sahraeian SM, Yoon BJ (2011) PicXAA-R: efficient structural alignment of multiple RNA sequences using a greedy approach. BMC Bioinformatics 12(1):S38
8. Moretti S, Wilm A, Higgins DG, Xenarios I, Notredame C (2008) R-Coffee: a web server for accurately aligning noncoding RNA sequences. Nucleic Acids Res 36:W10–W13
9. Wilm A, Higgins DG, Notredame C (2008) R-Coffee: a method for multiple alignment of non-coding RNA. Nucleic Acids Res 36:e52
10. Hamada M, Sato K, Kiryu H, Mituyama T, Asai K (2009) CentroidAlign: fast and accurate aligner for structured RNAs by maximizing expected sum-of-pairs score. Bioinformatics 25:3236–3243
11. Bauer M, Klau GW, Reinert K (2007) Accurate multiple sequence-structure alignment of RNA sequences using combinatorial optimization. BMC Bioinformatics 8:271
12. Dalli D, Wilm A, Mainz I, Steger G (2006) STRAL: progressive alignment of non-coding RNA using base pairing probability vectors in quadratic time. Bioinformatics 22:1593–1599
13. Lindgreen S, Gardner PP, Krogh A (2007) MASTR: multiple alignment and structure prediction of non-coding RNAs using simulated annealing. Bioinformatics 23:3304–3311
14. Meyer IM, Miklos I (2007) SimulFold: simultaneously inferring RNA structures including pseudoknots, alignments, and trees using a Bayesian MCMC framework. PLoS Comput Biol 3:e149
15. Miyazawa S (1995) A reliable sequence alignment method based on probabilities of residue correspondences. Protein Eng 8:999–1009
16. Do CB, Mahabhashyam MS, Brudno M, Batzoglou S (2005) ProbCons: probabilistic consistency-based multiple sequence alignment. Genome Res 15:330–340
17. Kiryu H, Tabei Y, Kin T, Asai K (2007) Murlet: a practical multiple alignment tool for structural RNA sequences. Bioinformatics 23:1588–1598
18. Sahraeian SM, Yoon BJ (2010) PicXAA: greedy probabilistic construction of maximum expected accuracy alignment of multiple sequences. Nucleic Acids Res 38:4917–4928
19. Hamada M, Kiryu H, Iwasaki W, Asai K (2011) Generalized centroid estimators in bioinformatics. PLoS ONE 6:e16450

Chapter 15

De Novo Discovery of Structured ncRNA Motifs in Genomic Sequences

Walter L. Ruzzo and Jan Gorodkin

Abstract

De novo discovery of "motifs" capturing the commonalities among related noncoding structured RNAs is among the most difficult problems in computational biology. This chapter outlines the challenges presented by this problem, together with some approaches towards solving them, with an emphasis on an approach based on the CMfinder program as a case study. Applications to genomic screens for novel *de novo* structured ncRNAs, including structured RNA elements in untranslated portions of protein-coding genes, are presented.

Key words CMfinder, Mutual information, ncRNA discovery, ncRNA gene, ncRNA motif, Riboswitch

1 Introduction

De novo discovery of "motifs" capturing the commonalities among related noncoding structured RNAs is among the most difficult problems in computational biology. The fundamental problem is that structurally constrained RNAs evolve while conserving structure, not sequence. Thus, computationally expensive structure prediction and/or structure-based search algorithms somehow must be part of the equation, to winnow rare homologous ncRNAs from the chaff of large genomes. Successful approaches to date combine sensitive algorithms for motif discovery with homology search and exploit prior biological knowledge wherever possible.

This chapter outlines this problem. Our goal is threefold—to illustrate the challenges presented by this problem, to point out some approaches that have been partially successful in addressing them, and to highlight areas where further improvements are especially desirable. As a particular case study, we emphasize an approach based on the CMfinder program [1], successfully used for discovery of functional noncoding RNA elements in prokaryotes (e.g., [2, 3]) and of strong candidates in vertebrates (e.g., [4]).

Jan Gorodkin and Walter L. Ruzzo (eds.), *RNA Sequence, Structure, and Function: Computational and Bioinformatic Methods*, Methods in Molecular Biology, vol. 1097, DOI 10.1007/978-1-62703-709-9_15,

Our outline is certainly not comprehensive. Other methods have been applied for genome-wide screens using different alignment strategies (ranging from making use of sequence-based alignments to complete realignment by Sankoff-style approaches), different scoring schemes, different exploitation of biological knowledge, etc. For more, we refer interested readers to other chapters in this volume and to any of the many excellent review articles that have appeared recently, e.g., [5–13].

2 Problem Overview

The comparative method, also known as covariation analysis, is one of the most powerful tools available for the elucidation of RNA secondary structures; *see*, e.g., Chapter 16, or [14, 15]. The key point is that mutations destroying a structurally important RNA base pair by altering one of its nucleotides may be repaired by a *compensating* change to its partner; observations of such changes between homologs thus highlight the base-paired regions in these molecules. The technical challenges in exploiting this are to find and align (sufficiently many) homologs and to extract the structural signal from them. Having sequences at appropriate evolutionary distances is critical to this, since very similar sequences, although easily recognizable and alignable, will exhibit few compensating substitutions; conversely, highly diverged sequences may exhibit many examples of compensatory changes, but may be difficult to find and difficult to align (either in sequence or structure).

In short, for genome-scale discovery of structured RNAs, success hinges on (a) selecting the right input data, (b) aligning it well (with attention to potential RNA secondary structure), and (c) inferring the important shared features of the alignment, including structure. Implicit in this is that (d) *local* alignment is critical, since exact boundaries of structured elements are unlikely to be known a priori, (e) entire input sequences may need to be discarded, for essentially the same reason, and (f) an iterative scheme involving genome-scale homology search for additional elements matching the discovered motif is valuable, since the initial input is unlikely to be complete, and additional examples will allow construction of more accurate motif models, iteratively amplifying discovery.

As a case study, this chapter focuses on one approach to the automation of this process, based on the CMfinder algorithm [1]. We do not suggest that this approach is the last word on the problem—it is not. But it is one of several available successful starting points. Many other tools may substitute for particular steps in the process, and we hope improvements will be made to all of them, and that the discussion below will help illustrate what may work and what needs improving.

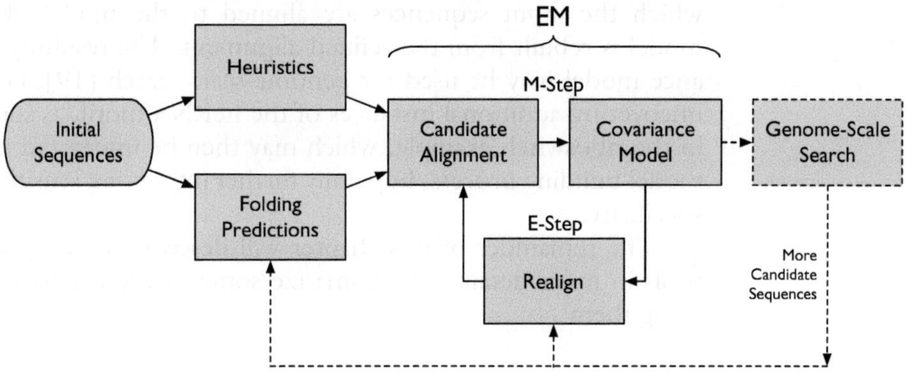

Fig. 1 Overview of the CMfinder discovery pipeline

As an example application, consider the problem of riboswitch discovery. *Riboswitches*, first discovered in 2002, are domains typically found in 5′ untranslated regions of messenger RNAs, where they conditionally control gene expression based on the presence of specific small molecules [16]. Ideally, one would like to have a tool that, given one or more genomes, would identify all riboswitches in them. Unfortunately, this is well beyond the state of the art; functional RNAs are simply too diverse and/or rare to be mechanically identified *de novo* amidst the vast bulk of non-RNA-containing genome sequence, at least by current methods. However, we can let discovery be guided by the known biology. Given that riboswitches are typically *cis*-acting elements with well-conserved secondary structures that regulate specific biochemical pathways in phylogenetically related prokaryotes, one might hope to find examples by examining upstream sequences extracted from orthologous genes in specific bacterial clades. Furthermore, given that multiple copies of a specific type of riboswitch sometimes regulate multiple steps in a particular biochemical pathway (steps *not* performed by orthologous enzymes), search based on the motif identified upstream of some enzyme in a pathway may well reveal additional paralogous riboswitch instances upstream of other enzymes in the same pathway, thus refining the motif model and providing further support for the significance and function of the RNA.

Figure 1 outlines the process. Input sequences that might contain a functional RNA motif, say 1,000 base pair sequences upstream of the start codons of orthologous enzymes in some prokaryotic clade, are gathered. A collection of smart heuristics applied to these input sequences, based on both sequence conservation and single-sequence structure prediction (Chapter 4, [17, 18]), results in a candidate alignment, from which a consensus RNA secondary structure prediction and covariance model (Chapters 5, 8 and 9, [19]) are built. This initial model is refined by an Expectation-Maximization-like (EM-like) [20, 21] iteration in which the input sequences are aligned to the model, then the model is rebuilt from the refined alignment. The resulting covariance model may be used for genome-scale search [19], hopefully uncovering additional instances of the ncRNA motif, as suggested in the riboswitch example, which may then be integrated into the model building process, hopefully further improving sensitivity and specificity.

The remainder of this chapter will describe these algorithmic tools in more detail and summarize some of the results obtained using them.

3 The CMfinder Algorithm

CMfinder builds on the DNA motif finding program MEME [22] in using an EM framework to search for motif instances embedded in a simple background, but replaces MEME's ungapped position weight matrix motif models with Covariance Models (CMs, [19, 23, 24], and Chapters 5, 8 and 9) to describe RNA motifs. As described above, it contains three main components: initial heuristic alignment, covariance model inference (the "M-step"), and CM-based realignment (the "E-step"). We will describe each in turn. (As usual, the synopsis below omits certain important details, but hopefully captures some of the more interesting features of the methods.)

3.1 Heuristic Alignment

This step identifies the approximate location and structure of a motif. A key issue is the tradeoff between accuracy and efficiency, but noting that the motif will be refined later, alignment errors are tolerable, provided that good alignments are well represented.

To start, strong candidates, i.e., segments with potentially stable secondary structure, are identified by using a single-sequence folding program [18] to compute the minimum free energy of all subsequences of the input. Candidates are ranked by their free energy, scaled by sequence length.

Local regions of sequence conservation are found by BLAST search, and pairs of candidates from different sequences are selected for comparison, using the structural alignment heuristic sketched below, if they are compatible with these "BLAST anchors." This heuristic improves accuracy by preventing obvious misalignments, as well as saving time by reducing the number of structural alignments calculated. The overall initial alignment consists of a central "consensus" candidate along with its nearest match in each other sequence. This process is repeated on the unselected candidates to find several initial alignments as seeds for the EM iteration.

Candidates are compared using the tree-edit algorithm of Hofacker et al. [18] (also *see* Chapter 12), modified to compare both unpaired and paired nucleotides. This improves discrimination among RNAs with relatively simple structures. This is a simpler comparison heuristic than those used in Carnac [25], ComRNA [26], or Sankoff-style algorithms (Chapter 13 or [27–29]), for example. Its drawbacks include potentially inaccurate secondary structure prediction and the simplified edit distance model, but it is relatively fast (approximately quadratic in the length of the candidates).

3.2 Model Inference

The key component of the M-step of the EM iteration is to (re-)build the covariance model to maximize the likelihood of the input data given the model. Much of this process is similar to analogous procedures from COVE [23] and CMbuild from Infernal [19]: given the estimated motif positions in the current alignment, and Dirichlet priors, maximum likelihood estimates for the transition and emission probabilities for the CM state machine are obtained, basically by estimating transitions/emissions observed on the training data. One technical innovation that we outline below is a Bayesian formulation of consensus secondary structure prediction that smoothly blends thermodynamic structure prediction with mutual information (defined in, e.g., Chapter 1, Eq. 3). The former is especially useful when sequence identity is high/covariation is low and the later is valuable when the opposite is true.

For $1 \leq i \leq l$, let L_i be the ith column of the current (length l) alignment $D = (L_1, L_2, \ldots, L_l)$ of n sequences, and let $\sigma = (\alpha, \beta)$ be the consensus secondary structure for D, where α is the set of indices of unpaired columns and β is the set of pairs of indices of base paired columns. The goal is to find the structure $\hat{\sigma}$ that maximizes $P(D, \sigma)$, the joint likelihood of the alignment and structure. Assuming independence of columns and column pairs,

$$P(D|\sigma) = \prod_{k \in \alpha} P(L_k) \prod_{(i,j) \in \beta} P(L_i L_j)$$

$$= \prod_{1 \leq k \leq l} P(L_k) \prod_{(i,j) \in \beta} \frac{P(L_i L_j)}{P(L_i) P(L_j)}.$$

The likelihood of an observed column is $P(L_i) = \prod_{x \in \{A,C,G,U\}} (p_x)^{n_x}$, where p_x is the probability of observing x in a row of L_i, and n_x is the observed number of such rows (ignoring gaps and assuming sequences are independent, i.e., a deep "star" phylogeny). Noting that n_x/n is the maximum likelihood parameter estimate for p_x (based on a multinomial model of the observed data L_i), and using an analogous expression for $P(L_i L_j)$, one can show that the term $I_{ij} = \log \frac{P(L_i L_j)}{P(L_i) P(L_j)}$ is proportional to the mutual information between columns i and j. The optimal structure $\hat{\sigma}$ maximizes $\sum_{(i,j)\in\beta} I_{ij}$. This approach is used by COVE [23] and works well in large, phylogenetically diverse datasets, but less well when covariance is limited, as with a few closely related sequences.

CMfinder introduces an informative prior on structures. If s_i is the prior probability that column i is single stranded, and p_{ij} the prior that columns i and j are base paired, then $P(\sigma) = \prod_{k\in\alpha} s_k \prod_{(i,j)\in\beta} p_{ij}$, and $P(D,\sigma)$ becomes:

$$P(D,\sigma) = P(D|\sigma)P(\sigma) = \prod_{1\le k\le l} P(L_k)s_k \prod_{(i,j)\in\beta} \frac{P(L_iL_j)}{P(L_i)P(L_j)} \frac{p_{ij}}{s_is_j}.$$

The maximum likelihood structure σ maximizes $K = \sum_{(i,j)\in\beta} K_{ij}$ where

$$K_{ij} = \log\left(\frac{P(L_iL_j)}{P(L_i)P(L_j)} \frac{p_{ij}}{s_is_j}\right) = I_{ij} + \log\frac{p_{ij}}{s_is_j},$$

and a simple dynamic programming algorithm can choose a compatible, pseudoknot-free set of base pairs maximizing K.

CMfinder's prior on structures is based on a thermodynamic model. For each sequence, calculate the partition function P_{ij} (Chapter 4, [18, 30]), which estimates the probability of forming base pair i, j, averaged over all possible structures. The column pairing probabilities p_{ij} are estimated by averaging the partition functions of the aligned sequences, and s_i is estimated as $1-\sum_j p_{ij}$.

Since p_{ij} and s_i are data-dependent, they are not "priors" in a strict Bayesian sense. However, the mutual information and the partition function look at the same data in different ways: mutual information measures the conservation of covarying base pairs in the particular sequences from an evolutionary perspective, while the partition function uses a thermodynamic model that is generically applicable to all RNAs. Combining them leverages both approaches: the energy model dominates when there is little mutual information and conversely mutual information dominates when the thermodynamic predictions are ambiguous. RNAalifold [31] uses a similar approach, calculating a linear combination of free energy and mutual information. (Since the partition function is proportional to $\exp(\Delta E/kT)$, free energy and log probabilities are comparable quantities.)

Conceptually, any of the many other RNA alignment/folding programs might be substituted for the particular method outlined here; *see*, e.g., Chapters 7, 13 and 14 and the references therein.

3.3 Realignment

The purpose of the E-step in the EM framework is to find the expected values of hidden parameters, which, in this problem, define the position (if any) of the motif instance in each input sequence as well as its alignment to the motif consensus. These values implicitly weight the candidates considered in the M-step, so that model rebuilding emphasizes good matches over poor ones. CMfinder accomplishes this by "scanning" each input sequence using the covariance model to identify the highest scoring subsequences of each input sequence, and aligning each to the model via the highest-probability path through the model (variously called the Viterbi or CYK algorithm; *see* Chapters 5, 8 and 9).

As noted earlier, an important aspect of the overall pipeline is the ability to use the model to search for additional matches in new genome sequences, and to incorporate them into the model. This fits naturally in the scheme outlined above—any high-scoring matches found in a genome scan can be aligned to the model by exactly the same process given in the previous paragraph, and (after calculating their partition functions [18, 30]) the model rebuilt as described in Subheading 3.2. This feature has proven highly effective at discovering large and diverse RNA families from a small set of related sequences.

3.4 Motif Scoring

One further piece of the puzzle is scoring—how does a novel "motif" stack up against ones discovered from *random* genomic sequences? A variety of approaches have been proposed to answer this question. One important class of solutions is *shuffling*—whatever scoring approach is taken, score the "real" motif by that method and compare it to similarly generated scores from a large number of sequences formed by randomly shuffling the nucleotide sequence of "real" motifs. An important *caveat* here is that, since the stacking energy in RNA helices is nucleotide-dependent, the average *dinucleotide* composition matters, so appropriate shuffling procedures must preserve these quantities. Altschul and Erickson [32] demonstrate how to shuffle single sequences while exactly preserving dinucleotide statistics, and methods that approximately achieve this goal for multiple sequence alignments are available [33–35].

Alternatively, a phylogenetically informed approach is possible—given a phylogenetic tree capturing the observed motif instances, together with branch lengths and estimated rates of indels, nucleotide substitutions (in unpaired regions) and nucleotide-pair substitutions (in paired regions), are the changes observed in the inferred motif more likely according to a model that accounts for base pairing or one that treats all columns as independent? A number of programs incorporate such models for phylogenetic inference (e.g., RAxML [36], phase [37]) and others directly use them for motif scoring (e.g., pfold [38, 39], EvoFold [40], pscore [41]). *See* also Chapter 16.

A large number of other approaches have been explored, including the scoring scheme used in RNAalifold [31, 42], and the SVM regression scheme used in RNAz [43].

4 Applications to *De Novo* Discovery of ncRNA Motifs

Our main application of interest is discovery of structured RNA motifs. These can be part of novel ncRNA genes or, for example, regulatory structures in untranslated regions of mRNAs. As with all methods, they are benchmarked on existing data, which, sadly, are sparse. It is therefore important to recognize that evaluating and comparing methods of this sort is fraught with difficulties, and conclusions depend strongly on specific benchmark data and/or evaluation criteria. Nevertheless, it seems safe to say that the CMfinder pipeline has been successful in several contexts.

The original CMfinder paper [1] demonstrated good results on a variety of Rfam [44–47] families, in terms of accurate recovery of the accepted consensus structure, robustness to inclusion of extraneous flanking sequences and to a declining proportion of motif-bearing sequences in its input. Yao et al. [2] examined extension of the method to incorporate results of genome-scale homolog search as outlined above, and applied it on a broad scale in 44 Firmicute bacteria. The motivating problem was riboswitch discovery. As suggested in Subheading 2, the starting points for motif discovery were datasets containing a few hundred nucleotides of unaligned noncoding sequence upstream of the start codons of genes containing orthologous protein domains, as identified by NCBI's Conserved Domain Database (CDD) [48]. Of the 13 *cis*-regulatory Rfam families present in the clade, 11 (mostly riboswitches) appeared among the 50 top-ranked motifs produced by this automated process, and the resulting models achieved greater than 75% sensitivity and specificity both in identifying family members, and in identifying paired nucleotides in them. The results also showed good rejection of negative controls (permuted alignments) and good recovery of RNA motifs known in the literature but not then in Rfam. Additionally, it discovered a number of novel elements such as ribosomal protein leaders that were consequently added to Rfam. Weinberg et al. [3] carried this analysis into other phyla, characterizing 22 strong candidate *cis*-regulatory RNAs, of which at least 5 were subsequently verified to be riboswitches [49–53]. Extensions of these techniques also have been used for the discovery of other riboswitch and ncRNA candidates; *see*, e.g., [54, 55].

In another direction, Torarinsson et al. [4] applied CMfinder to the human genome. Specifically, they ran it on MULTIZ alignment blocks [56] provided by the UCSC browser [57] within the pilot ENCODE regions of the 17-way human genome alignments [58]. In this context, the alignments were used only to indicate orthology—the detailed nucleotide-level alignments were ignored by CMfinder. The scan identified several thousand candidates, albeit with a high estimated false discovery rate (again based on shuffled alignments). The candidates showed highly significant enrichment for co-occurrence with "indel purified segments" [59]—noncoding segments that appear to be under purifying selection to exclude indels. Functional characterization of the candidates was not attempted, but of a small number of candidates selected for experimental follow-up, most were clearly expressed, usually in a tissue-specific manner.

One very interesting aspect of the results relates to alignment. Whole-genome alignments, as expected, strive to optimize nucleotide identity among aligned positions of different genomes. As explained earlier, however, *covariation* between aligned positions is a crucial indicator of RNA structure. Thus, it is expected that evidence for shared RNA structures is blunted in sequence-based alignments, but the impact of this bias is unclear. As one benchmark, Gardner et al. [60] report that all tested alignment methods exhibit a dramatic degradation in the quality of sequence-based multiple alignments of structured RNAs when sequence identity falls below ≈60%. Given that the majority of input alignments considered in [4] fall below this threshold, the effect of these misalignments on genome-scale RNA structure prediction may be quite significant. To further examine this issue, the data from [4] reproduced in Fig. 2 plots the percentage realignment within CMfinder candidates versus the percent sequence identity in the input alignment. It clearly shows the expected trend that candidates found in alignments with lower average identity tend to be more extensively realigned by CMfinder than those from high-identity blocks. More interesting is the extent of the effect. Approximately one-quarter of the candidates were realigned more than 50%, and significant adjustment occurs even in high-identity alignments, where one might have expected that most compensatory changes would be bracketed by well-conserved patches, constraining the changes to be correctly aligned.

To further emphasize the impact of alignment, Torarinsson et al. [61] report thousands of candidate RNA structures shared between human and mouse in regions that whole-genome alignment tools refuse to align. In short, conservation of secondary structure is potentially seriously underestimated by studies based on sequence-only multiple alignments.

As noted earlier, the above approach is only one of many possibilities that have been tried. Both sequence-based and

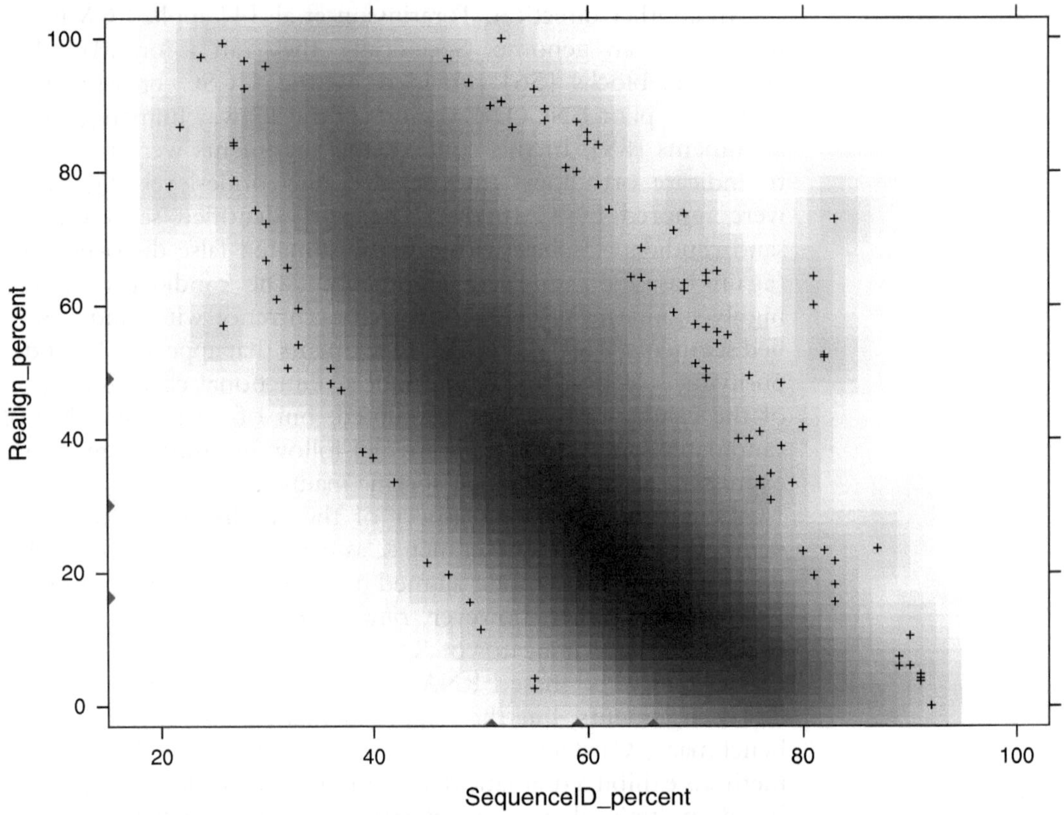

Fig. 2 Realignment versus sequence identity, based on ncRNA candidates from [4]. *Shading* represents a smoothed density estimate for the scatter plot; *pluses* mark the 1% of the data points comprising the lowest density regions. *Triangles* on the axes mark quartiles in the corresponding data

structure-based alignments have been used, with sequences from as few as two species to many dozens. In addition to the SCFG/CM approach outlined above, various groups have based structure inference on pair-models, on folding energy, on the powerful (but computationally expensive) Sankoff dynamic programming approach and heuristic approximations to it, on machine learning approaches, and combinations of these. Some approaches attempt to exploit phylogeny, others do not. Some have attempted clustering or other broader integration of results. For example, Lu et al. [62] combined RNA structure prediction with modENCODE data including RNAseq to characterize ncRNAs in *Caenorhabditis elegans*. Another recent approach makes implicit prediction through an SVM-based classifier tackling the problem of distinguishing structured from nonstructured UTRs [63]. Amidst this great diversity of approaches, one point of agreement is that these methods all identify hundreds to thousands of novel ncRNA candidates. Table 1 summarizes a sample of these results.

Table 1
Some genomic ncRNA scans

Method[a]	Organisms[b]	Num[c]	Notes[d]
QRNA [71] (pair-SCFG)	Bacteria	275	2-way, Seq, SW, [72]
	Yeasts	92	2-way, Seq, SW, [73]
RNAz [43] (Energy, SVM)	*Ciona*	2,109	2-way, Seq, SW, [74]
	Worm	2,366	2-way, Seq, SW, [75]
	Vertebrates/ PhastCons	35,985	8-way, Seq, SW, [76]
	Vertebrates/ ENCODE	3,707	28-way, Seq, SW, [77]
EvoFold [40] (phylo-SCFG)	Vertebrates/ PhastCons	48,479	8-way, Seq, SW, [40]
	Vertebrates/ ENCODE	4,986	28-way, Seq, SW, [77]
Dynalign [29] (Sankoff, SVM)	Bacteria	995	2-way, Seq, SW, H, [78]
FOLDALIGN [79, 80] (Sankoff)	Human-mouse	1,297	2-way, Str, [61]
CMfinder [1] (SCFG, EM)	Bacteria	1,466	*n*-way, Str, [2]
	Vertebrates/ ENCODE	6,587	17-way, Str, [4]
EvoFam [70] (SCFG)	Vertebrates/ PhastCons	220	41-way, Seq, [70]

[a]Name, references, and core methodology

[b]Genomes screened: ENCODE means pilot ENCODE regions (≈1% of the human genome, plus orthologous regions of other vertebrates); PhastCons means PhastCons conserved regions (≈5% of the human genome, plus orthologous regions of other vertebrates)

[c]Number of predicted RNA structures; where possible, numbers quoted are from a "stringent" set reported in the original paper, e.g., "$p > 0.9$" for RNAz and Dynalign, "Top 50%" for EvoFold ENCODE scan, and partially hand-curated for EvoFam. It is important to note that follow-up validation of predictions has been limited and estimated false discovery rates are often high

[d]*Notes*: *2-way* pairwise alignments, others were multiple alignments (CMfinder bacterial scan used varying numbers of sequences; RNAz generally selects five sequences if more are available), *Seq* sequence-based alignments, *Str* local structure-based alignments, *SW* limited length, sliding windows, *H* HMM-based alignment constraints

5 Conclusions and Future Prospects

To summarize, key challenges in *de novo* discovery of novel noncoding RNA genes (and other functional ncRNA elements, like riboswitches) center on the problems of (a) finding a sufficiently large set of putative representatives, sufficiently enriched with *actual* representatives, and (b) successfully inferring the consensus features of the family from these candidates, given that the candidate set is almost certainly of suboptimal diversity, with ill-defined borders, and contaminated with non- and atypical examples. "Suboptimal diversity" includes situations where sequence conservation is too high, potentially causing failures due to lack of covariation and/or spurious inclusion of "consensus" features that are absent from the broader family. It also includes situations where sequence conservation is too low, causing poor alignments and/or reduced enrichment of actual examples.

The case studies presented above outline some approaches to answering these challenges. Of these, we single out three general features that seem noteworthy and broadly applicable. First is the importance of exploiting prior knowledge where possible, especially for the purposes of generating candidate sequence sets on which to apply RNA motif discovery algorithms. The selection of upstream sequences of genes containing orthologous conserved domains for riboswitch discovery is one such example. A second important strategy is the integration of discovery with search—exposing additional examples of any motif helps both in accurately characterizing the motif and potentially in the ultimate functional characterization of the RNA. Again, this strategy has proved very useful in riboswitch discovery in prokaryotes, and we expect it to play an increasingly important role in analysis of mammalian genomes, although their larger sizes pose significant computational difficulties. Thirdly, we think that iterative approaches will continue to play important roles in these analyses. The enormous computational cost of "exact" algorithms even for idealized versions of these problems (e.g., the Sankoff algorithm) seem to necessitate use of heuristic approximations, but iterative refinement of the initial solution (e.g., as in CMfinder's EM-like approach, or the integration of search with discovery as mentioned above) builds in some tolerance for errors that will inevitably be made in the early stages, thus allowing them to be fast enough to be feasible.

Many challenges remain. Two issues are paramount among these. First, all genome-scale approaches to date, as assessed by the sophisticated methods mentioned in Subheading 3.4, have been plagued with high false discovery rates and/or low sensitivity. Better scoring, more realistic null models, and deeper exploitation of phylogenetic information might all help, as would more refined motif inference. As one example of the latter, improvements to

CMfinder's initialization might be attainable by augmenting single sequence folding with pairwise structural alignments, inclusion of suboptimal structural alignments, and/or phylogenetic information. The second major issue is that computational cost remains very high for the algorithms in use today. For example, one of the smaller projects mentioned above, the CMfinder Firmicutes scan, used about 1 year of computer time, and genome-scale screens in vertebrates have used hundreds. Obviously, parallelization is possible, computer costs continue to decline, and, independently, algorithms get faster. Dramatic improvements in covariance model search speed, for example, have been obtained in the past few years [19, 64–67]. Nevertheless, these problems are still at the margins of affordability. Genome-scale clustering of results [68–70] is another area that is potentially very compute-intensive and has received relatively little attention to date. We hope for progress on all of these fronts.

Acknowledgements

This work is supported by the Danish Council for Independent Research (Technology and Production Sciences), the Danish Council for Strategic Research (Programme Commission on Strategic Growth Technologies), as well as the Danish Center for Scientific Computing.

References

1. Yao Z, Weinberg Z, Ruzzo WL (2006) CMfinder—a covariance model based RNA motif finding algorithm. Bioinformatics 22(4):445–452. PMID:16357030
2. Yao Z, Barrick J, Weinberg Z, Neph S, Breaker R, Tompa M, Ruzzo WL (2007) A computational pipeline for high-throughput discovery of *cis*-regulatory noncoding RNA in prokaryotes. PLoS Comput Biol 3(7):e126. PMID:17616982
3. Weinberg Z, Barrick JE, Yao Z, Roth A, Kim JN, Gore J, Wang JX, Lee ER, Block KF, Sudarsan N, Neph S, Tompa M, Ruzzo WL, Breaker RR (2007) Identification of 22 candidate structured RNAs in bacteria using the CMfinder comparative genomics pipeline. Nucleic Acids Res 35:4809–4819. PMID:17621584
4. Torarinsson E, Yao Z, Wiklund ED, Bramsen JB, Hansen C, Kjems J, Tommerup N, Ruzzo WL, Gorodkin J (2008) Comparative genomics beyond sequence-based alignments: RNA structures in the ENCODE regions. Genome Res 18:242–251. PMID:18096747
5. Gorodkin J, Knudsen B (2000) RNA informatik. Naturens Verden 11–12:2–9
6. Eddy SR (2001) Non-coding RNA genes and the modern RNA world. Nat Rev Genet 2(12):919–929. PMID:11733745
7. Eddy SR (2002) Computational genomics of noncoding RNA genes. Cell 109(2):137–140. PMID:12007398
8. Bompfünewerer AF, Flamm C, Fried C, Fritzsch G, Hofacker IL, Lehmann J, Missal K, Mosig A, Müller B, Prohaska SJ, Stadler BMR, Stadler PF, Tanzer A, Washietl S, Witwer C (2005) Evolutionary patterns of non-coding RNAs. Theory Biosci 123(4):301–369. PMID:18202870
9. Mattick JS, Makunin IV (2006) Non-coding RNA. Hum Mol Genet 15(1):R17–R29. PMID:16651366
10. Bompfünewerer AF, Backofen R, Bernhart SH, Flamm C, Fried C, Fritzsch G, Hackermüller J, Hertel J, Hofacker IL, Missal K, Mosig A, Prohaska SJ, Rose D, Stadler PF, Tanzer A, Washietl S, Will S (2007) RNAs everywhere: genome-wide annotation of structured RNAs.

J Exp Zoolog B Mol Dev Evol 308:1–25. PMID:17171697
11. Gorodkin J, Hofacker IL, Torarinsson E, Yao Z, Havgaard JH, Ruzzo WL (2010) *De novo* prediction of structured RNAs from genomic sequences. Trends Biotechnol 28:9–19 (Feature Review). PMID:19942311
12. Gorodkin J, Hofacker IL (2011) From structure prediction to genomic screens for novel non-coding RNAs. PLoS Comput Biol 7(8):e1002100. PMID:21829340
13. Washietl S, Will S, Hendrix DA, Goff LA, Rinn JL, Berger B, Kellis M (2012) Computational analysis of noncoding RNAs. Wiley Interdiscip Rev RNA 3(6):759–778. PMID:22991327
14. Pace NR, Thomas BR, Woese CR (1999) Probing RNA structure, function, and history by comparative analysis. In: Gesteland RF, Cech TR, Atkins JF (eds) The RNA world, Chap. 4. Cold Spring Harbor Laboratory, Cold Spring Harbor, pp 113–141
15. Shang L, Xu W, Ozer S, Gutell RR (2012) Structural constraints identified with covariation analysis in ribosomal RNA. PLoS One 7(6):e39383. PMID:22724009
16. Barrick JE, Breaker RR (2007) The distributions, mechanisms, and structures of metabolite-binding riboswitches. Genome Biol 8(11):R239. PMID:17997835
17. Zuker M (1989) Computer prediction of RNA structure. Methods Enzymol 180:262–288. PMID:2482418
18. Hofacker IL, Fontana W, Stadler PF, Bonhoeffer LS, Tacker M, Schuster P (1994) Fast folding and comparison of RNA secondary structures. Monatshefte für Chemie 125:167–188
19. Nawrocki EP, Kolbe DL, Eddy SR (2009) Infernal 1.0: inference of RNA alignments. Bioinformatics 25(10):1335–1337. PMID:19307242
20. Durbin R, Eddy SR, Krogh A, Mitchison G (1998) Biological sequence analysis: Probabilistic models of proteins and nucleic acids. Cambridge University Press, Cambridge
21. Do CB, Batzoglou S (2008) What is the expectation maximization algorithm? Nat Biotechnol 26(8):897–899. PMID:18688245
22. Bailey TL, Elkan C (1995) The value of prior knowledge in discovering motifs in MEME. In: Proceedings of the third international conference on intelligent systems for molecular biology. AAAI, Menlo Park, pp 21–29. PMID:7584439
23. Eddy SR, Durbin R (1994) RNA sequence analysis using covariance models. Nucleic Acids Res 22(11):2079–2088. PMID:8029015
24. Sakakibara Y, Brown M, Hughey R, Mian IS, Sjölander K, Underwood RC, Haussler D (1994) Stochastic context-free grammars for tRNA modeling. Nucleic Acids Res 22(23):5112–5120. PMID:7800507
25. Touzet H, Perriquet O (2004) CARNAC: folding families of related RNAs. Nucleic Acids Res 32(Web server issue):W142–W145. PMID:15215367
26. Ji Y, Xu X, Stormo GD (2004) A graph theoretical approach for predicting common RNA secondary structure motifs including pseudoknots in unaligned sequences. Bioinformatics 20(10):1591–1602. PMID:14962926
27. Sankoff D (1985) Simultaneous solution of the RNA folding, alignment and protosequence problems. SIAM J Appl Math 45:810–825
28. Gorodkin J, Heyer LJ, Stormo GD (1997) Finding the most significant common sequence and structure motifs in a set of RNA sequences. Nucleic Acids Res 25(18):3724–3732. PMID:9278497
29. Mathews DH, Turner DH (2002) Dynalign: an algorithm for finding the secondary structure common to two RNA sequences. J Mol Biol 317(2):191–203. PMID:11902836
30. McCaskill JS (1990) The equilibrium partition function and base pair binding probabilities for RNA secondary structure. Biopolymers 29:1105–1119. PMID:1695107
31. Hofacker IL, Fekete M, Stadler PF (2002) Secondary structure prediction for aligned RNA sequences. J Mol Biol 319(5):1059–1066. PMID:12079347
32. Altschul SF, Erickson BW (1985) Significance of nucleotide sequence alignments: a method for random sequence permutation that preserves dinucleotide and codon usage. Mol Biol Evol 2(6):526–538. PMID:3870875
33. Babak T, Blencowe BJ, Hughes TR (2007) Considerations in the identification of functional RNA structural elements in genomic alignments. BMC Bioinformatics 8:33. PMID:17263882
34. Gesell T, Washietl S (2008) Dinucleotide controlled null models for comparative RNA gene prediction. BMC Bioinformatics 9:248. PMID:18505553
35. Anandam P, Torarinsson E, Ruzzo WL (2009) Multiperm: shuffling multiple sequence alignments while approximately preserving dinucleotide frequencies. Bioinformatics 25:668–669. PMID:19136551
36. Stamatakis A (2006) RAxML-VI-HPC: maximum likelihood-based phylogenetic analyses with thousands of taxa and mixed models. Bioinformatics 22(21):2688–2690. PMID:16928733
37. Gowri-Shankar V, Rattray M (2007) A reversible jump method for Bayesian phylogenetic inference with a nonhomogeneous substitution model. Mol Biol Evol 24(6):1286–1299. PMID:17347157

38. Knudsen B, Hein J (1999) RNA secondary structure prediction using stochastic context-free grammars and evolutionary history. Bioinformatics 15(6):446–454. PMID:10383470
39. Knudsen B, Hein J (2003) Pfold: RNA secondary structure prediction using stochastic context-free grammars. Nucleic Acids Res 31(13):3423–3428. PMID:12824339
40. Pedersen JS, Bejerano G, Siepel A, Rosenbloom K, Lindblad-Toh K, Lander ES, Kent J, Miller W, Haussler D (2006) Identification and classification of conserved RNA secondary structures in the human genome. PLoS Comput Biol 2:e33. PMID:16628248
41. Yao Z (2008) Genome scale search of noncoding RNAs: bacteria to vertebrates. Ph.D. thesis, Department of Computer Science and Engineering, University of Washington
42. Bernhart SHF, Hofacker IL, Will S, Gruber AR, Stadler PF (2008) RNAalifold: improved consensus structure prediction for RNA alignments. BMC Bioinformatics 9:474. PMID:19014431
43. Washietl S, Hofacker IL, Stadler PF (2005) Fast and reliable prediction of noncoding RNAs. Proc Natl Acad Sci USA 102:2454–2459. PMID:15665081
44. Griffiths-Jones S, Bateman A, Marshall M, Khanna A, Eddy SR (2003) Rfam: an RNA family database. Nucleic Acids Res 31(1):439–441. PMID:12520045
45. Griffiths-Jones S, Moxon S, Marshall M, Khanna A, Eddy SR, Bateman A (2005) Rfam: annotating non-coding RNAs in complete genomes. Nucleic Acids Res 33(Database issue):121–124. PMID:15608160
46. Gardner PP, Daub J, Tate JG, Nawrocki EP, Kolbe DL, Lindgreen S, Wilkinson AC, Finn RD, Griffiths-Jones S, Eddy SR, Bateman A (2009) Rfam: updates to the RNA families database. Nucleic Acids Res 37(Database issue):D136–D140. PMID:18953034
47. Gardner PP, Daub J, Tate J, Moore BL, Osuch IH, Griffiths-Jones S, Finn RD, Nawrocki EP, Kolbe DL, Eddy SR, Bateman A (2011) Rfam: Wikipedia, clans and the "decimal" release. Nucleic Acids Res 39(Database issue):D141–D145. PMID:21062808
48. Marchler-Bauer A, Anderson JB, Cherukuri PF, DeWeese-Scott C, Geer LY, Gwadz M, He S, Hurwitz DI, Jackson JD, Ke Z, Lanczycki CJ, Liebert CA, Liu C, Lu F, Marchler GH, Mullokandov M, Shoemaker BA, Simonyan V, Song JS, Thiessen PA, Yamashita RA, Yin JJ, Zhang D, Bryant SH (2005) CDD: a Conserved Domain Database for protein classification. Nucleic Acids Res 33(Database issue):D192–D196. PMID:15608175
49. Weinberg Z, Regulski EE, Hammond MC, Barrick JE, Yao Z, Ruzzo WL, Breaker RR (2008) The aptamer core of SAM-IV riboswitches mimics the ligand-binding site of SAM-I riboswitches. RNA 14:822–828. PMID:18369181
50. Regulski EE, Moy RH, Weinberg Z, Barrick JE, Yao Z, Ruzzo WL, Breaker RR (2008) A widespread riboswitch candidate that controls bacterial genes involved in molybdenum cofactor and tungsten cofactor metabolism. Mol Microbiol 68:918–932. PMID: 18363797
51. Sudarsan N, Lee ER, Weinberg Z, Moy RH, Kim JN, Link KH, Breaker RR (2008) Riboswitches in eubacteria sense the second messenger cyclic di-GMP. Science 321(5887):411–413. PMID:18635805
52. Wang JX, Lee ER, Morales DR, Lim J, Breaker RR (2008) Riboswitches that sense S-adenosylhomocysteine and activate genes involved in coenzyme recycling. Mol Cell 29:691–702. PMID:18374645
53. Meyer MM, Roth A, Chervin SM, Garcia GA, Breaker RR (2008) Confirmation of a second natural preQ1 aptamer class in Streptococcaceae bacteria. RNA 14:685–695. PMID:18305186
54. Weinberg Z, Wang JX, Bogue J, Yang J, Corbino K, Moy RH, Breaker RR (2010) Comparative genomics reveals 104 candidate structured RNAs from bacteria, archaea, and their metagenomes. Genome Biol 11(3):R31. PMID:20230605
55. Weinberg Z, Perreault J, Meyer MM, Breaker RR (2009) Exceptional structured noncoding RNAs revealed by bacterial metagenome analysis. Nature 462(7273):656–659. PMID:19956260
56. Blanchette M, Kent WJ, Riemer C, Elnitski L, Smit AFA, Roskin KM, Baertsch R, Rosenbloom K, Clawson H, Green ED, Haussler D, Miller W (2004) Aligning multiple genomic sequences with the threaded blockset aligner. Genome Res 14(4):708–715. PMID:15060014
57. Kent WJ, Sugnet CW, Furey TS, Roskin KM, Pringle TH, Zahler AM, Haussler D (2002) The human genome browser at UCSC. Genome Res 12(6):996–1006. PMID:12045153
58. ENCODE Project Consortium et al (2007) Identification and analysis of functional elements in 1% of the human genome by the ENCODE pilot project. Nature 447(7146):799–816. PMID:17571346
59. Lunter G, Ponting CP, Hein J (2006) Genome-wide identification of human functional DNA using a neutral indel model. PLoS Comput Biol 2(1):e5. PMID:16410828
60. Gardner PP, Wilm A, Washietl S (2005) A benchmark of multiple sequence alignment

programs upon structural RNAs. Nucleic Acids Res 33(8):2433–2439. PMID:15860779
61. Torarinsson E, Sawera M, Havgaard JH, Fredholm M, Gorodkin J (2006) Thousands of corresponding human and mouse genomic regions unalignable in primary sequence contain common RNA structure. Genome Res 16(7):885–889. Erratum: Genome Res 16:1439, 2006. PMID:16751343
62. Lu ZJ, Yip KY, Wang G, Shou C, Hillier LW, Khurana E, Agarwal A, Auerbach R, Rozowsky J, Cheng C, Kato M, Miller DM, Slack F, Snyder M, Waterston RH, Reinke V, Gerstein MB (2011) Prediction and characterization of noncoding RNAs in *C. elegans* by integrating conservation, secondary structure, and high-throughput sequencing and array data. Genome Res 21(2):276–285. PMID:21177971
63. Chen XS, Brown CM (2012) Computational identification of new structured cis-regulatory elements in the 3′-untranslated region of human protein coding genes. Nucleic Acids Res 40(18):8862–8873. doi: 10.1093/nar/gks684. PMID:22821558
64. Weinberg Z, Ruzzo WL (2004) Faster genome annotation of non-coding RNA families without loss of accuracy. In: RECOMB04: Proceedings of the eighth annual international conference on computational molecular biology. ACM, San Diego, pp 243–251. http://doi.acm.org/10.1145/974614.974647
65. Weinberg Z, Ruzzo WL (2004) Exploiting conserved structure for faster annotation of non-coding RNAs without loss of accuracy. Bioinformatics 20(1):i334–i341. PMID:15262817
66. Weinberg Z, Ruzzo WL (2006) Sequence-based heuristics for faster annotation of non-coding RNA families. Bioinformatics 22(1):35–39. PMID:16267089
67. Sun Y, Buhler J, Yuan C (2012) Designing filters for fast-known ncRNA identification. IEEE/ACM Trans Comput Biol Bioinformatics 9(3):774–787. PMID: 22084145
68. Will S, Reiche K, Hofacker IL, Stadler PF, Backofen R (2007) Inferring noncoding RNA families and classes by means of genome-scale structure-based clustering. PLoS Comput Biol 3(4):e65. PMID:17432929
69. Tseng HH, Weinberg Z, Gore J, Breaker RR, Ruzzo WL (2009) Finding non-coding RNAs through genome-scale clustering. J Bioinformatics Comput Biol 7:373–388. PMID:19340921
70. Parker BJ, Moltke I, Roth A, Washietl S, Wen J, Kellis M, Breaker R, Pedersen JS (2011) New families of human regulatory RNA structures identified by comparative analysis of vertebrate genomes. Genome Res 21(11):1929–1943. PMID:21994249
71. Rivas E, Eddy SR (2001) Noncoding RNA gene detection using comparative sequence analysis. BMC Bioinformatics 2(1):8. ISSN 1471-2105. PMID:11801179
72. Rivas E, Klein RJ, Jones TA, Eddy SR (2001) Computational identification of noncoding RNAs in *E. coli* by comparative genomics. Curr Biol 11(17):1369–1373. PMID:11553332
73. McCutcheon JP, Eddy SR (2003) Computational identification of non-coding RNAs in *Saccharomyces cerevisiae* by comparative genomics. Nucleic Acids Res 31(14):4119–4128. PMID:12853629
74. Missal K, Rose D, Stadler PF (2005) Non-coding RNAs in *Ciona intestinalis.* Bioinformatics 21(2):ii77–ii78. PMID:16204130
75. Missal K, Zhu X, Rose D, Deng W, Skogerbø G, Chen R, Stadler PF (2006) Prediction of structured non-coding RNAs in the genomes of the nematodes *Caenorhabditis elegans* and *Caenorhabditis briggsae.* J Exp Zoolog B Mol Dev Evol 306(4):379–392. PMID:16425273
76. Washietl S, Hofacker IL, Lukasser M, Hüttenhofer A, Stadler PF (2005) Mapping of conserved RNA secondary structures predicts thousands of functional noncoding RNAs in the human genome. Nat Biotechnol 23(11):1383–1390. PMID:16273071
77. Washietl S, Pedersen JS, Korbel JO, Stocsits C, Gruber AR, Hackermüller J, Hertel J, Lindemeyer M, Reiche K, Tanzer A, Ucla C, Wyss C, Antonarakis SE, Denoeud F, Lagarde J, Drenkow J, Kapranov P, Gingeras TR, Guigo R, Snyder M, Gerstein MB, Reymond A, Hofacker IL, Stadler PF (2007) Structured RNAs in the ENCODE selected regions of the human genome. Genome Res 17(6):852–864. PMID:17568003
78. Uzilov AV, Keegan JM, Mathews DH (2006) Detection of non-coding RNAs on the basis of predicted secondary structure formation free energy change. BMC Bioinformatics 7:173. PMID:16566836
79. Havgaard JH, Lyngsø RB, Stormo GD, Gorodkin J (2005) Pairwise local structural alignment of RNA sequences with sequence similarity less than 40%. Bioinformatics 21(9):1815–1824. PMID:15657094
80. Havgaard JH, Torarinsson E, Gorodkin J (2007) Fast pairwise structural RNA alignments by pruning of the dynamical programming matrix. PLoS Comput Biol 3:1996–1908. PMID:17937495

Chapter 16

Phylogeny and Evolution of RNA Structure

Tanja Gesell and Peter Schuster

Abstract

Darwin's conviction that all living beings on Earth are related and the graph of relatedness is tree-shaped has been essentially confirmed by phylogenetic reconstruction first from morphology and later from data obtained by molecular sequencing. Limitations of the phylogenetic tree concept were recognized as more and more sequence information became available. The other path-breaking idea of Darwin, natural selection of fitter variants in populations, is cast into simple mathematical form and extended to mutation-selection dynamics. In this form the theory is directly applicable to RNA evolution in vitro and to virus evolution. Phylogeny and population dynamics of RNA provide complementary insights into evolution and the interplay between the two concepts will be pursued throughout this chapter. The two strategies for understanding evolution are ultimately related through the central paradigm of structural biology: sequence ⇒ structure ⇒ function. We elaborate on the state of the art in modeling both phylogeny and evolution of RNA driven by reproduction and mutation. Thereby the focus will be laid on models for phylogenetic sequence evolution as well as evolution and design of RNA structures with selected examples and notes on simulation methods. In the perspectives an attempt is made to combine molecular structure, population dynamics, and phylogeny in modeling evolution.

Key words Evolution of structure, Multiple structures, Phylogeny, Quasispecies concept, Sequence-structure mappings, Sequence evolution, Simulations

1 Evolutionary *Thinking* in Mathematical Language

James Watson, one of the researchers who discovered the structure of the DNA double helix, begins his text book *Molecular Biology of the Gene* in 1965 with an euphoric introduction to evolution to underline its general importance. Today, almost 50 years after the discovery of the double helix and more than 200 years after Charles Darwin's birthday in 1809, evolution is still at the center of biological thinking: the statement of geneticist Theodosius Dobzhansky [1], *"Nothing in biology makes sense except in the light of evolution"*, is frequently cited and yet, Paul Griffiths [2] has given voice to the widespread feeling that an evolutionary perspective is indeed necessary, but it must be a *forward-looking* perspective

Jan Gorodkin and Walter L. Ruzzo (eds.), *RNA Sequence, Structure, and Function: Computational and Bioinformatic Methods*, Methods in Molecular Biology, vol. 1097, DOI 10.1007/978-1-62703-709-9_16,

allowing for a general understanding of the evolutionary process, not only a *backward-looking perspective* dealing with the specific evolutionary history of the species.

Here, we shall review current mathematical models of evolution while focusing in particular on two mechanistic aspects of evolution, phylogeny and population dynamics with respect to structure evolution, trying to work out how they are related. The first section introduces the concept of the phylogenetic tree and continues with an attempt to translate Charles Darwin's thoughts on natural selection into mathematical language with the knowledge of his contemporaries in mathematics. Subheading 2 elaborates on the state of the art in modeling phylogeny and continues with a review of evolutionary dynamics based on reproduction and mutation. Subheading 3 then deals with the translation of results from theory to applications in order to demonstrate the practical usefulness of evolutionary models. The notes (Subheading 4) will be concentrating on computation and simulation methods of phylogenetic aspects of the evolutionary process. In the prospects Subheading 5 we shall discuss the perspectives afforded by attempts to combine molecular structure and phylogeny in modeling evolution, and mention future developments in the experimental determination of fitness landscapes in the sense of Sewall Wright's metaphor [3].

Darwin's centennial book *The Origin of Species* [4] appeared in November 1859 and represents a masterpiece of abstraction and reduction, since the principle of evolution built upon only three factors—multiplication, variation, and (natural) selection in size-constrained populations—had been extracted from an overwhelming wealth of observations, but contains not a single mathematical expression and only one graphic illustration representing a *tree of life*. Figure 1 shows Darwin's conviction that all life on earth is related and that the pattern of relatedness is shaped like a tree. Darwin had drawn this sketch already in *Notebook B*, more than 20 years earlier, adding the words "I think" on its side. Darwin's fundamental concept of the tree of descendance of species is introduced by illustrating the present-day concept of phylogenetic trees. How might Charles Darwin have formulated his theory had he been a mathematician? We do not know, but we shall try to cast Darwin's natural selection in simple equations that were known at his time.

Phylogeny and population dynamics provide complementary insights into evolution and the interplay of the two concepts will be pursued throughout this chapter. The two strategies for understanding evolution are ultimately related through the central paradigm of structural biology (Fig. 2): sequences $\mathsf{S} = (s_1 s_2 \dots s_l)$ fold into structures Y and the structures encode the functions of molecules or organisms as expressed, for example, by fitness values. Both relations can be viewed as mappings, the first one

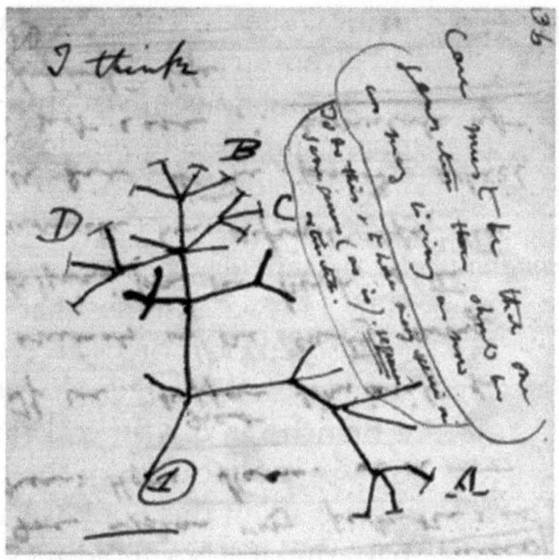

Fig. 1 *An evolutionary tree by Charles Darwin.* The ancestral species is at position "1." Extant species are denoted by endpoint and letters, and the remaining pendant edges represent extinctions. On the margin of his sketch of a tree Darwin had written, "I think," before expanding his idea in *The Origin of Species* [4]: "The affinities of all the beings of the same class have sometimes been represented by a great tree. I believe this simile largely speaks the truth. The green and budding twigs may represent existing species; and those produced during each former year may represent the long succession of extinct species..." (*First Notebook on Transmutation of Species*, 1837, syndics of Cambridge University Library)

$$\Phi: (\mathcal{Q},d_{\mathrm{H}}) \Rightarrow (\mathcal{Y},d_{\mathrm{Y}}) \qquad \Psi: (\mathcal{Y},d_{\mathrm{Y}}) \Rightarrow \mathbf{R}^1$$

$$\mathrm{S} \longrightarrow \mathrm{Y} = \Phi(\mathrm{S}) \longrightarrow f = \Psi(\mathrm{Y})$$

sequence structure function

Fig. 2 *Sequence, structure, and function in structural biology.* The relation between sequence, structure, and function is modeled by two successive mappings, ˇ from sequences into structures and ¯ from structures into real numbers being a quantitative measure of function

from sequence space[1] $\mathcal{Q}$ into structure space $\mathcal{Y}$, and the second one from structure space into the real numbers, $\Phi : (\mathcal{Q}, d_{\mathrm{H}}) \to (\mathcal{Y}, d_{\mathrm{Y}})$ and $\Psi : (\mathcal{Y}, d_{\mathrm{Y}}) \to \mathbb{R}^1$, respectively. Sequence space and structure space are metric spaces with the Hamming distance d_{H} and the structure distance d_{Y} as metrics. In evolution, phylogeny and selection are related by these mappings from sequence into function, in particular into fitness values f. Both mappings are many to one and hence cannot be inverted in strict mathe-

[1]Sequences are ordered strings of elements s_k ($k = 1, \ldots, l$), which in case of RNA are chosen from the alphabet $\mathcal{A}$ = {A,U,C,G}. The notions of sequence space and structure or shape space are essential for the definition of structure and function as results of mappings.

matical sense. It is nevertheless possible to construct pre-images of given structures in sequence space as well as pre-images of given functions in structure space. Their inversions then become important tools in the design of biomolecules—as we shall discuss in Subheading 2.3.

1.1 The Phylogenetic Tree

In present-day phylogenetic research—following Darwin's ideas (Fig. 1)—we assume that n sequences S_n are related by an (unrooted or rooted) tree $\mathcal{T}$ whose leaves represent the aligned sequences. The tree $\mathcal{T} = (V, E)$ consists of a vertex set V and a branch set $E \in V \times V$ [5], where the lengths of the branches of $\mathcal{T}$ are a measure of the extent of evolutionary change (Fig. 3). Above all, in this book we are interested in the change nucleotide patterns of RNA molecules along a phylogenetic tree. The vertex set V contains the taxon set S, which maps one-to-one onto the leaf set.

Definition 1: A phylogenetic tree $\mathcal{T}$ is a connected, undirected, acyclic graph whose leaves are labeled bijectively by the taxon set S.

1. An unrooted phylogenetic tree $\mathcal{T}$ has no vertices of degree two.
2. A rooted phylogenetic tree $\mathcal{T}$ has an internal vertex, which may have degree two and which forms the *root* of the tree.
3. A star tree is a phylogenetic tree $\mathcal{T}$ with one internal vertex, which may have a cardinality degree of the taxon set S or in other words, all taxa have one common ancestor.

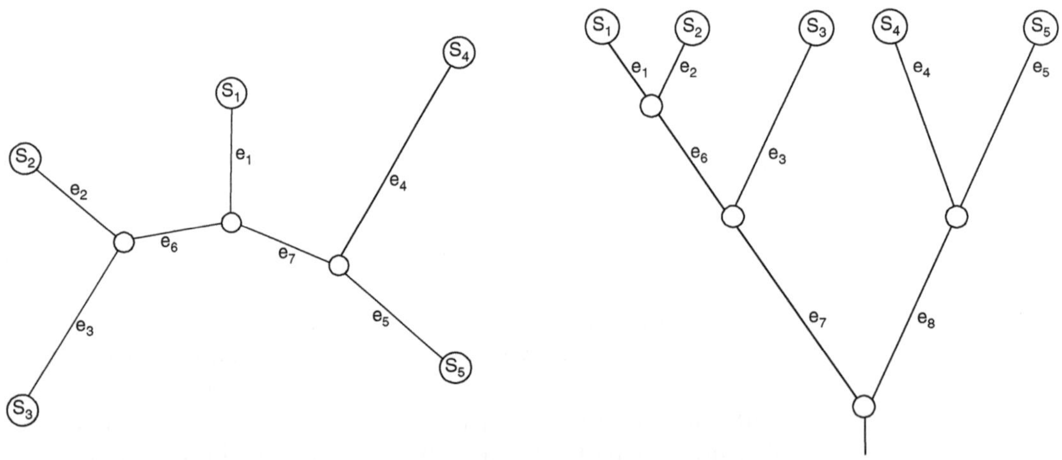

Fig. 3 *Phylogenetic trees showing relatedness of sequences (S_n).* The numbers e_i symbolize the lengths of the branches. The distance between any pair of sequences can be computed by adding up the lengths of the connecting branches. Lhs: Unrooted tree of five sequences (S_1 to S_5): this tree does not contain a node corresponding to the ancestor of all five sequences. Rhs: Rooted tree

The tree-length $\Lambda_{\mathcal{T}}$ is the sum of the branch lengths.

$$\Lambda_{\mathcal{T}} = \sum_{e \in E} \lambda_e \tag{1}$$

where $\lambda_e > 0$ represents the *length* of a branch $e \in E$. In commonly used phylogenetic methods the branch length is measured in numbers of substitutions per site. This *genetic distance d* estimated from nucleotide sequences are generally based on models of sequence evolution—as we shall discuss in Subheading 2.1.1. By contrast, the Hamming distance of two sequences, $d_{\mathrm{H}}(\mathsf{S}_i, \mathsf{S}_j)$, is the smallest number of substitutions, extensions, and deletions required to convert the two sequences into each other.

Charles Darwin's great foresight concerning the existence of a *tree of life* has been confirmed by phylogenetic reconstruction of the evolution of multicellular eukaryotes by making extensive use of data from molecular sequencing [6]. The idea of a single root of the tree of life and a common universal ancestor, however, turned out to be much less clear. Although only few doubts were raised regarding the single, non-recurring origin of life on Earth, the nature of the primordial prokaryote remained in the realm of speculation [7, 8], and for a long time it remained undecided whether such an ancestor had been a single species or a clan of genetically strongly interacting clones. As more and more sequence information became available on excessive horizontal gene transfer (*see* [9] and Subheading 1.2) among early prokaryotic organisms, it became clear that the tree concept becomes more and more obscure the further one approaches the distant past [10, 11]. The current reconstruction of the early history of life on Earth by a plethora of genomic data is not supportive of a single tree of life for archaebacterial, eubacterial, and primitive eukaryotic species but seems to be converging towards a scenario with multiple species rapidly exchanging genetic information [12–14].

1.2 Limitations of the Phylogenetic Tree Concept

Given the massive amount of sequence data observed today, a general goal in phylogeny is to reconstruct the evolutionary history of patterns of contemporary organisms, typically in the form of a phylogenetic tree. Accumulations of mutations by the copying process and environmental factors manifest the changes in the sequences called *substitutions*. Thus, given the vertical transmissions in time, the discrete character of mutations and a well-defined alphabet, the biopolymer sequences represent a unique memory of the phylogenetic past. The sequences may either come from DNA, protein, RNA, or other character-based molecules with linear arrangements of monomers from several classes. Misinterpretations of data, however, are possible, e.g., through time-heterogenous processes. Moreover, present day sequence-based phylogenies of

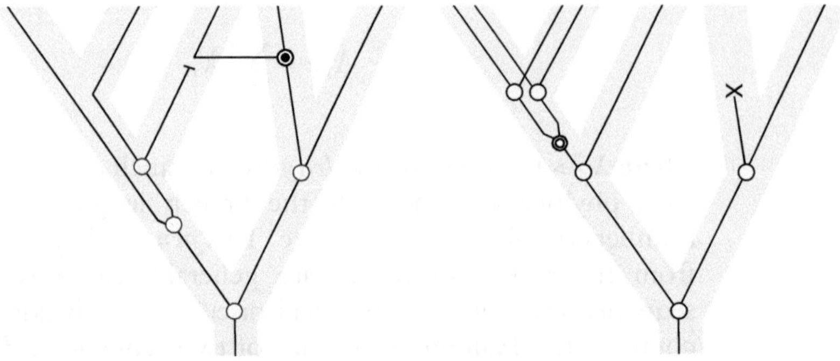

Fig. 4 *Limitations in phylogeny*. Lhs: trees for individual characters (inner tree) can differ from the species tree (outer gray tree). On the left branches, random sorting of ancestral polymorphism at subsequent speciation events. On the right branches horizontal gene transfer, i.e., the lateral transfer of individual genes or sequences between species. Rhs: a duplication event on the left branches and an instance of gene loss (x) on the right branch, both of which may lead to misconceptions of a species tree

organisms are based on many different genes, which can lead to controversial interpretations of evolutionary relationships between the organisms. Evolution of genes is different from species evolution, and they should not be confused. Indeed, there are various reasons why phylogenetic inference is not always straightforward and potentially leads to misconceptions (*see* Fig. 4), examples of problem sources are random sorting of ancestral polymorphism [15], horizontal gene transfer [16, 17], and gene duplication or gene loss [18]. As mentioned before, horizontal gene transfer obscures phylogeny and the question arises of how to define species trees for a set of prokaryotic taxa and subsequently, of how such a species tree can be inferred? In current research, phylogenetic networks offer an alternative to phylogenetic trees (A good introduction to this problematic is given by [19]). In addition, the fields of phylogenetic analysis and population genetics will come closer together in the near future as complete genomic sequences for large numbers of individuals, strains, and species will become available thanks to advanced sequencing technologies.

However, the reader should note that the application of phylogenetic methods goes beyond the reconstruction of phylogenetic trees for organisms. The phylogenetic analysis of molecular sequences using phylogenetic trees is an established field and several books (e.g., [5, 20]) offer detailed descriptions of the different approaches. In this chapter, we shall present some applications of phylogenetic substitution models in the context of RNA research and comparative genomics in Subheading 3. Just as phylogenetic analysis, population genetics is a mature discipline that has been addressed in a number of books, including [21, 22]. Both phylogeny and population genetics remain active and continually evolving areas of research. In an attempt to interconnect the two disciplines for future RNA research, we shall focus on the question of how phylogeny and population genetics are related to structure evolution on the pages to come.

1.3 The Mathematics of Darwin's Selection in Populations

In 1838 the Belgium mathematician Pierre-François Verhulst published a kinetic differential equation, which presumably was not known to Darwin. The Verhulst or logistic equation describes population growth in ecosystems with finite resources:

$$\frac{dN}{dt} = N \cdot r \left(1 - \frac{N}{K}\right) \quad \text{and}$$

$$N(t) = N(0) \frac{K}{N(0) + \left(K - N(0)\right) e^{-rt}} \,. \tag{2}$$

The number of individuals **I** in the population or the population size at time t is denoted by $[\mathbf{I}]_t = N(t)$, $N(0)$ is the population size at time $t = 0$, r is the growth parameter or *Malthusian* parameter named after the English economist Robert Malthus, and K is the *carrying capacity*, the maximal population size that can be sustained by the ecosystem. The interpretation is straightforward: Populations grow by reproduction and the population size increases proportionally to the number of individuals already present times the Malthusian parameter $N(t) \cdot r$, growth requires resources and this is taken into account by the third factor $\left(1 - N(t)/K\right)$, which approaches zero when the population size reaches the carrying capacity and no further growth is possible. The solution of Verhulst equation is called the *logistic curve*. In the early phase of growth, $N(t) \ll K$ (Fig. 5; upper plot, red curve), the logistic curve represents unlimited exponential growth, the effect of finite resources is significant at populations sizes in the range of 20% saturation, $N(t) = 0.2\,K$, and larger.

The Verhulst model of constrained growth is dealing with *multiplication*, the first factor of Darwin's principle. *Selection* follows straightforwardly from a partitioning into n subpopulations [23], $\mathbf{I}_j$ $(j = 1, \ldots, n)$ with $[\mathbf{I}_j] = N_j$ and $N(t) = \sum_{j=1}^{n} N_j(t)$. Each variant has a specific growth parameter or *fitness value* denoted by f_j; $j = 1, \ldots, n$ and the result is the *selection equation* (3) describing the evolution of the population:

$$\frac{dN_j}{dt} = N_j \left(f_j - \frac{N}{K}\,\phi(t)\right) ; \; j = 1, \ldots, n$$

$$\text{with} \quad \phi(t) = \frac{1}{N(t)} \sum_{i=1}^{n} f_i \, N_i(t) \,. \tag{3}$$

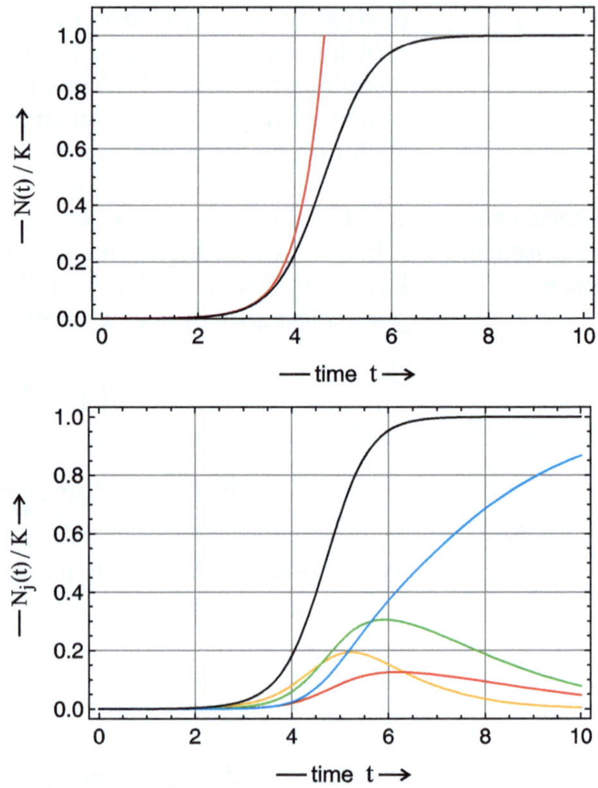

Fig. 5 *Solution curves of the logistic equations. Upper plot*: The *black curve* illustrates growth in population size from a single individual to a population at the carrying capacity of the ecosystem. The *red curve* represents the results for unlimited exponential growth, $N(t) = N(0)\exp(rt)$.
Parameters: $r = 2$, $N(0) = 1$, and $K = 10{,}000$.
Lower plot: Growth and internal selection is illustrated in a population with four variants. Color code: N black, N_1 yellow, N_2 green, N_3 red, N_4 blue.
Parameters: fitness values
$f_j = (1.75, 2.25, 2.35, 2.80)$, $N_j(0) = (0.8888, 0.0888, 0.0020, 0.0004)$, $K = 10{,}000$.
The parameters were adjusted such that the curves for the total populations size $N(t)$ coincide (almost) in both plots

Equation 3 is solved by means of normalized variables $x_i(t) = N_i(t)/N(t)$:

$$\frac{dx_i}{dt} = x_i\left(f_j - \phi(t)\right), \; j = 1, \ldots, n;$$

$$x_i(t) = \frac{x_i(0) \cdot \exp(f_j\, t)}{\sum_{i=1}^{n} x_i(0) \cdot \exp(f_i\, t)}, \; j = 1, \ldots, n. \tag{3a}$$

The size of the subpopulations is obtained through multiplication by the total population size

$$N_i(t) = N(t) \cdot x_i(t) = N(t) \cdot \frac{N(0) \cdot \exp(f_j\, t)}{\sum_{i=1}^{n} N_i(0) \cdot \exp(f_i\, t)} ;$$

$$j = 1, \ldots, n , \tag{3b}$$

and hence, the knowledge of the time dependence of population size is required. It is obtained by means of the integral $\Phi(t) = \int_0^t \phi(\tau) d\tau$

$$N(t) = N(0) \frac{K}{N(0) + \big(K - N(0)\big)\, e^{-\Phi(t)}} . \tag{2a}$$

Herein $\exp\big(-\Phi(t)\big)$ replaces $\exp(-rt)$ in the solution of the Verhulst equation. The course of selection in the variables N_j and x_j is essentially the same and the restriction to constant population size, $N = K$, was done only in order to simplify the analysis. Typical solution curves are shown in Fig. 6. The interpretation of the solution curves (3b) is straightforward: For sufficiently long time the sum in the denominator is dominated by the term containing the exponential with the highest fitness value, $f_m = \max\{f_j; j = 1, \ldots, n\}$, the consequence is that all variables except x_m vanish and the fittest variant is selected: $\lim_{t\to\infty}[X_m] = N$. Finally, the time derivative of the mean fitness, $d\phi(t)/dt = \mathrm{var}\{f\} \geq 0$, encapsulates optimization in Darwinian evolution: $\phi(t)$ is optimized during natural selection (*see*, e.g., [23]). The expressions for selection are rather simple, derivation and analysis are both straightforward, and everything needed was standard mathematics at Darwin's time.

Mechanisms creating new variants are not part of the nineteenth century model of evolution. Recombination and mutation were unknown and new variants appear spontaneously in the population like the *deus ex machina* in the ancient antique theater. Figure 6 illustrates selection at constant population size and the growth of a spontaneously created advantageous variant. Eventually the simple mathematical model described here encapsulates all three preconditions of Darwin's natural selection and reflects the state of knowledge of the evolutionists in the second half of nineteenth century.

2 Mathematical Models

Historically, the first synthesis of Darwin's theory and Mendelian genetics was performed through mathematical modeling of evolution by the three scholars of population genetics, Ronald Fisher, J.B.S. Haldane, and Sewall Wright. Modeling of evolution and many other topics in biology was and still is mainly based on ordinary differential equations (ODEs) for essentially two reasons:

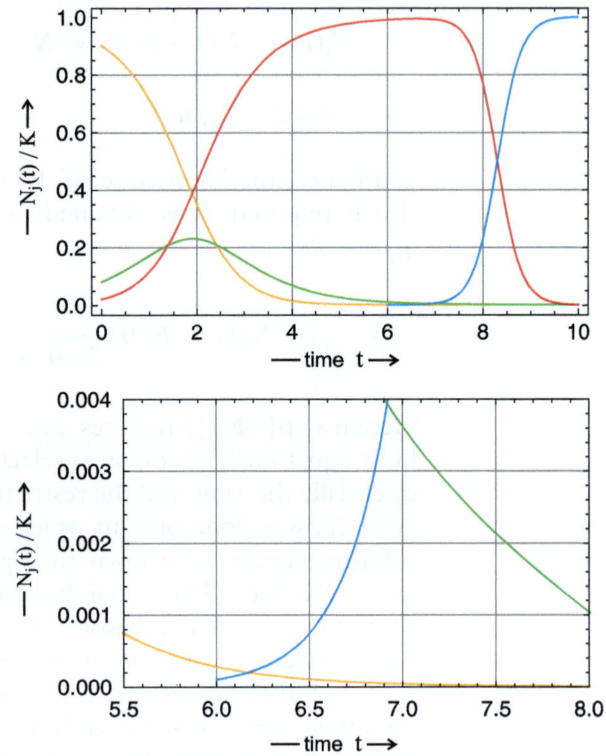

Fig. 6 *Solution curve of the selection equation* (3). The system is studied at constant maximal population size, $N = K$, and the plots represent calculated changes of the variant distributions with time. The *upper plot* shows selection among three species $\mathbf{I}_1$ (*yellow*), $\mathbf{I}_2$ (*green*), and $\mathbf{I}_3$ (*red*), and then the appearance of a fourth, fitter variant $\mathbf{I}_4$ (*blue*) at time $t = 6$, which takes over and becomes selected thereafter. The *lower plot* presents an enlargement of the upper plot around the point of spontaneous creation of the fourth species ($\mathbf{I}_4$). Parameters: fitness values
$f_j = (1, 2, 3, 7)$;
$x_j(0) = (0.9, 0.08, 0.02, 0)$
and $x_4(6) = 0.0001$

(i) ODEs are at least moderately well suited for handling the problems and (ii) handling and analyzing ODEs are based on 350 years experience from physics and mathematics. Partial differential equations (PDEs) are used for the description of migration and spreading of populations in space, in particular in ecological and epidemiological models, and for modeling morphogenesis. Apart from ODE and PDE models difference equations are used to describe discrete generations, and stochastic modeling by means of Markov processes is applied to analyze problems in biology. Markov models are particularly important in phylogeny.

Classical phylogeny was based on morphological comparisons which, although successful, escaped quantitative analysis and

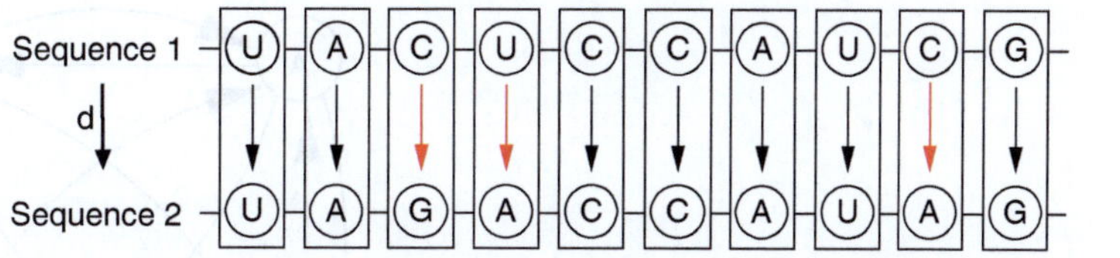

Fig. 7 *Sequence evolution*. The evolution of sequences is described by substitutional changes of single positions during a certain time span Δt, measured in number of substitution per site d. For independent-sites models we assume that positions evolve independently (as indicated by the framed boxes) and according to the same process (see also the closely related uniform error rate model in Subheading 2.3.4)

handling. The advent of molecular biology and the fast growing facilities for determination of biopolymer sequences and structures changed the scene entirely. Sequencing was developed first for proteins and later for nucleic acids, it laid down the basis for sequence comparisons and initiated the field of molecular evolution. Since 20 years a true explosion of available DNA sequences is taking place and the dramatically increasing demands for electronic storage facilities and retrieval as well as the requirements for new tools for data analysis are challenging computer science and led to the development of the novel discipline of bioinformatics. The enormous increase in computing power and data storage capacities, eventually, revolutionized also biological modeling and simulation, computational biology became a field of research in its own right. Discrete mathematics, in particular combinatorics and graph theory, are now indispensable for biological modeling not only in molecular phylogeny but also for nucleic acid structure predictions and several other fields.

2.1 Phylogenetic Models

Sequence structuring during a time span Δt is described by the evolution of sequences along a tree through single nucleotide substitutions at independent positions (Fig. 7). Explicit formal sequence evolution models are most prominently analyzed by maximum likelihood approaches [e.g. 24–27]. Clearly, the development of such models requires more or less justified assumptions concerning the evolutionary process.

2.1.1 Commonly Used Independent-Site Sequence Evolution Models

Molecular sequence data provide a recording of characters at a given instant, e.g., at time t_0, and contain no direct information on the rate at which they are evolving. During evolution, mutation and natural selection act upon molecules that are integrated in an organism, and molecules have no knowledge of their previous history. Such a *lack of memory* is one of the basic assumptions of phylogeny, and in the theory of stochastic processes this property

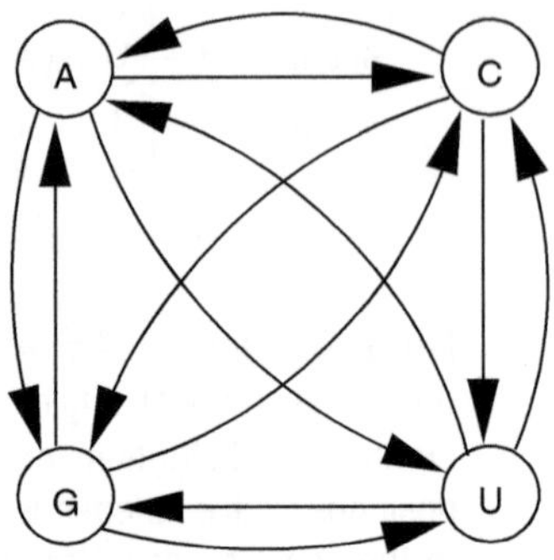

Fig. 8 *Standard models*. Independent evolution of sites, the Markov property, and continuous time are assumed, and the $|\mathcal{A}| \times |\mathcal{A}|$ instantaneous rate matrix is defined by $\mathbf{Q} = \{(\mathbf{Q})_{ij};\ i,j = 1,\ldots,|\mathcal{A}|\}$, and matrix, where $|\mathcal{A}| = \kappa$ is the number of character states. This chapter will mainly consider RNA evolution with $\mathcal{A} = \{A, C, G, U\}$, hence $|\mathcal{A}| = \kappa = 4$. For binary and nucleotide sequences, amino acid sequences in proteins and codon sequences in genes, the alphabet size is 2, 4, 20, and 64, respectively

is attributed to *Markov processes*. In other words, the evolutionary future depends on the current only state and not on any previous or ancestral state. Under the standard assumption of a *stationary, time homogeneous, and reversible Markovian substitution process*, the probability of nucleobase B_j at position s_k can be computed for every positive Δt, under the condition of nucleobase B_i having been at this position in the initial state ($t(0) = t_0$), which evolves into B_j within the time span $\Delta t = t - t_0$. To define such a process one only has to specify an instantaneous rate matrix $\mathbf{Q}$, in which each entry $q_{ij} > 0$ stands for the rate of change from state B_i to state B_j during an infinitesimal period of time dt, illustrated in Fig. 8. The diagonal elements of matrix $\mathbf{Q}$ are defined such that the individual rows add up to zero by

$$q_{ii} = -\sum_{j=1, j\neq i}^{\kappa} q_{ij} \quad \text{and} \quad \sum_{j=1}^{\kappa} q_{ij} = 0\,. \tag{4}$$

This definition is readily interpreted: q_{ij} is the rate determining the change at a given position s_k from B_i to B_j and $\sum_{j,j\neq i} q_{ij}$ is the rate for a change from B_i to any other nucleobase at this position. The diagonal element $q_{ii} = -\sum_{j\neq i} q_{ij}$, the so-called waiting time, thus determines the speed at which B_i is replaced by another nucleobase.

The rate matrix $\mathbf{Q}$ can be formulated in terms of time-dependent probabilities, which are elements of a probability matrix $\mathbf{P}(t)$: $P_{ij}(t)$ denotes the probability of finding B_j at time t at a given position s_k if B_i was the nucleobase at time t_0 at this position.

From the definition of the matrix $\mathbf{Q}$ for an infinitesimal dt and time homogeneity of the process follows

$$\mathbf{P}(t) = e^{\mathbf{Q}t} \cdot \mathbf{P}(0) = e^{\mathbf{Q}t}, \tag{5}$$

since $\mathbf{P}(0)$ is the unit matrix for the definition of probabilities used above. Equation 5 is a rather formal relation that cannot be evaluated analytically, because the exponential function of a general matrix can only be expressed as an infinite series of matrix products or computed numerically in terms of the eigenvalues and eigenvectors of the matrix $\mathbf{Q}$. In order to turn the probabilities into a dynamical model we define a normalized vector of nucleobase frequencies, $\boldsymbol{\pi}(t) = \big(\pi_1(t), \ldots, \pi_\kappa(t)\big)'$ with $\sum_{i=1}^{\kappa} \pi_i(t) = 1$,[2] which describes the nucleobase distribution at time t. The time development is therefore expressed in the form of a differential equation

$$\frac{d\boldsymbol{\pi}}{dt} = \mathbf{Q} \cdot \boldsymbol{\pi} \quad \text{and} \quad \boldsymbol{\pi}(t) = e^{\mathbf{Q}t} \cdot \boldsymbol{\pi}^{(0)}. \tag{6}$$

The equation on the rhs is identical with Eq. 5 and again solutions are accessible only through numerical computation of the eigenvalue problem for $\mathbf{Q}$ (*see* also Subheading 2.3.4). Independently of the initial values $\boldsymbol{\pi}^{(0)} = \boldsymbol{\pi}(0)$ the nucleobase distribution converges towards a stationary distribution $\bar{\boldsymbol{\pi}} = (\bar{\pi}_1, \ldots, \bar{\pi}_{|\mathscr{A}|})'$. The instantaneous rate matrix $\mathbf{Q}$ can be multiplied by an arbitrary factor $f > 0$ that can be absorbed in time $\tau = t \cdot f^{-1}$, since time and rate are confounded and only their product can be inferred without extrinsic information [24]. We typically scale time such that the expected rate of substitutions per site is

$$-\sum_i \pi_i q_{ii} = 1. \tag{7}$$

Due to the multiple substitutions we never observe the number of substitutions per site d. We rather observe the number of differences p, that is computed as

$$p = 1 - \sum_i \pi_i p_{ii(t)}. \tag{8}$$

The most common models are based both on independence along sites and on the assumptions stated in Table 1. It is worth noticing that not every stationary process is reversible. Reversibility, however, is a convenient and reasonable assumption underlying the

[2] For convenience we define $\boldsymbol{\pi}$ as column vector but to save space we write it as a transposed row vector $\boldsymbol{\pi}'$.

Table 1
Assumptions of the commonly used nucleotide substitutions models, of which some are collected in Table 2

1.	Each site of the sequence evolves independently
2.	The substitution process has no memory of past events (Markov property)
3.	The process remains constant through time (homogeneity)
4.	The process starts at equilibrium (stationarity)
5.	Substitutions occur in continuous time

Table 2
***Instantaneous rate matrices*. The table contains six different rate matrices that are based on the assumptions summarized, for convenience, in Table 1. The first model was developed by Jukes and Cantor [25], and is specified by a single free parameter since it assumes equal mutation rates between all nucleobases. The other models ordered by increasing number of parameters are K80: Motoo Kimura's two parameter model distinguishes between transitions and transversions [26], and the more general single substitution models, which consider different base compositions, in particular HKY: the Hasegawa–Kishino–Yano model [33], TN93: the Tamura–Nei model [34], F81: the Felsenstein 81 model [24], and GTR: the general time reversible model [35]. The diagonal elements (*) are given by Eq. 4**

	A	C	G	T	A	C	G	T	A	C	G	T
	JC69				K80				HKY			
A	*	α	α	α	*	β	α	β	*	$\beta\pi_C$	$\alpha\pi_G$	$\beta\pi_T$
C	α	*	α	α	β	*	β	α	$\beta\pi_A$	*	$\beta\pi_G$	$\alpha\pi_T$
G	α	α	*	α	α	β	*	β	$\alpha\pi_A$	$\beta\pi_C$	*	$\beta\pi_T$
T	α	α	α	*	β	α	β	*	$\beta\pi_A$	$\alpha\pi_C$	$\beta\pi_G$	*
	TN93				F81				GTR			
A	*	$\beta\pi_C$	$\alpha_1\pi_G$	$\beta\pi_T$	*	π_C	π_G	π_T	*	$a\pi_C$	$b\pi_G$	$c\pi_T$
C	$\beta\pi_A$	*	$\beta\pi_G$	$\alpha_2\pi_T$	π_A	*	π_G	π_T	$a\pi_A$	*	$d\pi_G$	$e\pi_T$
G	$\alpha_1\pi_A$	$\beta\pi_C$	*	$\beta\pi_T$	π_A	π_C	*	π_T	$b\pi_A$	$d\pi_C$	*	$f\pi_T$
T	$\beta\pi_A$	$\alpha_2\pi_C$	$\beta\pi_G$	*	π_A	π_C	π_G	*	$c\pi_A$	$e\pi_C$	$f\pi_G$	*

most commonly used models. A stationary Markov process is said to be time reversible if,

$$\pi_i q_{ij} = \pi_j q_{ji} \tag{9}$$

Under the reversibility assumption, the twelve nondiagonal entries of rate matrix **Q** can be described by six symmetric terms, the so-called exchangeability terms, and four stationary frequencies. Thus, the general time reversible model (GTR) of Table 2 has eight free parameters. The further assumption that the rate of substitution is the same for all nucleotides can be relaxed by including rate heterogeneity. This in turn assumes that many of the complexities of molecular evolution are primarily manifested as differences in the relative rate for sites changes, while all other aspects of the evolutionary process are maintained.

Rate Heterogeneity

Each site has a defined probability of evolving at a given rate, independently of its neighbors. The distribution of relative rates *r* is frequently assumed to follow a gamma distribution [28, 29]. Further developments, breaking the distribution into a pre-specified number of categories, make the model more computationally efficient [30]. Van de Peer et al. [31] used empirical pair-wise methods to infer site-specific rates of alignment positions. Based on this idea, Meyer et al. [32] introduced a maximum likelihood framework for estimating site-specific rates from pairs of sequences with an iterative extension in order to compute site-specific rates and the phylogenetic tree simultaneously. In the sense of reflecting different selective constraints at different sites hidden Markov models (HMM) are used to assign rates of change to each site, according to a Markov process that depends on the rate of change at the neighboring site. These approaches model site dependencies through shared rate parameters while still assuming independent changes at the different sites [e.g. 36]. As we are interested in the evolution of RNA structure, including different constraints such as energy values, it makes sense to focus on relaxing the assumption of independent evolving sites.

2.1.2 From Independent to Jointly Modeling Substitution Events

For many years, various authors have attempted to overcome the assumption of independence across sites. In the simplest cases, a dependence structure with joint substitution events is modeled. A prominent example are codon models, which consider the dependence of neighboring sites within a codon [37, 38] rather than mononucleotide models in protein coding regions. The assumption of independence among sites is naturally relaxed if the whole nucleobase triplet of a codon is taken as a unit of evolution. Evidently, this leads to larger rate matrices and makes computation more demanding. For protein coding regions, however, such models are comparatively more realistic.

In the same way, Schöniger and von Haeseler [39] have suggested modeling joint substitution events in RNA helical regions. Clearly, the nucleotides in stem regions of RNA molecules cannot evolve independently of their base-pairing counterparts. Given the frequencies of admissible base-paired doublets, it is quite likely that substitution at a non-base-pairing doublet (Fig. 9b) will lead to a base-paired doublet within a relatively short time interval, as we would expect in the case of so-called compensatory mutations (*see* Fig. 9). The sites are classified into two categories: (i) helical regions and (ii) loops, joints, and free ends—assuming, however, a fixed RNA secondary structure. While the units of loop regions are mononucleotides, as in the conventional independent models

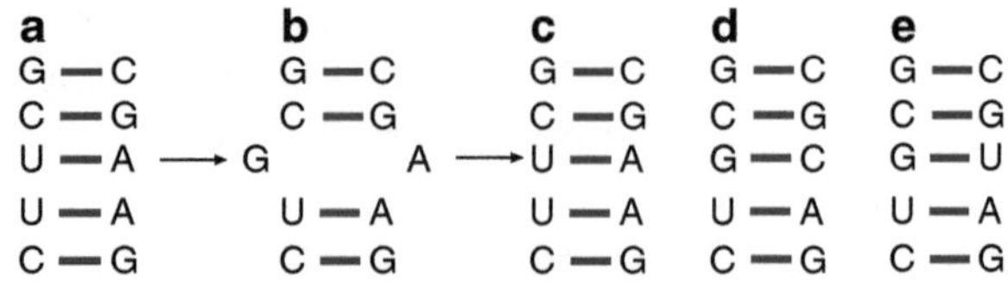

Fig. 9 *Compensatory mutation in an RNA double helix.* A substitution within a double helical region destroys a base pair and creates an internal loop: (a)→(b). Further substitution at the internal loop may restore the double helical structure either by reverting the mutation, (b)→(c), or by creating other base pairs at this position (b)→(d) and (b)→(e). The gain in fitness accompanying base pair formation is mirrored by high probabilities for such events, which is tantamount to relatively short waiting times (see text)

(Table 2), the units of helical regions are doublets (base-pairs). In helical regions, the state space is extended to all 16 pair combinations, $\mathsf{B}_i, \mathsf{B}_j \in \mathscr{B} = \mathscr{A} \times \mathscr{A} = \{\mathrm{AA}, \mathrm{AC}, \mathrm{AG}, \ldots, \mathrm{GU}, \mathrm{UU}\}$. Correspondingly, the F81 model (Table 2) is extended to a 16×16 instantaneous rate matrix by taking into account the stationary frequencies $\bar{\boldsymbol{\pi}}_{\mathscr{B}} = \{\bar{\pi}_{AA}, \bar{\pi}_{AC}, \bar{\pi}_{AG}, \ldots, \bar{\pi}_{GU}, \bar{\pi}_{UU}\}$ for the usual restriction that only one substitution per unit time is admissible.

Several other attempts to integrate site interactions of RNA base pairing into a Markov model of sequence evolution have been made [40–44]. Muse [40] introduced a new pairing parameter λ to account for the effects of forming or destroying a base pair. The six possible base pairs $\mathscr{B} = \{\mathrm{AU}, \mathrm{UA}, \mathrm{GC}, \mathrm{CG}, \mathrm{GU}, \mathrm{UG}\}$ are explicitly considered in a 6×6 instantaneous rate matrix $\mathbf{Q}$ [42], or augmented by one state for all mismatch pairs in a 7×7 matrix [43]. Commonly, the simultaneous substitution rate of both nucleobases in a base pair is assumed to be zero. However, models that allow doublet substitutions also exist [45]. Since Motoo Kimura's early works on the rates of nucleobase substitutions [26], the observation of compensatory mutations has been thought to explain the conservation of structure. Among further studies on the rate of compensatory mutations we mention [46, 47] here. Rate matrices for RNA evolution were also determined empirically from a large sample of related RNA sequences [48]. A more recent investigation [49] has shown that different secondary structure categories evolve at different rates.

2.1.3 Context-Dependent Substitutions

Relaxing the assumption of independently evolving sequence fragments through the introduction of overlapping dependencies makes the models substantially more involved (Fig. 10). Nevertheless, accounting for stacking interactions between pairs of adjacent residues in RNA helices and other energetic constraints in RNA structure (*see*, e.g., Chapters 2 and 4) would definitely improve the relation between substitutions and RNA structure. Not unexpectedly, empirical studies have found indications that

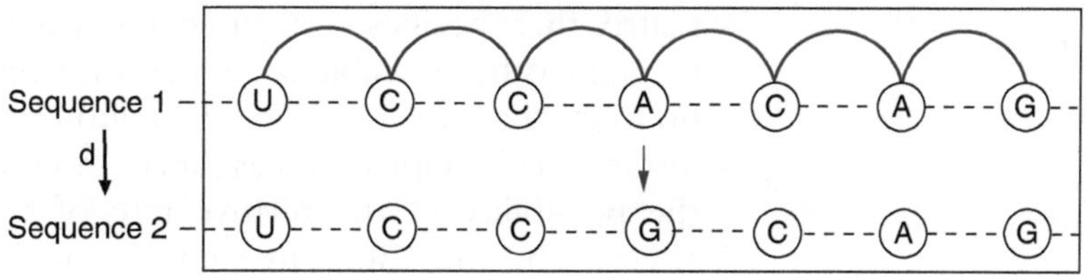

Fig. 10 *Context-dependent substitutions*. The sketch indicates that the rate for the substitution of the central nucleobase, A→G, depends on three further nucleobases on each side (the framed box comprises all seven nucleobases). Modeling global context dependency specifies rates of change from every possible sequence to every other possible sequence with the usual restriction of no more than one position being allowed to change in a particular instant. This dependency structure leads to rate matrices of high dimensionality, which require special approaches to overcome this problem (see text for details and Subheading 2.3.4 for an alternative approach)

the assumption of independent evolution of sites including the dependency on flanking sequence patterns is too restrictive and does not match the experimental data [e.g. 50, 51]. Pioneering work of context-dependent substitutions with respect to the CpG-deamination process has been done by Jens Jensen and Anne-Mette Pedersen [52] using a Markov model of nucleotide sequence evolution in which the instantaneous substitution rates at a site were allowed to depend on the states of a neighboring site at the instant of the substitution. So far they were able to consider pairs of sequences only. The model consists of a first component that depends on the type of change while the second component considers the CpG-deamination process. The model was later extended by taking into account arbitrary reversible codon substitutions with more flexibility concerning CpG-deamination [53]. In these approaches, inference is obtained using Markov chain Monte Carlo (MCMC) of expectation maximization (EM) based pseudo-likelihood estimation while phylogenetic reconstruction is still a problem [54]. Context-reducing models were developed that allow for context dependent substitutions but can be approximated by simple extensions of Joe Felsenstein's original framework. These models impose certain limitations on the independence of sites but allow for exact inference without too much additional cost in computation efforts (an example is the approach described in [55]). Nevertheless, some of these models are analytically solvable as was recently shown [56] for a special subcase of a model suggested by Tamura [34] and extended by the inclusion of CpG doublets. These works take up earlier ideas on the solvability of this class of models suggested in [57] and formally proven later [58].

Global context dependency models related to considering protein structure were first developed in Jeffrey Thorne's group using a Bayesian MCMC [59]. They defined an instantaneous rate matrix that specifies rates of change from every possible sequence to every other possible sequence with the common restriction of no more than one position being allowed to change at a particular instant. This approach has been modified in order to explore the possibility of the relative rate of sequence evolution being affected by the Gibbs free energy of RNA secondary structure [60]. The results are slightly ambiguous because medium size and large RNA molecules have natural or *expected* structures that differ substantially from the minimum free energy (mfe) structures and they may be constrained by structure and function rather than by thermodynamic stability (*see* Subheading 2.3.1). Taking into account the possible dependence of structure on all sites, the rate matrix $\mathbf{Q}$ is $4^l \times 4^l$ for sequence or chain length l, and any full analysis of $\mathbf{Q}$ is not feasible even if l is extremely small (for $l = 10$ the dimension of $\mathbf{Q}$ is already $4^{10} \times 4^{10}$). To overcome this high dimensionality a sequence path approach is used [52, 61].

It is a well-known fact in optimization theory that too many additional parameters for modeling add noise and imply a risk of overfitting the data. This is equally true for site dependence models. Careful strategies for model building are required in order to balance additional parameters with the available empirical information. Such models call for extensive analysis of their mathematical behavior, for example stability and convergence in a state of equilibrium, and for investigations into their ability to predict the statistical properties of biological sequences. As a consequence, computer simulation seems to be the most appropriate tool for identifying the necessary parameters for modeling site dependence.

A general simulation framework that takes into account site-specific interactions mimicking sequence evolution with various complex overlapping dependencies among sites has been introduced by [62]. This framework is based on the idea of applying different substitution matrices at each site ($\mathbf{Q}_k; k = 1, \ldots, l$) defined by the interactions with other sites in the sequence (Fig. 11). To this end, a neighborhood system $\mathcal{N} = (N_k)_{k=1,2,\ldots,l}$ is introduced such that $N_k \subset \{1, \ldots, l\}, k \notin N_k$ for each k and if $i \in N_k$ then $k \in N_i$ for each i, k. N_k contains all sites that interact with site s_k. With n_k they denote the cardinality of N_k, i.e., the number of sites that interact with s_k. $\mathbf{Q} = \{\mathbf{Q}_k; k = 1, \ldots, l\}$ thus constitutes a collection of possibly different substitution models $\mathbf{Q}_k$ acting on the sequence and an annotation of correlations among sites. The approach allows for modeling evolution of nucleotide sequences along a tree for user-defined systems of neighborhoods and instantaneous rate matrices.

2.2 Time in Phylogeny and Evolution

Evolutionary dynamics are commonly described in realtime or, in other words, the time axis in Figs. 5 and 6 corresponds to physical time. The morphological reconstruction of biological evolution makes extensive use of the fossil record and hence is

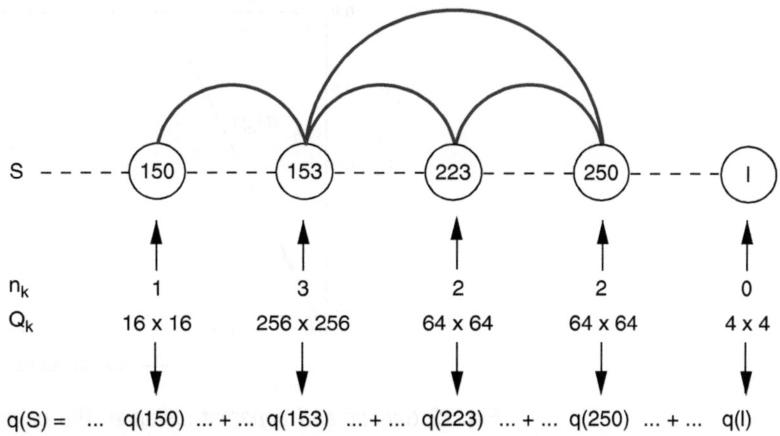

Fig. 11 *A ribozyme domain with overlapping dependencies.* Example of a sequence S with overlapping dependencies on site 153. Such dependencies occur, for example, in ribozyme domains [63]. The substitution rate for the whole sequence $q(S)$ is obtained as the sum of the rates of each site $q(S) = \sum_{k=1}^{l} \mathbf{Q}_k(B_i^{(k)}, B_j^{(k)})$. The nucleobase instantaneous substitution rate depends on the states of the neighborhood system of this site at the instant of the substitution, described in the instantaneous rate matrix $\mathbf{Q}_k$. The dimension of $\mathbf{Q}_k$ depends on the number of neighbors n_k taking into account at this site k

anchored in stratigraphy and other methods of palaeontological time determination. Phylogenetic reconstructions, on the other hand, are counting substitutions and thus not dependent on absolute timing. Without losing generality we assume that substitutions are related to realtime by some monotonously increasing function $g(t)$. It is important to distinguish between a genetic distance $d(g)$, i.e., the number of substitutions per site, and the observable distance $p(g)$. Because of double and multiple substitutions that remain undetectable by sequence comparisons, $d \geq p$ is always fulfilled (Fig. 12). Indeed, $p(g)$ converges to some saturation value $\lim_{g\to\infty} p(g) = p_\infty$. At this point the two sequences carry no more information on their phylogenetic relatedness. Assuming independent substitution events, p depends on chain length l and in the limit of long chains p and d become equal: $\lim_{l\to\infty} p(l) = d$.

The function $d(g)$, relating substitutions to real time, clearly depends on the evolutionary model, in particular on the rate matrix of substitutions (Table 2). In the simplest possible case, constituted by the Jukes–Cantor model, we have $p(g) = 3(1 - e^{-4g/3})/4$ and $d(g) = g$, at small values of d we have $p \approx d$ as required for the model, and the saturation value is $p_\infty = 3d/4$. For more models we refer to the monograph [64, p.265ff.]. As an example of an analysis of very ancient RNA genes that are already close to the saturation limit we mention the attempt to construct a single phylogenetic tree for all tRNAs [65].

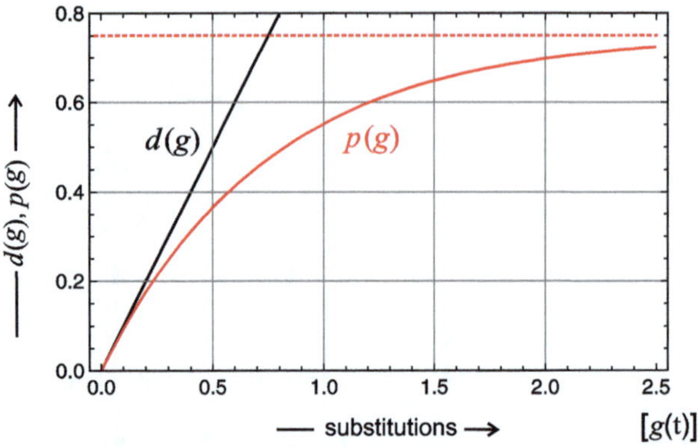

Fig. 12 *Genetic and apparent distance.* The genetic distance $d(g)$, with $g(t)$ being a monotonously increasing function of realtime, counts all substitutions whereas the observed distance $p(g)$ is reduced by double and multiple substitutions at the same site and hence $d \geq p$. The Jukes–Cantor model [25] is used

Historically, the first and also the most spectacular attempt to relate phylogeny and time was the hypothesis of a molecular clock of evolution [66–68], which is tantamount to the assumption of a linear relation $g = \lambda t$: The numbers of mutational changes in informational macromolecules, proteins and nucleic acids, are constant through time and possibly independent of the particular lineage in the sense of a universal λ value. The molecular clock hypothesis seemed to be in excellent agreement with the neutral theory of evolution [69], although the lack of influence of the generation time has always been kind of a mystery. More and more accurate data, however, have shown that the pace of the molecular clock varies with the particular protein under consideration. In addition, the molecular clock was found to tick differently in different lineages and for comparisons of sequences that diverged recently or a long time ago [70]. Although the usefulness of the molecular clock as an independent source of timing has never been seriously questioned by molecular evolutionary biologists [71], a deeper look at data and models revealed many deviations from the naïve clock hypothesis [72–74].

2.3 Evolution and Design of RNA Structures

The choice of RNA as model system has several reasons: (i) a large percentage of the free energy of folding comes from Watson–Crick base pairing, and base paring follows a logic that is accessible to combinatorics, (ii) a great deal but definitely not nearly all principles of RNA function can be derived from easy to analyze and predict secondary structures, (iii) evolution of RNA molecules can be studied in vitro and this fact found direct application in the design of molecules with predefined properties, and (iv) viroids, RNA viruses, their life cycles, and their evolution can be studied at molecular resolution and provide insight into a self-regulated evolutionary process. The knowledge on molecular details of RNA-based entities with partly autonomous life cycles provides direct access to a world of increasing complexity from molecules to viroids, from viroids to viruses, and so on—all being now an enormous source of molecular data that wait to be converted into comprehensive and analyzable models of, for example, host–pathogen interactions and their evolution. Nowhere else is the relation between phylogeny and functional evolution—sometimes even within one host—so immediately evident as with RNA viruses.

2.3.1 Notion of Structures

Notion, analysis, and prediction of RNA structures are discussed in other chapters of this edited volume, in particular, Chapter 1 (Introduction), Chapter 2 by Christian Zwieb, and Chapter 4 by Ivo Hofacker. Here, we want to concentrate on one aspect of RNA structures that became clear only within the last two decades: In the great majority of natural, evolutionary selected RNA molecules we are dealing with sequences forming a single stable structure, whereas randomly chosen sequences generically form a great variety of metastable suboptimal structures in addition to the minimum free energy structure [75]. Important exceptions of the one sequence-one structure paradigm are RNA switches fulfilling regulatory functions in nature [76–79] and synthetic biology [80]. Such riboswitches are multiconformational RNA molecules, which are involved in posttranscriptional regulation of gene expression. The conformational change is commonly induced by ligand binding or ribozymic RNA cleavage. Multitasking by RNA molecules clearly imposes additional constraints on genomics sequences and manifests itself through a higher degree of conservation in phylogeny.

2.3.2 Sequences with Multiple Structures

The extension of the notion of structure to multiconformational molecules is sketched in Fig. 13. The RNA molecule forms a well-defined minimum free energy (mfe) structure—being a perfect single hairpin (lhs of the figure). In addition to the mfe structure, the sequence S like almost all RNA sequences[3] forms a great variety of other, less stable structures called *suboptimal structures* (shown in the middle of the figure). Structures, mfe and suboptimal structures, are related through transitions, directly or via intermediates, which in a simplified version can be represented by means of a barrier tree [81, 82] shown on the rhs of the figure. Kinetic folding introduces a second time scale into the scenario

[3] Exceptions are only very special sequences, homopolynucleotides, for example.

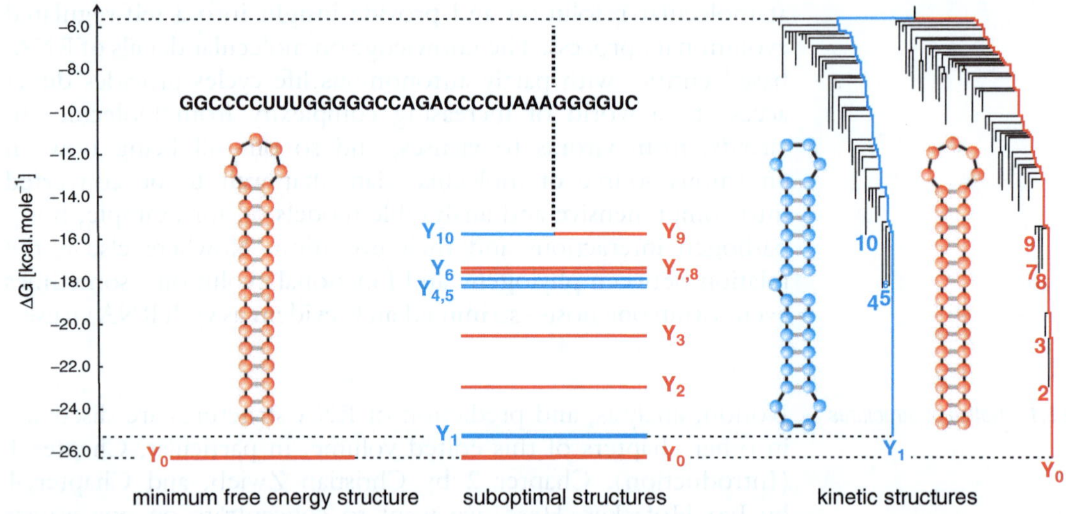

Fig. 13 *RNA secondary structures viewed by thermodynamics and folding kinetics.* An RNA sequence S of chain length l = 33 nucleotides has been designed to form two structures: (i) the single hairpin mfe structure,Y_0 (*red*) and (ii) a double hairpin metastable structure, Y_1 (*blue*). The Gibbs free energy of folding (ΔG) is plotted on the ordinate axis. The leftmost diagram shows the minimum free energy structure Y_0 being a single long hairpin with a free energy of $\Delta G = -26.3$ kcal/mole. The plot in the middle contains, in addition, the spectrum of the ten lowest suboptimal conformations classified and color coded with respect to single hairpin shapes (*red*) and double hairpin shapes (*blue*). The most stable—nevertheless metastable—double hairpin has a folding free energy of $\Delta G = -25.3$ kacl/mole. The rightmost diagram shows the barrier tree of all conformations up to a free energy of $\Delta G = -5.6$ kcal/mole where the energetic valleys for the two structures merge into one basin containing 84 structures, 48 of them belonging to the single hairpin subbasin and 36 to the double hairpin subbasin. A large number of suboptimal structures has free energies between the merging energy of the subbasins and the free reference energy of the open chain ($\Delta G = 0$)

of molecular evolution.[4] Based on Arrhenius theory of chemical reaction rates,

$$k = A \cdot e^{-E_a/RT} , \tag{10}$$

the height of the barrier, E_a determines the reaction rate parameter k and thereby the half life of the conformation $t_{1/2} = \ln 2/k$. In Eq. 10, A is the pre-exponential factor of the reaction, R is the gas constant, and T the absolute temperature in degree Kelvin. The two structures shown in Fig. 13 are connected by a lowest barrier of 20.7 kcal/mole that depending on the pre-exponential factors implies half lives of days or even weeks for the two conformations. In a conventional experiment with a time scale of hours the two conformations would appear as two separate entities. Barriers,

[4]Timescale number one is the evolutionary process itself. In order to be relevant for evolutionary dynamics the second timescale has to be substantially faster than the first one.

nevertheless, can be engineered to be much lower and then an equilibrium mixture of rapidly interconverting conformations may be observed. Several constraints are required for the conservation of an RNA switch, and the restrictions of variability in sequence space are substantial.

The comparison of the two dominant structures Y_0 and Y_1 in Fig. 13 provides a straightforward example for the illustration of different notions of stability: (i) thermodynamic stability, which considers only the free energies of the mfe structures—Y_0 in Fig. 13 is more stable than Y_1 since it has a lower free energy, $\Delta G(\mathsf{Y}_0) < \Delta G(\mathsf{Y}_1)$, (ii) conformational stability, which can be expressed in terms of suboptimal structures or partition functions within a basin or a subbasin of an RNA sequence—a conformationally stable molecule has no low lying suboptimal conformations that can be interconverted with the mfe structure at the temperature of the experiment; for kinetic structures separated by high barriers the partition functions are properly restricted to individual subbasins [83], and (iii) mutational stability that is measured in terms of the probability with which a mutation changes the structure of a molecule (*see* Fig. 14 and Subheading 2.3.3). All three forms of stability are relevant for evolution and phylogeny, but mutational stability and the spectrum of mutational effects—adaptive, neutral, or deleterious—are most important.

2.3.3 Mapping Sequences onto Structures

RNA secondary structures provide a simple and mathematically accessible example of a *realistic* mapping of biopolymer sequences onto structures [75, 84]. Here modeling will be restricted to the assignment of a single structure to each sequence (lhs diagram in Fig. 13). Apart from a few exceptions, experimental information on conformational free energy surfaces and fitness landscapes is rather very limited but the amount of available data is rapidly growing. *Realistic* means here that the neighborhood relations are similar to the observations in reality: (i) realistic landscapes are *rugged*, since nearest neighbor, i.e., Hamming distance one ($d_{\mathrm{H}} = 1$), sequences may have entirely different or very similar properties, and (ii) realistic landscapes are characterized by *neutrality* in the sense that different sequences may lead to similar or identical structures and may have indistinguishable properties for evolution. Both features are observed in mappings of RNA sequences onto secondary structures.

The mapping in the forward direction, i.e., from sequence space onto a space of structures called *shape space*, is formalized by

$$\Phi : \left(\mathcal{Q}_l^{(\kappa)}, d_{\mathrm{H}}\right) \Longrightarrow (\mathcal{Y}_l, d_{\mathrm{Y}}) \quad \text{or} \quad \mathsf{Y} = \Phi(\mathsf{S}) . \qquad (11)$$

As said in the introduction, both sequence and structure space are metric spaces having the Hamming distance (d_{H}) and an

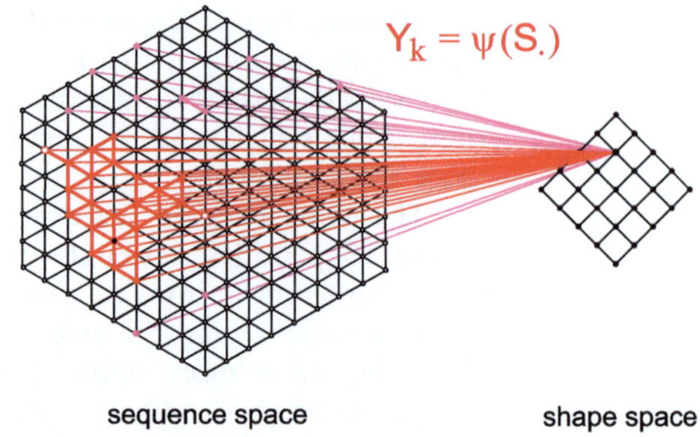

Fig. 14 *A sketch of the mapping of RNA sequences onto secondary structures.* The points of sequence space (here 183 on a planar hexagonal lattice) are mapped onto points in shape space (here 25 on a square lattice) and, inevitably, the mapping is many to one. All sequences $S_.$ folding the same mfe structure form a neutral set, which in mathematical terms is the preimage of Y_k in sequence space. Connecting nearest neighbors of this set—these are pairs of sequences with Hamming distance $d_{\mathrm{H}} = 1$—yields the neutral network of the structure, G_k. The network in sequence space consists of a giant component (*red*) and several other small components (*pink*). On the network the stability against point mutations varies from $\lambda = 1/6$ (*white points*) to $\lambda = 6/6 = 1$ (*black point*). We remark that the two-dimensional representations of sequence are used here only for the purpose of illustration. In reality, both spaces are high-dimensional—the sequence space of binary sequences $\mathcal{Q}_l^{(2)}$, for example, is a hypercube of dimension and that of natural four-letter sequences $\mathcal{Q}_l^{(4)}$ an object in 3 dimensional space

appropriate structure distance (d_{Y}) as metrics. The superscript "κ" is the size of the nucleobase alphabet ($\kappa = 4$ for natural sequences). The subscript "l" indicates that, for simplicity, we restrict the considerations here to sequences of the same lengths l. A set $\Gamma(\mathsf{Y}_k)$ containing all sequences folding into the same structure Y_k is denoted as *neutral set*, it represents the preimage of the structure in sequence space and it is defined by

$$\Gamma_k = \Gamma(\mathsf{Y}_k) = \Phi^{-1}(\mathsf{Y}_k) = \{\mathsf{S}_j | \mathsf{Y}_k = \Phi(\mathsf{S}_j)\} . \quad (12)$$

The neutral network G_k is the graph obtained from the set Γ_k by connecting all pairs of nodes with Hamming distance $d_{\mathrm{H}} = 1$ by an edge (*see* Fig. 14). The probability of a change in structure as a consequence of a mutation of the sequence $\mathsf{S}_j \in \Gamma_k$ is measured by the local degree of neutrality

$$\hat{\lambda}_k(\mathsf{S}_j) = \frac{\#\ \text{neutral mutations of}\ \mathsf{S}_j}{\#\ \text{all mutations of}\ \mathsf{S}_j} . \quad (13)$$

Accordingly, $\hat{\lambda}_k(\mathsf{S}_j) = 0$ implies that every mutation of S_j leads to a new structure and $\hat{\lambda}_k(\mathsf{S}_j) = 1$ expresses the fact that the structure Y_k is formed by all single point mutants of S_j, which might be characterized as a *mutation resistent* sequence. An average of the local degree of neutrality is taken over the entire neutral set Γ_k to yield $\lambda_k = \sum_{\mathsf{S}_j \in \Gamma_k} \hat{\lambda}_k(\mathsf{S}_j)/|\Gamma_k|$, the (mean or global) degree of neutrality, which is a measure for the mutational stability of the structure Y_k as discussed in Subheading 2.3.2. Typical λ_k-values for common RNA secondary structures—as found in nature or obtained by evolution experiments with RNA molecules—lie in the range $0.2 < \lambda_k < 0.3$, implying that roughly 25% of all mutations are neutral with respect to structure. The degree of neutrality for complete three-dimensional structures is not known and additional definitions are required, because spatial structures are points in a continuous rather than discrete shape space. Coarse graining of structures is necessary in order to be able to distinguish alike from different.

2.3.4 Chemical Kinetics of Evolution

In order to introduce mutations into selection dynamics Manfred Eigen [85] conceived a kinetic model based on stoichiometric equations, which handle correct replication and mutation as parallel reactions

$$(\mathsf{A}) + \mathsf{S}_i \xrightarrow{\mathrm{Q_{ji}f_i}} \mathsf{S}_j + \mathsf{S}_i; \; i,j = 1,\ldots,n\,. \qquad (14)$$

In normalized coordinates (14) corresponds to a differential equation of the form

$$\frac{dx_j}{dt} = \sum_{i=1}^{n} Q_{ji} f_i \, x_i - x_j \, \phi(t); \; j = 1,\ldots,n \; ; \; \sum_{i=1}^{n} x_i = 1\,. \qquad (15)$$

The finite size constraint $\phi(t) = \sum_{i=1}^{n} f_i x_i$ is precisely the same as in the mutation-free case (3), and the same techniques can be used to solve the differential equation [86, 87]. Solutions of (15) are derived in terms of the eigenvectors of the $n \times n$ matrix $\mathrm{W} = \mathrm{Q} \cdot \mathrm{F} = \{W_{ji} = Q_{ji} f_i\}$. The mutation frequencies are subsumed in the matrix $\mathrm{Q} = \{Q_{ji}\}$ with Q_{ji} being the probability that S_j is obtained as an error copy of S_i. The fitness values are the elements of a diagonal matrix $\mathrm{F} = \{F_{ij} = f_i \cdot \delta_{i,j}\}$. *Exact* solution curves and stationary mutant distributions are obtained by numerical computation. The stationary populations have been called *quasispecies* since they represent the genetic reservoirs of asexually reproducing species.

A quasispecies exists if every sequence in sequence space can be reached from every other sequence along a finite-length path of single point mutations in the mutation network.[5] In addition the quasispecies contains all mutants at nonzero concentrations ($\bar{x}_i > 0 \, \forall \, i = 1, \ldots, n$). In other words, after sufficiently long time a kind of *mutation equilibrium* is reached at which all mutants are present in the population. In the absence of neutral variants the quasispecies consists of a *master sequence*, the fittest sequence $\mathsf{S}_m : \{f_m = \max(f_i; i = 1, \ldots, n)\}$, and its mutants, S_j ($j = 1, \ldots, n, i \neq m$), which are present at concentrations that are, in essence, determined by their own fitness f_j, the fitness of the master f_m and the off-diagonal element of the mutation matrix Q_{jm} that depends on the Hamming distance from the master sequence $d_{\mathrm{H}}(\mathsf{S}_j, \mathsf{S}_m)$ $\left(see\,Eq.\,18\right)$.

The Error Threshold

Application of perturbation theory neglecting back mutations at zeroth order yields analytical approximations for the quasispecies [85, 89], which are valid for sufficiently small mutation rates [90], $Q_{ji} << \{Q_{ii}, Q_{jj}\} \, (i \neq j)$:

$$\bar{x}_m \approx \frac{Q_{mm} - \sigma_m^{-1}}{1 - \sigma_m^{-1}} \quad \text{and} \quad \frac{\bar{x}_j}{\bar{x}_m} \approx \frac{W_{jm}}{W_{mm} - W_{jj}}, \; j = 1, \ldots, n; j \neq m$$

$$\text{with } \sigma_m = f_m \,/\, \bar{f}_{-m} \text{ and } \bar{f}_{-m} = \sum_{=1, \neq m}^{n} f_j \bar{x}_j \,/\, (1 - \bar{x}_m) \,. \tag{16}$$

The superiority σ_m is a measure of the advantage in fitness the master has over the rest of the population, and $\bar{f}_{-m}$ is the mean fitness of this rest.[6]

In order to gain basic insight into evolutionary dynamics we introduce the uniform error rate model: The *point mutation rate* $p(s_k)$ is expressed in terms of a (mean) mutation rate per site and reproduction event that is assumed to be independent of the site s_k and the particular sequence S_i. Further simplification is introduced by the use of binary rather than four-letter sequences.[7] Then, diagonal and off-diagonal elements of matrix Q are of the simple form

[5]By definition of fitness values, $f_i \geq 0$, and mutation frequencies, $Q_{ji} \geq 0$, W is a non-negative matrix and the reachability condition boils down to the condition: $\mathrm{W}^k \gg 0$, i.e., there exists a k such that W^k has exclusively positive entries and Perron–Frobenius theorem applies [88].

[6]An exact calculation of $\bar{f}_{-m}$ is difficult because it requires knowledge of the stationary concentrations of all variants in the population: $\bar{x}_i$; $x = 1, \ldots, n$. For computational details *see* [23, 89, 91, 92].

[7]It should be noted that artificially synthesized two letter (DU; D = 2,6-diamino-purine) ribozymes have perfect catalytic properties [93].

$$Q_{ii} = (1-p)^l \quad \text{and}$$

$$Q_{ji} = (1-p)^{l-d_{\mathrm{H}}(\mathrm{S}_j,\mathrm{S}_i)}\, p^{d_{\mathrm{H}}(\mathrm{S}_j,\mathrm{S}_i)} = (1-p)^l \left(\frac{p}{1-p}\right)^{d_{\mathrm{H}}(\mathrm{S}_j,\mathrm{S}_i)}, \tag{17}$$

and insertion into Eq. 16 yields the three parameter (l, p, σ) expression

$$\bar{x}_m \approx \frac{(1-p)^l - \sigma_m^{-1}}{1-\sigma_m^{-1}} \quad \text{and} \quad \bar{x}_j \approx \left(\frac{p}{1-p}\right)^{d_{\mathrm{H}}(\mathrm{S}_j,\mathrm{S}_m)} \frac{f_m}{f_m - f_j}\, \bar{x}_m\,. \tag{18}$$

Equation 18 provides a quantitative estimate for the concentrations of mutants: For given p, $\bar{x}_j$ is the larger the smaller the Hamming distance from the master, $d_{\mathrm{H}}(\mathrm{S}_j, \mathrm{S}_m)$, is and the smaller the larger the difference in fitness, $f_m - f_j$ is. The stationary concentration of the master sequence, $\bar{x}_m$, vanishes at some critical mutation rate $p = p_{\mathrm{cr}}$ called the *error threshold* [85, 90, 91, 94, 95][8]:

$$p_{\mathrm{cr}} = 1 - \sigma^{-1/l} \quad \Longrightarrow \quad p_{\max} \approx \frac{\ln \sigma}{l} \quad \text{and} \quad l_{\max} \approx \frac{\ln \sigma}{p}\,. \tag{19}$$

Figure 15 compares a quasispecies calculated by the perturbation approach with the exact solution and shows excellent agreement up to the critical mutation rate $p = p_{\mathrm{cr}}$. This agreement is important because quantitative applications of quasispecies theory to virology are essentially based on Eq. 16 (*see* Subheading 2.3.7 and [96]).

At constant chain length l the error threshold defines a maximal error rate for evolution, $p \leq p_{\max}$, and at constant reproduction accuracy p the length of faithfully copied polynucleotides is confined to $l \leq l_{\max}$ [95, 97]. The first limit of a maximal error rate $p_{\max}$ has been used in pharmacology for the development of new antiviral strategies [98], and the second limit entered hypothetical modeling of early biological evolution where the accuracy limits of enzyme-free replication confine the lengths of polynucleotides that can be replicated faithfully [99].

Exact Solutions and Fitness Landscapes

Exact solutions of Eq. 15 computed numerically for a *single-peak fitness landscape*, $f_m = f_0$ and $f_j = f_n \,\forall j = 1, \ldots, n; j \neq m$, show a remarkably sudden change in the population structure at the critical mutation rate, $p = p_{\mathrm{cr}}$, that manifests itself in three observations (Fig. 15):

[8] Zero or negative concentrations of sequences clearly contradict the exact results described above and are an artifact of the perturbation approach. Nevertheless, the agreement between the exact solutions and the perturbation results up to the error threshold as shown in Fig. 15 is remarkable.

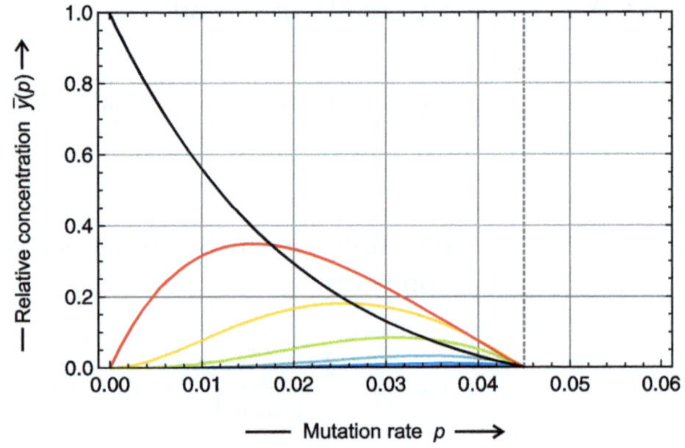

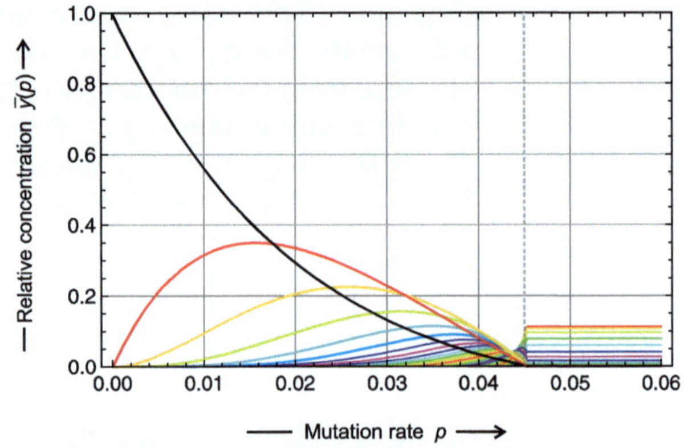

Fig. 15 *The quasispecies as a function of the point mutation rate p.* The plot shows the stationary mutant distribution of sequences of chain length $l = 50$ as a function of the point mutation rate p. The upper part contains the approximation by perturbation theory according to Eq. 16 and is compared with the exact results presented in the lower part of the figure. Plotted are the relative concentration of entire mutant classes: $\bar{y}_0$ (*black*) is the master sequence, $\bar{y}_1$ (*red*) is the sum of the concentrations of all one error mutants of the master sequence, $\bar{y}_2$ (*yellow*) that of all two error mutants, $\bar{y}_3$ (*green*) that of all three error mutants, and so on. In the perturbation approach the entire population vanishes at a critical mutation rate p_{cr} called the error threshold (which is indicated by a broken *gray line* at $p_{\mathrm{cr}} = 0.04501$) whereas a sharp transition to the uniform distribution is observed with the exact solutions. Choice of parameters: $f_m = 10$, $f_i = 1 \, \forall i = 1, \ldots, n; i \neq m$

(i) the concentration of the master sequence $\bar{x}_m$, becomes very small—zero in the perturbation approach (16),

(ii) an abrupt change in the population structure that sharpens with increasing chain length l in a phase transition like manner,

(iii) a transition to the uniform distribution, $\bar{x}_i = \kappa^{-l} \, \forall \, i = 1, \ldots, \kappa^l$, which is the exact solution at the mutation rate $p = \tilde{p} = \kappa^{-1}$.

The transition occurs already at small mutation rates far away from $\tilde{p}$ ($p_{\mathrm{cr}} = 0.045$ versus $\tilde{p} = 0.5$ in Fig. 15). Undoubtedly, evolution is not possible at point mutation rates $p > p_{\mathrm{cr}}$, since the existence of a uniform distribution implies that reproduction is operating but inheritance is not: Too many errors are made in the copying process and no individual sequence can be stably maintained over many generations.

Early works [100] have shown that the occurrence of error thresholds depends on the distribution of fitness values in sequence space. On *simple landscapes*, which are distinguished from *realistic*, complex landscapes, all genotypes of equal distance to the master have identical fitness values. Some smooth and simple fitness landscapes, in particular the additive, $f_m = f_0$ and $f_j = f_0 - \alpha \cdot d_{\mathrm{H}}(\mathsf{S}_j, \mathsf{S}_m)\,\forall\, j \neq m$, and the multiplicative landscape, $f_m = f_0$ and $f_j = f_0 \cdot \beta^{\,d_{\mathrm{H}}(\mathsf{S}_j,\mathsf{S}_m)}\,\forall\, j \neq m$, with $\alpha > 0$ and $\beta < 1$—which are both highly popular in population genetics—exhibit a gradual change from the homogeneous population, $\bar{x}_m = 1$, at $p = 0$ to the uniform distribution at $p = \tilde{p}$ in contrast to the single-peak landscape that shows the error threshold phenomenon (Fig. 15).

As said above (Subheading 2.3.3), two features are characteristic for ***realistic landscapes***: (i) ruggedness and (ii) neutrality [84, 101, 102]. Ruggedness can be modeled by assigning fitness differences at random within a band of fitness values with adjustable width d. The highest fitness value is assigned to the master sequence, $f_m = f_0$ and all other fitness values are obtained by means of the equation

$$f(\mathsf{S}_j) = f_n + 2d(f_0 - f_n)\left(\eta_j^{(s)} - 0.5\right),\; j = 1, \ldots, \kappa^l;\, j \neq m\,, \quad (20)$$

where $\eta_j^{(s)}$ is the j-th output random number from a pseudorandom number generator with a uniform distribution of numbers in the range $0 \leq \eta_j^{(s)} \leq 1$ that has been started with the seed s.[9] Typical results for the stationary mutant distribution are shown in Fig. 16: At $d = 0$ the landscape becomes the single-peak landscape and all mutants in a given mutant class have the same stationary concentration, $\bar{x}_j(p)$, the error threshold manifests itself by the sharp decrease in the stationary concentration of the master, $\bar{x}_m(p)$. The introduction of ruggedness has two effects: (i) the stationary solutions of the sequences belonging to one class split and form a band, and (ii) the error threshold becomes sharper and moves towards smaller value of p_{cr}.

[9] The seed s indeed defines all details of the landscape that in turn is completely defined by s and the particular type of the pseudorandom number generator.

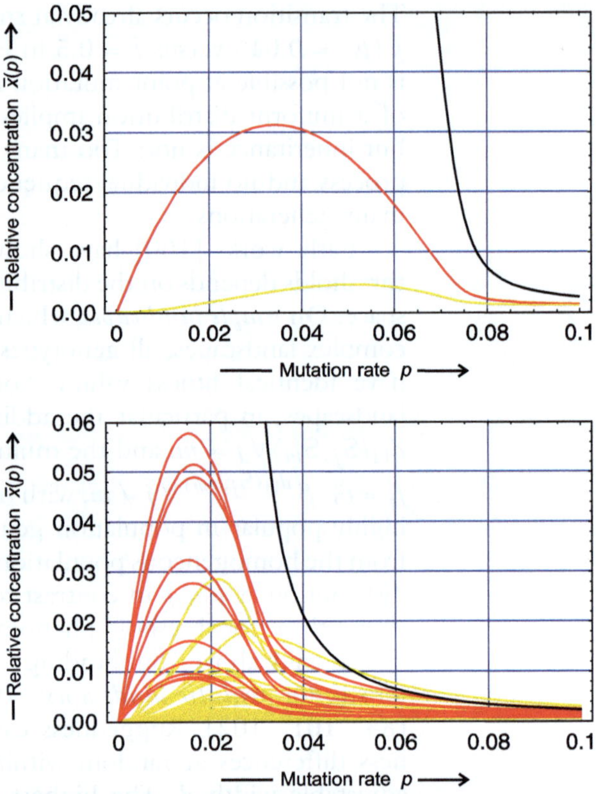

Fig. 16 *Quasispecies on model and realistic landscapes.* The plots show exact solution curves for individual variants in the stationary distribution as a function of the mutation rate, $\bar{x}_i(p)$. The chain length is $l = 10$ corresponding to 1,024 binary sequences on a fitness landscape defined by Eq. 20. The upper part of the figure was calculated with $d = 0$ corresponding to a single-peak landscape, whereas the lower part represents the results for a band width $d = 0.925$ close to full randomness of fitness values. Choice of parameters: $f_m = 2$, $f_i = 1 \, \forall i = 1, \ldots, n; i \neq m$, $s = 491$, $d = 0$ (*upper part*) and $d = 0.925$ (*lower part*); color code: master sequence (*black*), 10 individual single point mutants (*red*), and 45 individual double point mutants (*yellow*)

Neutrality

Neutrality can be introduced into random landscapes in straightforward way by means of a predefined degree of neutrality, λ. Then the fitness landscape is of the form

$$f(\mathrm{S}_j) = \begin{cases} f_0 & \text{if } \eta_j^{(s)} \geq 1 - \lambda \,, \\ f_n + \frac{2d}{1-\lambda}(f_0 - f_n)\left(\eta_j^{(s)} - 0.5\right) & \text{if } \eta_j^{(s)} < 1 - \lambda \,, \end{cases}$$

$$j = 1, \ldots, \kappa^l; \; j \neq m \,. \tag{21}$$

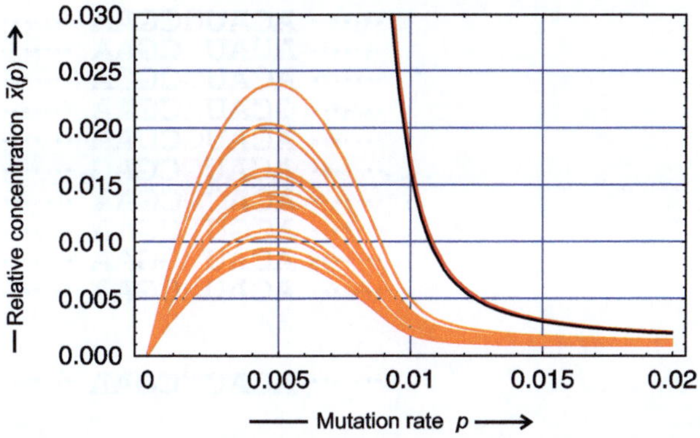

Fig. 17 *Quasispecies on a realistic landscape with neutral sequences.* Two neighboring sequences of highest fitness form a *master pair* of Hamming distance $d_{\mathrm{H}} = 1$, which is surrounded by 18 single point mutations. The plot shows the dependence of the joint quasispecies on the point mutation rate p. The concentrations of the two master sequences are practically identical for all mutations rates. Choice of parameters: $f_m = 1.1$, $f_n = 1.0$, $d = 0.5$, $\lambda = 0.1$; color code: master sequences (*black and red*), 18 individual single point mutants of both master sequences (*orange*)

with the two limiting cases: (i) lim $\lambda \to 0$ yielding the non-neutral random landscape (20) and (ii) lim $\lambda \to 1$ leading to the fully neutral case as modeled and analyzed by Motoo Kimura [69]. Evolution on neutral landscapes is described by neutral networks (Subheading 2.3.3) formed from sequences of identical fitness. Depending on the Hamming distance between neutral master sequences they form either a group of sequences coupled by selection dynamics or random selection takes place and only one sequence survives in the sense of Kimura's theory. The case of vanishing mutation rates, $\lim p \to 0$, has been analyzed for two neutral sequences [103], S_j and S_k and different Hamming distance $d_{\mathrm{H}}(\mathrm{S}_j, \mathrm{S}_k)$:

1. $d_{\mathrm{H}} = 1$: $\lim_{p \to 0} \frac{\bar{x}_j}{\bar{x}_k} = 1$ or $\lim_{p \to 0} \bar{x}_j = \lim_{p \to 0} = \bar{x}_k = 0.5$,
2. $d_{\mathrm{H}} = 2$: $\lim_{p \to 0} \frac{\bar{x}_j}{\bar{x}_k} = \alpha$ or $\lim_{p \to 0} \bar{x}_j = \alpha/(1+\alpha)$, $\lim_{p \to 0} \bar{x}_k = 1/(1+\alpha)$, with some value $0 \le \alpha \le 1$, and
3. $d_{\mathrm{H}} \ge 3$: $\lim_{p \to 0} \bar{x}_1 = 1$, $\lim_{p \to 0} \bar{x}_2 = 0$ or $\lim_{p \to 0} \bar{x}_1 = 0$, $\lim_{p \to 0} \bar{x}_2 = 1$.

In full agreement with the exact result we find that two fittest sequences of Hamming distance $d_{\mathrm{H}} = 1$ are selected as a strongly coupled pair with equal frequency of both members.

Numerical results show that strong coupling does not occur only for small mutation rates but extends over the whole range of p-values from $p = 0$ to the error threshold $p = p_{\mathrm{cr}}$ (Fig. 17).

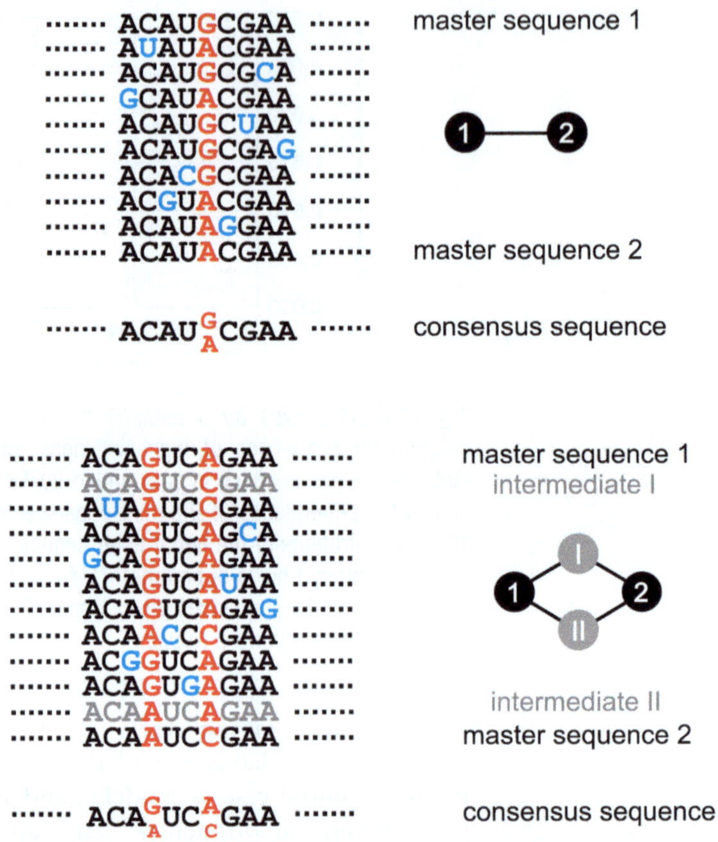

Fig. 18 *Quasispecies and consensus sequences in case of neutrality*. The upper part of the figure shows a sketch of sequences in the quasispecies of two fittest nearest neighbor sequences ($d_{\mathrm{H}} = 1$). The consensus sequence is not unique in a single position where both nucleotides appear with equal frequency. In the lower part the two master sequences have Hamming distance $d_{\mathrm{H}} = 2$ and differ in two positions. The two sequences are present at some ratio α that is determined by the fitness values of other neighboring sequences, and the nucleobases corresponding to the two master sequences appear with the same ratio α

Examples for case 2 are also found on random neutral landscapes and again the exact result for vanishing mutation rate holds up to the error threshold. The existence of neutral nearest and next nearest neighbors manifest itself by the lack of a unique consensus sequence of the population has an important consequence for phylogeny reconstruction (*see* Fig. 18). As shown in the sketch in Fig. 14 neutral networks may comprise several sequences and then, all neutral nearest neighbor sequences form a strongly coupled cluster in reproduction where the individual concentrations are determined by the largest eigenvector of the adjacency matrix of the network.

2.3.5 Advantages and Deficits of the Quasispecies Concept

The conventional synthetic theory of evolution considers reproduction of organisms rather than molecules [104]. In contrast, the kinetic theory of evolution [85, 92, 95] is dealing with the evolutionary processes at the molecular level. Correct reproduction and mutation are implemented as parallel reactions and various detailed mechanisms of reproduction can be readily incorporated into the kinetic differential equations. An extensively investigated example is RNA replication by means of an enzyme from the bacteriophage $Q\beta$ [105–108]. The detailed kinetic analysis reveals the condition for the applicability of the mutation selection model (15): The RNA concentration has to be smaller than the concentration of the replication enzyme. Provided the mechanism of reproduction is known it is straightforward to implement replication kinetics in a system of differential equations [109, pp. 29–75]. This is also true for epigenetic effects, which may require the use of delay differential equations. Furthermore, the kinetic theory rather than the conventional genetics approach is the appropriate basis for molecular evolution and, in particular, molecular phylogeny since these concepts deal with genes and genomes within organisms and not with the organisms themselves.

Explicit consideration of mutations as parallel reactions to error-free reproduction is manifested in the structure of stationary populations, no *ad hoc* assumptions are required for the appearance of mutants. In addition, high mutation rates as found, for example, in test tube evolution experiments [110–112] and in virus reproduction [96] do not represent a problem and are handled equally well as low mutation rates. The kinetic model relates evolutionary dynamics directly to fitness landscapes, which have a straightforward physical interpretation and can be measured. Unclear and ambiguous results as obtained with simple models of fitness landscapes demonstrate that an understanding of evolution is impossible without sufficient knowledge on the molecular basis of fitness. In simple systems fitness is a property that can be determined by the methods of physics and chemistry, and thus independently of evolutionary dynamics. The current explosion of harvested data provides a new source of molecular information that can be used for the computation of fitness values under sufficiently simple conditions.

As a starting point the phylogenetic approach focusses on the evolution of individual sites $s_k^{(i)}$ in a sequence $\mathsf{S}_i = \left(s_1^{(i)} s_2^{(i)} \cdots s_l^{(i)}\right)$ whereas the kinetic model considers full sequences initially as expressed by the mutation frequencies Q_{ji} and fitness values f_i. The independent site model of theoretical phylogeny is augmented by dependencies on other sites with increasing complexity eventually ending up at accounting all sites in the sequence. The kinetic approach progresses in opposite direction from sequences to smaller entities but eventually encounters the same parameter problem as the phylogenetic method. The uniform error model, for example, assumes independent mutation of and equal frequencies of mutation at all sites. A major difference between the two approaches concerns handling of mutations. In phylogeny the matrix **Q** is a rate matrix whereas the kinetic approach separates mutation expressed by the matrix Q and structural or functional effects represented by the fitness values in matrix F. This has the advantage that all fitness-related energetic effects, for example base pairing in double helical structure fragments, are entering F rather than Q. Of course, there are sequence-dependent effects on pure mutation like the occurrence of *hot spots* with higher mutation frequencies than at the other sites of the sequence, which commonly depend on RNA structure. Again taking into account site- and context-dependent point mutation rates increases the number of parameters enormously and considering to much detail encounters the same problem as in phylogeny. In the era of genomics the whole sequence approach appears more appropriate and the wealth of currently available data is more easily introduced into the kinetic approach.

The quasispecies is the stationary solution of a deterministic, differential equation-based model that, in principle, is bound to constant population size. Analysis of the basic ODEs, however, has shown that the results in relative concentrations (x_i; $\sum_{i=1}^{n} x_i = 1$) are generally valid as long as the population does neither die out ($\sum_{i=1}^{n} N_i = 0$) nor explode ($\sum_{i=1}^{n} N_i = \infty$) [92]. Then, the particle number are not normalizable and the structure of the population cannot be predicted from solutions of Eq. 15. Nevertheless, quasispecies may still exist for vanishing populations but any rigorous treatment has to start out from the original kinetic equations with population size being treated as a variable. Variants with zero fitness are compatible with the error threshold phenomenon [113, 114] but the prerequisites for the conventional calculation of the quasispecies are no longer fulfilled—not every sequence can be reached from every sequence by a finite chain of point mutations (Subheading 3.2.2).

Another question is hard to answer at present: Do the populations in nature ever reach a stationary state? In vitro evolution experiments can be carried out in such a way that stationarity or mutation equilibrium is achieved, but is this true also in nature in virus infections specific mutants appear also within the infected host but on the other hand the effect of infection and the course of disease is rather similar comparable hosts indicating that viruses are in a comparable state at the beginning of an infection.

Two other problems are quite general in biological modeling in particular on the molecular level: (i) Most models assume spatial homogeneity whereas cells are highly objects with limited diffusion, active transport, and spatial localization of molecular players, and (ii) many results are derived from differential equations, which are based on the use of continuous variables, and thereby it is implicitly assumed that populations are very large. In principle, the definition of continuous space and time requires infinite population size what is a reasonable and well-justified assumption in chemistry but not in biology where sample sizes may be very small. Examples are the often extremely small concentrations of regulatory molecules. Systematic studies on the very small bacterium *Mycoplasma pneumoniae* in the spirit of systems biology [115–117] have shown that in extreme cases only one molecule per hundreds or even thousands of bacterial cells is present at a given instant in the population. Also the assumption of a homogeneous space is questionable: There is very little free diffusion in real cells and even bacterial cells have a very rich spatial structure. Stochasticity plays an important role and discrete stochastic rather than deterministic continuous variables should be applied.

Finally we mention a general problem for evolutionary models. Using conventional modeling with ODEs all populations would extend over whole sequence space. A drastic example is the uniform distribution of sequences predicted at mutation rates above error threshold. Coverage of sequence space can never occur in a finite world: Even for small RNA molecules of tRNA size we would need a population size of $N = 10^{46}$ individuals in order to have one molecule for every possible sequence, whereas the largest populations in in vitro experiments with RNA hardly exceed $N = 10^{15}$ molecules. What we have instead are clones of sequences migrating through sequence space (*see*, for example, [101, 102, 118, 119]. Truncation of fitness landscapes has been suggested recently as a possible solution to the problem [120]. Alternatively, one could leave the full landscape and truncate populations through setting all concentrations less than one molecule per reaction volume equal to zero and eliminate the corresponding variables. New variables come into play when their concentration exceeds this truncation threshold similarly as in stochastic processes.

2.3.6 Evolutionary Optimization of Structure

In order to simulate the interplay between mutation acting on the RNA sequence and selection operating on the phenotypes, here the RNA structures, the sequence-structure map has to be an integral part of the model [101, 118, 121, 122]. The simulation tool starts from a population of RNA molecules and simulates chemical reactions corresponding to replication and mutation in a continuous stirred flow reactor (CSTR, *see* Subheading 4.2) and uses an algorithm developed by Daniel Gillespie [123, 124]. In target search problems the replication rate of a sequence S_k, representing its fitness f_k, is chosen to be a function of the structure[10] distance between the mfe structure formed by the sequence, $\mathsf{Y}_k = \Phi(\mathsf{S}_k)$ and the target structure Y_T,

$$f_k(\mathsf{Y}_k, \mathsf{Y}_T) = \frac{1}{\alpha + d_{\mathrm{S}}(\mathsf{Y}_k, \mathsf{Y}_T)/l}, \tag{22}$$

which increases when Y_k approaches the target (α is an adjustable parameter that was commonly chosen to be 0.1). A trajectory is completed when the population reaches a sequence that folds into the target structure. The simulated stochastic process has two absorbing barriers, the target and the state of extinction. For sufficiently large populations ($N > 30$ molecules) the probability of extinction is very small, for population sizes reported here, $N \geq 1{,}000$, extinction has been never observed.

A typical trajectory is shown in Fig. 19. In this simulation a homogenous population consisting on $N = 1{,}000$ molecules with the same random sequence and corresponding mfe structure is chosen as initial condition (Fig. 20). The target structure is the well-known clover leaf of phenyl-alanyl-transfer RNA ($\mathrm{tRNA^{phe}}$). The mean distance to target of the population decreases in steps until the target is reached [118, 121, 125]. Short adaptive phases are interrupted by long quasi-stationary epochs, the latter falling into two different scenarios:

(i) The structure is constant and we observe neutral evolution in the sense of Kimura [69]. The numbers of neutral mutations are proportional to the numbers of replications and the evolution of the population can be understood as a diffusion process on the corresponding neutral network [102].

(ii) The process during the stationary epoch involves several structures with identical replication rates and the evolutionary process is a kind of random walk in the space of these neutral structures.

The diffusion of the population on the neutral network is illustrated by the plot in the middle of Fig. 19 showing the width of the population as a function of time [75, 125]. The population width increases during the quasi-stationary epoch and sharpens almost instantaneously after a sequence that allows for the start of a new adaptive phase in the optimization process had been produced by mutation. The scenario at the end of the plateau corresponds to a *bottle neck* of evolution. The lower part of the figure shows a plot of the migration rate or drift of the population center and confirms the interpretation: The drift is almost always very slow unless the

[10]Several measures for the distance between structures can be applied. Here we have chosen the Hamming distance between the parentheses notation of structures, d_{S}.

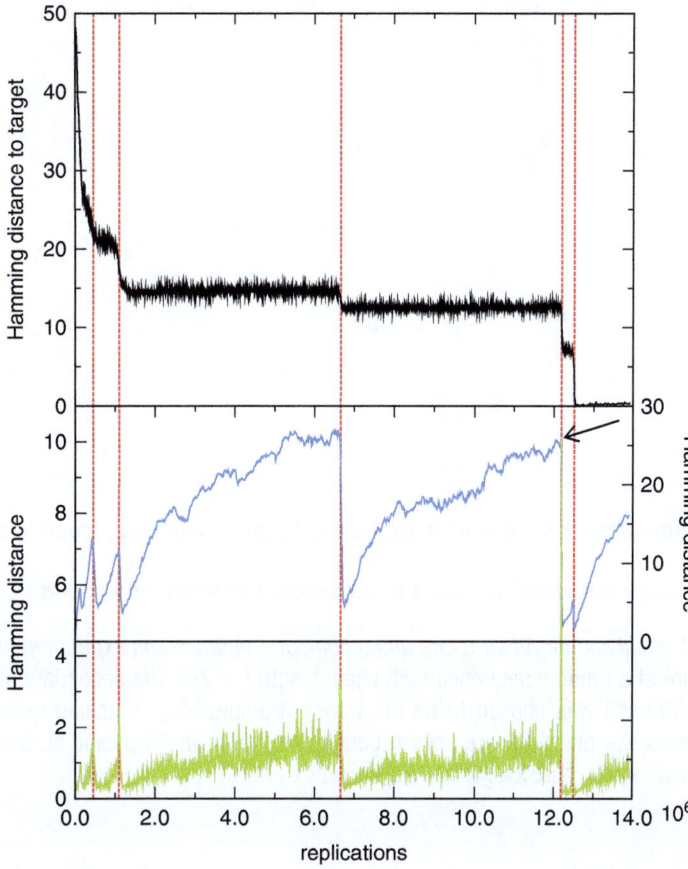

Fig. 19 *A trajectory of evolutionary structure optimization.* The topmost plot presents the mean distance to the target structure of a population of 1,000 molecules. The initial random structure and the target structure are shown in Fig. 20. The plot in the middle shows the width of the population in Hamming distance d_{H} and the plot at the bottom is a measure of the velocity with which the center of the population migrates through sequence space. Diffusion on neutral networks causes spreading on the population in the sense of neutral evolution [102]. A synchronization is observed at the end of each quasi-stationary plateau where a new adaptive phase in the approach towards the target is initiated. The synchronization is caused by a drastic reduction in the population width and a jump in the population center (the top of the peak at the end of the second long plateau is marked by a *black arrow*). A mutation rate of $p = 0.001$ was chosen, the replication rate parameter is defined in Eq. 22

population center "jumps" from one point in sequence space to another point from which the new adaptive phase is initiated. A closer look at the figure reveals the coincidence of the three events: (i) beginning of a new adaptive phase, (ii) collapse-like narrowing of the population, and (iii) jump-like migration of the population center.

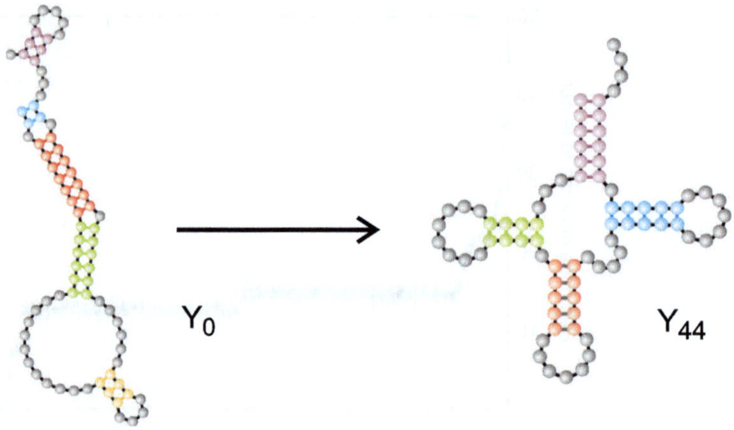

S_0: GUUAUGGGCGAUGAGGAGUAGAGAAAAAACCAAUCGGUGAAAGAUCUCGUGUGCCCAUUGCAUAGCCCGUACGGCA

S_{44}: GGGCAGAUAGGGCGUGUGAUAGCCCAUAGCGAACCCCCCGCUGAGCUUGUGCGACGUAUGCGCACCCUGUCCCGCU

Fig. 20 *Initial and target structure of the computer optimization experiment in Fig. 19*. Structure Y_0 is the mfe structure of a random sequence with chain length $l = 76$. The secondary structure of phenyl-alanyl-transfer RNA (tRNA^{phe}) was chosen to be the target structure Y_{44}. Stacking regions are shown in color. The two sequences, S_0 and S_{44}, are shown below; they differ in 46 positions. Structures were computed with the `Vienna RNA Package, Version 1.8.5`

2.3.7 From Sequences and Structures to Genotypes and Phenotypes

Genotypes or genomes are RNA or DNA sequences. The phenotype comprises structures and functions, both of which are dependent on the specific experimental or environmental setup. The simplest system capable of selection and mutation consists of RNA molecules in a medium that sustains replication. Then, genotype and phenotype are simply the RNA sequence and structure, respectively. In case of the Qβ evolution experiments the functional requirements are a result of the replication mechanism [105–107]. In order to be replicated the Qβ-virus RNA molecules must carry an accessible recognition site for binding to the enzyme Qβ-replicase [108, 126] and the replication involves a complementary or plus-minus strand copying mechanism with $f_j^{(+)}$ and $f_j^{(-)}$ being the replication rate parameters for plus-strand and minus strand synthesis. After internal equilibration the plus–minus ensemble grows exponentially with an overall fitness constant, which is the geometric mean of the fitness values of both strands: $f_j = \sqrt{f_j^{(+)} f_j^{(-)}}$. Thus, the phenotype in the case of Qβ-replication in the test tube is the ensemble consisting of both strands.

Outside plant cells viroids are *naked*, cyclic, and especially stable RNA molecules whose sequences are the viroid genotypes.

Viroid RNAs are multiplied through by the host cell machinery and carry specific recognition sites at which is initiated. Replication of *Potato spindle tuber viroid*, for example, starts predominantly at two specific sites with the closely related sequences GGAGCG**A** at position A_{111} and GGGGCG**A** at position A_{325} of the viroid RNA with a chain length of $l = 359$ nucleotides [127]—the two positions are almost on opposite sides of the cyclic RNA, 214 or 145 nucleotides apart. Viroid RNAs have many loops and bulges that serve two purposes: (i) They allow for melting of the viroid RNA since a fully double stranded molecule would be too stable to be opened, and (ii) they carry the recognition sites and motifs for replication and system trafficking [128, 129], which have also been studied on the 3D structural level [130]. Although viroid reproduction in nature requires a highly specific host cell and there was a common agreement that viroids in general are highly species specific, recent attempts to replicate *Avocado sunblotch viroid* in yeast cells have been successful [131]. The viroid phenotype is already quite involved: It has a structural component that guarantees high RNA stability outside the host cell, but at the same time the structure is sufficiently flexible in order to be opened and processed inside the cell. Like the Qβ RNA, viroid RNA carries specific recognition sites for initiation and control of the reproduction cycle.

Viruses, in essence, are like viroids but the complexity of the life cycle is increased by three important factors: (i) virus DNAs or RNAs carry genes that are translated in the host cell, yield virus specific factors, and accordingly viruses have genetic control on the evolution of these coding regions, (ii) the virus capsid may contain functional protein molecules, for example replicases, in addition to the virus-specific genetic material, and (iii) viruses are coated by virus-specific proteins or proteins and membranes (for a recent treatise of virus evolution *see* [96]).

Bacterial phenotypes are currently too complex for a comprehensive analysis at the molecular level. An exception is the particulary small and cell wall-free bacteria of the genus *Mycoplasma*. In particular, *Mycoplasma genitalium* is a parasitic bacterium and was considered to be the smallest organism for quite some time.[11] Its genome consists of one circular chromosome with 582,970 base pairs and 521 genes of which 482 encode for proteins. Extensive studies aiming at full systems biology of a cellular organism were performed on the some what larger species *Mycoplasma pneumoniae* with a genome size of $l \approx 816{,}000$ base pairs [115–117]. For larger organisms—ordinary bacteria of about tenfold size and small eukaryotic cells—flux balance analysis [132] rather than full

[11] The smallness record is currently hold by *Nanoarchaeum equitans* with a genome size of 490,885 base pairs.

molecular systems biology has been performed (an early example of flux balance analysis of *Escherichia coli* is found in [133]). Another approach to complex phenotypes is based on completely annotated genomes and information on gene interactions mainly coming from proteomics. A gene interaction network has been worked out recently for a small eukaryotic cell with roughly 6,000 genes [134] and it is overwhelmingly complex.

3 From Theory to Applications

In this section, we shall illustrate how the theory of evolution can be applied to problems of practical interest. We shall restrict ourselves to presenting only a few selected applications here. Even though the field has become so rich that several books could be filled with examples.

3.1 Applications of Phylogenetic Sequence Evolution Models

An in-depth knowledge of how sequences evolve may evidently help us to improve the reconstruction of phylogenetic trees based on sequences (Fig. 21). This classical application is described in Subheading 3.1.1. Regardless of this ongoing debate, sequence evolution models have in recent years also addressed the fields of RNA research more broadly. Here, we only mention two further applications: the value of phylogenetic simulation for distinguishing ancestral and functional correlations in Subheading 3.1.2 and the application of phylogenetic complex models to genomic ncRNA screens in Subheading 3.1.3.

3.1.1 Phylogenetic Tree Inference

There are currently four main methods of phylogenetic inference: methods based on the parsimonious principle, i.e., maximum parsimony [135], statistical methods such as maximum likelihood [24] or Bayesian inference [136], and distance-based methods like neighbor-joining [137]. For further examples of tree reconstruction methods and detailed descriptions we refer to [20].

Even though research has been done on the development of RNA base-pair substitution models with non-overlapping tuples, the results have not yet been widely adopted by the scientific community. One reason may be that a priori knowledge of molecular structure is necessary. The other reason could be that the improvements in phylogenetic inference afforded by these models are not significant enough in comparison to independent models with rate heterogeneity. The bias introduced in sequence analysis by ignoring heterogeneous rates among sites has been studied in population genetics [cf. 138] and phylogenetic reconstruction [139], where it has been shown that the inclusion of Γ-distribution usually improves the estimation of other evolutionary parameters, includ-

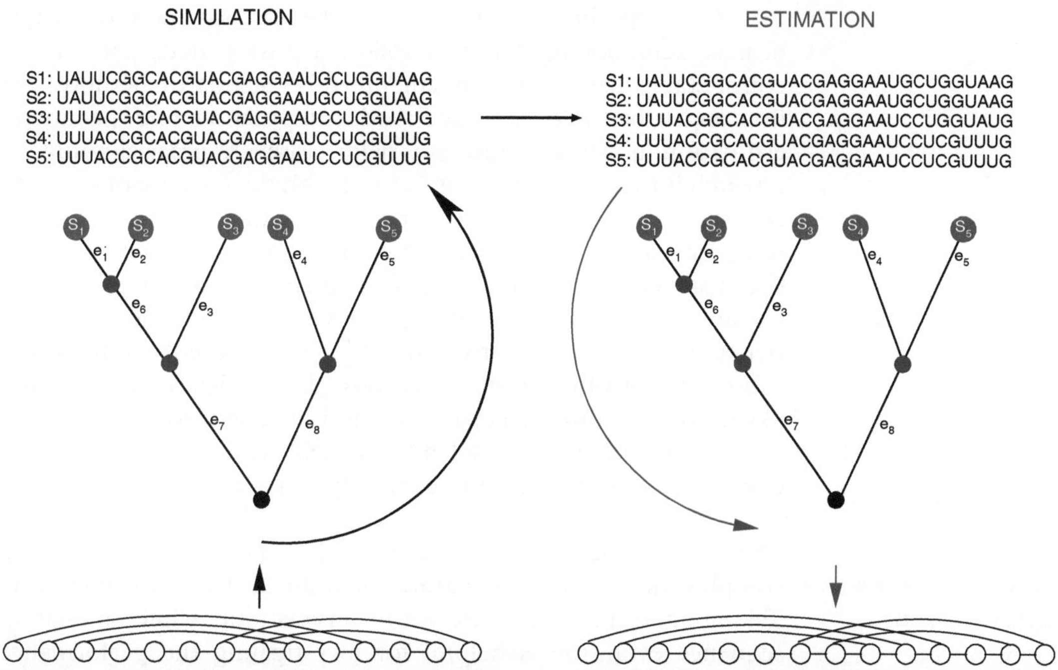

Fig. 21 *Phylogenetic simulations and estimations under constraints.* Whenever we analyze a set of homologous sequences, we have to take the evolutionary history of the observed sequences into account. Simulating sequence evolution while taking site-specific interactions into account is helpful for investigating the performance of both tree-building methods and structure prediction methods. Furthermore, these methods are of value in themselves because they contribute to our understanding of the interrelation between structure and substitution process

ing tree topology. Analyzing the efficiency of three reconstruction methods when sequence sites are not independent, Schöniger and von Haeseler [140] demonstrated that the inferred tree is not significantly affected by the presence of such correlations. In a comparative study Savill et al [45] have shown that models permitting a nonzero rate of double substitutions performed better than those restricting—as is usually done—the number of allowed substitutions to one per unit time. A software package, called `PHASE` [141], for phylogenetic inference of RNA sequences with a range of base-pairs substitution models is available. `MrBayes` [142] and `RAxML` [143] have also included RNA substitution models while two further simulation studies have been recently published [144, 145]. One study recommends the inclusion of RNA secondary structure during phylogenetic inference while the other notes that additional research is needed in order for the effects observed with mixed models in tree reconstruction to be fully understood. By contrast, it is well known that sequence quality can be improved by the use of secondary structure information, as described elsewhere in this volume.

Direct combination such as in the case of thermodynamic nearest neighbor models that follow a phylogenetic approach is considerably harder to implement [60]. Standard methods can no longer be used for likelihood computation and parameter estimation if Markov random fields arise. A great amount of research is therefore being carried out with the purpose of covering the required technical skills, even though these methods are still not practical on a wider scale. Neighbor-joining, which calculates the distances based on the ML estimated parameters, may be an alternative. For a parsimony study of the evolution of RNA structure we refer to, for example, [146]. A recent study on the robustness of phylogenetic methods [147], including maximum likelihood, maximum parsimony, and neighbor-joining, recommends tests such as model homogeneity and goodness of fit if confidence in the inferred trees is to be gained.

3.1.2 Phylogenetic Simulations: Ancestral and Functional Correlations

The models used for simulation need to be more accurate and complex descriptions of nature than those used for inference. Taking molecular structure into consideration while simulating sequence evolution is helpful for investigating the performance of both tree-building methods and structure prediction methods (Fig. 21). Furthermore, these simulations are of value in themselves because they contribute to our understanding of the intertwined relationship between structure and the substitution process. For example, the use of supervised sequence evolution allows us to control and study the extent of structural and sequence conservation as RNA structure stability or the influence of phylogentic diversity. Although methods using phylogeny and structure already exist, an explicit definition from a phylogenetic point of view was missing. A phylogenetic structure has since been introduced [148]. This definition is based on a simulation framework that includes a model which constitutes a range of different substitution models acting both along a sequence and an annotation of correlations among sites, a so-called neighborhood system as described in Subheading 2.1.3. A phylogenetic structure (PS) is then defined by a neighborhood system, a substitution model and a phylogenetic tree (Fig. 22). The substitution model specifies the evolutionary process of nucleotide evolution. However, the model is influenced by the neighborhood system that defines the interactions among sites in a sequence. The phylogenetic tree introduces an additional dependency pattern in the observed sequences. A PS appears in *a set of sequences at different instants*. These can be transformed, for example, into the minimum free energy secondary structure.

A realization of a PS is then a relational object at instant t. Figure 23 is an illustration of what has just been described, taking the example of a phylogenetic structure of Fig. 22 in relation to a minimum free energy (mfe) structure. For didactic reasons, a thermodynamically improbable neighborhood system with a

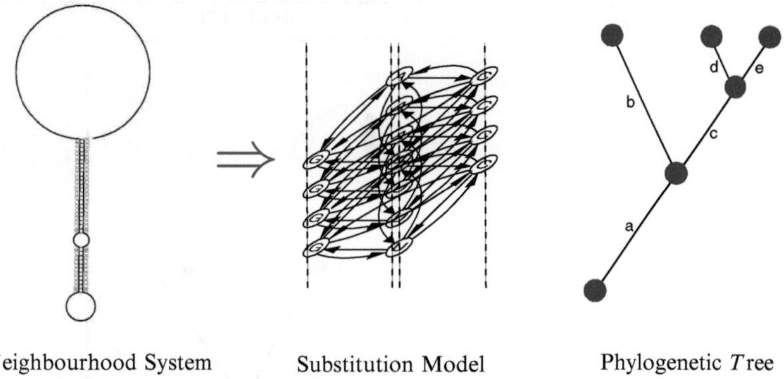

Fig. 22 *An example of a phylogenetic structure. Left:* an example of a thermodynamically improbable neighborhood system was chosen for didactic reasons. *Middle:* model **Q** constitutes a collection of possibly different substitution models *Right:* example of a phylogenetic tree with three extant taxa

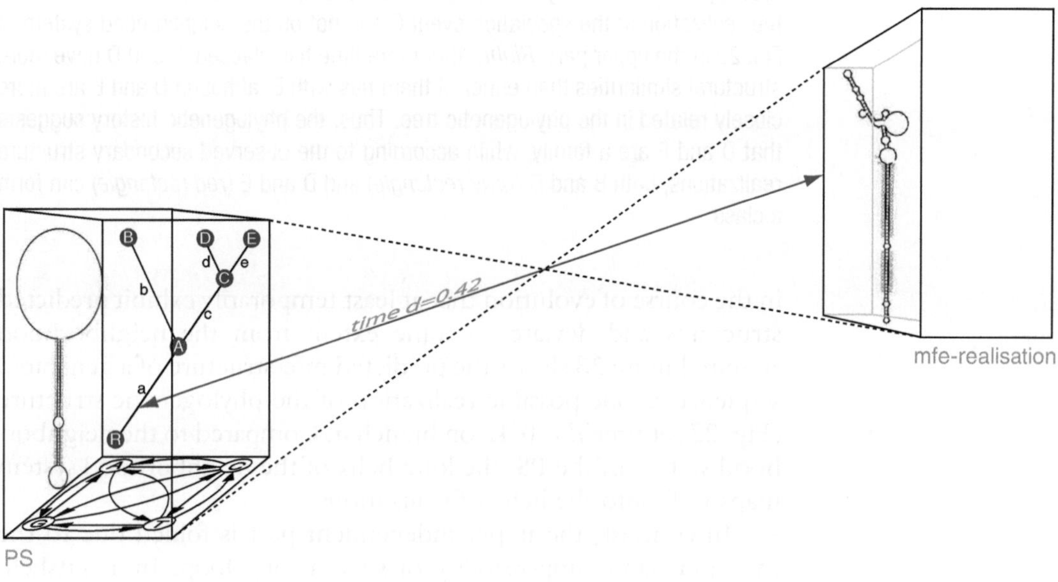

Fig. 23 *PS transformaion:* A Phylogenetic Structure (PS) of Fig. 22 is summarized on the left and is transformed at time d = 0.42 in a mfe realization and suboptimal mfe realizations on the right. By comparing the realization with the neighborhood system most base pairs are the same. However, due to the thermodynamical improbability of such a long loop, the upper independent part is folded

helix of such length and a long part of independent sites is used. However, this thermodynamically improbable neighborhood system influences the collection of different substitution models acting on the sequence. As part of this concept, it is possible to mimic sequence evolution under the structural constraint of the PS, e.g., under the condition of compensatory mutations with an extended Felsenstein model. Due to the stochastic nature of the substitution process, however, sequences will probably be observed

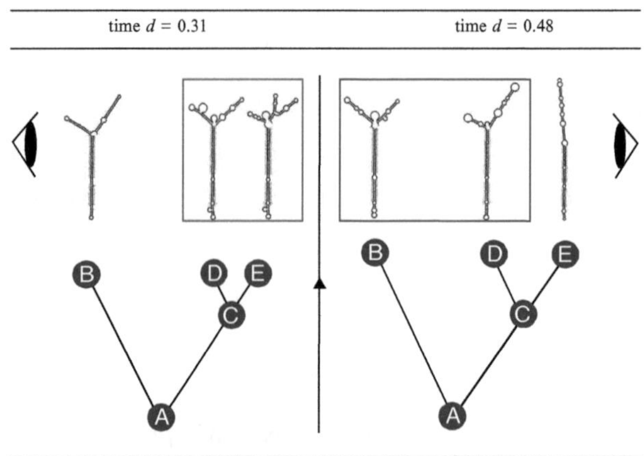

Fig. 24 *Observations of secondary structure on realization levels* while evolutionary history is neglected can mislead assignments of families, illustrated by *rectangles*. *Left*: shortly after a speciation event, we have similarities based on the realization at the speciation event C and not on the neighborhood system of Fig. 22 in the upper part. *Right*: after more time has elapsed, B and D have more structural similarities than either of them has with E, although D and E are more closely related in the phylogenetic tree. Thus, the phylogenetic history suggests that D and E are a family, while according to the observed secondary structure realizations, both B and D (*gray rectangle*) and D and E (*red rectangle*) can form a class

in the course of evolution that at least temporarily exhibit predicted structures and deviate to some extent from the neighborhood system. Figure 23 shows the predicted mfe-structure of a generated sequence as one possible realization of the phylogenetic structure (Fig. 22) at time $d = 0.42$ on branch a. Compared to the neighborhood system of the PS, the long helix of the neighborhood system maps well onto the helix of realizations.

In contrast, the upper independent part is folded due to the thermodynamic impossibility of such a long loop. In a nutshell, although the neighborhood system is thermodynamically improbable, it is transformed into possible thermodynamic realizations given an evolutionary history. Figure 24 illustrates the diversity of realizations of a PS as defined in Fig. 22: the stem region of the realization should be defined by the constraint through the neighborhood system and the substitution model, while the upper loop region has no site-specific interactions and should lead to different realizations through the mfe folding. In the upper part, however, we observe similar mfe folds between D and E at time $d = 0.31$, based on the previous ancestral state at C. With time, the differences between D and E become progressively more apparent. In the uppermost frame, the species B and D have more structural similarities than either has with E, although D and E are more closely related. Thus, the phylogenetic history suggests that D and

E are a family. However, according to the observed secondary structure realization B and D form a class at time $d = 0.48$.

Following a PS, we have to distinguish between different constraints for interactions observed among the sites. A structural constraint defines the evolutionary strength of structuring sequences at different instants t and can be differentiated into: ancestral constraint and neighborhood constraint. Neighborhood constraints are site-specific interactions acting on the sequence along the evolutionary process. In this sense, the interactions observable between the sites through this neighborhood constraint are called neighborhood or functional correlations. Following a PS, the evolution of nucleotides must also be taken into account. In phylogeny the states occurring at the internal nodes are important because of the likelihood of the state remaining unchanged after only a short period of time. This depends on the model; it is an ancestral constraint that defines the influence of ancestral nucleotide distribution at an alignment site and can be associated with observable ancestral correlations in sequences. In general, term associations, correlations, or dependencies are used to represent measurements from sequences via different estimation methods. If we estimate correlations from homologous sequence data, e.g., from an alignment, they are related through their evolutionary history and common ancestral states. Thus, ancestral as well as functional correlations may occur. So far, structure prediction methods have mostly been interested in predicting dependencies that result from neighborhood constraints.

3.1.3 Phylogenetic Background Models for Genomic Screens

The consensus structure prediction program `Pfold` [149] uses a 16×16 RNA base-pair substitution model combined with a context-free grammar similar to the genfinder programs `EvoFold` [150] and `QFold` [151]. For assessing the significance of predicted structures, e.g., estimating the false discovery rate in a genomic screen for ncRNAs, the genomic predictions should be compared to the results obtained from randomized data with the same dinucleotide content. In the case of single sequences, well-known and widely used algorithms are available for generating dinucleotide controlled random sequences either by shuffling or by first Markov chain simulations [152, 153]. One approach is to simulate the alignments of a given dinucleotide content [154]. `SISSIz` is based on a complex substitution model that captures the neighbor dependencies and other important alignment features except for the signal in question. In addition, it directly combines the phylogenetic null model with the `RNAalifold` [155] consensus folding algorithm, thus yielding a new variant of a thermodynamic structure-based RNA gene finding program that is not biased by the dinucleotide content. For further details on ncRNA gene finding programs we refer to, for example, Chapters 7, 9, 10, and 15.

3.2 Controlled Evolution and Evolutionary Design

Evolution under natural conditions is much too complex for modeling and the lack of detailed understanding at the molecular level is prohibitive for applications of evolution to solve practical design. Controlled environmental conditions and reduction of the complexity of the evolving unit, however, allow for the utilization of evolutionary methods in biotechnology. We choose here three prominent examples: (i) applied molecular evolution, (ii) lethal mutagenesis in viral pathology, and (iii) controlled bacterial evolution.

3.2.1 Evolutionary Design of Molecules

Experimental [112] and theoretical studies [85] showed that evolution through mutation and selection is not bound to the existence of cellular life but occurs readily with RNA molecules in test tube experiments (for details we refer to a recent review on RNA evolution in laboratory experiments [156]). The known mechanistic details of in vitro evolution encouraged the development of techniques exploiting the evolutionary principle for the production of designed molecules with given functions. All these methods are based on sequence variation and selection, whereby variation has become the easy part since the degree of sequence changes can be varied widely from mutation with low mutation rates to random synthesis of polynucleotides. Selection for predefined function is the tricky part of directed evolution and commonly represents a challenge for the intuition of the experimenter.

Selection by exponential enrichment (SELEX) has become a popular and routinely used method [157, 158]. Targets for binding are attached to the stationary phase of a chromatographic column and a solution containing a variety of RNA molecules is poured through the column. Molecules with high affinity to bind to the target, called *aptamers*, are first retained at the column, then washed out by another solvent, and the whole procedure is repeated several times with mutated samples derived from the best binders at this moment. Binding constants almost as large as those observed with the strongest natural aggregates can be obtained after some 20 selection cycles [159].

Evolutionary design has been applied to a great variety of other problems too. A view examples of recent review articles are given here: (i) directed evolution of nucleic acid enzymes [160], (ii) design of proteins by evolutionary methods [161, 162], and (iii) applications to design of low molecular weight compounds [163].

3.2.2 Virus Evolution: Error Thresholds and Lethal Mutagenesis

Virus evolution is a broad field with many aspects. The idea to relate the properties of viruses to the structures viral RNA is almost 40 years old: In a pioneering paper Charles Weissmann [108] related the life cycle of the RNA bacteriophage Qβ to the secondary structure of its RNA. The structural difference between newly synthesized and mature RNA is exploited as an important regulatory element. Mutations on the RNA genotype may have direct consequences for both RNA structures and viral life cycles [96]. Here only the practical aspect of developing antiviral medication is mentioned. Many antiviral drugs are powerful because they increase the mutation rate and drive virus populations to extinction but for a satisfactory molecular explanation of the mechanism the required information on the fitness landscape is still missing. Nevertheless, *lethal mutagenesis* is an important phenomenon and simplified models providing phenomenological explanations have been developed.[12]

Manfred Eigen and Esteban Domingo originally explained lethal mutagenesis caused by pharmaceutical compounds increasing the mutation rate through driving populations beyond the error threshold [98, 164]: At mutation rates above the error threshold replication becomes *random*[13] and the result is a complete loss of the genetic information and eventually the viral life cycle breaks down. Later on James Bull and Claus Wilke studied the dynamics of lethal mutagenesis in more detail and claimed that population extinction is a phenomenon independently of catastrophic error accumulation [165–167]. Recently, Francisco Montero and coworkers [114] modeled lethal mutagenesis by means of three-species replication-mutation kinetics with neglect of back mutation

$$\frac{dc_m}{dt} = \left(f_m(1-p)^l - \vartheta\right) c_m \,,$$

$$\frac{dc_k}{dt} = f_m(1-p)^d\left(1-(1-p)^{(l-d)}\right) c_m + \left(f_k(1-p)^d - \vartheta\right) c_k \,, \quad \text{and}$$

$$\frac{dc_j}{dt} = f_m\left(1-(1-p)^d\right) c_m + f_k\left(1-(1-p)^d\right) c_k - \vartheta\, c_j \,;\; c = \sum_{i=1}^{n} c_i \,, \tag{23}$$

where S_m is the master sequence with the concentration $[S_m] = c_m$ and the replication rate parameter f_m, c_k and f_k refer to the class of non-lethal mutants, and c_j to the class of lethal mutants, ϑ is the uniform degradation rate parameter for all sequences, l is the chain length, d the number of positions at which mutation yields a lethal

[12] An early paper [168] claimed that zero fitness values are incompatible with the existence of quasispecies and error threshold. The result, however, turned out to be an artifact of a rather naıve linear sequence space, since later works demonstrated that selection and mutation on realistic sequence spaces sustain error thresholds also in the presence of lethal variants [113, 114].

[13] *Random replication* expresses the fact that error accumulation destroys the relation between template and copy and inheritance is no longer possible.

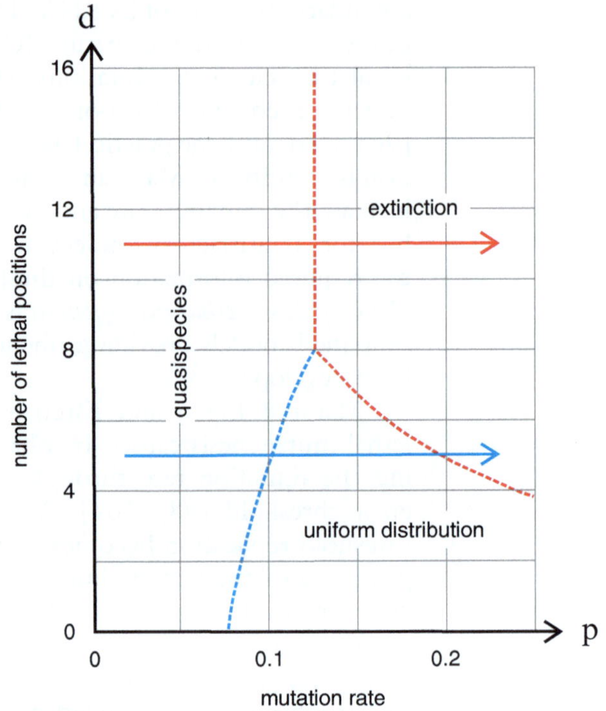

Fig. 25 *Quasispecies and lethal mutations.* The sketch shows the long time behavior of mutation-selection dynamics in the presence of lethal mutations. For a sufficiently large number of lethal positions (d8) in the virus genotype the quasispecies goes extinct at the extinction threshold (*red*) whereas for a smaller fraction of lethal variants two thresholds are observed: (i) an error threshold (*blue*) and an extinction threshold (*red*). The picture is redrawn from Tejero et al. [114], Fig. 2d; choice of parameters: chain length $l = 20$; kinetic parameters: $f_m = 15$, $f_k = 3$, $\vartheta = 1$

variant and eventually p the single point mutation rate. This model [114] is characterized by two features: (i) the concentration of the material consumed in the reproduction process is assumed to be constant, $[\mathrm{A}] = a_0$, and a_0 is absorbed in the fitness parameter f_i $(i = 1, \ldots, n)$, and (ii) a degradation rate ϑ is introduced for all species. In contrast to the selection-mutation equation (15) and the flow reactor discussed in Subheading 4.2, the model system Eq. 23 does not approach a stationary state but the total concentration c either grows infinitely or goes extinct. Figure 25 illustrates the result concerning lethal mutagenesis. There are two different scenarios of quasispecies development with increasing mutation rate p, which depend on the degree of lethality that is expressed in the number of lethal sites d: (i) at low lethality the quasispecies reaches first the error threshold at $p = p_{\mathrm{cr}}$, passes a range of p-values, and then becomes extinct at $p = p_{\mathrm{ext}}$, and (ii) at sufficiently high degree of lethality the error threshold merges with the extinction threshold and the quasispecies dies out directly at $p = p_{ext}$. It is worth noticing that the stability of the quasispecies against mutation increases with increasing degree of lethality corresponding to a shift of the error threshold towards higher mutation rate. Lethal mutagenesis is understood at the phenomenological level but when it comes to molecular details, more experimental data and a comprehensive molecular theory is required. Studies based on more realistic landscapes including lethal variants into model landscapes in the sense of (20) and (21) are still missing.

3.2.3 What We Learn from Bacterial Evolution

Controlled bacterial evolution has been and still is studied in a long time experiment by Richard Lenski. He and his groups started 12 parallel experiments derived from a single clonal ara^{-}[14] strain of *Escherichia coli* in February 1988. The original clone underwent an immediate revertant mutation to ara^{+}, and six ara^{+} and six ara^{-} population were chosen for the series experiment. Every day the cultures are propagated by transfer of samples into new growth medium that has been intentionally chosen as poor in glucose, probes are taken, isolated, and deep frozen at regular intervals of 500 generations [169, 170]. Until now the 12 populations have passed about 53,000 generations. Three findings are of direct relevance for this contribution: (i) the early adaptation to the changed environment [171], (ii) the phylogeny of controlled bacterial evolution [172], and (iii) contingency and repeatability in evolution [173].

Early adaptation to new environmental conditions occurs in steps rather than continuously [171] reminding of the course of structural optimization shown in Fig. 19. Although optimization of the phenotype exhibits punctuated appearance, mutations at the level of DNA sequences occur with a fairly constant rate per generation. The population evolves by forming clones that become in sequence space phylogenetic trees [172]. After 31,500 generations one of the twelve populations produced a mutant, which had the capability for citrate uptake from the growth medium [173].[15] This clone had an instantaneous advantage started to grow much faster and showed a dramatic increase in population size, because it had conquered a hitherto unexploited niche with a new nutrient. The main question to ask about contingency in evolution concerns the probability of the adaptive event: (i) was

---

[14]The variants ara^{+} and ara^{-} differ in a single point mutation and in the capacity to utilize arabinose as nutrient. In growth media free of arabinose the mutation $ara^{+} \leftrightarrow ara^{-}$ is neutral [173].

[15]Most *Escherichia coli* strains are unable to live on citrate buffer because they have no mechanism for uptake of citrate or citric acid into the cell. The growth medium used by Lenski et al. in the long time evolutions experiment contained citrate buffer for pH control.

it a highly improbable singular event or (ii) was it an ordinary event of common probability, which needed a preparation in the sense that the clone had to migrate into some region of sequence space before? In the first case there would be no chance to repeat the event, whereas in the second case the invention of a citrate channel should be repeatable if one started from some earlier isolate and *ran the tape a second time.* Richard Lenski and coworkers could find an answer: Scenario (ii) is what happens in the *Escherichia coli* experiments and indeed samples isolated as early as generation 20,000, i.e., 11,500 generations before the cit^{+} mutation had happened in the original population were able to develop advantageous citrate variants, whereas none was found with earlier isolates. The experiment is a beautiful demonstration of contingency in controlled evolution: Migration of the population in sequence space (in the sense of Fig. 19) sets the stage for mutation events.

4 Notes

In our notes we focus on simulation programs. The concept of *inverse folding* as used by the program `RNAinverse`, which has been described elsewhere, searches for sequences folding into a predefined structure. However, these programs do not take any phylogenetic relationship into account. The next Subheading 4.1 deals with sequence evolution models under constraints along phylogenetic trees (Fig. 21), including notes for both user and developer. In Subheading 4.2 we introduce a physical setup, the flow reactor, which is equally well suited for computer simulation and experimental implementation.

4.1 Phylogenetic Simulation Methods Under Constraints

Generating synthetic data is a significant task. Simulated data have to be generated with the same underlying parameters and statistics as the real data to which the tool will eventually be applied. Parameter estimation is a topic in its own right and we refer to other sources [174] as well as to Chapter 3. Given a sequence evolution model there are at least two ways of simulating the substitution of nucleotides along phylogenetic trees first, employing the matrix of substitution probabilities for any time interval second, using a rate matrix for an infinitesimally short time interval. The first approach requires the transition probability matrix, which is, for example, calculated by numerical computation of the eigenvalues and eigenvectors of the rate matrix **Q** (*see* Subheadings 2.1.1 and 2.3.4). If the number of substitution is large, the first probability matrix approach is faster since its computing time is independent of the number of substitutions. The second rate matrix approach, however, provides a way of simulating sequences under more complex models.

So far, different programs have been designed to simulate nucleotide sequences and protein sequences along a tree [175–182]. One of the most commonly used programs `Seq-Gen` [176] has implemented a wide range of independent nucleotide substitution models. The PHASE package [141] has implemented base-paired substitution models. It is, however, specifically designed for RNA sequences with secondary structure without taking any energy parameters into account. From a state-of-the-art perspective, the most important objective is to be flexible in terms of simplicity and complexity. We have to allow both general well-known structural constraints such as the mfe secondary structure and other constraints, e.g., from specific families, motifs or other RNA or Protein interactions. A method that focuses on a unifying framework for simulating sequence evolution with arbitrary complexity was therefore developed and implemented in the program `SISSI` (Simulating Site-Specific Interactions) [62]. Beside the model and the tree, the input file is a general neighborhood file for a user-defined neighborhood system or a `ct`-file. The neighborhood system can be transformed into another known structure file such as a grammar or motifs file. The framework also allows us to define a different substitution matrix for each site. Several other sequence simulators including indels exit such as `DAWG` [183] and `INDELible` [184]. However, none of these programs takes site-specific interactions into account. An algorithm for a maximum likelihood simulation program including an indel process and site-specific interactions is described in [148]. On the level of population genetics a simulation program for RNA macroevolution was developed [185].

By way of a last general note, the random generator should be chosen carefully. Furthermore, if a large number of simulations is run in fast succession, it is highly recommended to improve the resolution of the random number generator's automatic seeding by adding some milliseconds to it. Alternatively, most programs offer the option of specifying a seed for the random number generator. This is important for allowing the replication of results, e.g., while testing and debugging, or for repeating the simulation.

4.2 The Flow Reactor and Its Applications

Models based on differential equations yield solutions also in case of marginal stability. A famous example are the well-known oscillations of the Lotka–Volterra system [186, pp. 79–118]. Stochastic models commonly require a precise physical setup and a well-defined environment in order to yield stable and meaningful solutions. The flow reactor provides a defined and experimentally controllable environment for deterministic kinetics but it is also a suitable simulation tool for stochastic approaches to chemical reactions based on the *chemical master equation* [124, 187, 188]. Straightforward simulations are limited by the maximal population sizes ($< 10^6$) that can be handled in actual computations. These populations are too small for modeling chemical reactions and special techniques were developed that allow for separation of deterministic and stochastic components [189, 190]. For many biological applications, however, the tractable population sizes are sufficient.

The sketch presented in Fig. 26 shows an experimental setup that is suitable for analysis and computer simulation for various experimental implementations aiming at studies of evolution in the laboratory *see*, e.g., [191]. Two assumptions are commonly made: (i) the flow reactor is at thermal equilibrium with a controllable heat bath and (ii) the contents of the reactor is well stirred in order to guarantee spatial homogeneity. Non-equilibrium conditions are created by a flux of rate r that regulates influx and provides the source for the material consumed in the reactor and an outflux of reactor content to compensate for the change in volume. Thermodynamic equilibrium can be studied in the limit $(r \to 0, t \to \infty)$. The reactor in the sketch (Fig. 26) illustrates the optimization of RNA molecules through mutation and selection as described in Subheading 2.3.6 as well as Eq. 14. Further details can be found in the literature [102, 121, 122] and [109, pp. 9–17].

5 Prospects of Evolution

The present increase of molecular knowledge in the life sciences ranging from biopolymers to whole organisms, populations, and ecosystems is phenomenal. The advances in technology allow nowadays for harvesting data in large quantities that were unaccessible 20 years ago. Whole genome sequences as well as protein interaction maps for whole cells are now readily available. Still there is a long way to go from our present day data to a full understanding of cellular life. In particular, characterization of biomolecular structures and analysis of biochemical functions are indispensable for the bottom-up approach in the sense of systems biology. The impressive work on the mini-bacterium *Mycoplasma pneumoniae* [115–117] has demonstrated the need for biochemical analysis very clearly. The wealth of new data requires also a novel kind of theoretical biology [192], which sets the stage for modeling and computer simulation. As far as structure and phylogeny is concerned the new developments suggest to extend the initially presented paradigm of structural biology (Fig. 2) by introduction of the evolutionary aspect of structure and function being the target of selection, and the role of phylogeny that can be visualized as a coarse-grained mapping of evolution onto sequence space.

Phylogeny of RNA is particularly well suited for modeling evolution, because it comprises fairly simple systems like evolution in vitro of RNA molecules and allows for straightforward stepwise progression in complexity: molecules → viroids → RNA viruses

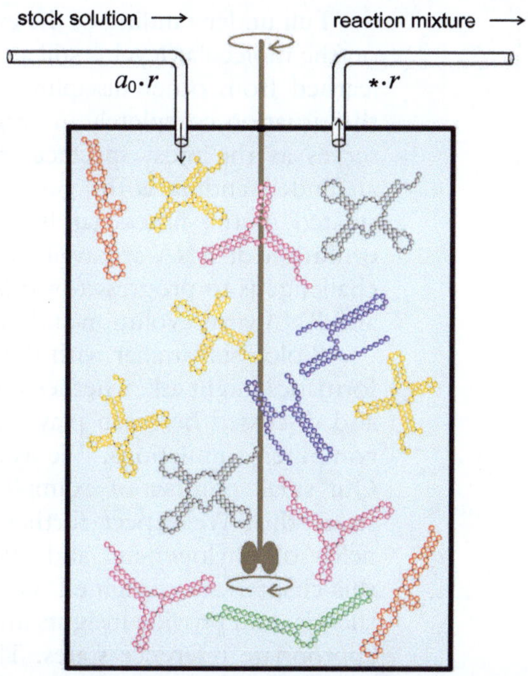

Fig. 26 *The flow reactor as a device for simulating deterministic and stochastic kinetics.* A stock solution containing all materials for RNA replication ($[\mathrm{A}] = a_0$) including an RNA polymerase flows continuously at a flow rate r into a well-stirred tank reactor (CSTR) and an equal volume containing a fraction of the reaction mixture ($[\star] = \{a, b, c_i\}$) leaves the reactor (for different experimental setups see Watts [191]). The population of RNA molecules in the reactor ($S_1, S_2, \ldots, S_n$ present in the numbers $N_1, N_2, \ldots, N_n$ with $N = \sum_{i=1}^{n} N_i$) fluctuates around a mean value, $N \pm \sqrt{N}$. RNA molecules replicate and mutate in the reactor, and the fastest replicators are selected. The RNA flow reactor has been used also as an appropriate model for computer simulations [102, 109, 121, 122]. There, other criteria for selection than fast replication can be applied. For example, fitness functions are defined that measure the distance to a predefined target structure and mean fitness increases during the approach towards the target [118]

and retroviruses. Sufficient data for comprehensive models of viroid or RNA virus life cycles are not yet available but there are no real technical obstacles for harvesting them and we can expect a lot of progress in the near future. Most RNA viruses mutate with high rates and retrieving phylogenies within and between hosts is a hot topic in clinical studies and in epidemiology. Apart from cancer viral infections are a field that is predestined for personal medicine since the spectrum of individual immunological responses to viral infections is rather broad. At present the role of RNA in the cell seems to be a never ending story adding more and more important features and regulatory tasks in genetics and epigenetics to this for a long time underestimated class of molecules.

Full understanding of the evolution of bacteria and eukaryotes on the molecular level is still a program for the future but we have learned from other disciplines that new techniques may change the situation completely in very short time—genome sequencing serves as the most spectacular example. The key towards such an understanding is the genotype–phenotype mapping as encapsulated in the fitness landscape. Fitness landscapes for in vitro evolution of RNA are available or at least accessible. The future challenge is to progress towards the more complex cases of viroid and RNA virus evolution and even further to free living organisms.

Biologists familiar with the quite sophisticated tools in bioinformatics might ask, whether simple models like the ones presented and discussed here can play a future role in the era of extensive computer simulations. We think, the answer is definitely yes. Our small number of examples have been sufficient to demonstrate this. We expect further progress to be made, when the fields of phylogenetic and evolutionary dynamics described in this chapter are combined. Only sufficiently simple concepts and theories can provide insights into complex systems and they define appropriate reference states. The complexity of the real world is then introduced similarly as in physics by means of intellectually comprehensible perturbations of the idealized cases. Admittedly, simple reference models are very often not yet known, complex networks may serve as a familiar example, and their development is an important future task for theorists in biology. Biology and chemistry are currently merging and there is legitimate hope that the common strategies of physicists and chemists will become more popular in biology.

Acknowledgement

The authors wish to express their gratitude to Carolin Kosiol for helpful discussions. T.G. is funded by a mobility fellowship of the Austrian genome research program GEN-AU and the GEN-AU project "Bioinformatics Integration Network III."

References

1. Dobzhansky T (1973) Nothing make sense except in the light of evolution. Am Biol Teach 35:125–129
2. Griffiths PE (2009) In what sense does 'nothing make sense except in the light of evolution'? Acta Biotheoretica 57:11–32
3. Wright S (1932) The roles of mutation, inbreeding, crossbreeding and selection in evolution. In: Jones DF (ed) Proceedings of the sixth international congress on genetics, vol 1. Brooklyn Botanic Garden, Ithaca, pp 356–366
4. Darwin C (1859) The origin of species, Murray edn. John Murray, London
5. Semple C, Steel MA (2003) Phylogenetics. Oxford lectures series in mathematics and its applications. Oxford University Press, Oxford
6. Goloboff PA, Catalano SA, Marcos Mirande J, Szumik CA, Salvador Arias J, Källersjö M, Farris JS (2009) Phylogenetic analysis of 73,060 taxa corroborates major eukayrotic groups. Cladistics 25:211–230
7. Ford Doolittle W (2000) Uprooting the tree of life. Sci Am 262(2):90–95

8. Woese C (1998) The universal ancestor. Proc Natl Acad Sci USA 95:6854–6859
9. Boto L (2010) Horizontal gene transfer in evolution: facts and challenges. Proc Roy Soc B 277:819–827
10. Ford Doolittle W (2010) The attempt on the life of the Tree of Life: science, philosophy, and politics. Biol Philos 25:455–473
11. Ford Doolittle W, Bapteste E (2007) Pattern pluralism and the Tree of Life hypothesis. Proc Natl Acad Sci USA 104:2043–2049
12. Bapteste E, O'Malley MA, Beiko RG, Ereshevsky M, Gogarten JP, Franklin-Hall L, Lapointe F-J, Dupré J, Dagan T, Boucher Y, Martin W (2009) Prokaryotic evolution and the tree of life are two different things. Biol Direct 4:34
13. Pace NR (2009) Mapping the tree of life: progress and prospects. Microbiol Mol Biol Rev 73:565–576
14. Sleator RD (2011) Phylogenetics. Arch Microbiol 193:235–239
15. Satta Y, Takahata N (2000) DNA archives and our nearest relative: the trichotomy problem revisited. Mol Phylogenet Evol 14(2): 259–275
16. Ford Doolittle W (1999) Phylogenetic classification and the unversal tree. Science 284:2124–2128
17. Kurland CG, Canback B, Berg OG (2003) Horizontal gene transfer: a critical view. Proc Natl Acad Sci USA 100:9658–9662
18. Rasmussen MD, Kellis M (2007) Accurate gene-tree reconstruction by learning gene- and species-specific substitution rates across multiple complete genomes. Genome Res 17:1932–1942
19. Huson DH, Rupp R, Scornavacca C (2010) Phylogenetic networks. Cambridge University Press, Cambridge
20. Felsenstein J (2004) Inferring phylogenies. Sinauer Associates, Sunderland
21. Hartl DL, Clark AG (1998) Principles of population genetics. Sinauer Associates, Sunderland
22. Hein J, Schierup MH, Wiuf C (2005) Gene genealogies, variation and evolution: a primer in coalescent theory. Oxford University Press, New York
23. Schuster P (2011) Mathematical modeling of evolution. Solved and open problems. Theor Biosci 130:71–89
24. Felsenstein J (1981) Evolutionary trees from DNA sequences: a maximum likelihood approach. J Mol Evol 17:368–376
25. Jukes TH, Cantor CR (1969) Evolution of protein molecules. In: Munro HN (ed) Mammalian protein metabolism, vol 3. Academic, New York, pp 21–123
26. Kimura M (1980) A simple method for estimating evolutionary rates of base substitutions through comparative studies of nucleotide sequences. J Mol Evol 16:111–120
27. Tavaré S (1986) Some probabilistic and statistical problems on the analysis of DNA sequences. Lec Math Life Sci 17:57–86
28. Uzzell T, Corbin KW (1971) Fitting discrete probability distributions to evolutionary events. Science 172(3988):1089–1096
29. Yang Z (1993) Maximum-likelihood estimation of phylogeny from DNA sequences when substitution rates differ over sites. J Mol Evol 42:587–596
30. Yang Z (1994) Maximum likelihood phylogenetic estimation from DNA sequences with variable rates over sites: approximative methods. J Mol Evol 39:306–314. doi:10.1007/BF00160154
31. Van de Peer Y, Neefs JM, De Rijk P, De Wachter R (1993) Reconstructing evolution from eukaryotic small-ribosomal-subunit RNA sequences: calibration of the molecular clock. J Mol Evol 37:221–232
32. Meyer S, von Haeseler A (2003) Identifying site-specific substitution rates. Mol Biol Evol 20:182–189
33. Hasegawa M, Kishino H, Yano T (1985) Dating of the human–ape splitting by a molecular clock of mitochondrial DNA. J Mol Evol 22:160–174
34. Tamura K, Nei M (1993) Estimation of the number of nucleotide substitutions in the control region of mitochondrial DNA in humans and chimpanzees. Mol Biol Evol 10:512–526
35. Rodriguez F, Oliver JL, Main A, Medina JR (1990) The general stochastic model of nucleotide substitution. J Theor Biol 142:485–501
36. Felsenstein J, Churchill GA (1996) A hidden markov model approach to variation among sites in rate of evolution. J Mol Evol 13:92–104
37. Goldman N, Yang Z (1994) A codon-based model of nucleotide substitution for protein-coding DNA sequences. Mol Biol Evol 11:725–736
38. Muse SV, Gaut BS (1994) A likelihood approach for comparing synonymous and nonsynonymous nucleotide substitution rates, with application to the chloroplast genome. Mol Biol Evol 11:715–724
39. Schöniger M, von Haeseler A (1994) A stochastic model for the evolution of autocorrelated DNA sequences. Mol Phylogenet Evol 3:240–247. doi: 10.1006/mpev.1994.1026

40. Muse SV (1995) Evolutionary analyses of DNA sequences subject to constraints on secondary structure. Genetics 139:1429–1439
41. Rzhetsky A (1995) Estimating substitution rates in ribosomal RNA genes. Genetics 141:771–783
42. Tillier ERM (1994) Maximum likelihood with multi-parameter models of substitution. J Mol Evol 39:409–417
43. Tillier ERM, Collins RA (1998) High apparent rate of simultaneous compensatory basepair substitutions in ribosomal RNA. Genetics 148:1993–2002
44. Tillier ERM, Collins RA (1995) Neighbor joining and maximum likelihood with RNA sequences: addressing the interdependence of sites. Mol Biol Evol 12:7–15
45. Savill NJ, Hoyle DC, Higgs PG (2000) RNA sequence evolution with secondary structure constraints: comparison of substitution rate models using maximum-likelihood methods. Genetics 157:399–411
46. Innan H, Stephan W (2001) Selection intensity against deleterious mutations in RNA secondary structures and rate of compensatory nucleotide substitutions. Genetics 159:389–399
47. Stephan W (1996) The rate of compensatory evolution. Genetics 144:419–426
48. Smith AD, Lui TWH, Tillier ERM (2004) Empirical models for substitution in ribosomal RNA. Mol Biol Evol 21:419–427
49. Smit S, Widmann J, Knight R (2007) Evolutionary rates vary among rRNA structural elements. Nucleic Acids Res 35:3339–3354
50. Bulmer M (1986) Neighboring base effects on substitution rates in pseudogenes. Mol Biol Evol 3:322–329
51. Morton BR (1995) Neighboring base composition and transversion/transition bias in a comparison of rice and maize chloroplast noncoding regions. Proc Natl Acad Sci USA 92:9717–9721
52. Jensen JL, Pedersen A-MK (2000) Probabilistic models of DNA sequence evolution with context dependent rates of substitution. Adv Appl Prob 32:449–467. (Pages in jstore: 499–517)
53. Christensen OF, Hobolth A, Jensen JL (2005) Pseudo-likelihood analysis of codon substitution models with neighbor-dependent rates. J Comput Biol 12:1166–1182
54. Baele G, Van de Peer Y, Vansteelandt S (2008) A model-based approach to study nearest-neighbor influences reveals complex substitution patterns in non-coding sequences. Syst Biol 57:675–692
55. Siepel A, Haussler D (2004) Phylogenetic estimation of context-dependent substitution rates by maximum likelihood. Mol Biol Evol 21:468–488
56. Bérard J, Gouéré JB, Piau D (2008) Solvable models of neighbor-dependent substitution processes. Math Biosci 211:56–88
57. Duret L, Galtier N (2000) The covariation between tpa deficiency, cpg deficiency, and g+c content of human isochores is due to a mathematical artifact. Mol Biol Evol 17(11):1620–1625
58. Arndt PF, Burge CB, Hwa T (2003) DNA sequence evolution with neighbor-dependent mutation. J Comput Biol 10:313–322
59. Robinson DM, Jones DT, Kishino H, Goldman N, Thorne JL (2003) Protein evolution with dependence among codons due to tertiary structure. Mol Biol Evol 20:1692–1704
60. Yu J, Thorne JL (2006) Dependence among sites in RNA evolution. Mol Biol Evol 23:1525–1537
61. Pedersen A-MK, Jensen JL (2001) A dependent rates model and MCMC based methodology for the maximum likelihood analysis of sequences with overlapping reading frames. Mol Biol Evol 18:763–776
62. Gesell T, von Haeseler A (2006) *In silico* sequence evolution with site-specific interactions along phylogenetic trees. Bioinformatics 22:716–722
63. Cate JH, Gooding AR, Podell E, Zhou K, Golden BL, Kundrot CE, Cech TR, Doudna JA (1996) Crystal structure of a group I ribozyme domain: principles of RNA packing. Science 273:2–33
64. Nei M, Kumar S (2000) Molecular evolution and phylogenetics. Oxford University Press, New York
65. Eigen M, Winkler-Oswatitsch R (1981) Transfer-RNA: the early adaptor. Naturwissenschaften 68:217–228
66. Kumar S (2005) Molecular clocks: four decades of evolution. Nat Rev Genet 6:654–662
67. Morgan GJ (1998) Emile Zuckerkandl, Linus Pauling and the molecular evolutionary clock. J Hist Biol 31:155–178
68. Takahata N (2007) Molecular clock: an *anti*-neo-Darwinian legacy. Genetics 176:1–6
69. Kimura M (1983) The neutral theory of molecular evolution. Cambridge Univeresity Press, Cambridge
70. Ayala FJ (1997) Vagaries of the molecular clock. Proc Natl Acad Sci USA 94:7776–7783
71. Bromham L, Penny D (2003) The modern molecular clock. Nat Rev Genet 4:216–224
72. Bromham L (2011) The genome as a life-history character: why rate of molecular evolution varies between mammal species. Proc Trans Roy Soc B 366:2503–2513

73. Ho SYW, Lanfear R, Bromham L, Phillips MJ, Soubrier J, Rodrigo AG, Cooper A (2011) Time-dependent rates of molecular evolution. Mol Ecol 20: 3087–3101
74. Roger AJ, Hug LA (2006) The origin and diversification of eukaryotes: problems with molecular phylogenies and molecular clock estimation. Proc Trans Roy Soc B 361: 1039–1054
75. Schuster P (2006) Prediction of RNA secondary structures: from theory to models and real molecules. Rep Prog Phys 69: 1419–1477
76. Gottesman S (2004) The small RNA regulators of *Escherichia coli*: roles and mechanisms. Annu Rev Microbiol 58:303–328
77. Mandal M, Boese B, Barrick JE, Winkler WC, Breaker RR (2003) Riboswitches control fundamental biochemical pathways in *Bacillus subtilis* and other bacteria. Cell 113:577–596
78. Serganov A, Patel DJ (2007) Ribozymes, riboswitches and beyond: regulation of gene expression without proteins. Nat Rev Genet 8:776–790
79. Winkler WC (2005) Metabolic monitoring by bacterial mRNAs. Arch Microbiol 183:151–159
80. Boyle PM, Silver PA (2009) Harnessing nature's toolbox: regulatory elements for synthetic biology. J Roy Soc Interface 6:S535–S546
81. Flamm C, Fontana W, Hofacker IL, Schuster P (2000) RNA folding at elementary step resolution. RNA 6:325–338
82. Wolfinger MT, Svrcek-Seiler WA, Flamm C, Hofacker IL, Stadler PF (2004) Efficient computation of RNA folding dynamics. J Phys A Math Gen 37:4731–4741
83. Mann M, Klemm K (2011) Efficient exploration of discrete energy landscapes. Phys Rev E 83:011113
84. Schuster P, Fontana W, Stadler PF, Hofacker IL (1994) From sequences to shapes and back: a case study in RNA secondary structures. Proc R Soc Lond B 255:279–284
85. Eigen M (1971) Selforganization of matter and the evolution of biological macromolecules. Naturwissenschaften 58: 465–523
86. Jones BL, Enns RH, Rangnekar SS (1976) On the theory of selection of coupled macromolecular systems. Bull Math Biol 38: 15–28
87. Thompson CJ, McBride JL (1974) On Eigen's theory of the self-organization of matter and the evolution of biological macromolecules. Math Biosci 21:127–142
88. Seneta E (1981) Non-negative matrices and markov chains, 2nd edn. Springer, New York
89. Eigen M, McCaskill J, Schuster P (1988) Molecular quasispecies. J Phys Chem 92:6881–6891
90. Swetina J, Schuster P (1982) Self-replication with errors - a model for polynucleotide replication. Biophys Chem 16:329–345
91. Eigen M, McCaskill J, Schuster P (1989) The molecular quasispecies. Adv Chem Phys 75:149–263
92. Eigen M, Schuster P (1978) The hypercycle. A principle of natural self-organization. Part B: the abstract hypercycle. Naturwissenschaften 65:7–41
93. Reader JS, Joyce GF (2002) A ribozyme composed of only two different nucleotides. Nature 420:841–844
94. Biebricher CK, Eigen M (2005) The error threshold. Virus Res 107:117–127
95. Eigen M, Schuster P (1977) The hypercycle. A principle of natural self-organization. Part A: emergence of the hypercycle. Naturwissenschaften 64:541–565
96. Domingo E, Parrish CR, Holland JJ (eds) (2008) Origin and evolution of viruses, 2nd edn. Elsevier, Academic, Amsterdam
97. Eigen M, Schuster P (1978) The hypercycle. A principle of natural self-organization. Part C: the realistic hypercycle. Naturwissenschaften 65:341–369
98. Domingo E (ed) (2005) Virus entry into error catastrophe as a new antiviral strategy. Virus Res 107(2):115–228
99. Eigen M, Schuster P (1982) Stages of emerging life - five principles of early organization. J Mol Evol 19:47–61
100. Wiehe T (1997) Model dependency of error thresholds: the role of fitness functions and contrasts between the finite and the infinite sites models. Genet Res Camb 69: 127–136
101. Fontana W, Schuster P (1998) Shaping space: the possible and the attainable in RNA genotype-phenotype mapping. J Theor Biol 194:491–515
102. Huynen MA, Stadler PF, Fontana W (1996) Smoothness within ruggedness. The role of neutrality in adaptation. Proc Natl Acad Sci USA 93:397–401
103. Schuster P, Swetina J (1988) Stationary mutant distribution and evolutionary optimization. Bull Math Biol 50:635–660
104. Maynard Smith J (1998) Evolutionary genetics, 2nd edn. Oxford University Press, Oxford
105. Biebricher CK, Eigen M, Gardiner WC Jr (1983) Kinetics of RNA replication. Biochemistry 22:2544–2559
106. Biebricher CK, Eigen M, Gardiner WC Jr (1984) Kinetics of RNA replication: plus-minus asymmetry and double-strand formation. Biochemistry 23:3186–3194

107. Biebricher CK, Eigen M, Gardiner WC Jr (1985) Kinetics of RNA replication: competition and selection among self-replicating RNA species. Biochemistry 24:6550–6560
108. Weissmann C (1974) The making of a phage. FEBS Lett 40:S10–S18
109. Phillipson PE, Schuster P (2009) Modeling by nonlinear differential equations. Dissipative and conservative processes. World scientific series on nonlinear science A, vol 69. World Scientific, Singapore
110. Biebricher CK (1983) Darwinian selection of self-replicating RNA molecules. In: Hecht MK, Wallace B, Prance GT (eds) Evolutionary biology, vol 16. Plenum Press, New York, pp 1–52
111. Biebricher CK, Gardiner WC Jr (1997) Molecular evolution of RNA *in vitro*. Biophys Chem 66:179–192
112. Spiegelman S (1971) An approach to the experimental analysis of precellular evolution. Quart Rev Biophys 4:213–253
113. Takeuchi N, Hogeweg P (2007) Error-thresholds exist in fitness landscapes with lethal mutants. BMC Evol Bio 7:15
114. Tejero H, Marín A, Moran F (2010) Effect of lethality on the extinction and on the error threshold of quasispecies. J Theor Biol 262:733–741
115. Güell M, van Noort V, Yus E, Chen W-H, Leigh-Bell J, Michalodimitrakis K, Yamada T, Arumugam M, Doerks T, Kühner S, Rode M, Suyama M, Schmidt S, Gavin A-C, Bork P, Serrano L (2009) Transcriptome complexity in a genome-reduced bacterium. Science 326:1268–1271
116. Kühner S, van Noort V, Betts MJ, Leo-Macias A, Batisse C, Rode M, Yamada T, Maier T, Bader S, Beltran-Alvarez P, Castaño-Diez D, Chen W-H, Devos D, Güell M, Norambuena T, Racke I, Rybin V, Schmidt A, Yus E, Aebersold R, Herrmann R, Böttcher B, Frangakis AS, Russell RB, Serrano L, Bork P, Gavin A-C (2009) Proteome organization in a genome-reduced bacterium. Science 326:1235–1240
117. Yus E, Maier T, Michalodimitrakis K, van Noort V, Yamada T, Chen W-H, Wodke JAH, Güell M, Martínez S, Bourgeois R, Kühner S, Raineri E, Letunic I, Kalinina OV, Rode M, Herrmann R, Gutiérez-Gallego R, Russell RB, Gavin A-C, Bork P, Serrano L (2009) Impact of genome reduction on bacterial metabolism and its regulation. Science 326:1263–1268
118. Fontana W, Schuster P (1998) Continuity in evolution. On the nature of transitions. Science 280:1451–1455
119. Peliti L, Derrida B (1991) Evolution in a flat fitness landscape. Bull Math Biol 53:355–382
120. Saakian DB, Biebricher CK, Hu C-K (2009) Phase diagram for the Eigen quasispecies theory with a trancated fitness landscape. Phys Rev E 79:041905
121. Fontana W, Schnabl W, Schuster P (1989) Physical aspects of evolutionary optimization and adaptation. Phys Rev A 40:3301–3321
122. Fontana W, Schuster P (1987) A computer model of evolutionary optimization. Biophys Chem 26:123–147
123. Gillespie DT (1976) A general method for numerically simulating the stochastic time evolution of coupled chemical reactions. J Comp Phys 22:403–434
124. Gillespie DT (2007) Stochastic simulation of chemical kinetics. Annu Rev Phys Chem 58:35–55
125. Schuster P (2003) Molecular insight into the evolution of phenotypes. In: Crutchfield JP, Schuster P (eds) Evolutionary dynamics – exploring the interplay of accident, selection, neutrality, and function. Oxford University Press, New York, pp 163–215
126. Biebicher CK, Luce R (1992) *In vitro* recombination and terminal elongation of RNA by Qβ-replicase. EMBO J 11:5129–5135
127. Fels A, Hu K, Riesner D (2001) Transcription of potato spindle tuber viroid by RNA polymerase II starts predominantly at two specific sites. Nucleic Acids Res 29:4589–4597
128. Ding B, Itaya A (2007) Viroid: a useful model for studying the basic principles of infection and RNA biology. Mol Plant Microbe Interact 20:7–20
129. Zhong X, Archual AJ, Amin AA, Ding B (2008) A genomic map of viroid RNA motifs critical for replication and systemic trafficking. Plant Cell 20:35–47
130. Zhong X, Leontis N, Qian S, Itaya A, Qi Y, Boris-Lawrie K, Ding B (2006) Tertiary structural and functional analyses of a viroid RNA motif by isostericity matrix and mutagenesis reveal its essential role in replication. J Virol 80:8566–8581
131. Delan-Forino C, Maurel M-C, Torchet C (2011) Replication of *avocado sunblotch viroid* in the yeast *Saccaromyces cerevisiae*. J Virol 85:3229–3238
132. Orth JD, Thiele I, Palsson BØ (2010) What is flux balance analysis? Nat Biotechnol 28:245–248
133. Edwards JS, Ibarra RU, Palsson BØ (2001) *In silico* predictions of *Escherichia coli* metabolic capabilities are consistent with experimental data. Nat Biotechnol 19:125–130
134. Costanzo M, Baryshnikova A, Bellay J, Kim Y, Spear ED, Sevier CS, Ding H, Koh JLY, Toufighi K, Mostafavi S, Prinz J, Onge RPSt, Van der Sluis B, Makhnevych T, Vizeacoumar

FJ, Alizadeh S, Bahr S, Brost RL, Chen Y, Cokol M, Deshpande R, Li Z, Li Z-Y, Liang W, Marback M, Paw J, San Luis B-J, Shuteriqi E, Tong AHY, van Dyk N, Wallace IM, Whitney JA, Weirauch MT, Zhong G, Zhu H, Houry WA, Brudno M, Ragibizadeh S, Papp B, Pál C, Roth FP, Giaver G, Nislow C, Troyanskaya OG, Bussey H, Bader GD, Gingras A-C, Morris QD, Kim PM, Kaiser CA, Myers CL, Andrews BJ, Boone C (2010) The genetic landscape of a cell. Science 317: 425–431
135. Fitch WM (1971) Toward defining the course of evolution: minimum change for a specific tree topology. Syst Zool 20:406–416
136. Rannala B, Yang Z (1996) Probability distribution of molecular evolutionary trees: a new method of phylogenetic inference. J Mol Evol 43:304–311
137. Saitou N, Nei M (1987) The neighbor-joining method: a new method for reconstructing phylogenetic trees. Mol Biol Evol 4:406–425
138. Aris-Brosou S, Excoffier L (1996) The impact of population expansion and mutation rate heterogeneity on DNA sequence polymorphism. Mol Biol Evol 13:494–504
139. Yang Z (1996) Maximum-likelihood models for combined analyses of multiple sequences data. J Mol Evol 42:587–596
140. Schöniger M, von Haeseler A (1995) Performance of the maximum likelihood, neighbor joining, and maximum parsimony methods when sequence site are not independent. Syst Biol 44:533–547
141. Hudelot C, Gowri-Shankar H, Rattray M, Higgs P (2003) RNA-based phylogenetic methods: application to mammalian mitochondrial RNA sequences. Mol Phylogenet Evol 28:241–252
142. Ronquist F, Huelsenbeck JP (2003) Mrbayes 3: Bayesian phylogenetic inference under mixed models. Bioinformatics 19(12):1572–1574. doi:10.1093/bioinformatics/btg180. URL http://bioinformatics.oxfordjournals.org/content/19/12/1572.abstract
143. Stamatakis A (2006) RAxML-VI-HPC: maximum likelihood-based phylogenetic analyses with thousands of taxa and mixed models. Bioinformatics 22:2688–2690
144. Keller A, Forster F, Muller T, Dandekar T, Schultz J, Wolf M (2010) Including RNA secondary structures improves accuracy and robustness in reconstruction of phylogenetic trees. Biol Direct 5:4
145. Letsch HO, Kck P, Stocsits RR, Misof B (2010) The impact of rrna secondary structure consideration in alignment and tree reconstruction: simulated data and a case study on the phylogeny of hexapods. Mol Biol Evol 27(11):2507–2521. doi:10.1093/molbev/msq140. URL http://mbe.oxfordjournals.org/content/27/11/2507.abstract
146. Caetano-Anolles G (2002) Tracing the evolution of RNA structure in ribosomes. Nucleic Acids Res 30:2575–2587
147. Thi Nguyen MA, Gesell T, von Haeseler A (2012) Imosm: intermittent evolution and robustness of phylogenetic methods. Mol Biol Evol 29:663–673. doi:10.1093/molbev/msr220. URL http://mbe.oxfordjournals.org/content/early/2011/09/22/molbev.msr220.abstract
148. Gesell T (2009) A phylogenetic definition of structure. PhD thesis, University of Vienna
149. Knudsen B, Hein J (2003) Pfold: RNA secondary structure prediction using stochastic context-free grammars. Nucleic Acids Res 31(13):3423–3428
150. Pedersen JS, Bejerano G, Siepel A, Rosenbloom K, Lindblad-Toh K, Lander ES, Kent J, Miller W, Haussler D (2006) Identification and classification of conserved rna secondary structures in the human genome. PLoS Comput Biol 2(4):e33. doi:10.1371/journal.pcbi.0020033. URL http://dx.plos.org/10.1371/journal.pcbi.0020033
151. Rivas E, Eddy SR (2001) Noncoding RNA gene detection using comparative sequence analysis. BMC Bioinformatics 2:8
152. Altschul SF, Erickson BW (1985) Significance of nucleotide sequence alignments: a method for random sequence permutation that preserves dinucleotide and codon usage. Mol Biol Evol 2(6):526–538
153. Clote P (2005) An efficient algorithm to compute the landscape of locally optimal RNA secondary structures with respect to the Nussinov-Jacobson energy model. J Comp Biol 12:83–101
154. Gesell T, Washietl S (2008) Dinucleotide controlled null models for comparative rna gene prediction. BMC Bioinformatics 9:248–264
155. Hofacker IL, Fekete M, Stadler PF (2002) Secondary structure prediction for aligned RNA sequences. J Mol Biol 319(5): 1059–1066. doi:10.1016/S0022-2836(02)00308-X
156. Joyce GF (2007) Forty years of *in vitro* evolution. Angew Chem Int Ed 46:6420–6436
157. Ellington AD, Szostak JW (1990) *In vitro* selection of RNA molecules that bind specific ligands. Nature 346:818–822
158. Tuerk C, Gold L (1990) Systematic evolution of ligands by exponential enrichment: RNA ligands to bacteriophage T4 DNA polymerase. Science 249:505–510

159. Klussmann S (ed) (2006) The aptamer handbook. Functional oligonucleotides and their applications. Wiley-VCH, Weinheim
160. Joyce GF (2004) Directed evolution of nucleic acid enzymes. Annu Rev Biochem 73:791–836
161. Brakmann S, Johnsson K (eds) (2002) Directed molecular evolution of proteins or how to improve enzymes for biocatalysis. Wiley-VCH, Weinheim
162. Jäckel C, Kast P, Hilvert D (2008) Protein design by directed evolution. Annu Rev Biophys 37:153–173
163. Wrenn SJ, Harbury PB (2007) Chemical evolution as a tool for molecular discovery. Annu Rev Biochem 76:331–349
164. Eigen M (2002) Error catastrophe and antiviral strategy. Proc Natl Acad Sci USA 99:13374–13376
165. Bull JJ, Ancel Myers L, Lachmann M (2005) Quasispecies made simple. PLoS Comput Biol 1:450–460
166. Bull JJ, Sanjuán R, Wilke CO (2007) Theory for lethal mutagenesis for viruses. J Virol 81:2930–2939
167. Summers J, Litwin S (2006) Examining the theory of error catastrophe. J Virol 80: 20–26
168. Wagner GP, Krall P (1993) What is the difference between models of error thresholds and Muller's ratchet. J Math Biol 32:33–44
169. Lenski RE, Rose MR, Simpson SC, Tadler SC (1991) Long-term experimental evolution in *Escherichia coli*. I. Adaptation and divergence during 2,000 generations. Am Nat 38:1315–1341
170. Lenski RE, Travisano M (1994) Dynamics of adaptation and diversification: a 10,000-generation experiment with bacterial populations. Proc Natl Acad Sci USA 91:6808–6814
171. Elena SF, Cooper VS, Lenski RE (1996) Punctiated evolution caused by selection of rare beneficial mutants. Science 272:1802–1804
172. Papadopoulos D, Schneider D, Meies-Eiss J, Arber W, Lenski RE, Blot M (1999) Genomic evolution during a 10,000-generation experiment with bacteria. Proc Natl Acad Sci USA 96:3807–3812
173. Blount ZD, Christina Z, Lenski RE (2008) Historical contingency an the evolution of a key innovation in an experimental population of *Escherichia coli*. Proc Natl Acad Sci USA 105:7898–7906
174. Steel M (2005) Should phylogenetic models be trying to "fit an elephant"? Trends Genet 21:307–309
175. Schöniger M, von Haeseler A (1995) Simulating efficiently the evolution of DNA sequences. Comput Appl Biosci 11:111–115
176. Rambaut A, Grassly NC (1997) Seq-Gen: an application for the Monte Carlo simulation of DNA sequence evolution along phylogenetic trees. Comput Appl Biosci 13: 235–238
177. Grassly NC, Adachi J, Rambaut A (1997) PSeq-Gen: an application for the Monte Carlo simulation of protein sequence evolution along phylogenetic trees. Comput Appl Biosci 13:559–560
178. Yang Z (1997) PAML: a program package for phylogenetic analysis by maximum likelihood. Comput Appl Biosci 13:555–556
179. Stoye J, Evers D, Meyer F (1998) Rose: generating sequence families. Bioinformatics 14:157–163
180. Nicholas JS, Hoyle DC, Higgs PG (2000) RNA sequence evolution with secondary structure constraints: comparison of substitution rate models using maximum-likelihood methods. Genetics 157: 399–411
181. Tufféry P (2002) CS-PSeq-Gen: simulating the evolution of protein sequence under constraints. Bioinformatics 18:1015–1016
182. Pond SLK, Frost SDW, Muse S (2005) HyPhy: hypothesis testing using phylogenies. Bioinformatics 21:676–679
183. Cartwright RA (2005) DNA assembly with gaps (Dawg): simulating sequence evolution. Bioinformatics 21(Suppl 3):i31–38
184. Fletcher W, Yang Z (2009) INDELible: a flexible simulator of biological sequence evolution. Mol Biol Evol 26:1879–1888
185. Guo S, Kim J (2009) Large-scale simulating of RNA macroevolution by an energy-dependent fitness model (Preprint)
186. Murray JD (2002) Mathematical biology I: an introduction, 3rd edn. Springer, New York
187. Gardiner CW (2009) Stochastic methods. A handbook for the natural and social sciences. Springer series in synergetics, 4th edn. Springer, Berlin
188. Gillespie DT (1992) A rigorous derivation of the chemical master equation. Phys A 188:404–425
189. van Kampen NG (1961) A power series expansion of the master equation. Can Chem Phys 39:551–567
190. van Kampen NG (1976) The expansion of the master equation. Adv Chem Phys 34:245–309
191. Watts A, Schwarz G (eds) (1997) Evolutionary biotechnology – from theory to experiment. Biophysical chemistry, vol 66/2–3. Elesvier, Amsterdam, pp 67–284
192. Brenner S (1999) Theoretical biology in the third millenium. Philos T Roy Soc Lond B 354:1963–1965

Chapter 17

The Art of Editing RNA Structural Alignments

Ebbe Sloth Andersen

Abstract

Manual editing of RNA structural alignments may be considered more art than science, since it still requires an expert biologist to take multiple levels of information into account and be slightly creative when constructing high-quality alignments. Even though the task is rather tedious, it is rewarded by great insight into the evolution of structure and function of your favorite RNA molecule. In this chapter I will review the methods and considerations that go into constructing RNA structural alignments at the secondary and tertiary structure level; introduce software, databases, and algorithms that have proven useful in semiautomating the work process; and suggest future directions towards full automatization.

Key words RNA structural alignment, Comparative analysis, Secondary structure, Tertiary structure, Software, Databases, Algorithms

1 Introduction

The strength of comparative analysis at the secondary structure level was first discovered for tRNA and later applied systematically to elucidate the structure of the ribosomal RNA (reviewed by [1]). With the introduction of biochemical methods to probe RNA structure, the secondary and tertiary structure of several RNA molecules began to emerge [2]. A data revolution came with the determination of the full atomic resolution structures of the ribosome by X-ray crystallography [3, 4] that both reconciled all the previous biochemical data and provided many new insights into the complexity and design principles of the ribosome and the greater understanding and appreciation of RNA as a structural and multifunctional molecule.

Databases for the major functional RNA species (Table 1) have been created and curated as sequence data has become available, e.g., for the ribosome [5], ribonuclease P [6], transfer-messenger RNA [7, 8], and the signal recognition particle [9]. Projects like the Rfam database have taken up the task of providing RNA structural alignments for all known RNA structures and automating

Jan Gorodkin and Walter L. Ruzzo (eds.), *RNA Sequence, Structure, and Function: Computational and Bioinformatic Methods*, Methods in Molecular Biology, vol. 1097, DOI 10.1007/978-1-62703-709-9_17,

Table 1
Databases for RNA structural alignment

Name	URL
Rfam	http://rfam.sanger.ac.uk
RNP database	http://rnp.uthscsa.edu/
Ribosomal database project (RDP)	http://rdp.cme.msu.edu/
Comparative RNA website (CRW)	http://www.rna.ccbb.utexas.edu
The tmRNA website	http://www.indiana.edu/~tmrna/
The Rnase P database	http://jwbrown.mbio.ncsu.edu/RNaseP/
Viral RNA structure database	http://rna.tbi.univie.ac.at/cgi-bin/virusdb.cgi

Table 2
Editors for RNA structural alignment

Name	URL
4SALE	http://4sale.bioapps.biozentrum.uni-wuerzburg.de/
ARB	http://www.arb-home.de/
BioEdit	http://www.mbio.ncsu.edu/bioedit/bioedit.html
Boulder ALE	http://boulderale.sourceforge.net/
Construct	http://www.biophys.uni-duesseldorf.de/construct3/
JalView	http://jalview-rnasupport.blogspot.com/
RALEE	http://personalpages.manchester.ac.uk/staff/sam.griffiths-jones/software/ralee/
Ribostral	http://rna.bgsu.edu/ribostral
S2S	http://www.bioinformatics.org/S2S/
SARSE	http://sarse.ku.dk/

the update of alignments as new sequences become available [10]. With the availability of increasing amounts of 3D structural data, the challenge has been to represent these data as part of the 2D structural alignment maps. With the classification of non-Watson-Crick base pairs that are geometrically similar (isosteric) [11], it has now become possible to identify and annotate isosteric base changes in the alignments. Initiatives have been taken to make a new data format that can contain several levels of information [12] and to collect the knowledge of how to construct RNA structure alignments in an ontology [13].

An easily overlooked workhorse for the construction of RNA structural alignment databases has been dedicated software for sequence-structure editing (Table 2). Several editors have been developed for this task, e.g., DCSE [14], SEQPUP, BioEdit, GDE [15], Jalview [16], Construct [17], ARB [18], RALEE [19], 4SALE [20], SARSE [21], S2S [22], and Boulder ALE [23]. These structural alignment editors have several features in common

Table 3
Algorithms for RNA structural alignment

Name	URL
Pfold	http://www.daimi.au.dk/~compbio/pfold/
RNAalifold	http://rna.tbi.univie.ac.at/cgi-bin/RNAalifold.cgi
Pcluster	http://sarse.kvl.dk/
FoldalignM	http://foldalign.ku.dk/
StrAl	http://www.biophys.uni-duesseldorf.de/stral/
WAR	http://genome.ku.dk/resources/war/
Dynalign	http://rna.urmc.rochester.edu/dynalign.html
Stemloc	http://biowiki.org/StemLoc
QRNA	http://selab.janelia.org/software.html
Consan	http://selab.janelia.org/software/consan/

including annotation of stem regions and base changes, dot plots, secondary structure drawings, and so on. On other aspects they differ: 4SALE and SARSE allow integration of algorithms for extended analysis; S2S and Boulder ALE are currently the only editors that support the annotation of isosteric and non-Watson-Crick base pairs. The complexity of the data and visualization of features on large alignments often make it necessary to identify patterns by human eyeballing. By inspecting a multiple alignment, the expert can make corrections based on experimental data with close consideration of structure and function. Semiautomated updates of Rfam have been done using SARSE and recently Boulder ALE has been used to update a set of alignments using 3D structural data and consideration of noncanonical isosteric base pairs [24].

Algorithms for RNA secondary structure prediction are useful tools for facilitating the construction of RNA structural alignments (Table 3), and some of these have been collected in packages like the RNAdbtools package [25] and the Vienna package [26]. The dream would be to be able to predict the structure of two sequences separately, observe the common structure, and align based on this. Unfortunately, the low information content of a single sequence does not make this possible and thus this approach often fails. In cases where many closely related sequences exist, programs like Pfold and RNAalifold [27–29] can be employed to predict the secondary structure based both on folding rules and on the phylogenetic information available in the alignment. These methods however depend on simple sequence alignment software and start to fail for pairwise distances above 30% and fail miserably above 60% [30]. Methods for simultaneous folding and aligning sequences have been developed, e.g., FOLDALIGN [31], PMcomp [32], Dynalign [33, 34], and Stemloc [35, 36].

However, these methods are very calculation expensive and thus applicable to only relatively small sequence regions. More efficient heuristic methods have been introduced: RNAcast [37], CMfinder [38], and FOLDALIGNM [39]. For a comparison of the various approaches, see Torarinsson et al. [39].

Since most of the genomic sequences are transcribed and a plethora of noncoding RNAs have been discovered, the RNA structural alignment task has in principle been extended to include the whole genome. This is also true since the role of noncoding RNA elements is often only understood fully in the genomic context of coding regions, protein binding sites, splice sites, and so on. Complex examples of this are observed for RNA viruses that are packed with RNA structural elements, overlapping protein binding sites, and coding regions. The ability to make detailed RNA structural alignments has and will continue to be important for molecular biology in several different ways: (1) The identification of novel RNA structures will inform experimental work. (2) It will continue to improve classification and taxonomy for a more detailed view of evolutionary history. (3) Structural analysis across broad phylogenies will help in understanding the mechanisms that drive diversity and evolution, e.g., insertion, deletion, and recombination. (4) The observation of the allowed structural changes for preserving molecular structure and function provides us with design principles that can be used in attempts to design novel molecules for application in bionanotechnology and synthetic biology.

2 Basic Principles

2.1 Editing Procedures

The editing of sequence-structure alignments refers to the manual process of moving base symbols around horizontally to correct their position in the alignment. This manual process is done by selecting blocks of base symbols and using the mouse or keyboard to drag the symbols horizontally in the alignment where gap characters allow it. There are two ways to start a structural alignment: (1) The known structure of the reference sequence is used as the initial pairing mask and the other sequences are moved until they fit this pairing mask. (2) To start a structural alignment from scratch, a prediction is made for one or more sequences and the common pairing mask is used. Coloring of the sequences in relation to the pairing mask facilitates this process, e.g., red to show that base pairs does not fit and green to show that base pairs fit.

2.2 Structure Representations

When editing structural alignments, several visual aids are useful. The main abstraction levels are defined by the primary, secondary, tertiary, and quaternary structure of molecules or their complexes. The sequence alignment can be annotated with various

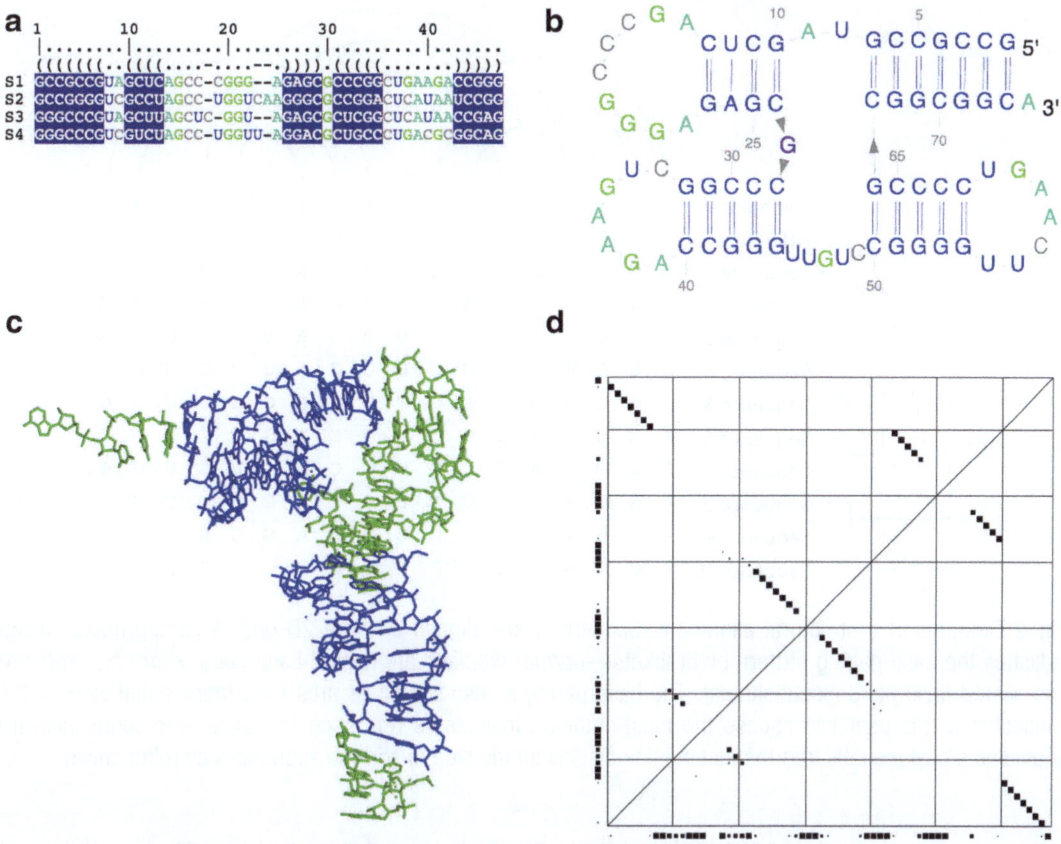

Fig. 1 Representations of RNA structure at primary, secondary, and tertiary level. (**a**) Partial structural alignment of tRNA sequences. (**b**) Secondary structure of tRNA with only nested base pairs shown. (**c**) Atomic model of the tRNA molecules with secondary structure regions indicated in *blue*. *Green regions* contain non-nested base pairs

information, e.g., color coding to highlight the base paired and unpaired regions (Fig. 1a). Secondary structure drawings are highly useful for a more direct viewing of the helix and loop regions (Fig. 1b). The structural alignment and secondary structure are representations of the 3D atomic model that can either be based on biophysical data or inferred from biochemical or phylogenetic data (Fig. 1c). Several additional plots are useful for displaying the base pair connectivity along the backbone, e.g., dot plots (Fig. 1d) and circular arch diagrams.

2.3 The Multidimensional Alignment

The annotated sequence alignment is the most efficient way to display all the data related to the structure and phylogeny of an RNA molecule. In Fig. 2 several of the common features of a structural alignment are shown. The sequences are listed as rows and are often sorted according to their phylogeny (indicated here by a phylogenetic tree). The sequence for which most information

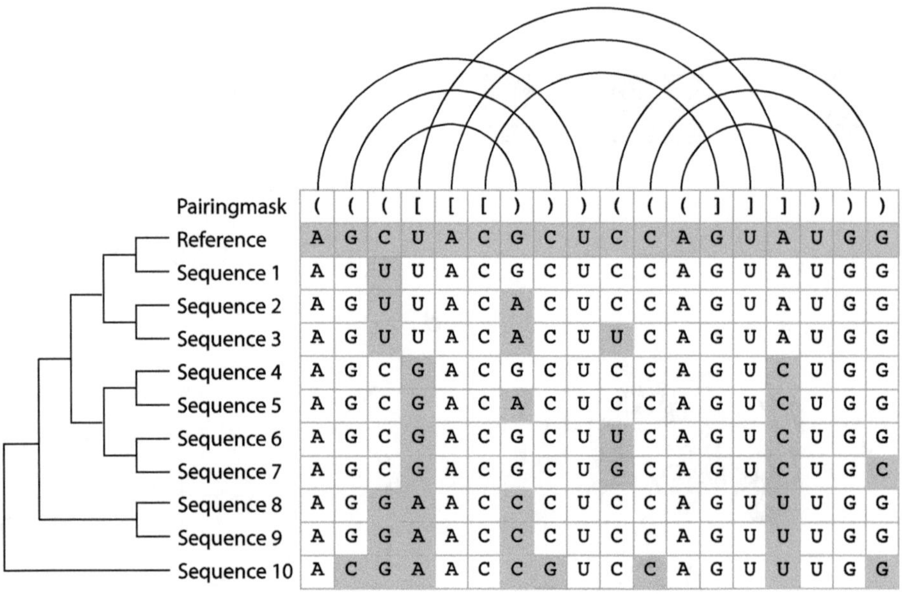

Fig. 2 Elements of a structural alignment. Sequences are aligned up on a 2D grid. A pairing mask on top indicates the base pairing pattern by brackets—*normal brackets* are nested base pairs, *sharp brackets* are non-nested base pairs (pseudoknots). The base paring is also shown as arcs for a more visual view of the connections. It is useful to choose the most studied molecule as reference sequence. The sequences are related as a phylogenetic tree that is useful to display on the side to sort the sequences by relatedness

is available is often chosen as the reference sequence. The common structure of the aligned sequences is indicated by a pairing mask that in its simplest form contains a set of nested brackets (and) that define the base pairing patterns between columns. Sharp brackets are used to indicate non-nested base pairs [and] also called pseudoknots. In S2S < and > are used to indicate non-Watson-Crick base pairs. The pairing mask can also be based on alphanumerical symbols that help in annotating stem regions and pseudoknots.

2.4 Types of Base Pair Changes

When bases change within an RNA structure, they have different meaning (Table 4). When a base pair changes from G-C to G-U, it is called a conservatory base change since the G-U wobble base pair is accepted in the RNA helix. When a base pair changes from G-C to A-U, it is called a coordinated base change since both positions change to preserve the base pair. If the base pairs changes from G-C to A-C, it is called a mismatch since it disrupts the standard Watson-Crick base pair. In some cases base changes that look like a mismatch might be the result of changes between isosteric non-Watson-Crick base pairs (Table 5). These base changes can be isosteric (conserve geometry), be non-isosteric but geometrically allowed, or be disallowed/not observed [40]. Other changes that can happen are insertion or deletion of one or more nucleotides.

Table 4
Types of base pairs

Type	Short	Symbol
Cis Watson-Crick/Watson-Crick	cWW	─●─
Trans Watson-Crick/Watson-Crick	tWW	─○─
Cis Watson-Crick/Hoogsteen	cWH	●─■
Trans Watson-Crick/Hoogsteen	tWH	○─□
Cis Watson-Crick/Sugar Edge	cWS	●─▶
Trans Watson-Crick/Sugar Edge	tWS	○─▷
Cis Hoogsteen/Hoogsteen	cHH	─■─
Trans Hoogsteen/Hoogsteen	tHH	─□─
Cis Hoogsteen/Sugar Edge	cHS	■─▶
Trans Hoogsteen/Sugar Edge	tHS	□─▷
Cis Sugar Edge/Sugar Edge	cSS	─▶─
Trans Sugar Edge/Sugar Edge	cSS	─▷─
Bifurcated	bif	

Table 5
Types of base changes

Type	Examples
Covariation	G─●─C to A─●─U
Conservatory	G─●─C to G─●─U
Mismatch	G─●─C to A x C
Insertion	G─●─C to A─●─G C─●─U
Deletion	G─●─C to - -
Isosteric	A─●─U to G─●─C A─○─A to G─○─C G□─▷A to A□─▷A A■─▶A to A■─▶C
Non-isosteric but geometrically possible	U─●─A to U○─□A
Not possible	G─■─C to A x U

3 Applications

The main application of constructing RNA structural alignments is to gain insight into the biological function of the particular molecule. This insight comes through comparing sequences and structures and thus observing their evolutionary changes. Deeper insight into how organisms change their basic function is gained when changes in structural elements in relation to important sequence motifs are observed.

The identified covariation patterns have additional applications as data sets for training algorithms to recognize these patterns. Apart from standard covariation observed for secondary structure, base triples and tertiary interactions can sometimes be deciphered. With the increasing amount of structural data for RNA molecules, it becomes feasible to investigate the covariation of non-Watson-Crick base pairs and provide sequence alignments based on 3D structural data [24].

Another application of structural sequence analysis is to extract motifs and rules for the design of RNA nanostructures [41]. In this case the tertiary motifs are of particular interest since they allow geometrically interesting molecular architectures to be built [42]. Structural alignment of the tertiary motifs provides information about the possible sequences that can be used to design a particular motif and gives rise to a set of motifs and rules to follow when designing a novel RNA structure [43–45].

4 Software Usage Examples

In this section the SARSE and S2S editors are described in more details as examples of RNA alignment based on secondary and tertiary structure, respectively.

4.1 SARSE: Semiautomated RNA Structure Editor

The SARSE program was developed for curating existing RNA databases using various tools to assist the process. The editor can be downloaded at http://sarse.ku.dk/ but has not been updated for a number of years. The standard sequence editor window (Fig. 3b) has a list of sequence names next to an alignment of sequences. The sequence viewer can be split in two (vertically or horizontally) which allows easy viewing of the two sides of a helix. The pairing mask shown at the top of the alignment defines the selection of base pairs by the mouse. Selections in one window will simultaneously select the base pairing partners in the other window. The SARSE editor also provides an overview window that displays the colors of the alignment in a zoomable map. Clicking this map moves the main window to the corresponding position. An additional window called "History" tracks all the editing steps done to the alignment. Clicking on an element of the list will take you back to this point in your editing history. The history window thus both documents your editing and provides an extended undo button.

A particular usage of the editor to ease the identification of misaligned structures is illustrated in Fig. 3. Figure 3a shows the comparison of two pairing masks that define base pairs as misaligned (red), aligned (green), or nonoverlapping (yellow). These annotations can become especially useful when the alignment consists of a structure prediction for each sequence or if more structural groupings exist as shown in Fig. 3b. To aid the overview of large alignments, a plotting scheme named "ali-stem plot" is used to identify areas of misalignment and nonoverlapping structure (Fig. 3c, d). The ali-stem plot has a representation of the alignment in the top rectangle and underneath a triangle with lines connecting the base pairing positions of the alignment. In contrast to a normal dot plot, the ali-stem plot shows both sides of the

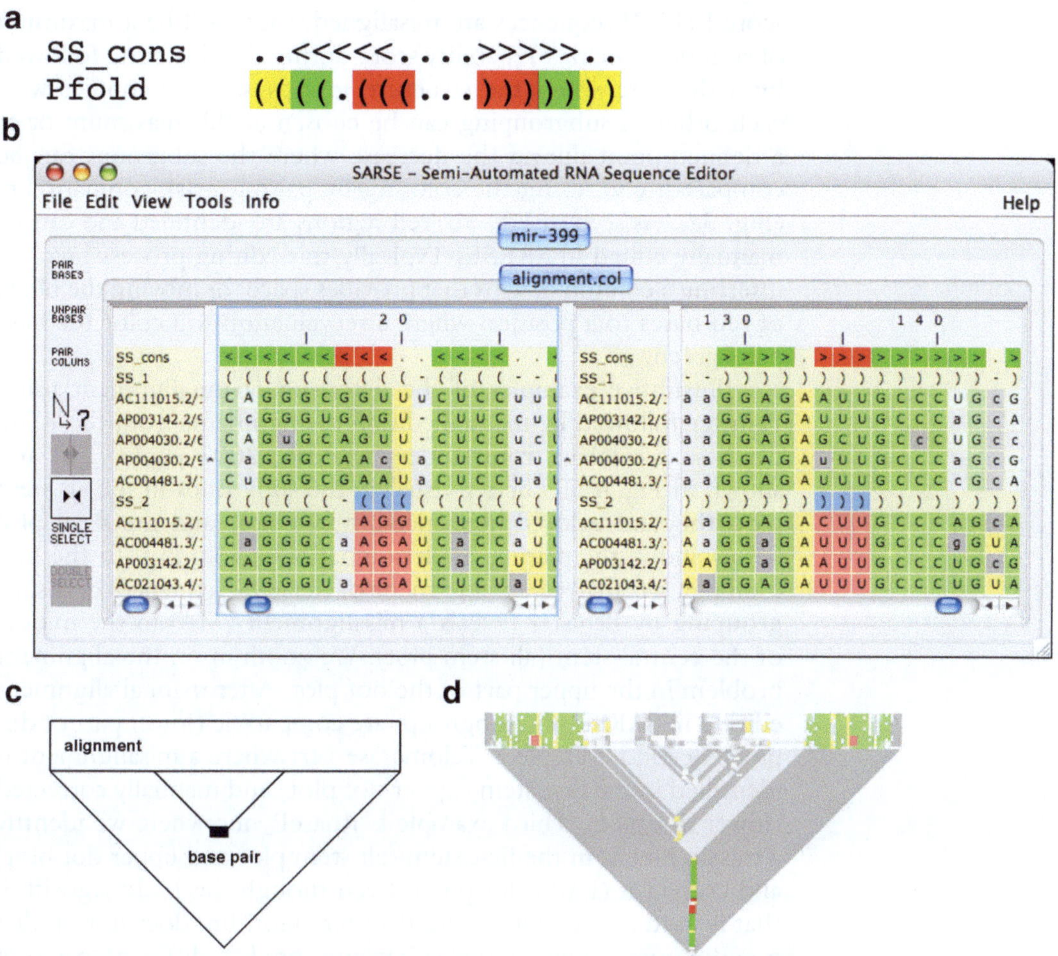

Fig. 3 SARSE, semiautomated RNA sequence editor. (**a**) Comparison between two pairing masks "SS_cons" from Rfam and a pairing mask predicted by Pfold. SS_cons is taken as the reference pairing mask and the Pfold pairing mask is colored in relation to this reference: *Red* indicates nonoverlapping base pairs, *green* indicates overlapping base pairs, and *yellow* indicates non-annotated base pairs. (**b**) Screenshot of the SARSE editor using the above coloring scheme. An alignment of sequences is shown in split view that makes it possible to see two parts of the stem at position 11–25, 129–142. The alignment has been split into two groups by the Pcluster algorithm which makes it possible to spot a misalignment of group two at position 17–20, 134–136. The colors are more transparent depending on the reliability of Pfold prediction. (**c**) A simple plotting scheme "ali-stem plot" shows the alignment on top and the base pair connections in a *triangle* below. (**d**) The alignment above shown as ali-stem plot creates a fast overview

base pair in the lower rectangle which makes it possible to make asymmetric color annotations on the helices.

The SARSE program was originally presented together with a clustering algorithm called Pcluster. Pcluster uses the RNA structure prediction algorithm Pfold to predict a structure for each individual sequence and then uses a greedy algorithm to group sequences based on obtaining a high summed base pair reliability score [21]. If sequences are misaligned, there will be a maximum of summed base pair reliability score during the clustering followed by a decrease when the structure groups start to conflict with each other. A subgrouping can be chosen at this maximum or at a defined point during the decrease where the subgroups can be compared, e.g., using the coloring by pairing mask compatibility (Fig. 3a). As seen in Fig. 3b, red regions are identified and can be manually edited in SARSE. Typically this editing process involves inserting a column of gaps that provides space for moving the block of red bases to a position where a reevaluation will color the base pairs green.

Illustrative examples of the detection of misalignment using Pcluster followed by manual editing of alignments from the Rfam database are taken from Andersen et al., 2007 paper [21] and shown in Fig. 4. To better visualize the subgroups in the ali-stem plot, they are colored here in rainbow colors and the stem plot below shows the pairing mask compatibility coloring. In the first example, we analyze an alignment of mir-399 sequences and subgrouping by Pcluster reveals a misalignment (MA) in the middle of the central stem (ali-stem plot). We zoom in on the alignment problem in the upper part of the dot plot. After manual alignment editing in SARSE, the subgroups are compatible (lower part of dot plot). Second example is Telomerase-vert where a misalignment is identified in the first stem (upper dot plot) and manually corrected (lower dot plot). Third example is RnaseP_nuc where we identify a misalignment in the first stem (ali-stem plot and upper dot plot) and correct it (lower dot plot). Even though the Pfold algorithm that is used as the basis of the Pcluster algorithm does not predict pseudoknots, some of the subgroups predict different parts of the pseudoknot (annotated in the dot plot). Similar observations are observed when clustering an alignment of tmRNA sequences, where several of the pseudoknots in this molecule are also identified in this way. We zoom in on one misaligned part of a pseudoknot (upper dot plot) and realign it (lower dot plot). The identification/prediction of pseudoknots by structural clustering shows that the two stems of a pseudoknot are considered by the Pfold algorithm as two mutually exclusive structures. The alignments and curation can be evaluated by the increase in structure score that is gained by the editing procedure. Likewise, the alignments can be ranked by mismatch and overlapping and nonoverlapping summed scores, which makes it easy to identify the alignments that need curation. See example at http://sarse.ku.dk/Rfam_sarse/.

4.2 S2S: Sequence to Structure

The S2S program [22] was the first to introduce a framework for working with multiple alignments based on tertiary structure derived from PDB files. It can be downloaded at http://www.bioinformatics.org/S2S/ and has a graphical user interface that makes it easy to display, manipulate, and interact with data at the primary, secondary, and tertiary structure level.

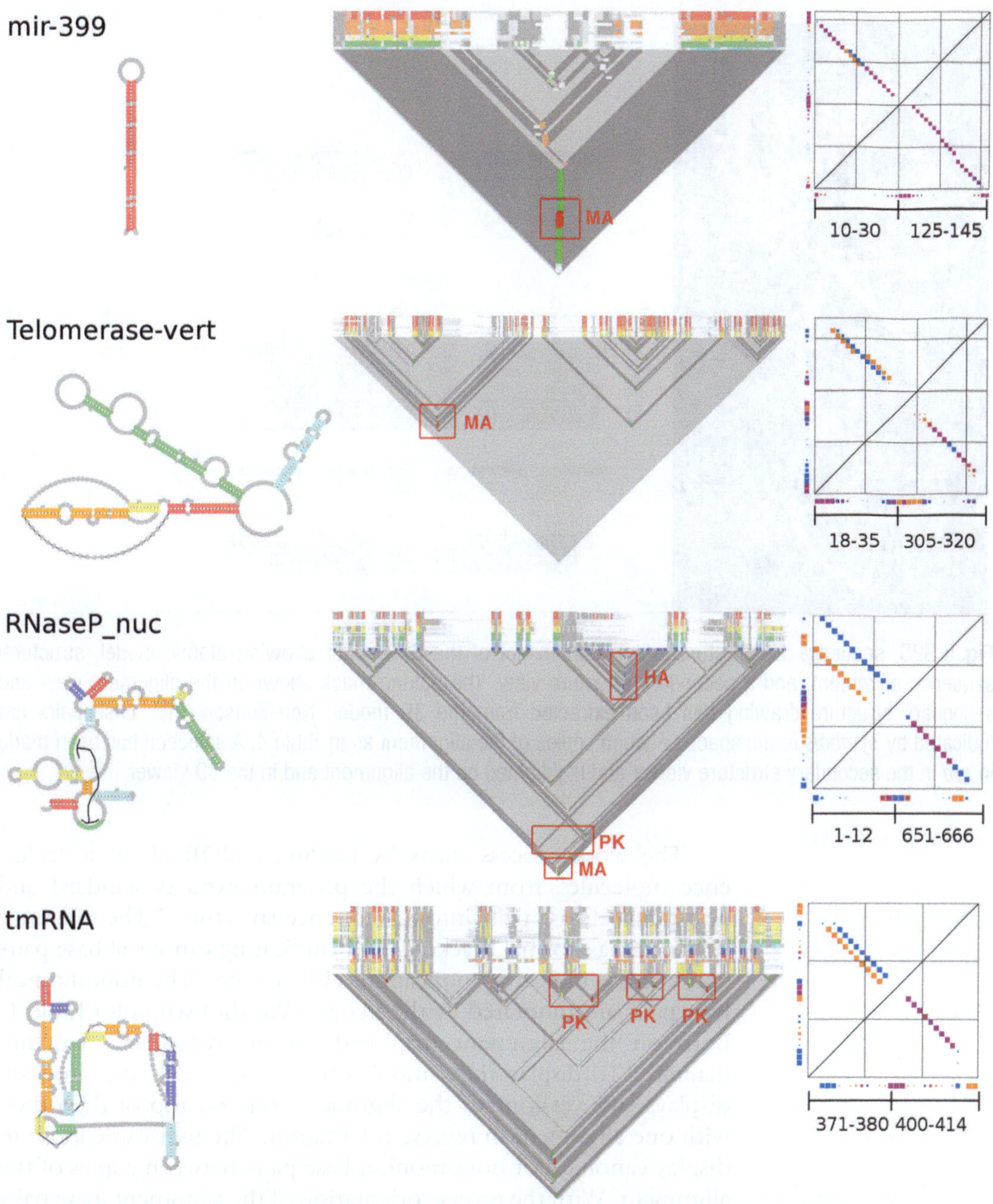

Fig. 4 Examples of detection and correction of alignment errors. Secondary structure drawings of the analyzed RNAs are shown to the *left*. Ali-stem plots are shown in the *middle*. Alignments are subgrouped by the Pcluster algorithm. Misalignments are *boxed* and marked "MA." Pseudoknots are marked "PK." Dot plots are shown to the *right* with misalignment shown in *upper triangle* and correction shown in *lower triangle*. Zoom-in regions are indicated below dot plot

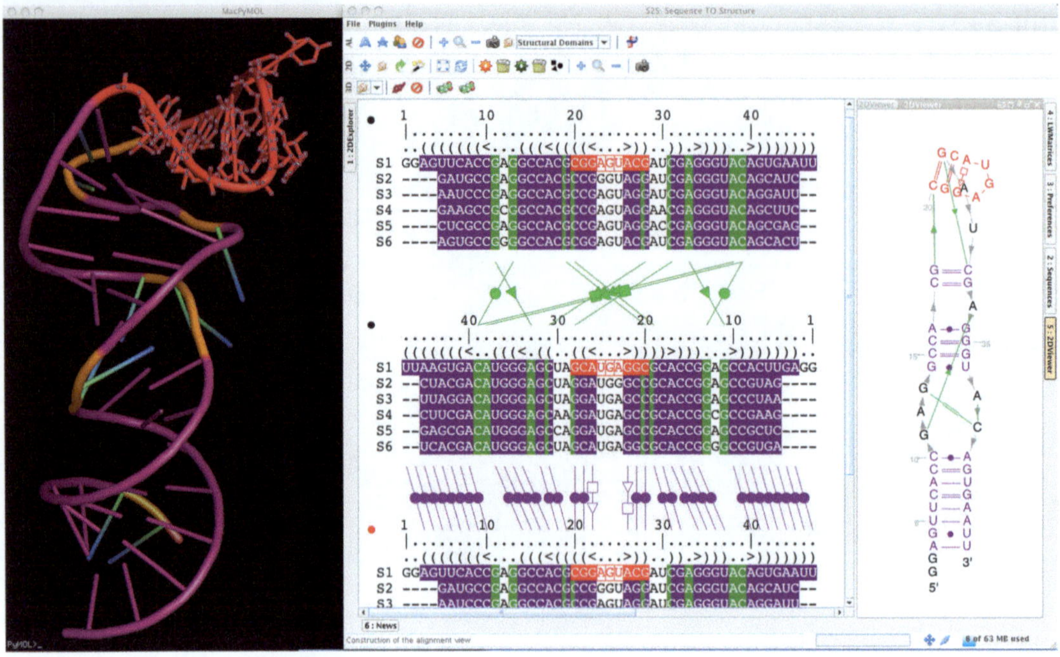

Fig. 5 S2S, sequence to structure editor. Screenshot of the S2S editor showing atomic model, structural sequence alignment, and a secondary structure view. The pairing mask shown in the alignment view and secondary structure drawing has been extracted from the 3D model. Non-Watson-Crick base pairs are indicated by *symbols* in the space between copies of the alignment as in Table 4. A selection has been made in *red* in the secondary structure viewer and highlighted on the alignment and in the 3D viewer

The work process starts by loading a PDB file as a "reference molecule" from which the program extracts standard and nonstandard base pairs into a "reference structure." The reference structure has normal brackets (and) indicating canonical base pairs and < and > indicating noncanonical base pairs. The noncanonical base pairs are annotated by the Leontis-Westhof symbols (Table 4) both on the alignment map and on the secondary structure diagram. To display the symbols on the alignment, the program displays two versions of the alignment, one on top of the other, with one alignment in reverse orientation. The user can choose to display canonical or noncanonical base pairs between copies of the alignment. With the reverse orientation of the alignment, base pairs are displayed next to each other like in the stem of a helix (Fig. 5c).

Next step is to import sequences to be aligned to your "reference molecule." The sequences to be imported can either be prealigned by a sequence alignment program or just a list of sequences. To display the sequences, go to the Preferences panel and change the number of sequences displayed. In the Preferences panel, it is also possible to choose to display canonical and/or noncanonical base pairs, which makes it easy to focus on the noncanonical base pairs and compare base changes between two columns.

Structural conservation of the imported sequences is shown according to the "reference structure." S2S uses two different colors to display the isosteric secondary and tertiary base interactions and two brighter colors to show when base changes are non-isosteric but geometrically possible. A white box is used to indicate base combinations that have not been observed for this type of base pair as defined by the "reference structure." These annotations are defined by isostericity matrices [46] that can be modified by the user. The program furthermore displays base triples directly on the alignment by splitting the color of the box of a base into two parts, which allows nine possible types of annotation of bases involved in triple interactions.

S2S has several ways to navigate and edit the alignment and secondary structure. By clicking a button next to the alignment, you can choose the range and type of symbols shown between the two versions of the alignment. Selections are very easy since the alignment and secondary structure viewer are linked and the selection mode can be set to work at residue, interaction, and domain level. The manual editing of the alignment is done by keyboard shortcuts to insert gaps and move residues horizontally, and sequences can be rearranged by drag and drop. The direct link to the secondary structure in the 2D panel is a very strong feature and by using a 3D viewer like PyMol makes fully informed structural editing possible.

5 Prospects

The main goal of the editor/curator of an RNA database is to make high-quality RNA structure and sequence alignments available for further experimental and/or theoretical studies. There are still no algorithms to solve this automatically, which might be caused by the multidimensional nature of this problem. The problem will eventually be solved and automated but the road towards this preferable situation will need to be supported by an integrated framework and graphical user interface linking all the involved data types, e.g., phylogeny, sequence alignments, secondary structure, 3D structures, and ways to look at relations between complexes of molecules. To my mind the system has to be visual and interactive to allow the researcher to edit and modify the data and do experiments to gain further insight. In addition detailed annotations of functions at the DNA, RNA, and protein level are required to understand the evolutionary pressures that have shaped the molecule being studied. This is especially important for RNA structures that have overlapping transcription factor binding sites, splice sites, regulatory protein binding sites, protein coding regions, and so on.

Sequence-structure alignments are a main way of organizing and classifying biomolecular data. Apart from the sequence data explosion, high-throughput techniques are being developed for obtaining secondary and tertiary structure data. Integrating these data in sequence alignment databases is a main challenge. Algorithms for prediction and classification will have to be scalable due to the high data volumes. Many current algorithms are *NOT* scalable and thus novel and faster methods have to be developed. As illustrated by the subgrouping studies using Pcluster and SARSE, parting alignments into regions makes it possible to make local predictions where the prediction method works. The divide-and-conquer strategy might be a way to use existing algorithms to analyze parts of large data sets as part of analysis pipelines directed towards automated construction of sequence-structure alignments. Pipelines are being devised but still require manual editing steps [24].

A future direction is to get better data concerning RNA motifs and study the 3D structural evolution of RNA molecules in excessive detail. For recurrent structural RNA motifs, the alignments do not have to be between homologous sequences—one could, e.g., align a common motif from RNase P and the ribosome. These structural alignments do not make sense in an evolutionary context, but show what sequences can adopt a certain motif and provide information about how the structural context affects the motif. These types of investigations will allow the extraction of the design principles that have been employed by evolution and inform novel rational designs of biomolecules for use in nanotechnology and synthetic biology applications.

References

1. Pace NR, Thomas BC, Woese CR (1999) Probing RNA structure, function, and history by comparative analysis. The RNA world, 2nd edn. Gesteland RF, Cech TR, Atkins JF (eds.), Cold Spring Harbor Laboratory Press, Cold Spring Harbor, New York, pp 113–141
2. Ehresmann C et al (1987) Probing the structure of RNAs in solution. Nucleic Acids Res 15(22):9109–9128
3. Ban N, Nissen P, Hansen J, Moore P, Steitz T (2000) The complete atomic structure of the large ribosomal subunit at 2.4 A resolution. Science 289(5481):905–920
4. Wimberly BT et al (2000) Structure of the 30S ribosomal subunit. Nature 407(6802): 327–339
5. Olsen GJ, Larsen N, Woese CR (1991) The ribosomal RNA database project. Nucleic Acids Res 19 Suppl:2017–2021
6. Brown JW (1999) The ribonuclease P database. Nucleic Acids Res 27(1):314
7. Zwieb C, Larsen N, Wower J (1998) The tmRNA database (tmRDB). Nucleic Acids Res 26(1):166–167
8. Williams KP, Bartel DP (1998) The tmRNA website. Nucleic Acids Res 26(1):163–165
9. Larsen N, Zwieb C (1993) The signal recognition particle database (SRPDB). Nucleic Acids Res 21(13):3019–3020
10. Griffiths-Jones S, Bateman A, Marshall M, Khanna A, Eddy SR (2003) Rfam: an RNA family database. Nucleic Acids Res 31(1): 439–441
11. Leontis NB, Westhof E (1998) Conserved geometrical base-pairing patterns in RNA. Q Rev Biophys 31(4):399–455
12. Waugh A et al (2002) RNAML: a standard syntax for exchanging RNA information. RNA 8(6):707–717
13. Brown JW et al (2009) The RNA structure alignment ontology. RNA 15(9): 1623–1631

14. De Rijk P, De Wachter R (1993) DCSE, an interactive tool for sequence alignment and secondary structure research. Comput Appl Biosci 9(6):735–740
15. De Oliveira T, Miller R, Tarin M, Cassol S (2003) An integrated genetic data environment (GDE)-based LINUX interface for analysis of HIV-1 and other microbial sequences. Bioinformatics 19(1):153–154
16. Clamp M, Cuff J, Searle SM, Barton GJ (2004) The Jalview Java alignment editor. Bioinformatics 20(3):426–427
17. Luck R, Graf S, Steger G (1999) ConStruct: a tool for thermodynamic controlled prediction of conserved secondary structure. Nucleic Acids Res 27(21):4208–4217
18. Ludwig W et al (2004) ARB: a software environment for sequence data. Nucleic Acids Res 32(4):1363–1371
19. Griffiths-Jones S (2005) RALEE—RNA ALignment editor in Emacs. Bioinformatics 21(2):257–259
20. Seibel PN, Muller T, Dandekar T, Schultz J, Wolf M (2006) 4SALE—a tool for synchronous RNA sequence and secondary structure alignment and editing. BMC Bioinformatics 7:498
21. Andersen ES et al (2007) Semiautomated improvement of RNA alignments. RNA 13(11):1850–1859
22. Jossinet F, Westhof E (2005) Sequence to Structure (S2S): display, manipulate and interconnect RNA data from sequence to structure. Bioinformatics 21(15):3320–3321
23. Stombaugh J, Widmann J, McDonald D, Knight R (2011) Boulder ALignment Editor (ALE): a web-based RNA alignment tool. Bioinformatics 27(12):1706–1707
24. Widmann J et al (2012) RNASTAR: an RNA STructural Alignment Repository that provides insight into the evolution of natural and artificial RNAs. RNA 18(7):1319–1327
25. Gorodkin J, Zwieb C, Knudsen B (2001) Semi-automated update and cleanup of structural RNA alignment databases. Bioinformatics 17(7):642–645
26. Hofacker IL (2003) Vienna RNA secondary structure server. Nucleic Acids Res 31(13):3429–3431
27. Knudsen B, Hein J (1999) RNA secondary structure prediction using stochastic context-free grammars and evolutionary history. Bioinformatics 15(6):446–454
28. Knudsen B, Hein J (2003) Pfold: RNA secondary structure prediction using stochastic context-free grammars. Nucleic Acids Res 31(13):3423–3428
29. Hofacker IL, Fekete M, Stadler PF (2002) Secondary structure prediction for aligned RNA sequences. J Mol Biol 319(5):1059–1066
30. Knudsen B, Hein J (2003) Pfold: RNA secondary structure prediction using stochastic context-free grammars. Nucleic Acids Res 31(13):3423–3428
31. Havgaard JH, Lyngso RB, Stormo GD, Gorodkin J (2005) Pairwise local structural alignment of RNA sequences with sequence similarity less than 40%. Bioinformatics 21(9):1815–1824
32. Hofacker IL, Bernhart SH, Stadler PF (2004) Alignment of RNA base pairing probability matrices. Bioinformatics 20(14):2222–2227
33. Mathews DH, Turner DH (2002) Dynalign: an algorithm for finding the secondary structure common to two RNA sequences. J Mol Biol 317(2):191–203
34. Mathews DH (2005) Predicting a set of minimal free energy RNA secondary structures common to two sequences. Bioinformatics 21(10):2246–2253
35. Holmes I (2005) Accelerated probabilistic inference of RNA structure evolution. BMC Bioinformatics 6(1):73
36. Holmes I, Rubin GM (2002) Pairwise RNA structure comparison with stochastic context-free grammars. Pac Symp Biocomput: 163–174
37. Reeder J, Giegerich R (2005) Consensus shapes: an alternative to the Sankoff algorithm for RNA consensus structure prediction. Bioinformatics 21(17):3516–3523
38. Yao Z, Weinberg Z, Ruzzo WL (2006) CMfinder—a covariance model based RNA motif finding algorithm. Bioinformatics 22(4):445–452
39. Torarinsson E, Havgaard JH, Gorodkin J (2007) Multiple structural alignment and clustering of RNA sequences. Bioinformatics 23:926–932
40. Leontis NB, Stombaugh J, Westhof E (2002) The non-Watson-Crick base pairs and their associated isostericity matrices. Nucleic Acids Res 30(16):3497–3531
41. Andersen ES (2010) Prediction and design of DNA and RNA structures. New Biotechnol 27(3):184–193
42. Jaeger L, Chworos A (2006) The architectonics of programmable RNA and DNA nanostructures. Curr Opin Struct Biol 16(4):531–543
43. Geary C, Baudrey S, Jaeger L (2008) Comprehensive features of natural and in vitro selected GNRA tetraloop-binding receptors. Nucleic Acids Res 36(4):1138–1152

44. Jaeger L, Verzemnieks EJ, Geary C (2009) The UA handle: a versatile submotif in stable RNA architectures. Nucleic Acids Res 37(1): 215–230
45. Geary C, Chworos A, Jaeger L (2011) Promoting RNA helical stacking via A-minor junctions. Nucleic Acids Res 39(3):1066–1080
46. Leontis N, Stombaugh J, Westhof E (2002) The non-Watson-Crick base pairs and their associated isostericity matrices. Nucleic Acids Res 30(16):3497–3531

Chapter 18

Automated Modeling of RNA 3D Structure

Kristian Rother, Magdalena Rother, Pawel Skiba, and Janusz M. Bujnicki

Abstract

This chapter gives an overview over the current methods for automated modeling of RNA structures, with emphasis on template-based methods. The currently used approaches to RNA modeling are presented with a side view on the protein world, where many similar ideas have been used. Two main programs for automated template-based modeling are presented: ModeRNA assembling structures from fragments and MacroMoleculeBuilder performing a simulation to satisfy spatial restraints. Both approaches have in common that they require an alignment of the target sequence to a known RNA structure that is used as a modeling template. As a way to find promising template structures and to align the target and template sequences, we propose a pipeline combining the ParAlign and Infernal programs on RNA family data from Rfam. We also briefly summarize template-free methods for RNA 3D structure prediction. Typically, RNA structures generated by automated modeling methods require local or global optimization. Thus, we also discuss methods that can be used for local or global refinement of RNA structures.

Key words RNA structure, Structure prediction, Homology modeling, Comparative modeling, *De novo* modeling, Template search, Structure refinement

1 Introduction

The discrepancy between the number of known sequences and structures has been quoted many times as the driving force for 3D modeling of macromolecular structures. RNA structures are no exception here, and the field has seen tremendous progress over the last few years. It seems that RNA bioinformatics is setting out to repeat the success story of protein structure prediction. The major breakthroughs of the last years include *de novo* folding of RNA structures up to 50 nucleotides long starting from sequence alone [1, 2] and template-based modeling of the *S. cerevisiae* 80S ribosome based on cryo-EM data [3] *in silico*. What challenges remain? To get an idea what could be achieved, it is worth to peek into the scientific record of protein structure prediction.

Jan Gorodkin and Walter L. Ruzzo (eds.), *RNA Sequence, Structure, and Function: Computational and Bioinformatic Methods*, Methods in Molecular Biology, vol. 1097, DOI 10.1007/978-1-62703-709-9_18,

1.1 Prediction of 3D Structures: Lessons from the Protein World

Over the last 20 years, prediction of protein structures has evolved from the first optimistic attempts into a widely established and mature field with useful applications. Protein structure models can be used to suggest residues for mutation in functional assays, to determine surface residues for chemical cross-linking experiments, and to obtain starting structures for molecular replacement in crystallography and for fitting them into low-resolution electron density maps from cryo-EM or SAX experiments. The 3D structure modeling methods are complemented by automatic model quality prediction (MQAP) tools like ProQ [4] or MetaMQAP [5] that can assess the likelihood whether the structural model is likely to be correct globally as well as to predict the accuracy of individual residues. 3D structure prediction is usually accompanied by prediction of protein order and disorder that can provide information which parts of a protein can be expected to have a stable structure at all [6]. Independent benchmarks including the biannual CASP experiment [7] and the continuous LiveBench experiment [8] help to evaluate the current state of software for structure prediction. An impressive outcome of these efforts includes the ModBase repository [9], containing ten million models of protein 3D structure built by the MODELLER program. For a similar data collection, SAHG, 42,500 models covering the entire human proteome have been constructed [10] using an automatic pipeline involving BLAST, PSI-BLAST, different alignment methods, disorder prediction, and MODELLER [11].

The field of RNA structure poses unique challenges that cannot be compared directly to those from the protein world: the number of known structured RNAs is much lower than that of proteins; there are many RNA molecules active through base pairing, e.g., miRNA, a phenomenon unparalleled in proteins; and many RNAs like riboswitches and aptamers can have highly dynamic structures that can hardly be described by a single conformation. Nevertheless, there are goals imaginable for RNA that are yet to be reached: there is currently no repository that attempts to gather a complete set of structured RNA for a given species. The question whether a given RNA has a stable 3D structure is focused mainly on the development of methods for 2D structure prediction. Only recently a procedure to predict the quality of RNA models has been developed [12], and there is no long-term benchmark being performed. Using theoretical RNA models to support experimental studies cannot be considered a standard method, and the relevant scientific hypotheses need to be formulated carefully for each individual case, depending on the amount of experimental data and the expected accuracy of the model.

Is it realistic to borrow ideas for RNA modeling from methods developed for proteins? After all, there are considerable differences: in proteins, secondary structure is formed by the main chain and in RNA by base pairing of the side chains. The base pairs itself are highly interchangeable by structurally equivalent (isosteric) canonical and noncanonical pairs [13]. The tertiary structure of RNA is dominated by base stacking as the main stabilizing force, leading to coaxial stacks. Any method for RNA structure prediction needs to take these interactions into account. On the other hand, RNA and proteins have many building principles in common: both are linear polymers consisting of a limited number of building blocks that determine the folding of the structure. There is considerable flexibility of the backbone that allows for secondary structures being formed. Repeating secondary structural elements assemble to higher-order structures and promote stable 3D structures. These similarities allow the approaches that were used successfully for proteins to be applied to building RNA models (review in ref. 14).

1.2 Template-Based vs. Template-Free Modeling

There are three groups of methods that have been demonstrated to be useful for predicting RNA and protein structures alike: template-based, template-free, and hybrid methods [14]. In template-based modeling (also termed homology or comparative modeling), a known 3D structure (of another, related protein) is modified to obtain a model with a new sequence but the same overall architecture. The success of template-based modeling stands and falls with the choice of a proper template, and finding the right template is a separate and challenging problem. Template-free modeling applies physical and/or statistical rules to generate a folded structure from an unfolded one. This procedure is generally more calculation-intensive than template-based modeling and is currently limited to small- and medium-sized RNA molecules. The hybrid knowledge-based approach samples conformations by assembling small fragments derived from known structures and uses a statistics-based scoring function to evaluate the generated candidates. While it still requires considerable computational power, the knowledge-based approach overcomes the main drawback of the physical methods, which may waste a lot of time exploring conformations that are far away from any real structure. In this chapter, we describe two methods for template-based modeling in detail, including how to find and prepare a template structure and how to refine the resulting model. We also briefly review methods for template-free modeling.

2 Automated Template-Based Modeling of RNA 3D Structure

The basic idea of template-based modeling is to use an existing structure (the template) and a set of instructions that allow to construct a similar model with an altered sequence (the target).

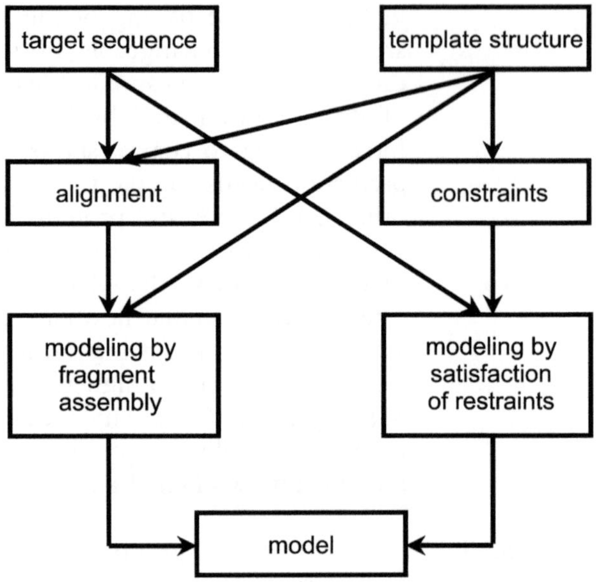

Fig. 1 Two alternative ways of template-based modeling. Models can be built by fragment-based assembly or satisfaction of restraints. In both cases, a target sequence and a template structure are necessary

These instructions can be represented by a pairwise alignment of the target and template sequence. In this minimal definition, the pros and cons of template-based modeling become apparent. On one hand, using a template structure saves a lot of time compared to the calculation-intensive sequence folding methods, which have to perform a search of the vast conformational space. On the other hand, a sufficiently similar structure (typically of an evolutionarily related molecule) and biologically relevant sequence alignment are absolutely necessary for the template-based method to work. Technically, two ideas to build a model automatically have been proposed (*see* Fig. 1): the first is to copy the template structure and modify it until it matches the target sequence. The second is to derive a set of restraints from the template and use them for a fast nonphysical simulation that folds the target chain. In the protein modeling field, these approaches can be represented by Swiss-MODEL. [15] and MODELLER [16], respectively. Both ideas have been recently implemented also for RNA modeling and are available in the ModeRNA [17] and MacroMoleculeBuilder [18] programs. (These and other software tools we discuss are summarized in Table 1.)

2.1 Template-Based Modeling by Fragment Assembly

ModeRNA implements the idea of copying and editing the template (review on this idea: [19]). As mentioned above, in this procedure, ModeRNA is conceptually similar to successful programs for protein structure modeling such as Swiss-MODEL [20]. As an input, ModeRNA takes a PDB structure file and a FASTA

Table 1
Examples of software for template-based and template-free modeling, template search, and refinement of RNA models

Program	Summary
ModeRNA	Program for template-based modeling by fragment replacement. It critically depends on the template structure and target-template alignment provided by the user. It replaces parts of the template structure by fragments of various sizes, according to a sequence alignment. Can model posttranscriptionally modified nucleosides and has many functions for analyzing and manipulating RNA structures. Can use multiple templates and secondary structure restraints provided by the user and can assemble a model from user-defined fragments
MMB/RNABuilder	Program for template-based modeling by satisfying constraints and restraints. In order to obtain a biologically relevant model, it requires a set of restraints (base pairs, base stacking, distances, etc.) and runs a simulation on an unfolded RNA chain or preliminary model. Is capable of limiting the simulation to parts of the structure and adjust many parameters like the strength of restraints or repulsion of VdW spheres
MC-Fold|MC-Sym	Program for template-free RNA modeling, which assembles 3D structures from a library of "Nucleotide Cyclic Motifs," i.e., fragments in which all nucleotides are circularly connected by covalent, pairing, or stacking interactions. It implements two energy functions, one based on nonbonded terms (van der Waals and stacking interactions) from the AMBER package and another one based on statistics of the experimentally determined structures
FARNA/FARFAR	FARNA is a program for template-free RNA modeling, which assembles an RNA 3D structure from short linear fragments, using a knowledge-based energy function, which takes into account preferences of the backbone and side-chain conformations and of base-pairing and base-stacking interactions, derived from experimentally determined RNA structures. Fragments for the assembly of RNA structure were taken from the large ribosomal subunit of Haloarcula marismortui (PDB code 1ffk). FARFAR is an extension of FARNA, which uses a full-atom refinement in order to optimize the RNA structures generated by FARNA. The full-atom energy function is supplemented with harmonic constraints placed between Watson–Crick edge atoms in the two residues that are assumed to form each bounding canonical base pair and a term to approximately describe the screened electrostatic interactions between phosphates. It also includes terms derived from earlier work on proteins: a potential for weak carbon hydrogen bonds, an alternative orientation-dependent model for desolvation based on occlusion of protein moieties. The search of the conformational space is carried out based on Monte Carlo dynamics
SimRNA	Program for template-free RNA modeling that uses a coarse-grained representation of the nucleotide chain (three or five pseudoatoms per nucleotide residue). Bonded and nonbonded terms in its scoring function are based entirely on database statistics. The search of the conformational space is carried out based on Monte Carlo dynamics
DMD	Program for template-free RNA modeling that uses a coarse-grained representation of the nucleotide chain (three pseudoatoms per nucleotide residue). It uses experimentally tabulated energy values to parametrize base pairing and base stacking as well as to estimate loop entropy. It also uses an explicit representation of hydrogen bonding to enforce base pair formation and an additional term for phosphate–phosphate repulsion. It uses the Discrete Molecular Dynamics approach for searching the conformational space

Table 1
(continued)

ParAlign	Program for searching RNA sequences in a large dataset based on similarity. Can be used to find template candidates
Infernal	Program that calculates alignments for RNA families. Requires a set of sequences and a covariance model (CM). Can be used to create target-template sequence alignment. Also can generate CMs from a multiple sequence alignment (used in Rfam)
PYMOL	Molecular viewer with a set of functions to edit structures. Allows the manipulation of bonds, angles, and torsions as well as manual superposition of groups of molecules. Has a scripting interface based on Python 2.3
Chimera	Molecular viewer that can be used to edit structures. Provides a set of functions to analyze structures, e.g., calculate clashes and hydrogen bonds, and to use AMBER tools to, e.g., solvate a molecule or add counterions. Has a scripting interface using Python 2.6
Coot	Program for modeling protein and nucleic acid structures into electron density maps. It supports manual editing by, e.g., fitting rotamers of a single residue or constructing idealized RNA helices. Supports assembly of single nucleotide building blocks. Has a Python scripting interface using an embedded interpreter
AMBER	Molecular Dynamics package. Uses the AMBER force field that supports simulations of nucleotides. An extension for handling nucleotide modifications is available
Zephyr	Graphical interface for the GROMOS Molecular Dynamics package. Uses the AMBER force field. Accelerates calculations by using graphical processors
HyperChem	Commercial molecule editor that provides six procedures for minimization of structures

alignment. The alignment is decomposed into elementary operations such as copying parts from the template structure that are identical in the target and substituting bases by adding small fragments for individual nucleotides. In this context it is noteworthy that ModeRNA is capable of modeling not only the four standard residues (A, G, C, U) but also >100 posttranscriptionally modified ones, which is unique in the RNA modeling field. Insertions and deletions in the alignment are modeled by inserting fragments of appropriate length from a library of more than 100,000 fragments built from known RNA structures. Fragments are selected based on spatial compatibility with the rest of the molecule, i.e., geometrical match between the termini of the inserted fragment and the anchor points in the framework, and the absence of steric clashes, but no physical energy function is used to assess the resulting structure. As a result, the models may exhibit local steric problems such as distorted backbone, which can make further refinement necessary.

ModeRNA offers a variety of functions for manipulating RNA structure via a scripting interface using the Python language. These functions can be used to add custom fragments or entire secondary structure elements (see below) and manipulate individual residues in a structure without the need to provide a template and alignment. Other functions are devoted to analyzing a structure, e.g., reporting base pairs, base stacking, unusual geometry, or detecting interatomic clashes. Last but not least, ModeRNA can "clean up" PDB structures to make them usable as templates; remove water molecules, ions, and ligands; and standardize the nomenclature of atoms. Summarizing, ModeRNA is a comparative modeling package with a multipurpose structure editing functionality. ModeRNA is freely available for Linux, Windows, and Mac OS platforms.

Models of tRNA, rRNA, a riboswitch, and a group I intron have been modeled successfully. A major benchmark was performed on a set of 99 tRNA structures, using each other as templates. The resulting 9,675 models had an average RMSD of 5.6 Å, an average discrepancy index DI [21] of 0.62, and an average GDT_TS of 0.5.

2.2 Template-Based Modeling by Satisfaction of Spatial Restraints

Alternatively, the template structure can be used to generate a set of spatial descriptors for the structure. These so-called restraints are used to guide the folding of the target sequence in order to obtain a model. The clear classification of base-pairing and stacking interactions in RNA makes them ideal restraints that describe RNA structure on a high level. Other structural properties that may be used as restraints are backbone torsions [22, 23], relative positioning of bases [24], or simply interatomic distances and angles. An according modeling program implements restraints as elastic rubber bands connecting parts of the molecule and simulates movements of the molecule in such a way that restraints get satisfied. In contrast to template-free prediction methods, restraint-based modeling is mostly data-driven (e.g., the folding is driven by restraints instead of the physical energy), although some physical terms like repulsion of VdW spheres may be included.

One program that uses spatial restraints for RNA 3D structure modeling is MacroMoleculeBuilder (MMB, formerly known as RNABuilder) [25]. It allows the user to specify a target sequence, base-pairing and stacking interactions, and distance restraints. From this information, MMB is able to construct an RNA model, effectively threading the target RNA chain along the backbone of one or more structural templates. Generally, the model will be the better, the more knowledge a user provides. Additionally, MMB can rigidify residues or entire domains of the molecule. In practice, this is useful to perform the modeling in several stages, to relax the structure of elements not represented in the template, or simply to save calculation time. There is however a chance that the RNA chain gets caught in a local minimum during simulation and tangles up. MMB can perform simulated annealing and rescale the potential in order to resolve such situations. The restraints are given to MMB as a script file, allowing highly customized simulations.

MMB uses the Simbody/Molmodel [26] internal coordinate library to represent a full-atomic and a coarse-grained model in parallel. By default, no forces act on the atoms unless constraints or restraints are specified. Forces are resolved by a harmonic potential that becomes inverse at distances higher than a cutoff value. MMB is freely available for Linux systems.

A sample application of MMB/RNABuilder was to construct a model for the *Azoarcus* group I intron based on a template structure from the *Twort* phage and adding two domains and according tertiary interactions from a second template from *Tetrahymena*. The final model reached an overall RMSD of 4.4 Å, the core being very accurate with a RMSD of 2.7 Å (this model could be reproduced with ModeRNA with a similar accuracy).

It needs to be pointed out that although a program like MMB does not use a target-template sequence alignment explicitly, the base pairing/stacking/distance restraints from the template need to be assigned to defined residues in the target. With that, the information encoded in the input of MMB is equivalent to that in a sequence alignment, although users have very good control which kinds of restraints they want to use and which they don't. One key advantage of modeling with spatial restraints is that experimental data that provides partial information on the structure (e.g., distance restraints from FRET measurements, information on pairing, or accessibility of particular residues from chemical probing or hydroxyl radical experiments) integrates well with the mechanics used in the according software [18].

2.3 Template Search

Finding an appropriate template is essential for the success of any method for template-based modeling. In proteins, the main criterion for target-template correspondence is sequence similarity resulting from evolutionary conservation. In RNA, one has to consider the correspondence of base pairs, in particular the conservation of canonical base pairs from a potential template in the target sequence. For this reasons, running BLAST over the PDB, as it is often used for proteins, is not sufficient to find good structural templates unless the sequences are highly similar. Figure 2 gives an overview of a pipeline for template search, utilizing information from RNA families of template candidates are from in order to take base pair correspondence into account. When there is no template or alignment known, the target sequence can be searched against the Rfam database [27], e.g., with the ParAlign program [28], in order to identify families similar to the target. For each identified family, representatives (exemplars) with a known 3D structure can be aligned to the target sequence, e.g., using Infernal [29] and a covariance model corresponding to the family of the template candidate. The covariance models can be derived, e.g., from sequence alignments in the Rfam database and represent

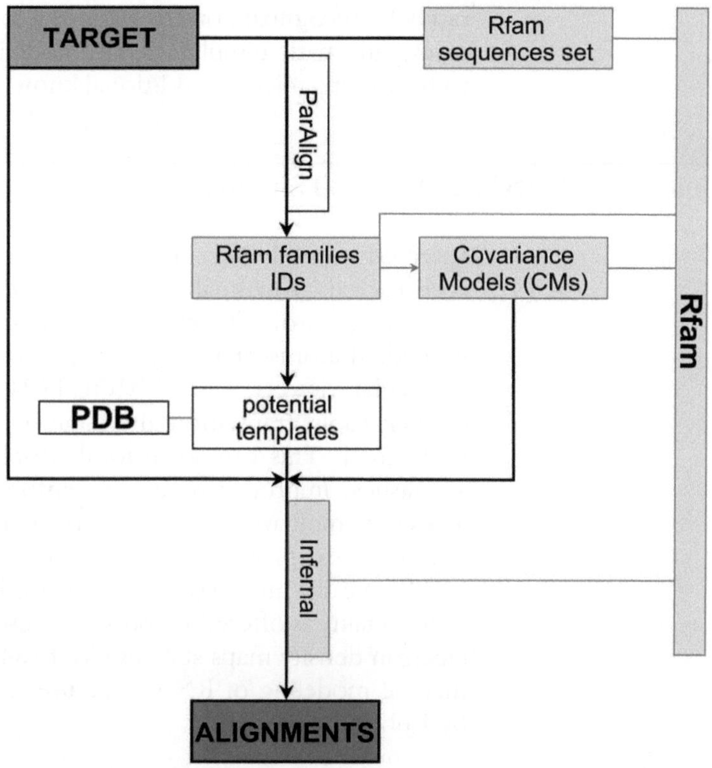

Fig. 2 Pipeline for template search. First, the ParAlign program is used to find families in Rfam that have a similarity with the target sequence (*top*). For each family, a list of corresponding PDB structures is prepared (*middle*). Finally, each potential template is aligned to the target sequence by the Infernal program, using a covariance model (*bottom*)

correlated base exchanges over the entire family. Alternatively, RNA sequence alignment methods such as R-Coffee [30] can be used to align the target sequence to the user-defined template. Our studies have shown that this approach may not be accurate enough for large and highly diverged RNAs, but it may be feasible for short and similar sequences [17].

There are many RNAs, for which a template exists, but it has low sequence identity to the target, and hence its identity and/or a biologically relevant target-template alignment is hard to find. Recently, Capriotti et al. have found a strong correlation between structural and sequence similarity [31]. However, below 60% identity this correlation weakens, making reliable alignments harder to obtain. They identified the twilight zone for RNA (the region where comparative modeling becomes difficult) between 40 and 70% identity. Also, below 70% sequence identity the tertiary interactions of homologous residues may differ between the template and the target, and such structural variations can rarely be recognized on the basis of a sequence alignment. In these cases, automatic template-based modeling is not sufficient and has to be accompanied by additional knowledge-based modeling steps.

3 Interactive Modeling of RNA 3D Structure

Interactive modeling means that parts of an RNA model are constructed by manually selecting or constructing pieces, ranging from single atoms to entire structural elements. On the level of individual atoms or residues, structures can be manipulated with molecular viewers, e.g., PyMOL [32] or Chimera [33]. In most of such tools, translation and rotation of groups of atoms can be performed. This can solve local problems with a structure such as clashes, improper bonds, or enforce two bases to be paired. These manipulations need to be performed carefully, though, because these programs typically do not support the maintenance of chemically and biologically reasonable structures. A similar functionality is offered by programs used for building models into electron density maps such as Coot [34]. More information about manual modeling of RNA structures can be found in Chapter 17 by Ebbe S. Andersen.

3.1 Editing Structural Elements

When larger elements need to be edited in an RNA structure, manual modeling becomes inconvenient. An obvious solution is to reuse secondary structural elements or parts thereof taken from known structures. Effectively, this applies the idea of template-based modeling on a smaller scale. The ModeRNA software can treat secondary structural elements as individual building blocks and combine them into a single model according to users' demands. To do so, up to four anchor residues in the model are superimposed with the according ends of a fragment to connect the backbones of the respective strands. In Fig. 3, some common operations are listed. They include (a) elongating or shortening a helix by insertion of an idealized helical fragment, (b) replacing a local motif (e.g., a loop) by a different structural element such as a longer or shorter loop based on superimposing two residues at its termini, (c) adding a base pair in a standard geometry to a single residue, and (d) inserting a secondary structural element such as a bulge or an internal loop into an existing model. A practical example is the modeling of a tertiary interaction in the *Azoarcus* group I intron using a structural fragment from *Tetrahymena*, which was successfully carried out with both the MMB and ModeRNA programs.

A number of web resources are available that describe motifs from RNA structures: FR3D [13] focuses on the classification of two-residue interactions, FRABASE [35] annotates secondary

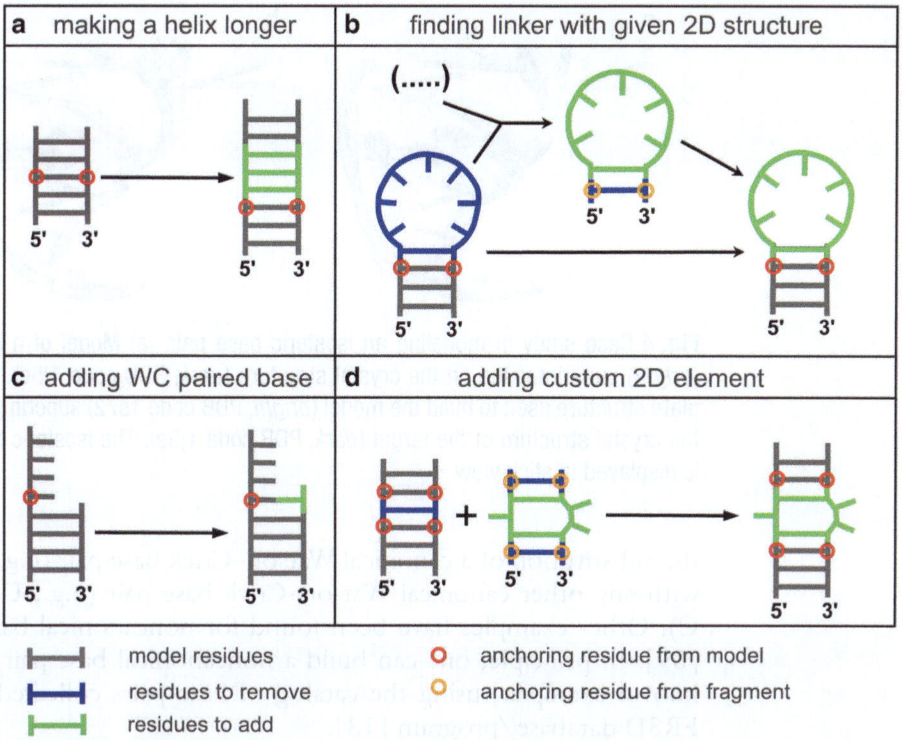

Fig. 3 Typical fragment-based manipulations of 3D models. (**a**) Inserting a helical stem between two Watson–Crick base pairs. (**b**) Finding a linker between two nucleotide residues with a given secondary structure constraint for the linker. (**c**) Adding a canonical base pair to a single residue and connecting the RNA backbone. (**d**) Inserting a custom piece of RNA structure into a target molecule

structural elements, RNAJunction contains 12,000 junctions and kissing loop structures [36], and SCOR contains manually annotated motifs of many different types [37]. The idea of building an RNA structure from small fragments of 2–8 residues can be carried a step further: instead of using one large template in the first place, one could build the entire model by trying different combinations from a sufficiently large set of building blocks. This is exactly what *de novo* modeling methods do (see below).

3.2 Modeling Noncanonical Base Pairs/Isostericity

One important challenge for template-based modeling is the construction of proper noncanonical base pairs. The first step is to recognize correspondencies between the template and the alignment, where a base pair is replaced by another pair, which exhibits the same mutual orientation of the backbone, irrespectively of the base moieties, which is referred to as isostericity [39]. For a practical implementation of isostericity, a set of empirical matrices has been determined that describe which base pair types are structurally equivalent to which and can be substituted by each other without any major modification of the RNA backbone. One example is

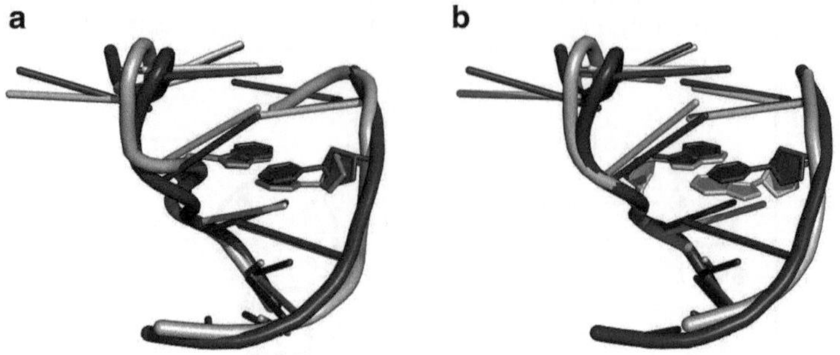

Fig. 4 Case study in modeling an isosteric base pair. (**a**) Model of a kink turn (*bright*) superimposed on the crystal structure (*dark*, PDB code 1j5e). (**b**) Template structure used to build the model (*bright*, PDB code 1s72) superimposed on the crystal structure of the target (*dark*, PDB code 1j5e). The isosteric base pair is displayed in stick view

the substitution of a canonical Watson–Crick base pair (e.g., A–U) with any other canonical Watson–Crick base pair (e.g., U–A, G–C). Other examples have been found for noncanonical base pairs [39]. In principle, one can build a noncanonical base pair from a known exemplar, using the catalog of base pairs collected in the FR3D database/program [13].

In a preliminary study, we attempted to build a comparative model of the kink turn from the 30S ribosomal subunit from the bacterium *Thermus thermophilus* (reference PDB code 1J5E, residues 683–688/699–707) without using information from that structure. As a template, we chose a kink turn from the 50S ribosomal subunit from the bacterium *Haloarcula marismortui* (PDB code 1S72, residues 77–81/93–100) based on a comparative analysis of kink turn structures [40]. We focused our attention on the trans-Sugar–Hoogsteen pair 79G–98A. This was to be replaced by the pair 686U–704A in the model. According to the isostericity matrices by Stombaugh et al. [41], this replacement is isosteric with an isostericity index of 0.62, the values ranging from 0 (best) to 9 (worst). We have built a model for the target kink turn based on the template structure with the scripting interface of ModeRNA. To construct the correct noncanonical base pair, we took the U–A base pair exemplars from FR3D and superimposed them using different ribose and base atoms for superposition. Finally, the backbone repair procedure of ModeRNA was applied. The resulting model is shown in Fig. 4. The RMSD for the model built in this piloting study is 2.01 Å.

In the target, there is a trans-Watson–Hoogsteen pair in position 686U–704A. Our procedure was able to create such an interaction in this position. It can insert base pair exemplars with a specified type of noncanonical pair and allows to try base pair exemplars regardless of their base-pairing type, if the base pair type is not known. One challenge that remains is that the superposition in this approach does not necessarily keep base stacking intact (we tried other examples, and we were not necessarily successful). For this application, either an algorithmic solution that maintains stacking during insertion of a pair is required (a local simulation for satisfying restraints) or a protocol using fragments of two subsequent base pairs like those in MC-Sym can be applied. It is however unsure, whether such a procedure would guarantee perfect modeling of all isosteric replacements.

4 Template-Free Modeling of RNA 3D Structure

An alternative approach to 3D structure prediction, sometimes termed "ab initio" prediction, is based on the thermodynamic hypothesis formulated by Anfinsen for proteins, according to which the native structure of a macromolecule corresponds to the global minimum of the free energy of the system comprising the macromolecule [42]. Accordingly, physics-based methods model the process of folding by simulating the conformational changes of a macromolecule while it searches for the state of minimal free energy (review in ref. 43). The "score" of each conformation is calculated as the true physical energy based on the interactions within the macromolecule and between the macromolecule and the solvent [44]. There exist a number of software packages for simulation of macromolecular folding in atomic detail. Examples of all-atom simulations of RNA with general-purpose software packages such as AMBER or CHARMM include the folding of small RNA hairpins [45, 46] or modeling the interaction of "kissing loops" in the dimerization initiation site (DIS) of HIV [47]. Molecular dynamics simulations restrained by experimental data have been also used to model the conformational transitions of large macromolecular complexes involving both RNAs and proteins, such as the ribosome (review in ref. 48).

The ab initio approach is plagued by serious problems. In particular, a full-atom structural model of a macromolecule has a large number of degrees of freedom, which makes the search space enormous, and the function with which to calculate the energy of the system is very complex. As a result, both the sampling and energy calculations are very costly in terms of computational power required. Typically, the free energy landscape is extremely rugged, i.e., it possesses multiple local minima, and it is essentially impossible to perform an exhaustive evaluation of all these minima to identify the one with the globally lowest value. Thus, it is currently unfeasible to use this approach to fold RNA molecules larger than small hairpins, at least not in a fully automated way.

The number of degrees of freedom can be reduced by using coarse-grained models, which treat groups of atoms as united interaction centers, so that a smaller number of elements and interactions need to be considered (review in ref. 49). Additionally, the force field derived for the united interaction centers yields a much smoother energy surface than that for the all-atom energy function. As a result, many local energy minima are removed, in which the system could become trapped during the simulation. Another approach to template-free modeling involves the use of previously solved structures—as a source of structural fragments, parameters of the scoring function, or both. This type of structure prediction is often termed "*de novo* modeling," and should not be confused with the "ab initio" modeling, as it relies not on the "first principles" of physics, but on information from databases. The use of information of databases can be also combined with coarse graining of the representation.

Recently, a number of new methods have been developed that allow for coarse-grained folding simulations of RNA structure. YUP [50] and NAST [51] represent RNA by just one pseudoatom per nucleotide residue. Vfold [52], DMD [53], and the first version of SimRNA [54] represent RNA by three pseudoatoms per residue. CG [55] and the newest version of SimRNA represent RNA with five pseudoatoms per residue. Finally, HiRE-RNA [56] uses six or seven pseudoatoms for purine or pyrimidine residues, respectively. These methods use various scoring functions with different emphasis on "physical" and "statistical" elements.

De novo methods for protein structure prediction that use structural fragments and a scoring function at least partially derived from the analysis of the structures in the database are represented by FARNA/FARFAR [2, 38] and MC-Fold|MC-Sym [1] (Table 1).

As in the field of protein structure prediction, template-free methods can be useful for predicting tertiary structures for RNA sequences up to 100 nt long. Longer molecules are increasingly more difficult to model, even with the use of experimental data in the form of spatial restraints. Thus, the modeling of large RNA molecules has to rely on the use of templates.

5 Refinement of RNA 3D Structural Models

Essentially all models of RNA 3D structure contain errors. Modeling errors can range from a completely wrong global structure to minor local inaccuracies. Large global errors may result from using a wrong template in comparative modeling, using a good template but a wrong alignment, or from having a template-free method converge to a wrong minimum of the scoring function that does not correspond to the true native structure. Large errors can be also due to the use of wrong constraints (e.g., incorrectly predicted secondary structure) either at the stage of the template search or in template-free folding. However, even models that are globally correct (i.e., exhibit correct secondary structure and have native-like tertiary contacts as well as the global shape) often contain various local errors caused by various inaccuracies of the modeling protocols. Local errors often include improper bond lengths and angles, steric clashes, cavities in the structure, missing hydrogen bonds, and distorted base-pairing and stacking interactions.

To improve a structure resulting from any type of automated modeling, one can use a program for structure refinement. The primary goal is to optimize local properties of a model (geometry, pairing and stacking interactions, hydrogen bonds, and electrostatic interactions), while keeping the overall architecture unchanged. Generally, such a program applies a more rigid set of rules than the original modeling program. Ideally, the refinement program takes all physical properties of the molecule into account that were not considered during the modeling, in addition to those already included in the modeling procedure. Eventually a full physical force field can be applied, but only to find a local energy minimum, rather than to sample all possible conformations in search for a global energy minimum.

Recently, RASP, a fine-grained force field to evaluate RNA models or assist in their improvement, has been developed [12]. This method was reported to be accurate in distinguishing models close to the native structure generated by the FARFAR program [2].

5.1 Local Refinement

Most models constructed from templates need local refinement, even those that are based on correct alignments to correct templates. Often small inaccuracies are introduced, e.g., when indels are being modeled from the fragment library in ModeRNA, but no fragment fits ideally to the insertion site, or when MMB fails to satisfy two restraints simultaneously and tries to make a compromise. Many of these errors are local and can be repaired by taking only fragments of structure into consideration, without editing the entire structure.

Backbone breaks are relatively easy to repair. As long as the residues to be connected are not too far from each other, finding a backbone conformation connecting them presents a simple task of the loop closure. For this, there are many algorithms such as CCD or FCCD. The CCD algorithm [57] is an iterative procedure that tries to optimize the torsion angles of a mobile chain of atoms. It attempts to minimize the RMSD between three terminal mobile atoms with three static atoms of a target site. One after another, the torsion angles are twisted by an angle increment θ that minimizes the RMSD of the three mobile and static atoms. The values for θ

are determined from the first derivative of the RMSD (θ) function. By repeating this procedure for all torsion angles, the RMSD is minimized and the mobile chain may eventually close the gap.

For longer chains or unfavorable starting conformations, the CCD algorithm may end up in local minima. This problem is avoided by its extension, the FCCD algorithm [58] that follows the same iterative approach as CCD—it repetitively alters bond after bond until a RMSD threshold is reached. The difference is that FCCD determines a rotation matrix for a part of the mobile chain directly and rotates it around a single atom at its end. As a consequence, FCCD not only changes the torsion angle but also the flat angle at the atom rotated, possibly creating values that are chemically unreasonable. The bond lengths are conserved by both procedures. When the mobile chain is just a few atoms long, such as the phosphodiester bridge between the O3' and C5' between two adjacent RNA residues, a simple minimization procedure can try to correct the flat angles after FCCD has closed the gap. An implementation of both the CCD and FCCD algorithms and the correction procedure are available within the ModeRNA program.

When several residues need to be corrected—e.g., to remove a clash—one can use the MMB program, which is capable of freezing most of the structure and move only a section in order to reach a relaxed conformation. This option is straightforward to use, but it has the disadvantage that the forces applied only take standard interactions into account, e.g., all Watson–Crick base pairs are expected to have the same ideal conformation.

5.2 Refinement of the Entire Model

When a model contains many local distortions, or the overall geometry is to be improved (e.g., to maximize the hydrogen bonding), a global refinement is required. This is often the case with structures generated by *de novo* modeling methods that use a reduced representation of the RNA. Recently, Bernauer et al. have developed a knowledge-based potential for RNA which can be used for RNA refinement [59]. The potential is available in a coarse-grained form, allowing for fast simulations of RNA molecules, e.g., for folding, and an all-atom form for optimizing structural details. The authors report that their potential outperforms the RNA potential in Rosetta [2]. The only disadvantage of this method that we can see is that it is not available as a stand-alone program yet.

Another important group of physics-based methods usable for refinement are Molecular Dynamics (MD). They are generally too calculation-intensive to fold longer RNA, but can be useful for optimizing a model achieved by other means. Three force fields have been widely used for molecular dynamics of nucleic acids: GROMOS [60], AMBER [61], and CHARMM27 [62]. The precision of MD methods can be measured by quantum mechanical calculations for small sets of 30–50 atoms. It has been shown that the AMBER force field represents stacking and hydrogen bonding interactions well enough to reproduce conformations from high-resolution crystal structures and to explore reaction mechanisms of ribozymes [63, 64].

The force fields typically have more problems with polarizable groups, in particular the phosphate group, and polarization induced by divalent cations [65]. This means that the MD is less accurate for RNA backbones. Problems with sampling backbone conformations in DNA have been reported [66], but subtle backbone features have been successfully reproduced as well [67]. In the most recent edition of the GROMOS force field, the representation of backbone torsion angles has been improved, their accuracy now being close to that of experimental structures [68].

The use of MD software is costly, not only in terms of computation time, but also because a lot of effort has to be put into preparing an RNA model to make it acceptable for an MD package. Several steps are necessary to refine a structure, including the following:

- Atom and residue names have to comply to a rigorous standard imposed by the MD package.
- The structural model needs to be placed in a virtual water box (unless an implicit solvent model is used, which can lead to inaccurate electrostatic calculations, especially for a highly charged molecule like RNA).
- The water molecules overlapping with the RNA need to be removed, and the coordinates of water molecules need to be relaxed in order to close small cavities left by removal of water molecules near the RNA surface.
- Counterions need to be added to the solvent and the solvent needs to be relaxed again.
- The minimization needs to be executed for the entire structural ensemble.
- Finally, the energy output needs to be evaluated, and coordinates of the minimized structure without water and ions need to be extracted.

Executing these steps with proper parameters using one of the MD packages mentioned above easily requires an effort that exceeds that of the original modeling. This technical barrier is lowered by the OpenMM Zephyr program [69]. It provides a graphical interface to the GROMACS MD package that allows performing all the preparation steps automatically. It uses graphical processors (GPUs) to perform the calculation and can thus be used with acceptable performance on powerful office computers and does not require access to a computing cluster or a supercomputing facility. Zephyr uses the AMBER force field. Another program that is straightforward to use is the commercial software HyperChem (www.hyper.com). It features several molecular mechanics options, including an AMBER force field and six procedures for energy minimization.

One problem common to all force fields mentioned so far is that they apply physically realistic electrostatic interactions, while the typical model is not presented in a realistic environment, because of improper ion positioning. As a consequence, the repulsion of negative charges on phosphate groups may lead to a steady distortion or unfolding of the structure during minimization. In reality, the negative charges are stabilized by positively charged counterions (Mg^{2+}, K^{+}, Na^{+}) that have strong effects on RNA folding [70, 71]. One part of these ions is bound to specific binding sites on the RNA surface, others bind through a hydration shell or are displaced in vicinity of the RNA. Consequently, the exact locations of ions in RNA models are not known, and including them in the abovementioned minimization procedures is troublesome for that reason. Finally, if the distortions present in a preliminary model are too large, the force field may be unable to find a folded minimum energy conformation and unfolds the structure instead. Also, if the model is in a conformation confined by energy barriers (>5 kcal/mol), an MD simulation may not be able to move the system to an alternative state [72].

It must be noted that most force fields are capable of handling only the four basic ribonucleosides (A, U, G, C). However, many RNA molecules contain chemically modified residues. The AMBER force field has been extended to allow for the simulation of dynamics, energy minimization, and refinement of some of the noncanonical nucleosides [73]. However, in general the refinement of modified nucleosides and their contacts with the rest of the RNA molecule in the context of a 3D structure is a difficult problem and has not been automated yet in any of the modeling packages.

Summarizing, an ideal procedure for refining RNA models does not exist yet. To improve the current situation, it would be worthwhile to have a number of improvements, such as a reasonable simplification of existing force fields that prevents the phosphates from drifting apart through electrostatic repulsion while not keeping them completely immobile, as well as to improve the handling of the modified versions of the canonical ribonucleosides.

Acknowledgements

We thank Sam Flores for fruitful discussion. The development of modeling tools in the Bujnicki laboratory has been supported by the grants from the Polish Ministry of Science and Higher Education (MNiSW; grant HISZPANIA/152/2006), the Foundation for Polish Science (FNP, grant TEAM/2009-4/2), Deutsche

Forschungsgemeinschaft (DFG, grant SPP 1258), and by the 6th Framework Programme of the European Commission (EC FP6, Network of Excellence EURASNET, contract number LSHG-CT-2005-518238). J.M.B. has been supported by the European Research Council (ERC, StG grant RNA + P = 123D) and by the "Ideas for Poland" fellowship from the FNP.

References

1. Parisien M, Major F (2008) The MC-Fold and MC-Sym pipeline infers RNA structure from sequence data. Nature 452:51–55
2. Das R, Karanicolas J, Baker D (2010) Atomic accuracy in predicting and designing noncanonical RNA structure. Nat Methods 7: 291–294
3. Armache JP, Jarasch A, Anger AM et al (2010) Cryo-EM structure and rRNA model of a translating eukaryotic 80S ribosome at 5.5-A resolution. Proc Natl Acad Sci USA 107:19748–19753
4. Wallner B, Elofsson A (2003) Can correct protein models be identified? Protein Sci 12: 1073–1086
5. Pawlowski M, Gajda MJ, Matlak R, Bujnicki JM (2008) MetaMQAP: a meta-server for the quality assessment of protein models. BMC Bioinforma 9:403
6. Noivirt-Brik O, Prilusky J, Sussman JL (2009) Assessment of disorder predictions in CASP8. Proteins 77(Suppl 9):210–216
7. CASP-1 (1995) Special issue. Proteins 23
8. Bujnicki JM, Elofsson A, Fischer D, Rychlewski L (2001) LiveBench-1: continuous benchmarking of protein structure prediction servers. Protein Sci 10:352–361
9. Pieper U, Webb BM, Barkan DT et al (2011) ModBase, a database of annotated comparative protein structure models, and associated resources. Nucleic Acids Res 39:D465–D474
10. Motono C, Nakata J, Koike R et al (2011) SAHG, a comprehensive database of predicted structures of all human proteins. Nucleic Acids Res 39:D487–D493
11. Eswar N, Eramian D, Webb B, Shen MY, Sali A (2008) Protein structure modeling with MODELLER. Methods Mol Biol 426:145–159
12. Capriotti E, Norambuena T, Marti-Renom MA, Melo F (2011) All-atom knowledge-based potential for RNA structure prediction and assessment. Bioinformatics 27:1086–1093
13. Sarver M, Zirbel CL, Stombaugh J, Mokdad A, Leontis NB (2008) FR3D: finding local and composite recurrent structural motifs in RNA 3D structures. J Math Biol 56:215–252
14. Rother K, Rother M, Boniecki M, Puton T, Bujnicki JM (2011) RNA and protein 3D structure modeling: similarities and differences. J Mol Model 17(9):2325–2336
15. Guex N, Peitsch MC (1997) SWISS-MODEL and the Swiss-PdbViewer: an environment for comparative protein modeling. Electrophoresis 18:2714–2723
16. Sali A, Blundell TL (1993) Comparative protein modelling by satisfaction of spatial restraints. J Mol Biol 234:779–815
17. Rother M, Rother K, Puton T, Bujnicki JM (2011) ModeRNA: a tool for comparative modeling of RNA 3D structure. Nucleic Acids Res 39:4007–4022
18. Flores SC, Altman RB (2010) Turning limited experimental information into 3D models of RNA. Rna 16:1769–1778
19. Krieger E, Nabuurs SB, Vriend G (2003) Homology modeling. Methods Biochem Anal 44:509–523
20. Bordoli L, Kiefer F, Arnold K, Benkert P, Battey J, Schwede T (2009) Protein structure homology modeling using SWISS-MODEL workspace. Nat Protoc 4:1–13
21. Parisien M, Cruz JA, Westhof E, Major F (2009) New metrics for comparing and assessing discrepancies between RNA 3D structures and models. RNA 15:1875–1885
22. Wadley LM, Keating KS, Duarte CM, Pyle AM (2007) Evaluating and learning from RNA pseudotorsional space: quantitative validation of a reduced representation for RNA structure. J Mol Biol 372:942–957
23. Richardson JS, Schneider B, Murray LW et al (2008) RNA backbone: consensus all-angle conformers and modular string nomenclature (an RNA Ontology Consortium contribution). RNA 14:465–481
24. Sykes MT, Levitt M (2005) Describing RNA structure by libraries of clustered nucleotide doublets. J Mol Biol 351:26–38
25. Flores SC, Wan Y, Russell R, Altman RB (2010) Predicting RNA structure by multiple template homology modeling. Pac Symp Biocomput:216–227

26. Schmidt JP, Delp SL, Sherman MA, Taylor CA, Pande VS, Altman RB (2008) The Simbios National Center: Systems Biology in Motion. Proc IEEE Inst Electr Electron Eng 96:1266–1280
27. Gardner PP, Daub J, Tate JG et al (2009) Rfam: updates to the RNA families database. Nucleic Acids Res 37:D136–D140
28. Saebo PE, Andersen SM, Myrseth J, Laerdahl JK, Rognes T (2005) PARALIGN: rapid and sensitive sequence similarity searches powered by parallel computing technology. Nucleic Acids Res 33:W535–W539
29. Nawrocki EP, Kolbe DL, Eddy SR (2009) Infernal 1.0: inference of RNA alignments. Bioinformatics 25:1335–1337
30. Wilm A, Higgins DG, Notredame C (2008) R-Coffee: a method for multiple alignment of non-coding RNA. Nucleic Acids Res 36:e52
31. Capriotti E, Marti-Renom MA (2010) Quantifying the relationship between sequence and three-dimensional structure conservation in RNA. BMC Bioinformatics 11:322
32. DeLano WL (2002) The PyMOL Molecular Graphics System, Version 1.5.0.4 Schrödinger, LLC
33. Pettersen EF, Goddard TD, Huang CC et al (2004) UCSF Chimera–a visualization system for exploratory research and analysis. J Comput Chem 25:1605–1612
34. Emsley P, Cowtan K (2004) Coot: model-building tools for molecular graphics. Acta Crystallogr D Biol Crystallogr 60:2126–2132
35. Popenda M, Szachniuk M, Blazewicz M et al (2010) RNA FRABASE 2.0: an advanced web-accessible database with the capacity to search the three-dimensional fragments within RNA structures. BMC Bioinformatics 11:231
36. Bindewald E, Hayes R, Yingling YG, Kasprzak W, Shapiro BA (2008) RNAJunction: a database of RNA junctions and kissing loops for three-dimensional structural analysis and nanodesign. Nucleic Acids Res 36:D392–D397
37. Tamura M, Hendrix DK, Klosterman PS, Schimmelman NR, Brenner SE, Holbrook SR (2004) SCOR: structural classification of RNA, version 2.0. Nucleic Acids Res 32:D182–D184
38. Das R, Baker D (2007) Automated *de novo* prediction of native-like RNA tertiary structures. Proc Natl Acad Sci U S A 104:14664–14669
39. Leontis NB, Stombaugh J, Westhof E (2002) The non-Watson-Crick base pairs and their associated isostericity matrices. Nucleic Acids Res 30:3497–3531
40. Lescoute A, Leontis NB, Massire C, Westhof E (2005) Recurrent structural RNA motifs, Isostericity Matrices and sequence alignments. Nucleic Acids Res 33:2395–2409
41. Stombaugh J, Zirbel CL, Westhof E, Leontis NB (2009) Frequency and isostericity of RNA base pairs. Nucleic Acids Res 37:2294–2312
42. Anfinsen CB (1973) Principles that govern the folding of protein chains. Science 181:223–230
43. Hardin C, Pogorelov TV, Luthey-Schulten Z (2002) Ab initio protein structure prediction. Curr Opin Struct Biol 12:176–181
44. Scheraga HA (1996) Recent developments in the theory of protein folding: searching for the global energy minimum. Biophys Chem 59:329–339
45. Zuo G, Li W, Zhang J, Wang J, Wang W (2010) Folding of a small RNA hairpin based on simulation with replica exchange molecular dynamics. J Phys Chem B 114:5835–5839
46. Deng NJ, Cieplak P (2010) Free energy profile of RNA hairpins: a molecular dynamics simulation study. Biophys J 98:627–636
47. Sarzynska J, Reblova K, Sponer J, Kulinski T (2008) Conformational transitions of flanking purines in HIV-1 RNA dimerization initiation site kissing complexes studied by CHARMM explicit solvent molecular dynamics. Biopolymers 89:732–746
48. Sanbonmatsu KY, Tung CS (2007) High performance computing in biology: multimillion atom simulations of nanoscale systems. J Struct Biol 157:470–480
49. Tozzini V (2010) Multiscale modeling of proteins. Acc Chem Res 43:220–230
50. Tan RKZ, Petrov AS, Harvey SC (2006) YUP: a molecular simulation program for coarse-grained and multiscale models. J Chem Theory Comput 2:529–540
51. Jonikas MA, Radmer RJ, Laederach A et al (2009) Coarse-grained modeling of large RNA molecules with knowledge-based potentials and structural filters. RNA 15:189–199
52. Cao S, Chen SJ (2009) A new computational approach for mechanical folding kinetics of RNA hairpins. Biophys J 96:4024–4034
53. Ding F, Sharma S, Chalasani P, Demidov VV, Broude NE, Dokholyan NV (2008) Ab initio RNA folding by discrete molecular dynamics: from structure prediction to folding mechanisms. RNA 14:1164–1173
54. Rother K, Rother M, Boniecki M et al (2012) Template-based and template-free modeling of RNA 3D structure: inspirations from protein structure modeling. In: Leontis NB, Westhof E (eds) RNA 3D structure analysis and prediction. Springer, Berlin
55. Xia Z, Gardner DP, Gutell RR, Ren P (2010) Coarse-grained model for simulation of RNA three-dimensional structures. J Phys Chem B 114:13497–13506
56. Pasquali S, Derreumaux P (2010) HiRE-RNA: a high resolution coarse-grained energy model for RNA. J Phys Chem B 114:11957–11966

57. Canutescu AA, Dunbrack RL Jr (2003) Cyclic coordinate descent: a robotics algorithm for protein loop closure. Protein Sci 12: 963–972
58. Boomsma W, Hamelryck T (2005) Full cyclic coordinate descent: solving the protein loop closure problem in Calpha space. BMC Bioinforma 6:159
59. Bernauer J, Huang X, Sim AY, Levitt M (2011) Fully differentiable coarse-grained and all-atom knowledge-based potentials for RNA structure evaluation. RNA 17:1066–1075
60. Christen M, Hunenberger PH, Bakowies D et al (2005) The GROMOS software for biomolecular simulation: GROMOS05. J Comput Chem 26:1719–1751
61. Case DA, Cheatham TE 3rd, Darden T et al (2005) The Amber biomolecular simulation programs. J Comput Chem 26:1668–1688
62. Foloppe N (2000) D. MA. All-atom empirical force field for nucleic acids: I. Parameter optimization based on small molecule and condensed phase macromolecular target data. J Comput Chem 21:86–104
63. Sefcikova J, Krasovska MV, Spackova N, Sponer J, Walter NG (2007) Impact of an extruded nucleotide on cleavage activity and dynamic catalytic core conformation of the hepatitis delta virus ribozyme. Biopolymers 85:392–406
64. Sefcikova J, Krasovska MV, Sponer J, Walter NG (2007) The genomic HDV ribozyme utilizes a previously unnoticed U-turn motif to accomplish fast site-specific catalysis. Nucleic Acids Res 35:1933–1946
65. Ditzler MA, Otyepka M, Sponer J, Walter NG (2010) Molecular dynamics and quantum mechanics of RNA: conformational and chemical change we can believe in. Acc Chem Res 43:40–47
66. Fadrna E, Spackova N, Stefl R, Koca J, Cheatham TE, 3rd, Sponer J (2004) Molecular dynamics simulations of Guanine quadruplex loops: advances and force field limitations. Biophys J 87:227–242
67. Ditzler MA, Sponer J, Walter NG (2009) Molecular dynamics suggest multifunctionality of an adenine imino group in acid-base catalysis of the hairpin ribozyme. Rna 15:560–575
68. Soares TA, Hunenberger PH, Kastenholz MA et al (2005) An improved nucleic acid parameter set for the GROMOS force field. J Comput Chem 26:725–737
69. Eastman P, Pande V (2010) OpenMM: A Hardware-Independent Framework for Molecular Simulations. Computing in Science & Engineering 12:34–39
70. Draper DE, Grilley D, Soto AM (2005) Ions and RNA folding. Annu Rev Biophys Biomol Struct 34:221–243
71. Draper DE (2008) RNA folding: thermodynamic and molecular descriptions of the roles of ions. Biophys J 95:5489–5495
72. Reblova K, Fadrna E, Sarzynska J et al (2007) Conformations of flanking bases in HIV-1 RNA DIS kissing complexes studied by molecular dynamics. Biophys J 93:3932–3949
73. Aduri R, Psciuk BT, Saro P, Taniga H, Schlegel HB, SantaLucia J (2007) AMBER Force Field Parameters for the Naturally Occurring Modified Nucleosides in RNA. J Chem Theory Comput 3:1464–1475

Chapter 19

Computational Prediction of RNA–RNA Interactions

Rolf Backofen

Abstract

We describe different tools and approaches for RNA–RNA interaction prediction. Recognition of ncRNA targets is predominantly governed by two principles, namely the stability of the duplex between the two interacting RNAs and the internal structure of both mRNA and ncRNA. Thus, approaches can be distinguished into different major categories depending on how they consider inter- and intramolecular structure. The first class completely neglects the internal structure and measures only the stability of the duplex. The second class of approaches abstracts from specific intramolecular structures and uses an ensemble-based approach to calculate the effect of internal structure on a putative binding site, thus measuring the accessibility of the binding sites.

Since accessibility-based approaches can handle only one continuous interaction site, two addition types of approaches were introduced which predict a joint structure for the interacting RNAs. Since this problem is NP-complete, the approaches can handle only a restricted class of joint structures. The first are co-folding approaches, which predict a joint structure that is nested when the both sequences are concatenated. The last and most complex class of approaches impose only the restriction that they discard zipper-like structures. Finally, we will discuss the use of conservation information in RNA-target prediction.

Key words RNA–RNA interaction, RNA target prediction, Accessibility, Inter- and intramolecular base pairs, Joint secondary structure

1 General Principles of RNA–RNA Interactions

Many non-coding RNAs (ncRNAs) are known to regulate target RNAs by base pair interactions. Examples range from prokaryotic antisense RNAs and small RNAs (sRNAs) to eukaryotic small interfering RNAs (siRNAs), microRNAs (miRNAs), and small nucleolar RNAs (snoRNAs).

Recognition of ncRNA targets is predominantly governed by two principles. First, the target specificity is determined by the stability of the duplex that can be formed between the two interacting RNAs. The requirements on the duplex strength may depend on the biological class, e.g., miRNA interaction in plants compared to mammals. In plants, miRNAs cause target cleavage by forming a

Jan Gorodkin and Walter L. Ruzzo (eds.), *RNA Sequence, Structure, and Function: Computational and Bioinformatic Methods*, Methods in Molecular Biology, vol. 1097, DOI 10.1007/978-1-62703-709-9_19,

nearly perfect duplex, i.e., they exhibit almost full complementary to the target, as it is also the case for siRNAs. For mammalian miRNAs, the preferred mode of action is translational inhibition, which allows for a less stable duplex and thus provides a more flexible interaction pattern.

The second component influencing the efficiency of RNA–RNA interaction is the internal structure of both mRNA and ncRNA. Here, different aspects have to be considered. On one hand, secondary structures might block a putative binding site, which can be assessed by the accessibility of the binding site. There are several papers that investigated the influence and importance of this parameter for functional interaction sites (see also Subheading 3.2). On the other hand, there are known cases where internal structure does not lead to in-functional interactions, but instead blocks the formation of a longer duplex that would be preferred in case there was no internal structure. A well-known example is the interaction between the antisense RNA CopA and its target CopT, which could form a full duplex. However, due to internal structure, the full CopA-CopT duplex does not form. Instead, there is an intermediate kissing loop complex that is sufficient for the biological function (*see*, e.g., [1]).

2 RNA–RNA Interaction Prediction Approaches Neglecting Intramolecular Structure

Computational prediction of a target for a specific ncRNA is based on the search for complementary regions between both RNAs. However, the "strength" of complementarity of these regions is measured in many different ways. For the evaluation of the complementarity, solely sequence-based methods like `BLAST` [2] can be used to search for long stretches of complementarity. Furthermore, it is important to consider also the non-Watson Crick G–U base pairs as realized in `GUUGle` [3]. The individual base pair model used by `TargetRNA` [4] can also be considered as a pure sequence-based approach. Here, A–U and G–C base pairs are given the same score. For this reason, [5] introduced a similar model, where the scoring of individual base pairs was inspired by the strength of the base pair, which is especially important for genomes with low GC-content such as *Listeria* [6]. The main advantage of these approaches is their simplicity; the computational costs usually grow at most geometrically with the input length. Another advantage is that one can easily calculate the significance of the matches, which will be discussed later.

The next step of complexity are approaches that do not score the all interaction base pairs independently, but use instead a scoring of stacked base pairs and internal loops as also employed in folding of single RNAs. This thermodynamic scoring of duplexes between two RNA can be considered as a restricted and specialized version of full RNA secondary structure prediction (like in `Mfold` [7] and `RNAfold` [8]). The first approach using this idea was `RNAhybrid` [9] (also implemented as `RNAduplex` in the *Vienna RNA Package* [10]), which main application is the prediction of miRNA targets. Here, the scoring of a base pair (i, k), where i is a position in the mRNA, and k is a position in the ncRNA, depends on the consecutive base pair (i', k') with $i' > i$ and $k < k'$ (assuming that both mRNA and ncRNA are annotated $5' \rightarrow 3'$, and that the interaction is anti-parallel as usual). If $i' = i + 1$ and $k' = k - 1$, then the two paired bases form a stacking, which is usually energetically favorable. Otherwise, the two base pairs close an internal loop or bulge. The energy parameters for this scoring are the same as in RNA secondary structure folding and represent free energies (in kcal/mol) that were derived from experimental data using the nearest neighbor model by [11]. Later, a similar approach was also used in `TargetRNA` for the prediction of bacterial sRNA targets. Both aforementioned approaches use a restriction on the length of internal loops in the mixed duplexes since long internal loops are energetically not favorable and increase the computational complexity (where the maximal loop length L contributes quadratically to the run time). `RNAplex` [12] uses an energy model similar to `RNAhybrid` and `RNAduplex`, but differs in the treatment of internal loops. In the standard energy model, explicit energy tables are used for small internal loops and large internal loops are evaluated using a logarithmic length-dependent term and an asymmetry penalty. In `RNAplex`, the length-dependent term is replaced by an affine gap penalty, which allows to avoid the quadratic factor introduced by the maximal loop length L. In addition, `RNAplex` introduces a penalty term proportional to the interaction length to mimic the effect of interaction site accessibility.

There are several advantages of this simple energy model. First, it models RNA–RNA interactions much more realistic than approaches based on sequence complementarity alone. Furthermore, it allows to take temperature into account, which is an important parameter when considering the stability of duplexes. Second, energy-based approaches are very fast since their computational complexity is comparable to simple local sequence alignment. Third, one can easily calculate the significance (i.e., *p*-values) of the hits, which is again due to similarity to local sequence alignment [9]. For local alignment, it is well-known that scores of optimal local alignments follow an extreme value distribution [13]. The similarity between sequence alignment and RNA–RNA duplexes becomes evident when we consider interaction prediction as matching of complementary sequences. Clearly, different scoring models are used, e.g., the energy model of `RNAhybrid` uses a special scoring of loop interactions compared to gap penalties in sequence alignment. However, experimental investigation showed that the assumption of extreme value-distributed scores can be carried over.

The extreme value distribution for gap-less local alignment can be derived roughly as described in [14, 15]. For the moment, we assume to have only one sequence and ignore the other sequence. We have at every sequence position the same probability p for a match, which corresponds to a coin toss model with probability p to observe "head." Then, a run of ℓ "heads" correspond to an alignment (also called patch) with score ℓ (where a match is scored with 1 and a mismatch is scored with $-\infty$). The expected number $E(\text{Score} \geq \ell)$ of matchings with Score $\geq \ell$ (or equivalently runs of at least length ℓ) in a sequence of length n is given by

$$E(\text{Score} \geq \ell) \approx np^{\ell}. \tag{1}$$

We assume that the length of the maximal run, i.e., alignment, is unique. Then, we obtain for the length (or score) R of this maximal run (or alignment)

$$np^{R} = 1 \quad \text{and hence} \quad R = \log_{\frac{1}{p}}(n). \tag{2}$$

When going back to the original problem of local gap-less alignment of two sequences (being equivalent to finding a local complementary matching region), we have the same match probability p for each pair of positions, leading to nm possible starting positions for a run of matches. The run of length ℓ starting at position (i,j) then occupies positions $(i,j),(i+1,j+1)\ldots(i+\ell-1,j+\ell-1)$. Thus, instead of Equation (2) we simply get

$$nmp^{R} = 1 \quad \text{and hence} \quad R = \log_{\frac{1}{p}}(nm). \tag{3}$$

This shows that the expected length of a maximal run (or score of the alignment) grows with the logarithm of the product of sequence lengths. Thus, in case of hybridization between two sequences of lengths n and m, the duplex formation score has to be normalized by $\log(nm)$ to be comparable, which is the first important insight.

Second, we want to derive the final distribution of scores (and henceforth p-values) from the expected number of runs. Equations 1 to 3 can be rewritten as $E(\text{Score} \geq \ell) \approx nmp^{\ell} = nme^{\ln(p)\ell} = nme^{-\lambda\ell}$ with $\lambda = -\ln(p)$, which provides the expected number of matching regions of Score $\geq \ell$ between two random sequences. Following [13], one has to correct this expected value to $E(\text{Score} \geq \ell) \approx Knme^{-\lambda\ell}$ to compensate for the fact that not all starting positions are independent.

Given that we know the expected number of matching regions with Score $\geq \ell$, the remaining question is what is the likelihood to see a matching region with Score $\geq \ell$ more than once? This is solved using the Poisson distribution, where the probability of observing an event n times, with the event being expected to be seen μ times, is given by

$$\frac{e^{-\mu}\mu^n}{n!}. \tag{4}$$

From our previous calculations, the event "Score $\geq \ell$" is expected to be seen $\mu = E(\text{Score} \geq \ell) = Knme^{-\lambda\ell}$ many times. Thus, using Equation 4 we can calculate the probability of observing Score $\geq \ell$ at least one time as 1– (the probability of observing it zero times), i.e., $1 - \frac{e^{-\mu}\mu^0}{0!}$ or

$$1 - e^{-Knme^{-\lambda\ell}}.$$

This formula represents an instance of an extreme value distribution $P(S \geq x) = 1 - e^{-e^{-(x-u)/s}}$ with location parameter u and scale parameter s, which are in our case $u = (\ln Kmn)/\lambda$ and $s = 1/\lambda$.

For the problem of defining p-values for hybridization energies, one can estimate the scale and location parameters by fitting the extreme value distribution to an empirical distribution generated from normalized hybridization scores for a large set of randomly generated sequences. These sequences are generated using the actual dinucleotide frequency of the mRNA space of interest. It is important to use dinucleotide instead of mononucleotide shuffling since the energy of duplexes depend on the dinucleotide frequencies due to base pair stacking. The scores have to be normalized according to length since longer putative target and ncRNA sequences will tend to have more negative energies.

All aforementioned approaches based on finding complementary regions have the main disadvantage of neglecting intramolecular base pairs. This can result in two consequences. First, these approaches could predict biologically implausible interactions where one of the interacting regions is sequestered in a stable intramolecular structure. Second, they tend to predict too long interactions since the scoring makes it usually more favorable to extend interactions if the effect of breaking intramolecular base pairs is ignored (*see* Fig. 1)

a

```
5'-...GUC                              CACGU AAAACAUAAG...-3'
          UGCAAACAC   ACAACGC
          ACGUUUGUG   UGUUGUG
3'-...UUA          U                   UUGUAGUGUUGGCAU...-5'
```

b

```
5'-...GUC                   C           A  A                ...-3'
         UGCAAACAC ACAACGC ACGU A  AA  C  UAAG   AAAA
         ACGUUUGUG UGUUGUG UGUA U  UU  G  AUUC   UUUU
3'-...UUA        U          U      G  G  G C      GG        ...-5'
```

Fig. 1 Comparison of (**a**) an experimentally verified RNA–RNA interaction (GcvB-*argT* in *Salmonella* taken from [46]) with (**b**) a typical prediction from an approach that neglects internal structure. Highlighted in gray is the part of the prediction that is not present in the validated interaction structure

3 RNA–RNA Interaction Prediction Approaches Considering Internal Structures

The drawbacks of the methods described in the previous section led to the introduction of several other approaches that consider the effects of internal structures in both mRNA and ncRNA. In principle, there are two main classes that can be distinguished (*see* Fig. 2). Approaches from the first class calculate in a first step for each possible region in every sequence the energy required to make the region accessibly, and then combine these energies with duplex energies in a second step. The second class of approaches calculates a common structure for the interacting RNAs in one step. The problem of directly predicting a common structure for two interacting RNAs is NP-complete in general. Hence, one has to impose restrictions on the class of considered structures to make the problem computationally feasible.

3.1 Concatenation Approaches

One of the first type of methods for prediction of RNA–RNA interactions incorporating the internal structure is provided by `pairfold` [16], `RNAcofold` [17], and the method presented by Dirks et al. as part of the `NUpack` package [18]. They predict a joint structure of mRNA and ncRNA by the concatenation of the two sequences using a special linker character. Then, a modified version of the usual RNA folding (like in `Mfold` [7] and `RNAfold` [8]) is applied. In the following, we will thus refer to this class of approaches as *concatenation approaches*. Basically, the recursive structure is the same as in single sequence structure prediction, but there is a special treatment of loops that contain the linker symbol. The reason is that an internal loop containing the linker element is in fact not an internal loop, but consists only of external bases. Technically, this is solved by numbering the first sequence from 1 to n_1 and the second sequence from $n_1 + 1$ to n_1+n_2 with $n_{1,2}$ being the sequence lengths. Then, in the recursion for a hairpin loop between position i and j, we have the linker

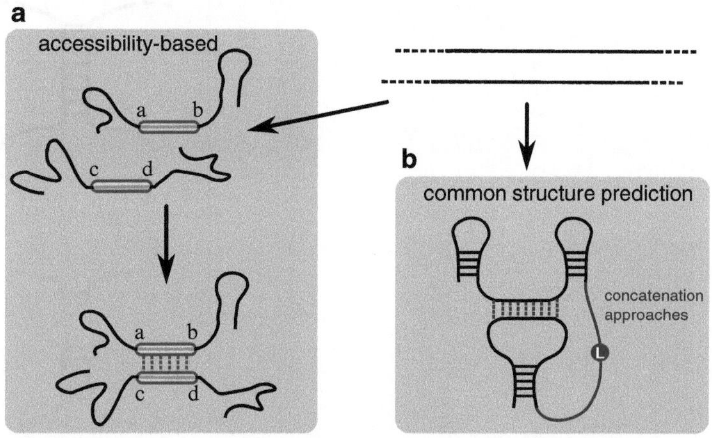

Fig. 2 Two main approaches for prediction of RNA–RNA interactions considering internal structure. (**a**) Accessibility-based approaches use an ensemble approach to determine unfolding energies of interaction sites in a first step, which are then combined with duplex energies. (**b**) In concatenation approaches, the two RNA sequences are joined by a linker symbol and a common secondary structure is predicted. The class of possible common structures has to be restricted to nested structures

symbol contained if $i \leq n_1 < j$. In this case, one may not add the hairpin contribution since all bases $i+1 \ldots j-1$ are external. Similarly, for an internal loop spanning $i \ldots i'$ on the left side and $j' \ldots j$ on the right side, all these bases between $i+1 \ldots i'-1$ and $j'+1 \ldots j-1$ are again external if $i \leq n_1 < i'$ or $j' \leq n_1 < j$.

As a result, these approaches can only predict joint structures that are nested in the two concatenated input sequences, since this is the crucial restriction for the recursive calculation of an RNA secondary structure in general. However, this restriction implies that important known interaction motifs like kissing hairpins (*see* Fig. 3) cannot be predicted. The reason is that those structures form pseudoknots when concatenating the two sequences.

An advantage of the concatenation approach is that all techniques regularly used in RNA secondary structure prediction can be transferred to this cofolding approach. Hence, it is also possible to calculate the partition function of all joint structures as well as base pair probabilities (intramolecular as well as base pairs between the two sequences) using a variant of McCaskill's approach [19]. The partition function Z_S for a sequence S (which might be composed of two sequences using a linker symbol) is the sum of all Boltzmann-weighted energies of all structures Q that sequence S can form, i.e.,

$$Z_S = \sum_{Q \text{ structure of } S} e^{-\frac{E(Q)}{RT}}.$$

Fig. 3 Kissing hairpin interaction between two RNA molecules

After calculation of the partition function, the Boltzmann probability of a specific secondary structure Q can be calculated by $\frac{e^{-\frac{E(Q)}{RT}}}{Z_S}$. Even more important, the probability of a base pair (i, j) (where i, j are positions in either of the sequences) can be calculated using a modification of the partition function computation to sum up the Boltzmann-weighted energies for all structures that contain the base pair, i.e., to calculate

$$Z_S^{(i,j)} = \sum_{Q \text{ contains } (i,j)} e^{-\frac{E(Q)}{RT}}.$$

The probability of the base pair (i, j) is then given by $Z_S^{(i,j)}/Z_S$. Furthermore, the concatenation approach allows for two interacting molecules A and B the calculation of the partition function for the single molecules A and B, for the homo-dimers AA and BB and for the hetero-dimer AB. With these partition functions, it is possible to calculate concentration-dependent melting temperatures, which can be used to estimate affinities.

An interesting application of the concatenation approach is presented in [20, 21]. It addresses the problem that most methods for prediction of bacterial sRNA targets restrict the search to a region around the start codon, but use many different settings concerning the length and position of this region. Here, a classification method was used to solve this problem. First, multiple overlapping regions were considered and the joint minimum free energy structure for all the overlapping regions was calculated using a simple concatenation approach (without a special treatment of the linker sequence). The minimum free energy structure of a region is then used to extract different features such as percent composition of bases in interior loops, bulge loops, etc. In addition, sequence features like the percentage of A+U bases were used, which is supposed to account for putative AU-rich binding sites of the protein Hfq that facilitates interaction formation. In total, 10 features for 1,000 overlapping sequences in the region +30 to −30 around the start codon were computed, which gives rise to a *secondary structure profile*. These features were then used to train two classifier (based on Naive-Bayes and SVM) on a data set of 46 positive samples and 86 negative samples.

3.2 Accessibility-Based Approaches

Since concatenation approaches can only predict nested joint structures, i.e., structures without pseudoknots, they cannot handle important structures like kissing hairpins. For that reason, another class of approaches has been introduced that can handle such interactions. The basic idea is not to predict a single joint structure (or an ensemble of joint structures), but to investigate first the ensemble properties of the single sequences that are important for a putative interaction. Basically, an interaction site must be accessible for binding the interaction partner (i.e., not covered by intramolecular base pairs). Thus, for any two positions $a < b$ in a sequence, one computes the free energy that is required to make the sequence stretch between a and b free of intramolecular base pairs. This can be accomplished by calculating the partition function $Z^{sg(a,b)}$ for the ensemble of structures that leave the putative interaction site ranging from a to b single-stranded (*see* Fig. 4). Now one calculates the *ensemble energy* of these structures by the formula $E^{sg(a,b)} = -RT\ln(Z^{sg(a,b)})$. Defining the energy of the ensemble of all structures by $E_{\text{all}} = -RT\ln(Z)$, where Z is the total partition function, we define the energy $ED(a, b)$ that is required to make the interaction site accessible as

$$ED(a, b) = E^{sg(a,b)} - E_{\text{all}}.$$

Note that this term is positive and thus can be considered as a penalty. Approaches like `RNAup` [22] and `IntaRNA` [23] use precalculated ED-values for all possible interaction regions to calculate a combined energy consisting of the ED-values of both interacting sequences and the energy of the duplex between them. Thus, the energy of an interaction of two regions $i \ldots i'$ of the first sequence and $k \ldots k'$ of the second sequence is calculated as the optimal duplex energy for this interaction regions, plus $ED^{\text{mRNA}}(i, i')$ and $ED^{\text{ncRNA}}(k, k')$. The ED-values for all regions in one sequence can be precalculated with basically the same complexity as the calculation of base pair probabilities in normal RNA folding using the `RNAplfold` approach [24, 25]. For the combined energy E, the recursion is in principal similar to pure hybridization approaches as `RNAhybrid` with the exception that

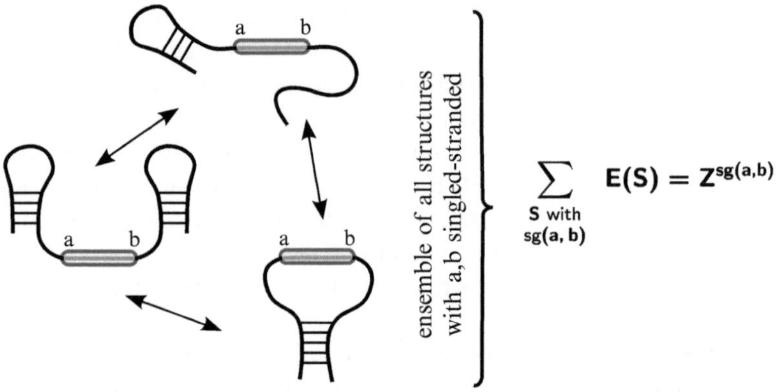

Fig. 4 Energy landscape and accessibility. Given a putative interaction site between positions *a* and *b*, there are several structures where this site is single-stranded, whereas others cover the interaction site. The latter ones cannot be adopted in a joint structure using *a*...*b* as an interaction site. The partition function $Z^{sg(a,b)}$ for the ensemble of structures with the subsequence between *a* and *b* being single-stranded is the sum of all Boltzmann-weighted energies for the structures with *blue ovals* (Color figure online)

all possible right end points have to be considered separately to to determine the optimal region for the ED-values. This approach, which is used in RNAup, leads to a quadratic overhead depending on the maximal length of the interaction site considered. The reason is simply that one has to consider all pairs of regions $i \ldots i'$ of the first sequence and $k \ldots k'$ of the second sequence, storing the energy in a four-dimensional matrix $C(i, i'; k, k')$. This gives rise to an algorithm with a time and space complexity of $O(n^4)$. In IntaRNA, this overhead is avoided by using a heuristic approach. To give an idea of this heuristic approach, let's first consider an optimization that reduces the space complexity of the approach to $O(n^2)$. The idea is to fix a right start point (i', k'), where i' is a position in the first sequence, and k' a position in the second. Then, one can calculate all interactions with this start point by using only a two-dimensional matrix $C^{i',k'}(i, j)$, which stores the best energy of an interaction between $i \ldots i'$ and $k \ldots k'$. The recursion is simply given by

$$C^{i',k'}(i,k) = \min_{\substack{i<j\leq i' \\ k<l\leq k'}} \begin{pmatrix} \text{Loop}(i,k,j,l) + C^{i',k'}(j,l) \\ -ED^{\text{mRNA}}(j,i') - ED^{\text{ncRNA}}(l,k') \\ +ED^{\text{mRNA}}(i,i') + ED^{\text{ncRNA}}(k,k') \end{pmatrix}.$$

Here, $\text{Loop}(i, k, j, l)$ denotes the energy for an internal loop closed by the intermolecular basepairs (i, k) and (j, l) (*see* Fig. 5). Once all interactions with the same right end have been calculated, the optimal interaction for that right end is stored, and the matrix

$$C^{i',k'}(i,k) = \min_{\substack{i<j\leq i'\\ k<l\leq k'}}$$

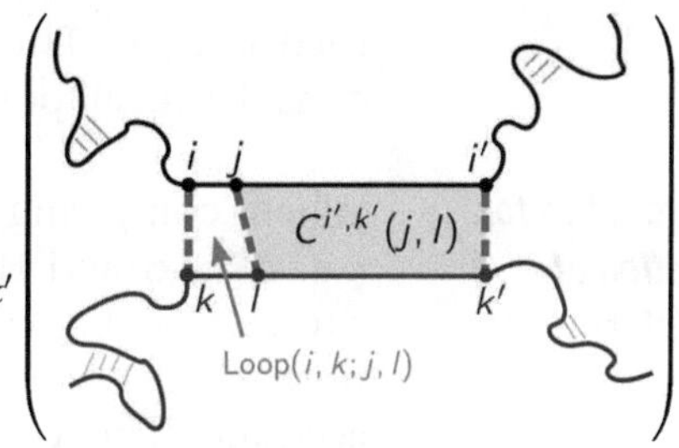

Fig. 5 The optimized recursion for `IntaRNA`

is reused for the next possible right ends. Thus, the algorithm has still $O(n^4)$ time complexity, but only $O(n^2)$ space. The heuristics of `IntaRNA` now does not consider all possible right ends, but uses only the best right ends for each possible left interaction end. Nevertheless, the prediction quality of `RNAup` and `IntaRNA` are basically equal since `IntaRNA` uses a seed condition in addition.

Although the main tools in this class assume only one contiguous region of interaction, the idea can be extended to several interaction sites, however with growing computational complexity. This idea was first exemplified in `biRNA` [26, 27]. It is based on the calculation of joint probabilities for pairs of unpaired regions. To this end, we have to recall that the ensemble energy $E^{sg(a,b)}$ can be used to calculate the probability of the ensemble of structures where region $a \ldots b$ is single-stranded. This probability is given by

$$P[sg(a,b)] = \frac{Z^{sg(a,b)}}{Z} = \frac{e^{-\frac{E^{sg(a,b)}}{RT}}}{e^{-\frac{E_{\text{all}}}{RT}}} = e^{-\frac{E^{sg(a,b)}-E_{\text{all}}}{RT}} = e^{-\frac{ED(a,b)}{RT}}.$$

Thus, we can use the previously calculated ED-values to calculate the probability that a region is single-stranded region or vice versa. Now to calculate the probability (and hence the ensemble energy difference) for a pair of regions a, b and c, d, one has to calculate the joint probability $P[sg(a,b) \wedge sg(c,d)]$. We can calculate this joint probability using conditional probabilities as follows:

$$P[sg(a,b) \wedge sg(c,d)] = P[sg(c,d)|sg(a,b)] \cdot P[sg(a,b)].$$

We already showed that $P[sg(a,b)]$ can be calculated from the ED-values. For determining $P[sg(c,d)|sg(a,b)]$, we can apply a modified version of the ED-algorithm from [28] to determine ED-values (and thus the associated probabilities) for every region $a \ldots b$ to calculate $P[sg(c,d)|sg(a,b)]$ for every region $c \ldots d$ in a single run of this algorithm. Since this involves calculation of the partition function, one has to consider the different classes of combinations for two single-stranded regions $sg(a,b)$ and $sg(c,d)$ (*see* [27] for details). Overall, the complexity is $O(n^3)$ for each $sg(a, b)$. Thus, we obtain the complexity $O(n^3 \cdot n \cdot w)$ when considering all possible regions $a \ldots b$ with length $b - a \leq w$.

3.3 Approaches for the Prediction of General Joint Structures

When comparing the concatenation approaches (RNAcofold, pairFold and NUpack) with the approaches using interaction site accessibility (RNAup and IntaRNA), it becomes obvious that both classes restrict the set of interactions that are taken into account. The concatenation approaches can only predict joint structures where the intermolecular base pairs are not covered by intramolecular base pairs. An equivalent condition is that the intermolecular base pairs between the two RNAs may only occur at external positions in each of the RNA. A position k is external in a structure of an RNA if there is no base pair (i,j) in the structure that covers k (i.e., where $i < k < j$). The approaches using accessibility on the other hand assume a single interaction site, which may not contain bases participating in intramolecular pairings.

These restrictions are due to the fact that the unrestricted problem (i.e., finding the best joint structure of two interacting RNAs without any restriction on the type of structures) is computationally a very hard problem. It was shown in [29] that the general problem is NP-complete, which means in practice that an exact algorithm would require exponential time.[1] He could also show that the problem can be solved in polynomial time if so-called zigzag structures (i.e., zipper-like structure consisting of inter- and intra-molecular structures) are not considered.

However, there are RNA–RNA interactions, e.g., between the sRNA OxyS and its target *fhlA*, that involve more than one kissing hairpin interaction (*see* Fig. 6). Such interactions can neither be predicted by concatenation approaches nor by approaches using accessibility of a single interaction site. For that reason, new methods have been introduced that extend the class of allowed joint structures. The IRIS tool [30] introduced a new recursive scheme that allowed to consider several kissing hairpins for the first time. It uses an energy model that maximizes the number of base pairs. Then [29] considered the extension to a more realistic energy model, which is inspired by the standard nearest neighbor energy model of single RNA sequence folding. Furthermore, they gave a precise definition of the class of structures treated by their approach.

---

[1]The precise definition of NP-complete is more complex. NP is a class of problems, which are currently believed to be different from the class P of problems that can be solved in polynomial time. Unless NP = P (which is believed to be very unlikely), there cannot be an algorithm that exactly solves the general interaction problem in polynomial time for *all* instances. However, there might be algorithms that solve the problem in reasonable time for most practical instances.

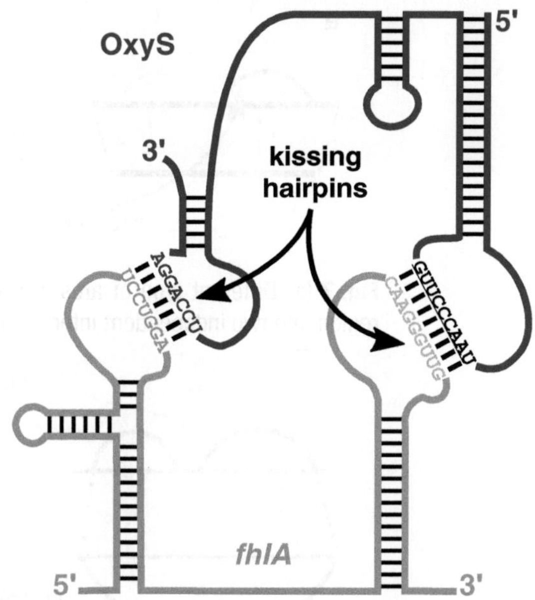

Fig. 6 Interaction model of the sRNA OxyS and its target mRNA *fhlA* as proposed by [47]. The Oxys-*fhlA* interaction involves two kissing hairpins between the interacting RNAs

Both approaches can handle the OxyS-*fhlA* interaction and both approaches predict a single structure with minimal free energy (mfe) (the structure with maximal number of base pairs in the case of IRIS). However, as already observed for single RNA folding, mfe structures are often wrong. The standard way to overcome this problem is to use a partition function variant as already described above for the concatenation approaches. Since one has to calculate the sum over all possible joint structures, it is necessary to reformulate the recursion equations such that every joint structure is decomposed in a unique way. This problem was solved independently by piRNA [31] and RIP [32]. Thus, both approaches allow to calculate important quantities like melting temperatures and base pair probabilities. The melting temperatures calculated by the algorithm are in good agreement with the experimentally measured ones as exemplified in [31] for the wild-type and three mutated constructs of the OxyS-*fhlA* interaction.

The basic recursion type for these approaches is as follows. In the following, we define an interaction arc as an intermolecular base pair between two RNA molecules. The intramolecular base pairs are called direct or indirect arc, depending on whether they are directly spanning an interaction arc or not (*see* Fig. 7a). The main idea of the algorithm is to combine a standard folding algorithm for each single sequence with a recursion that splits an interaction region between the two input sequence into two independent

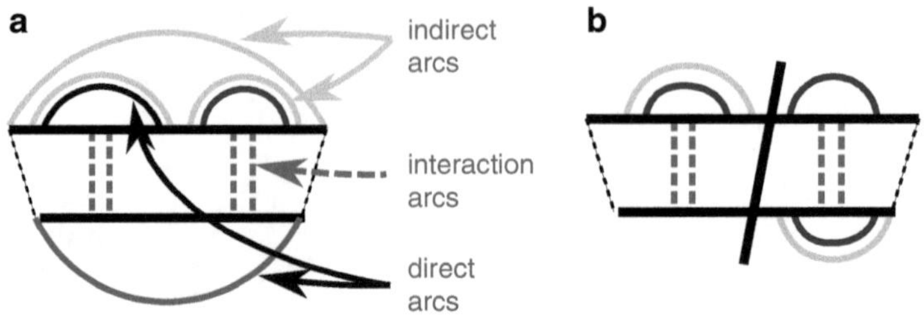

Fig. 7 (**a**) Different type of arcs in a joint structure. (**b**) Split of an interaction region into two independent interaction regions

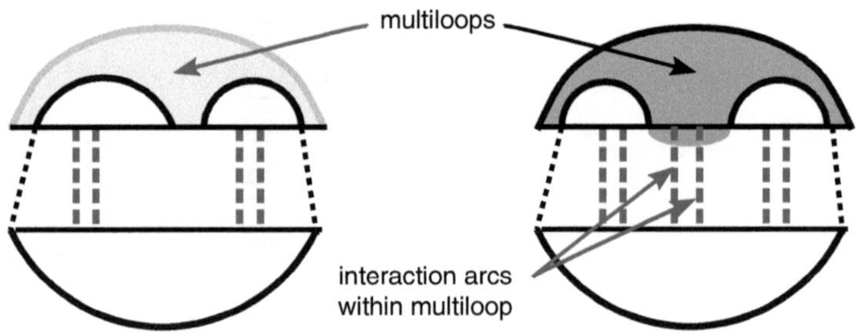

Fig. 8 Multiloops for (**a**) indirect and (**b**) direct arcs. Multiloops closed by indirect arcs are scored using the standard multiloop energies from single sequence folding. Multiloops closed by direct arcs require a different scoring since they include intermolecular base pairs (i.e., interaction arcs)

interaction regions (*see* Fig. 7b). This is only possible if there is no arc spanning this split. Thus, a split assumes that each spanning arc has already been recursively decomposed by the folding recursion for the single sequence. Furthermore, one has to consider all possible split points. Since we have $O(n^4)$ possible interaction regions, and we have to consider $O(n^2)$ splits for each region, the overall complexity is $O(n^6)$ time and $O(n^4)$ space. However, when only the prediction of minimal free energy interactions is of interest, one can use a technique called sparsification to reduce the time complexity to $O(n^4 * \phi(n))$, where $\phi(n)$ is a function that turns out to be $O(n)$ on average [33].

Finally, the energy model has to be adopted to account for all types of multiloops that have to be distinguished. The reason is that the combination of direct arcs and interaction arcs results in special multiloops that require a different scoring (*see* Fig. 8). For instance, the interaction arcs change the number of unpaired bases within the multiloop. To handle all multiloop cases, many different dynamic programming matrices must be introduced (*see* [31] for details).

All aforementioned tools for the prediction of a joint structure have still a very high computational complexity (the computation time is in the order of $O(n^6)$, where n is the length of the input sequence(s)). On one hand, a technique called sparsification, which has already been successfully applied to RNA secondary structure prediction and alignment [34–36], allowed to greatly improve both time and space requirement for the problem of predicting an optimal joint structure [27]. On the other hand, there were attempts to reduce the complexity by considering approximations to the original problem. A very intuitive way is to use accessibilities (as in `RNAup` and `IntaRNA`), but to allow more than one interaction site. The ED-value for measuring the energy required to make a site accessible can be calculated from the probability that this site is single-stranded. It should be immediately clear that these probabilities are not independent for different interaction sites. Thus, conditional probabilities have to be used instead (see Subheading 3.2 for details). Although it initially seemed to be too complex to be calculated, a Bayesian approximation of these conditional probabilities was introduced by [26] and [27]. This allowed a fast calculation of these conditional probabilities and resulted in a fast heuristic method to predict the specific (multiple) binding sites of two interacting RNAs.

4 Comparative RNA–RNA Interaction Prediction Approaches

The still existing problem of all aforementioned tools is their high false positive rate in genome-wide target predictions. A possible way to tackle this problem is the inclusion of additional biological knowledge (*see*, e.g., [37, 38]). However, another reasonable way is the use of comparative information to improve the prediction accuracy. This has both advantages and disadvantages. As shown in [39], conservation of interaction sites is not a general feature of all interactions, in contrast to accessibility since true interaction sites are significantly more accessible than random interaction sites. If an interaction site, however, is conserved, then this is a strong signal and drastically reduces the false positive rate.

Based on the observation that RNA–RNA interaction prediction can be considered as prediction of a common RNA secondary structure, it is more than natural to carry over the concept of RNA consensus structures to interaction prediction. For predicting a RNA consensus structure for a given alignment, there are mainly three strategies employed. The first strategy is to rely purely on phylogeny as realized in, e.g., `Pfold` [40]. The second strategy is to use thermodynamic folding and do a (more or less) average folding for all sequences in the alignment. `RNAalifold` [41] uses this strategy combined with a measurement of co-variation between the sequences. Finally, `PETfold` [42] unifies both strategies by using a maximum expected accuracy scoring to combine thermodynamic and phylogenetic information. Thus, for a consensus base pair

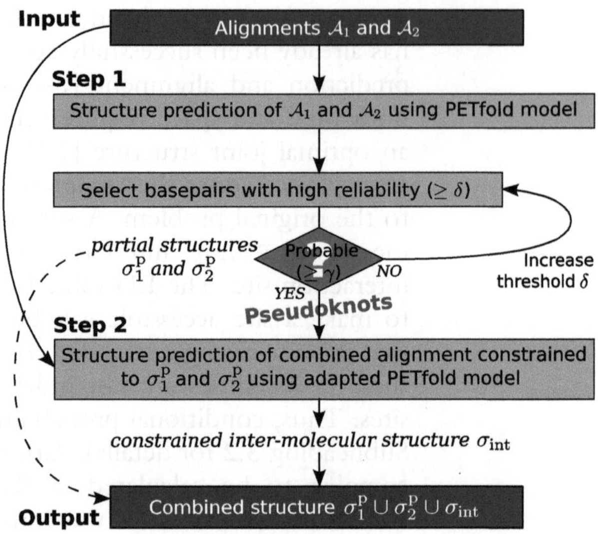

Fig. 9 The PETcofold pipeline according to [43]. In Step 1, highly reliable intramolecular base pairs are determined by PETfold and a set of highly reliable base pairs as selected. Furthermore, the partial structures consisting of the selected base pairs have to have a combined probability within a specified range. In Step 2, an intermolecular folding by an adapted PETfold using the constraints from Step 1 is computed. Finally, the structures from both steps are combined, which allows the introduction of pseudoknots

between two alignment columns, the phylogenetic reliability of this consensus base pair (as calculated by Pfold) is combined with the average base pair probability of the associated base pairs in the individual sequences.

PETcofold [43, 44] is an extension of PETfold to the problem of predicting a consensus interaction. The input consists of two alignments, one for the ncRNA, and the other for the putative target mRNA. However, a direct application of PETfold to the concatenated alignments would have required the use of a partition function version of the full joint structure prediction, which would have been too complex since these approaches typically require $O(n^6)$ time. Nevertheless, the RNAalifold consensus folding strategy was recently also applied to the full $O(n^6)$ joint structure prediction model in [45].

To reduce the time complexity, PETcofold uses a two step approach (*see* Fig. 9). The main idea is to rely on the concatenation approach RNAcofold to predict probabilities of interaction base pairs. However, this alone would have excluded important interaction motifs such as kissing hairpins as described before. Thus, PETcofold uses a hierarchical approach for the joint structure prediction. In the first step, highly reliable consensus base pairs are identified independently in the ncRNA alignment and mRNA alignment using PETfold. The position of these base pairs are considered to be generally bound and, thus, inaccessible for the interaction. In the second step, an adapted `PETfold` using `RNAcofold` instead of `RNAfold` is used with the condition that the previously identified reliable base pairs may not be used for the interaction. This is achieved by constraining all the associated position as single-stranded for `RNAcofold`.

Acknowledgments

I would to thank Andreas Richter for his careful reading and editing of the manuscript, and for many helpful comments and suggestions. This quite improved the manuscript.

References

1. Malmgren C, Wagner EG, Ehresmann C, Ehresmann B, Romby P (1997) Antisense RNA control of plasmid R1 replication. The dominant product of the antisense rna-mrna binding is not a full RNA duplex. J Biol Chem 272(19):12508–12512
2. Altschul SF, Gish W, Miller W, Myers EW, Lipman DJ (1990) Basic local alignment search tool. J Mol Biol 215(3):403–410
3. Gerlach W, Giegerich R (2006) GUUGle: a utility for fast exact matching under RNA complementary rules including G-U base pairing. Bioinformatics 22(6):762–764
4. Tjaden B, Goodwin SS, Opdyke JA, Guillier M, Fu DX, Gottesman S, Storz G (2006) Target prediction for small, noncoding RNAs in bacteria. Nucleic Acids Res 34(9): 2791–2802
5. Mandin P, Repoila F, Vergassola M, Geissmann T, Cossart P (2007) Identification of new noncoding RNAs in Listeria monocytogenes and prediction of mRNA targets. Nucleic Acids Res 35(3):962–974
6. Glaser P, Frangeul L, Buchrieser C, Rusniok C, Amend A, Baquero F, Berche P, Bloecker H, Brandt P, Chakraborty T, Charbit A, Chetouani F, Couve E, de Daruvar A, Dehoux P, Domann E, Dominguez-Bernal G, Duchaud E, Durant L, Dussurget O, Entian KD, Fsihi H, Garcia-del Portillo F, Garrido P, Gautier L, Goebel W, Gomez-Lopez N, Hain T, Hauf J, Jackson D, Jones LM, Kaerst U, Kreft J, Kuhn M, Kunst F, Kurapkat G, Madueno E, Maitournam A, Vicente JM, Ng E, Nedjari H, Nordsiek G, Novella S, de Pablos B, Perez-Diaz JC, Purcell R, Remmel B, Rose M, Schlueter T, Simoes N, Tierrez A, Vazquez-Boland JA, Voss H, Wehland J, Cossart P (2001) Comparative genomics of Listeria species. Science 294(5543):849–852
7. Zuker M (1994) Prediction of RNA secondary structure by energy minimization. Methods Mol Biol 25:267–294
8. Hofacker IL, Fontana W, Stadler PF, Bonhoeffer S, Tacker M, Schuster P (1994) Fast folding and comparison of RNA secondary structures. Monatshefte Chemie 125:167–188
9. Rehmsmeier M, Steffen P, Höchsmann M, Giegerich R (2004) Fast and effective prediction of microRNA/target duplexes. RNA 10(10):1507–1517
10. Gruber AR, Lorenz R, Bernhart SH, Neubock R , Hofacker IL (2008) The Vienna RNA websuite. Nucleic Acids Res 36(Web Server issue): W70–W74
11. Mathews DH, Sabina J, Zuker M, Turner DH (1999) Expanded sequence dependence of thermodynamic parameters improves prediction of RNA secondary structure. J Mol Biol 288(5):911–940
12. Tafer H, Hofacker IL (2008) RNAplex: a fast tool for RNA-RNA interaction search. Bioinformatics 24(22):2657–2663
13. Karlin S, Altschul SF (1990) Methods for assessing the statistical significance of molecular sequence features by using general scoring schemes. Proc Natl Acad Sci USA 87(6):2264–2268
14. Waterman MS (1995) Introduction to computational biology - maps, sequences and genomes. London, England
15. Pearson WR, Wood TC (2001) Statistical significance in biological sequence comparison. In: Balding DJ, Bishop M, Cannings C (eds) Handbook of statistical genetics. Chichester, UK, pp 39–65
16. Andronescu M, Chuan Zhang Z, Condon A (2005) Secondary structure prediction of interacting RNA molecules. J Mol Biol 345(5):987–1001

17. Bernhart SH, Tafer H, Mückstein U, Flamm C, Stadler PF, Hofacker IL (2006a) Partition function and base pairing probabilities of RNA heterodimers. Algorithms Mol Biol 1(1):3
18. Dirks RM, Bois JS, Schaeffer JM, Winfree E, Pierce NA (2007) Thermodynamic analysis of interacting nucleic acid strands. SIAM Rev 49(1):65–88
19. McCaskill JS (1990) The equilibrium partition function and base pair binding probabilities for RNA secondary structure. Biopolymers 29(6–7):1105–1119
20. Cao Y, Zhao Y, Cha L, Ying X, Wang L, Shao N, Li W (2009) sRNATarget: a web server for prediction of bacterial sRNA targets. Bioinformation 3(8):364–366
21. Zhao Y, Li H, Hou Y, Cha L, Cao Y, Wang L, Ying X, Li W (2008) Construction of two mathematical models for prediction of bacterial sRNA targets. Biochem Biophys Res Comm 372(2):346–350
22. Mückstein U, Tafer H, Hackermüller J, Bernhart SH, Stadler PF, Hofacker IL (2006) Thermodynamics of RNA-RNA binding. Bioinformatics 22(10):1177–1182
23. Busch A, Richter AS, Backofen R (2008) IntaRNA: efficient prediction of bacterial sRNA targets incorporating target site accessibility and seed regions. Bioinformatics 24(24):2849–2856
24. Bernhart SH, Hofacker IL, Stadler PF (2006b) Local RNA base pairing probabilities in large sequences. Bioinformatics 22(5): 614–615
25. Bompfünewerer AF, Backofen R, Bernhart SH, Hertel J, Hofacker IL, Stadler PF, Will S (2008) Variations on RNA folding and alignment: lessons from Benasque. J Math Biol 56(1–2):129–144
26. Chitsaz H, Backofen R, Cenk Sahinalp S (2009a) biRNA: Fast RNA–RNA binding sites prediction. In: Salzberg S, Warnow T (eds) Proc. of the 9th workshop on algorithms in bioinformatics (WABI), vol 5724 of Lecture notes in computer science. Springer, Berlin/Heidelberg, pp 25–36
27. Salari R, Backofen R, Cenk Sahinalp S (2010a) Fast prediction of RNA-RNA interaction. Algorithms Mol Biol 5:5
28. Mückstein U, Tafer H, Bernhart SH, Hernandez-Rosales M, Vogel J, Stadler PF, Hofacker IL (2008) Translational control by RNA–RNA interaction: Improved computation of RNA–RNA binding thermodynamics. In: Elloumi M, Küng J, Linial M, Murphy R, Schneider K, Toma C (eds) Bioinformatics research and development, vol 13 of Communications in computer and information science. Springer, Berlin/Heidelberg, pp 114–127
29. Alkan C, Karakoç E, Nadeau JH, Cenk Sahinalp S, Zhang K (2006) RNA-RNA interaction prediction and antisense RNA target search. J Comput Biol 13(2):267–282
30. Pervouchine DD (2004) IRIS: intermolecular RNA interaction search. Genome Inform 15(2):92–101
31. Chitsaz H, Salari R, Cenk Sahinalp S, Backofen R (2009b) A partition function algorithm for interacting nucleic acid strands. Bioinformatics 25(12):i365–i373
32. Huang FWD, Qin J, Reidys CM, Stadler PF (2009) Partition function and base pairing probabilities for RNA-RNA interaction prediction. Bioinformatics 25(20):2646–2654
33. Salari R, Möhl M, Will S, Cenk Sahinalp S, Backofen R (2010b) Time and space efficient RNA-RNA interaction prediction via sparse folding. In: Berger B (ed) Proc of RECOMB 2010, vol 6044 of Lecture notes in computer science. Springer, Berlin/Heidelberg, pp 473–490
34. Wexler Y, Ben-Zaken Zilberstein C, Ziv-Ukelson M (2006) A study of accessible motifs and rna folding complexity. In: Apostolico A, Guerra C, Istrail S, Pevzner PA, Waterman MS (eds) Proc. of the tenth annual international conferences on computational molecular biology (RECOMB'06), vol 3909 of Lecture notes in computer science. Springer, Berlin/Heidelberg, pp 473–487
35. Ziv-Ukelson M, Gat-Viks I, Wexler Y, Shamir R (2008) A faster algorithm for RNA cofolding. In: Crandall KA, Lagergren J (eds) WABI 2008, vol 5251 of Lecture notes in computer science. Berlin Heidelberg, pp 174–185
36. Backofen R, Tsur D, Zakov S, Ziv-Ukelson M (2009) Sparse RNA folding: Time and space efficient algorithms. In: Kucherov G, Ukkonen E (eds) Proc. 20th symp. combinatorial pattern matching, vol 5577 of LNCS. Springer, pp 249–262
37. Richter AS, Schleberger C, Backofen R, Steglich C (2010) Seed-based IntaRNA prediction combined with GFP-reporter system identifies mRNA targets of the small RNA Yfr1. Bioinformatics 26(1):1–5
38. Sonnleitner E, Gonzalez N, Sorger-Domenigg T, Heeb S, Richter AS, Backofen R, Williams P, Huttenhofer A, Haas D, Blasi U (2011) The small RNA PhrS stimulates synthesis of the Pseudomonas aeruginosa quinolone signal. Mol Microbiol 80(4):868–885
39. Richter AS, Backofen R (2012) Accessibility and conservation - general features of bacterial small RNA-mRNA interactions? RNA Biol 9(7):954–965
40. Knudsen B, Hein J (2003) Pfold: RNA secondary structure prediction using stochas-

tic context-free grammars. Nucleic Acids Res 31(13):3423–3428
41. Hofacker IL, Fekete M, Stadler PF (2002) Secondary structure prediction for aligned RNA sequences. J Mol Biol 319(5):1059–1066
42. Seemann SE, Gorodkin J, Backofen R (2008) Unifying evolutionary and thermodynamic information for RNA folding of multiple alignments. Nucleic Acids Res 36(20): 6355–6362
43. Seemann SE, Richter AS, Gesell T, Backofen R, Gorodkin J (2011) PETcofold: predicting conserved interactions and structures of two multiple alignments of RNA sequences. Bioinformatics 27(2):211–219
44. Seemann SE, Richter AS, Gorodkin J, Backofen R (2010) Hierarchical folding of multiple sequence alignments for the prediction of structures and RNA-RNA interactions. Algorithms Mol Biol 5:22
45. Li AX, Marz M, Qin J, Reidys CM (2011) RNA-RNA interaction prediction based on multiple sequence alignments. Bioinformatics 27(4):456–463
46. Sharma CM, Darfeuille F, Plantinga TH, Vogel J (2007) A small RNA regulates multiple ABC transporter mRNAs by targeting C/A-rich elements inside and upstream of ribosome-binding sites. Genes Dev 21(21):2804–2817
47. Argaman L, Altuvia S (2000) *fhlA* repression by OxyS RNA: kissing complex formation at two sites results in a stable antisense-target RNA complex. J Mol Biol 300(5):1101–1112

Chapter 20

Computational Prediction of *Micro*RNA Genes

Jana Hertel, David Langenberger, and Peter F. Stadler

Abstract

The computational identification of novel microRNA (miRNA) genes is a challenging task in bioinformatics. Massive amounts of data describing unknown functional RNA transcripts have to be analyzed for putative miRNA candidates with automated computational pipelines. Beyond those miRNAs that meet the classical definition, high-throughput sequencing techniques have revealed additional miRNA-like molecules that are derived by alternative biogenesis pathways. Exhaustive bioinformatics analyses on such data involve statistical issues as well as precise sequence and structure inspection not only of the functional *mature* part but also of the whole *precursor* sequence of the putative miRNA. Apart from a considerable amount of species-specific miRNAs, the majority of all those genes are conserved at least among closely related organisms. Some miRNAs, however, can be traced back to very early points in the evolution of eukaryotic species. Thus, the investigation of the conservation of newly found miRNA candidates comprises an important step in the computational annotation of miRNAs.

Topics covered in this chapter include a review on the obvious problem of miRNA annotation and family definition, recommended pipelines of computational miRNA annotation or detection, and an overview of current computer tools for the prediction of miRNAs and their limitations. The chapter closes discussing how those bioinformatic approaches address the problem of faithful miRNA prediction and correct annotation.

Key words miRNA, Machine learning, Homology, Structure conservation

1 Introduction

MicroRNAs (miRNAs) constitute an abundant class of short RNA molecules. With an average length of 22 nucleotides they are found throughout most eukaryotic clades. MiRNAs operate as post-transcriptional regulators of gene expression via specific complementary binding of target mRNA transcripts for cleavage or translational repression. Influencing the regulation of many protein-coding genes, miRNAs are presumably involved in most biological processes [1]. Different cell types and tissues vary significantly in the set of expressed miRNAs [2]. The expression of miRNAs also deviates substantially in numerous disease

Jan Gorodkin and Walter L. Ruzzo (eds.), *RNA Sequence, Structure, and Function: Computational and Bioinformatic Methods*, Methods in Molecular Biology, vol. 1097, DOI 10.1007/978-1-62703-709-9_20,

states which makes miRNA-based therapies an interesting field of research (*see*, e.g., [3] or [4]).

Canonical miRNAs are processed in a multi-step process from polymerase II transcripts (pri-miRNAs) that undergo splicing and polyadenylation. In the first step, hairpin-shaped precursors (pre-miRNAs) are excised, exported to the cytoplasm, and then cleaved by the type III RNAse Dicer to release the functional mature miRNA (miR), which then binds to Argonaut proteins. Thus, canonical miRNAs form one particular class of endogenous substrates of the RNAi machinery. In recent years, a variety of non-canonical processing pathways have been found to lead to small RNAs that function like the canonical miRNAs. This diversity of pathways is an important issue for the prediction of miRNA genes, in particular since it became clear that secondary structure of the precursor RNA determines the processing pathway in vivo [5]. Below we will briefly review the major types of the biogenesis of miRNAs and miRNA-like small RNAs.

A second important issue is the phylogenetic distribution of miRNAs and the patterns of their evolution. It appears that miRNAs have originated independently in several eukaryotic lineages as endogenous regulators that act via an RNAi-like pathway. The best understood classes are animal and plant miRNAs.

The canonical miRNAs of metazoa (except those in sponges [6]) evolve extremely slowly. At present, hundreds of miRNA families are compiled in the `Rfam` database [7]. Genomically, miRNAs are often located within introns of protein-coding genes—and are therefore co-transcribed together with their host gene. Alternatively, miRNAs are encoded in independent non-coding transcripts known as pri-miRNAs [8, 9]. These are often referred to as "intergenic," as the primary transcripts were mostly unknown in the early history of miRNA research. The pri-miRNAs are often poly-cistronic giving rise to multiple pre-miRNA hairpins, thus explaining the genomic clustering of many miRNAs [10, 11]. Drosha cleavage to produce pre-miRNA hairpins occurs co-transcriptionally on both independently transcribed and intron-encoded canonical pre-miRNAs [12].

The number of miRNAs in plants differs significantly to those in animals. While there are much fewer miRNAs known in plants, the size of their precursor sequence is much larger than those found in animal genomes [13]. The precursor and even the mature sequence of animal miRNAs are often diverged a lot within one miRNA family. In plants, on the other hand, not only the mature but also the whole precursor sequences of miRNAs of one family are highly similar. This may indicate that the expansion of plant miRNAs has a more recent origin while the origin of animal miRNAs is ancient [14, 15], suggesting that plant and animal miRNAs have been originated independently. The different functional modes of perfect target binding in plants compared to the target binding of animal miRNAs, where some mismatches are allowed, support this thesis.

Recently, miRNAs also have been discovered in the unicellular organism green algae *Chlamydomonas reinhardtii* [16, 17]; soil-living amoeba *Dictyostelium discoideum* [18] and sponges [6]. Although these RNAs are not recognizable as homologs of miRNA families conserved within either multicellular animals or multicellular plants, they share functional characteristics with typical miRNAs. In particular, it has been proven that they can direct cleavage of target mRNAs both in vitro and in vivo. The change of their expression patterns during gamete differentiation in algae and amoeba suggests a possible role of those miRNAs in the regulation of sexual reproduction. In sponges, a complete small RNA processing machinery for animals can be found. Nonetheless, the miRNAs of poriferans, cnidarians, and bilaterians show great variation in precursor sizes and mature miRNA sequences. Clearly, the emergence of miRNAs is not linked to the evolution of multicellularity. On the other hand, there is no evidence of a universally conserved miRNA. Together with the rapid innovation of miRNA families in both animal and plants, the most plausible hypothesis is still an independent origin in several lineages.

MiRNAs can also be found in virus. Here miRNAs are used to change processes of host cells or replicated viruses. In particular, these changes help to escape from the immune surveillance of their hosts. While no virus miRNA is conserved among different virus families or between viruses and their host species, their biogenesis is similar to those of plants and animals. For a more detailed review on viral miRNAs we refer to [19].

Canonical miRNA genes in both animals and plants are among the most slowly evolving sequences in animal genomes. On the other hand, there is an ongoing process of innovation of novel miRNAs, as well as proliferation of paralogs by local and non-local gene-duplications. The strong stabilizing selection of miRNAs, which is often a consequence of their involvement in critical processes in early development, renders the complete loss of a miRNA family a rather rare event. Thus, miRNAs have been advocated as powerful phylogenetic markers, *see*, e.g., [14, 20, 21]. In particular, this field of application requires a detailed understanding of the evolutionary patterns of miRNAs, and hence a reliable recognition of homology. It was recognized early in the history of miRNA research that the sequence of the mature miRNA remains nearly unchanged over extensive evolutionary time-scales, and even ancient paralogs often differ only by a couple of nucleotides. This has led to the definition of miRNA families and in particular to a nomenclature system that attempts to encode orthology at least approximately [22]. Several studies hence use near-identity of the mature sequence as the only criterion to define families [20, 23].

More distant homologs, however, are not always trivially recognizable. The selection pressure on duplicated copies of miRNAs is relaxed at least temporarily, sometimes allowing significant sequence divergence of paralogs. As a consequence, stringent comparisons of mature sequences alone are at times insufficient to establish family membership [24]. Detailed investigations of miRNA evolution thus consider the entire sequence of the precursor, requiring that the precursor of putative homologs is alignable in both sequence and secondary structure [14, 25]. This is crucial also in cases of "arm-switching", where the predominant mature product changes from one side to the other side of the precursor hairpin [26]. Different mature miRNA sequences thus not necessarily arise from evolutionarily unrelated precursors.

Several studies emphasize the importance of the seed region, canonically comprising nucleotides 2–8 of the mature sequence. Nearly identical seeds have been interpreted as indication of paralogy, *see* e.g. [27–29]. Similarity of seed regions also often goes hand-in-hand with related functions [30]. This is not a compelling argument for paralogy, however. The rapid innovation of novel miRNA families suggest that sequences with the same or similar seeds may just as well, or maybe even more easily, arise *de novo*, compared to a scenario in which ancient paralogs have lost all sequence similarity except for a perfectly conserved heptamer. Conversely, there are several examples where the 5′-end of the mature miRNA is shifted by one or two nucleotides within a family of easily recognizable paralogs. In addition, 5′ editing of the mature sequence, which also affects the seed region, seems to be a common phenomenon [20]. For the purpose of this contribution, therefore, we define miRNA families in terms of clearly recognizable sequence homology.

Until recently, novel miRNAs were typically detected from cloning experiments and analyzed individually. Next generation sequencing techniques have changed the situation [31]: massive amounts of data are being produced that require automatized analysis pipelines. Canonical miRNAs are by no means the only class of small RNAs detected by such approaches. An increasing number of alternative routes have been described in recent years [32]. The class of miRNAs, as it is currently annotated, thus, is far from homogeneous. Several recent papers emphasize this issue [33–35] in response to conflicting annotations. Most classes of structured "housekeeping RNAs," including tRNAs [36–39] and both classes of snoRNAs [40–43], as well as vault RNA and the Y5 RNA [44] can give rise to small RNAs, which in some cases have been included in `miRBase` [45]. The recently detected class of mirtrons provides further non-canonical precursors that, after splicing, are processed in the same way as canonical pre-miRNAs [46–49].

The computational prediction of miRNAs heavily relies on well-defined training sets of "true" miRNAs that are set

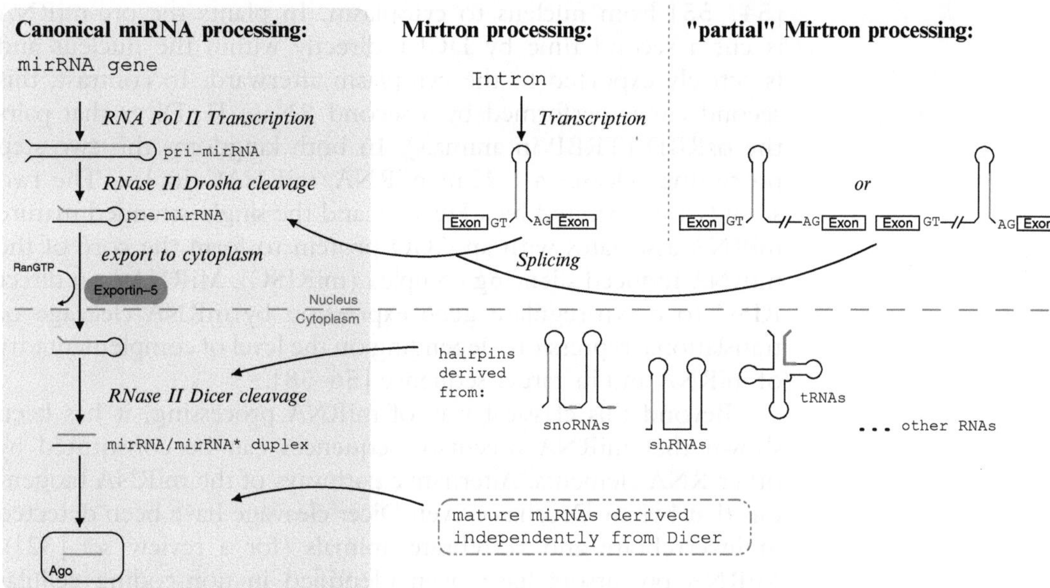

Fig. 1 MiRNA biogenesis pathways

apart by distinctive features from other classes of small RNAs. Since sequence and structure features are the result of selective constraints, it appears crucial to computational miRNA finding methods that the miRNAs used for training share all or at least parts of the processing pathway, *see* Fig. 1. In other words, miRNA gene prediction depends on a consensus in the definition of "miRNA" and a consistent annotation. In contrast, homology-based methods do not crucially depend on such a coherent classification. In the following paragraphs we therefore survey the main features of miRNAs.

Primary miRNA (pri-miRNA) genes are 5′capped, spliced, and polyadenylated polymerase II transcripts. After transcription the RNA molecule folds to a single- or multi stem-loop structure, such that each hairpin contains one miRNA [50]. Consequently, a mechanism that precisely recognizes and accurately excises the functional mature miRNAs is required. In animals miRNAs are processed in two major steps. First, the pri-miRNA is cleaved by the RNase II Drosha which is directed to the base of each single hairpin by the double-stranded RNA binding domain protein (dsRBD) DGCR8 in the nucleus of the cell [51, 52]. The released intermediate precursor sequence (pre-miRNA) shows the characteristic hairpin structure with a 2 nt 3′-overhang. In plants, Drosha itself is not available, though its functionality is provided by the Dicer-like protein DCL1 which also cuts the pri-miRNA and liberates the pre-miRNA(s) [53]. In animals the 2 nt overhang of the pre-miRNA is recognized by the export receptor Exportin-5 and is actively transported via a Ran-GTP-dependent mechanism

[54, 55] from nucleus to cytoplasm. In plants the pre-miRNA is cut a second time by DCL1 directly within the nucleus and is actively exported to the cytoplasm afterward. In contrast, this second cut is performed by a second RNase II, Dicer that pairs the dsRBD (TRBP in animals). In both kingdoms this two-step processing releases a ~22 nt miRNA/miRNA* duplex. The two strands are separated by a Helicase and the single-stranded mature miRNA associates with an AGO protein to form the core of the miRNA-induced silencing complex (miRISC). MiRNAs can direct RISC to down-regulate gene expression by mRNA cleavage or translational repression, depending on the level of complementarity of miRNA and its target sequence [56–58].

Beyond this classical way of miRNA processing, it has been shown that miRNA precursor sequences can be constituted by other RNA elements. Alternative pathways of the miRNA biogenesis that bypass Drosha or even Dicer cleavage have been detected in invertebrate and vertebrate animals (for a review *see* [32]). MiRNA precursors have been identified in non-coding cellular transcripts [10, 11] and in the 3′UTRs of mRNAs [59, 60]. Even transposable elements [61] and viral transcripts [62] can provide genes that act as miRNA precursors. In fruitfly and roundworm several debranched introns are processed by the intron pathway in the nucleus before the transcripts biogenesis converges with the canonical miRNA pathway during hairpin export to the cytoplasm [47, 48]. These miRNA precursors are termed *mirtrons*. In the ancient eukaryote *Giardia lamblia* it has first been shown that small nucleolar RNAs (snoRNAs) can also serve as precursor for miRNAs [63]. Such small RNAs that originate from snoRNAs and can function like miRNAs could also be verified in human for both functional H/ACA and C/D box snoRNAs [64–66].

Both, mirtron and snoRNA derived miRNA processing pathways are Drosha independent but require the activity of the RNase II Dicer to release the miRNA/miRNA* duplex . They exert their function embedded in the RISC complex just like canonical miRNAs.

Plant miRNAs are in some respects very similar to animal miRNAs, in particular in their size but also in their evolutionary conservation which is associated with a varying level of sequence conservation in particular of the mature sequence. There are also major differences, however, that have to be taken into account in computational tools. As in animals, plant miRNAs are processed from primary precursors that are ordinary pol-II transcripts, from which a hairpin-shaped pre-miRNA is excised. In plants this step is controlled by the Dicer-homolog DCL1, not by a homolog of Drosha. DCL1 is also responsible for the extraction of miRNA/miRNA* duplexes. As in animals, the mature miRNAs are incorporated into the RISC complex, *see* [67] for a detailed review. In contrast to animal miRNAs, their plant analogs often, but not always [68], form perfect helices with their targets, *see*, e.g., [69]. In fact, plant miRNAs may originate from inverted duplications of target gene sequences [70]. This makes it feasible to use the existence of putative targets already at the level of miRNA gene prediction. The second important difference is that the precursor hairpins can be much larger and more elaborated structured than their animal counterparts. Furthermore, they may host multiple mature miRNAs in the same stem-loop structure [71], apparently extracted by multiple DCL1 cuts. Analogous multiple Dicer cuts, however, are also observed in animal pre-miRNAs. The resulting moRNAs (miRNA offset RNAs) function like miRNAs at least in some cases [72], hence this may be less of a difference than previously thought.

In summary, all small RNAs that result from the above described pathways seem to act the same: partitioning in the Ago protein complex, directing this complex to the target gene, and finally down-regulating gene expression or repressing their translation. Despite likewise functionality, the variations in their biogenesis and their origin make it difficult for computer programs to detect *all* miRNAs in the transcriptome of a given (set of) species. Specific tools for specifically originated miRNAs are preferable, in particular when analyzing the result of next generation sequencing methods. Not only the fact that a small RNA with a mature miRNA-like length of ~22 nt is expressed and is incorporated into Ago complex defines a miRNA. It is recommended to analyze the whole precursor for the verification of such candidates. The mature part is simply too short and thus carries too few information which can be used for a reliable miRNA classification.

2 MiRNA Databases

The detection of novel miRNAs is based on the information that can be derived from already found genes of this ncRNA class. To date, the `miRBase` database is the most comprehensive collection of miRNAs annotated in plants, viruses, and animals. It stores information about genomic location, sequence, and secondary structure, length of the precursor and the mature sequences and lists related publications of each gene. The annotated miRNAs are retrieved continuously from single gene experiments, next generation sequencing methods, computational prediction and by replicating miRNAs that are sent to lineage specific databases like `Flybase` [73], `Wormbase` [74] , or the more general ncRNA database `Rfam` [7] database. Comprehensive data on plant miRNAs and their targets are compiled in miRBase and in MicroPC (μPC) [75].

This data can be used for the annotation of homologous miRNAs in so far unregarded species, or to find miRNAs *de novo* in any species. The first case relies on the sequences and structures as they are stored in the databases. For the latter case, this data needs to be processed statistically and miRNA-specific characteristics need to be extracted. Although plant pre-miRNAs are more heterogeneous than animal pre-miRNAs in both size and secondary structure [76], they can be efficiently discriminated from other RNAs by machine learning techniques. The problem with this collection of information is that it is not easily distinguishable in which way the annotated miRNA has been processed in the cell. All miRNAs, irrespective of their biogenesis, are combined under the homogeneous term "miRNA." Further problems arise with conflicting annotation of the same genomic location in different databases. Considering all these aspects, those databases can be regarded as helpful source of information. Nevertheless, it is essential for high quality computational prediction methods to carefully gather and analyze the data before creating reliable input datasets.

In the following common methods to predict miRNAs computationally from genomic sequences and deep sequencing data, *de novo* and by homology to already known miRNAs are summarized. In general, the available data of miRNAs is analyzed statistically by means of typical miRNA characteristics for the computational annotation of novel miRNA genes. In combination with similar values from other ncRNA classes or sequences that are expected to be no miRNAs, this data is used to classify unknown sequences. In bioinformatics this is mainly handled with machine learning techniques.

3 *De Novo* Computational MiRNA Prediction

De novo miRNA gene finding is a challenging task in bioinformatics. The characteristic secondary structure of the precursor, the highly conserved primary sequence of mature and precursor sequence as well as expression information encoded in small RNA sequence libraries are used altogether to find novel miRNA genes.

There are several starting points for the computational annotation of miRNAs in a certain organism. The most natural approach is the assignment of homologous miRNAs from related species. Here, either the complete precursor or the mature sequence(s) of known miRNAs are compared to the genome using local or semi-global alignment techniques, e.g., NCBI-`blast` [77] or `GotohScan` [78]. The resulting candidates should be verified further by confirming the conservation of precursor sequence and secondary structure over their complete length with respect to their homologs. In other words, true candidates are those that fit well in the alignment and share the consensus secondary structure, while the region of the mature miRNA(s) is not allowed to have more than a few substitutions. One obvious problem when using (only) the mature sequences as starting point is that one might miss miRNAs that have undergone an arm-switching in the requested species. Thus, the original mature region might have evolved too much and the query miRNA cannot be detected anymore.

After the annotation of homologs it remains to find novel or species specific miRNAs. The whole genome, or—if available whole genome wide alignments with related species—are scanned for putative miRNA candidates using machine learning approaches. In principal, the characteristics of precursor, its fold-back, and conservation information of both, sequence and structural issues are considered and evaluated to classify each candidate sequence/alignment. In particular, programs that apply this kind of artificial intelligence are trained based on a set of known miRNAs. Dependent on the implemented algorithm a dataset that represents sequences that are not miRNAs but show similar characteristics is created and used as opponent to the set of true miRNAs. A summary on available software can be found in Table 1.

A recent detailed analysis of features that are useful in machine learning approaches to plant miRNA prediction can be found in [79]. A variety of software tools for this purpose has been devised, which share the common architecture of miRNA gene finders, differing in the structure prediction algorithm used for hairpin detection, the definition of features used for classification, and the machine learning approach itself. While some tools, such as `PlantMiRNAPred` [80], are content with identifying precursors, most tools focus on identifying the mature miRNA. To this end, possible positions of a mature miRNAs are identified on the pre-miRNA structure. The machine learning algorithms are then fed with features of the precursor hairpin, the mature miRNA candidate, and, in most cases, some information on the flanking sequence. A simple filter is implemented in `miRcheck` [81]. SVM-based tools are `tripletSVM` [82] and `MiRPara` [83], while `HHMMiR` [84] and `NOVOMIR` [85] use Hidden Markov Models. In order to avoid artifacts from using sequence windows, `NOVOMIR` uses first `RNAlfold` [86] to scan for local secondary structures, and then employs `RNAshapes` [87] to extract stem-loop like regions.

In particular in earlier approaches, conservation in other plant species is used as an additional filter [88, 89]. Usually, sequence conservation is used as an additional feature of evaluation separate from the prediction tools themselves. An efficient tool for homology search geared specifically towards plant miRNAs is `microHARVESTER` [90].

Naturally, the resulting set of potential miRNAs is compared to existing annotation to reduce the number of false positives and avoid confusion in different annotations for the same sequence.

Table 1
Available software tools that can be used for the prediction of miRNAs

Tool	Ref.	Prediction	Basis	Species	Type
Support vector machine/random forest					
miRPara	[83]	mat	seq.	Animal & plant	s.a.
MiRenSVM	[114]	pre	seq.	Animal	s.a.
triplet-SVM	[82]	pre	seq./struct.	Animal	s.a.
microPred	[115]	pre	seq.	?	s.a.
PMirP	[116]	pre	seq.	Animals	s.a. + ws.
mir-abela	[98]	pre	seq.	Animal	s.a.
miRanalyzer	[106, 107]	pre	HTS	Animal	ws.
RNAmicro	[117]	pre	aln	Animal	s.a.
RNAz	[118]	ncRNA	aln	Animal	s.a.
Hidden Markov Model/Covariance Model					
SSCprofiler	[119]	pre	seq.	Vertebrata	s.a.
HMMMiR	[84]	pre	seq.	Animal & plant	s.a.
NOVOMIR	[85]	pre	seq.	Plant	s.a.
Infernal	[120]	ncRNA	seq.	Any	s.a.
Naive Bayes classifier					
BayesMiRNAfinc	[121]	pre	seq.	Human & mouse	s.a.
MatureBayes	[122]	mat	pre-miRNA	Animals	ws.
Other scoring methods					
V(ir)Mir	[123]	pre	seq.	Virus	s.a.
DIANA-mirExTra	[124]	mat	HTS	Human & mouse	ws.
miRcheck	[81]	seq.	Plant	s.a.	
ProMiR	[125]	pre	seq.	Human	s.a.
MiRDeep2	[105, 126]	pre	seq.	Animal, plant	s.a.
miRTRAP	[108]	pre	HTS	Animal	s.a.
ALPS	[112]	ncRNA	HTS	Any	s.a.
incRNA	[113]	ncRNA	HTS		s.a.
mirTools	[127]	mat	HTS	Animal	ws. using miRDeep
deepBase	[128]	ncRNA	HTs	Plant & animal	db. using miRDeep

(continued)

Table 1
(continued)

Tool	Ref.	Prediction	Basis	Species	Type
Support vector machine/random forest					
Homology					
`miroHARVESTER`	[90]	pre	seq.	Plant	s.a. + ws.
`Blast`	[77]	pre & mat	seq.	Any	s.a. + ws.
`GotohScan`	[78]	pre & mat	seq.	Any	s.a. + ws.

The columns list the name of the program (Tool), its reference publication, the type of predicted miRNA, the type of sequence data and that can be analyzed with this program, the set of species that can be handled, and how the authors have made their program available to other researchers

mat mature, *pre* precursor sequence, *seq.* sequence, *struc.* secondary structure, *aln* multiple sequence alignment, *s.a.* stand alone, *ws.* webserver, *db.* database

If small RNA libraries are available for the respective organism, this procedure can be supported by expression profiles which might represent mature miRNAs. An unknown position in the genome, that exhibits the characteristics of a miRNA and further shows expression in the estimated mature region(s), is a very likely candidate for a novel miRNA.

In plants, several approaches explicitly employ the complementarity to a putative target mRNA as a filtering step. This is in particular the case of EST-based studies such as [91]. An early tool of this type is `findMiRNA` [92]. In [93], the order of the filtering steps is turned around. Their "intra-genomic matching" approach first finds an exact complementary match with a putative target and then evaluates the sequence and secondary structure around a putative mature miRNA. Although the pri-miRNA transcripts are typically spliced, one frequently finds the pre-miRNA hairpins in a single exon. A notable exception is the miR444 family, which is specific to grasses [94, 95]. In all these cases, the exon–exon junction is located in or close to the loop of the stem-loop structure of the pre-miRNA, and the introns feature the canonical splice motif GU..AG. Based on these observations, `SplamiR` [96] searches for spliced miRNAs. Starting from a search for complementary sequences in genomic 30 kb windows with modified `blast` algorithm that accepts GU mismatches, candidate pairs are searched for splice sites with `GeneSplicer` [97] for possible introns. The spliced precursor is then scored with `miR-abela` [98].

As introduced at the beginning of this chapter, often miRNAs are located in genomic gene clusters together with other miRNAs, in introns or up-/down-stream of certain protein coding genes.

Therefore, it makes a lot of sense to take such syntenic information into account when filtering true from false predictions. Thus, homologous miRNAs that occur in the same genomic situation as known from related organisms are very likely to be true.

4 Computational miRNA Prediction Using Short RNA-seq

High-throughput sequencing of small RNAs has shown to offer possibilities to quantify miRNA expression. Low molecular weight RNA is isolated (~17–28 nt long molecules), ligated to the adapters, amplified, and sequenced following a small RNA preparation protocol, using an Illumina (Solexa) machine [99], the 454 pyrosequencing system [100], or the SOLID sequencing machine, to name the most commonly used. The short RNA sequences (reads) are then mapped back to a reference genome. Tools like `segemehl` [101], `BWA` [102], `SOAP2` [103], or `Bowtie` [102] can be used for this alignment step. Here, it is important to allow mismatches and multiple mappings, since miRNAs are well known to be targets of RNA editing [104] and commonly occur in multiple copies.

Small RNA sequencing (small RNA-seq) does not only give information on pure expression, but gives also the actual genomic region the RNA molecules have arisen from. By taking a deeper look to known miRNA loci, it is possible to observe a read pattern generated by the miRNA processing mechanism. Two high read stacks are positioned directly above the annotated miR and miR* regions, showing a specific distance with almost no reads in between. This gap belongs to the loop region which is not measured in the RNA-seq data, since the length of this region (~15 nt) cannot be measured after the size fractionating step (~17–28 nt). Most of the currently available methods use this short RNA reads in combination with secondary structure predictions to identify novel miRNA loci.

In 2008, `miRDeep` [105] was the first tool using high-throughput sequencing data to predict new miRNA candidates. This stand-alone application includes all steps from mapping the reads to a reference genome, clustering consecutive reads occurring in close genomic distance, elongating the region to fetch the whole precursor, calculating the secondary structure, and determining a probabilistic score. The score is calculated using information from relative positions of the reads within a predicted hairpin (miR/miR*), the 3′ 2 nt overhang of the assumed miR and miR* sequences, as well as secondary structure information, like the minimum free energy (mfe). Unfortunately, `miRdeep` does not allow errors in the alignment steps and discards reads mapping to multiple loci, loosing all edited miRNAs. The input data has to be a list of reads in `fasta`-format, excluding the color space outputs from a SOLID machine. In 2012, a new version of the software `miRDeep2` was published. It identifies miRNAs with better accuracy and shows an improvement of usability.

In 2009, a web server called `miRanalyzer` [106] was released. To predict new miRNAs with this online tool, only the output of a sequencing machine is needed. In contrast to `miRDeep`, where the user has to download the reference genomes, install several third-party tools, and use the own machine (which might be rather slow), `miRanalyzer` offers a one stop shop solution. A recent version of `miRanalyzer` [107] supports 34 species. The simplicity of a web server is a big benefit for researchers with no computational background. miRanalyzer uses a random forest machine learning approach to decide if a cluster of consecutive reads (distance < 30 nt) belongs to a miRNA or not. The used set of features consists not only of the relative position of the mapped reads on the hairpin, but also on a wide range of secondary structure information, like the mfe, the number of paired nucleotides, the number of bulges and the length of the loop, to name a few. Up to two mapping errors are allowed in the read alignments and the results of the prediction can be downloaded. Furthermore, a new version also allows mapping of data in the color space from the SOLID sequencer.

`miRTRAP` [108] is the most recent approach for pure miRNA prediction and was published in 2010. This tool incorporates currently found miRNA-offset RNAs (moRs) [109, 110] and takes the genomic context into account to eliminate false positive predictions. It evaluates the secondary structure, the position of the reads on the hairpin and discards candidates with read patterns that are incompatible with the miRNA biogenesis. `mirTRAP` is a stand-alone tool and cannot handle reads in color space.

The web server tool `DARIO` does not only deal with miRNA prediction, but also other classes of non-coding RNAs (ncRNAs) [111]. It uses processing patterns of mapped HTS data, to classify new ncRNAs. After clustering consecutive reads, stacks of them are merged to blocks, using `blockbuster` [110]. These blocks form specific patterns for different classes of short RNAs and were thus used to train a random forest classifier [42]. This machine learning approach is then used to identify three classes of the short RNAs miRNAs, tRNAs, and snoRNAs. `DARIO` is sequencing-platform independent, since it does not include the mapping of the reads. Two other tools with a comparable method are `ALPS` [112] and `incRNA` [113].

There are some other applications that predict miRNAs by using high-throughput sequencing, but these use previously published methods for prediction.

Table 1 presents an overview of these and other miRNA prediction tools.

5 Closing Remarks

Naturally, there are numerous limitations of computationally predicting novel genes—this is true not only for the prediction of novel miRNA genes. In this section we conclude by discussing some of these limitations in the field of miRNA prediction.

First of all, the quality of the genome assembly is a major point. Are there already chromosomes assembled, or still scaffolds or even shorter contigs? For low coverage and/or incomplete assembled genomes it is hard to verify or even classify feasible miRNA candidates. Often, the syntenic information cannot be inferred due to too short contigs/scaffolds. Duplicated sequence parts in the genome assembly further hamper the computational verification of miRNA candidates with the help of surrounding information. For example, if a hit is found in a multiply assembled region this does not mean, that the gene exists in multiple copies. However, this is a lot of manual post-processing which is crucially important for correct annotations!

The quality of computational prediction of novel miRNAs is also constrained by the current annotation of the respective and also of related species. The main question here is: is the annotation trustworthy, and if not, what can I do about it? There are several miRNAs annotated of which it is known, that they really are no miRNAs at all but they are still listed as miRNA and—what is extremely annoying—are used as examples for further miRNA prediction pipelines! In addition, the nomenclature of gene productions is not always unique. Genes that are named differently are not necessarily different genes! Such problems must be resolved by comparing such genes on sequence and structure basis if they are expected to be the same. This is in particular the case for annotated genes located in a sensible genomic distance to one of the putative miRNAs.

Another obstruction is the problem of pseudogenes. Especially, in the case of finding more distantly related miRNAs it is hard to say from the sequence-structure conservation only, if it might be a pseudogene or not. The lack of function can only be observed in specific experimental analysis. Also, it is not clear, if positively classified miRNA candidates that do not show expression in small RNA libraries are false predictions. They might also be pseudogenes or miRNAs that are expressed in different developmental states or tissues than those of which the respective library has been extracted from. It is therefore not recommended to discard candidates with no expression, on the other hand, if expression is proven it increases the probability of the candidate to be a true miRNA.

It is also a bad idea to discard miRNAs that map to other non-protein-coding RNAs in the first instance. As described in the introductory part, a lot of non-canonical miRNAs exist in eukaryotic genomes. For example, mapping a miRNA candidate to an annotated *sno*RNA or a 3′UTR of an mRNA should not lead to the conclusion that the miRNA prediction is wrong. Here, it is necessary to check whether this situation can be observed in other species, too. If the existing annotation does not give rise to an exceptional miRNA, this information can be obtained by simply sending the known sequence through the same miRNA prediction pipeline and analyze its outcome.

References

1. Bartel DP (2004) Micro RNAs: genomics, biogenesis, mechanism, and function. Cell 116:281–297
2. Lagos-Quintana M, Rauhut R, Yalcin A, Meyer J, Lendeckel W, Tuschl T (2002) Identification of tissue-specific microRNAs from mouse. Curr Biol 12:735–739
3. Trang P, Weidhaas JB, Slack FJ (2008) MicroRNAs as potential cancer therapeutics. Oncogene 27(2):S52–S57
4. Fasanaro P, Greco S, Ivan M, Capogrossi MC, Martelli F (2010) MicroRNA: emerging therapeutic targets in acute ischemic diseases. Pharmacol Ther 125:92–104
5. Cifuentes D, Xue H, Taylor DW, Patnode H, Mishima Y, Cheloufi S, Ma E, Mane S, Hannon GJ, Lawson ND, Wolfe SA, Giraldez AJ (2010) A novel miRNA processing pathway independent of Dicer requires Argonaute2 catalytic activity. Science 328:1694–1698
6. Grimson A, Srivastava M, Fahey B, Woodcroft BJ, Chiang HR, King N, Degnan BM, Rokhsar DS, Bartel DP (2008) Early origins and evolution of microRNAs and Piwi-interacting RNAs in animals. Nature 455:1193–1197
7. Griffiths-Jones S, Bateman A, Marshall M, Khanna A, Eddy SR (2003) Rfam: an rna family database. Nucleic Acids Res 31(1):439–441
8. Cai X, Hagedorn CH, Cullen BR (2004) Human microRNAs are processed from capped, polyadenylated transcripts that can also function as mRNAs. RNA 10:1957–1966
9. Pawlicki JM, Steitz JA (2008) Primary microRNA transcript retention at sites of transcription leads to enhanced microRNA production. J Cell Biol 182:61–76
10. Lau NC, Lim LP, Weinstein EG, Bartel DP (2001) An abundant class of tiny RNAs with probable regulatory roles in caenorhabditis elegans. Science 294:858–862
11. Lagos-Quintana M, Rauhut R, Lendeckel W, Tuschl T (2001) Identification of novel genes coding for small expressed RNAs. Science 294:853–858
12. Morlando M, Ballarino M, Gromak N, Pagano F, Bozzoni I, Proudfoot NJ (2008) Primary microRNA transcripts are processed co-transcriptionally. Nat Struct Mol Biol 15:902–909
13. Jones-Rhoades MW, Bartel DP, Bartel B (2006) MicroRNAS and their regulatory roles in plants. Annu Rev Plant Biol 57: 19–53
14. Hertel J, Lindemeyer M, Missal K, Fried C, Tanzer A, Flamm C, Hofacker IL, Stadler PF (2006) The expansion of the metazoan microRNA repertoire. BMC Genomics 7:25
15. Li A, Mao L (2007) Evolution of plant microRNA gene families. Cell Res 17:212–218
16. Zhao T, Li G, Mi S, Li S, Hannon GJ, Wang XJ, Qi Y (2007) A complex system of small RNAs in the unicellular green alga Chlamydomonas reinhardtii. Genes Dev 21:1190–1203
17. Molnár A, Schwach F, Studholme DJ, Thuenemann EC, Baulcombe DC (2007) miRNAs control gene expression in the single-cell alga Chlamydomonas reinhardtii. Nature 447:1126–1129
18. Hinas A, Reimegård J, Wagner EGH, Nellen W, Ambros VR, Söderbom F (2007) The small RNA repertoire of the unicellular amoeba *Dictyostelium discoideum*: microRNA candidates, small antisense RNAs that may be derived from longer transcripts and multiple classes of repeat-associated RNAs. Nucleic Acids Res 35:6714–6726
19. WanZhong J, Zhi L, ZhaoRong L (2008) Discoveries and functions of virus-encoded microRNAs. Chin Sci Bull 53(2):169–177
20. Wheeler BM, Heimberg AM, Moy VN, Sperling EA, Holstein TW, Heber S, Peterson

KJ (2009) The deep evolution of metazoan microRNAs. Evol Dev 11:50–68
21. Peterson KJ, Dietrich MR, McPeek MA (2009) MicroRNAs and metazoan macroevolution: insights into canalization, complexity, and the cambrian explosion. Bioessays 31:736–747
22. Ambros V, Bartel B, Bartel DP, Burge CB, Carrington JC, Chen X, Dreyfuss G, Eddy SR, Griffiths-Jones S, Marshall M, Matzke M, Ruvkun G, Tuschl T (2003) A uniform system for microRNA annotation. RNA 9:277–279
23. Grun D, Wang Y-L, Langenberger D, Gunsalus KC, Rajewsky N (2005) MicroRNA target predictions across seven drosophila species and comparison to mammalian targets. PLoS Comput Biol 1:e13
24. Huang Y, Gu X (2007) A bootstrap based analysis pipeline for efficient classification of phylogenetically related animal miRNAs. BMC Genomics 8:66
25. Heimberg AM, Sempere LF, Moy VN, Donoghue PJC, Peterson KJ (2008) MicroRNAs and the advent of vertebrate morphological complexity. Proc Natl Acad Sci USA 105:2946–2950
26. Griffiths-Jones S, Hui JLH, Marco A, Ronshaugen M (2011) MicroRNA evolution by arm switching. EMBO Rep 12:172–177
27. Grad Y, Aach J, Hayes GD, Reinhart BJ, Church GM, Ruvkun G, Kim J (2003) Computational and experimental identification of *C. elegans* microRNAs. Mol Cell 11:1253–1263
28. Lim LP, Lau NC, Weinstein EG, Abdelhakim A, Yekta S, Rhoades MW, Burge CB, Bartel DP (2003) The microRNAs of *Caenorhabditis elegans*. Genes Dev 17:991–1008
29. Hayes GD, Frand AR, Ruvkun G (2006) The mir-84 and let-7 paralogous microRNA genes of *Caenorhabditis elegans* direct the cessation of molting via the conserved nuclear hormone receptors NHR-23 and NHR-25. Development 133:4631–4641
30. Roush S, Slack F (2008) The let-7 family of micrornas. Trends Cell Biol 18(10):505–516
31. Buermans HJP, Ariyurek Y, van Gertjan O, den Dunnen JT, ’t Hoen PCA (2010) New methods for next generation sequencing based microRNA expression profiling. BMC Genomics 11:716
32. Miyoshi K, Miyoshi T, Siomi H (2010) Many ways to generate microrna-like small RNAs: non-canonical pathways for microRNA production. Mol Genet Genomics 284:95–103
33. Schopman NCT, Heynen S, Haasnoot J, Berkhout B (2010) A miRNA–tRNA mix-up: tRNA origin of proposed miRNA. RNA Biol 7:573–576
34. Langenberger D, Bartschat S, Hertel J, Hoffmann S, Tafer H, Stadler PF (2011) MicroRNA or not microRNA? In: BSB 2011. Lecture notes in computer science. Springer, Berlin, pp 1–9 (accepted)
35. Hansen TB, Kjems J, Bramsen JB (2011) Enhancing miRNA annotation confidence in miRBase by continuous cross dataset analysis. RNA Biol 8(3):378–383
36. Lee YS, Shibata Y, Malhotra A, Dutta A (2009) A novel class of small RNAs: tRNA-derived RNA fragments (tRFs). Genes Dev 23:2639–2649
37. Cole C, Sobala A, Lu C, Thatcher SR, Bowman A, Brown JW, Green PJ, Barton GJ, Hutvagner G (2009) Filtering of deep sequencing data reveals the existence of abundant Dicer-dependent small RNAs derived from tRNAs. RNA 15:2147–2160
38. Haussecker D, Huang Y, Lau A, Parameswaran P, Fire AZ, Kay MA (2010) Human tRNA-derived small RNAs in the global regulation of RNA silencing. RNA 16:673–695
39. Findeiß S, Langenberger D, Stadler PF, Hoffmann S (2011) Traces of post-transcriptional RNA modifications in deep sequencing data. Biol Chem 392:305–313
40. Kawaji H, Nakamura M, Takahashil Y, Sandelin A, Katayama S, Fukuda S, Daub C, Kai C, Jun Kawai J, Yasuda J, Carninci P, Hayashizaki Y (2008) Hidden layers of human small RNAs. BMC Genomics 9:157
41. Taft RJ, Glazov EA, Lassmann T, Hayashizaki Y, Carninci P, Mattick JS (2009) Small RNAs derived from snoRNAs. RNA 15:1233–1240
42. Langenberger D, Bermudez-Santana C, Stadler PF, Hoffmann S (2010) Identification and classification of small RNAs in transcriptome sequence data. Pac Symp Biocomput 15:80–87
43. Brameier M, Herwig A, Reinhardt R, Walter L, Gruber J (2011) Human box C/D snoRNAs with miRNA like functions: expanding the range of regulatory rnas. Nucleic Acids Res 39:675–686
44. Canella D, Praz V, Reina JH, Cousin P, Hernandez N (2010) Defining the RNA polymerase III transcriptome: genome-wide localization of the RNA polymerase III transcription machinery in human cells. Genome Res 20:710–721
45. Griffiths-Jones S (2004) The microrna registry. Nucleic Acids Res 32:D109–D111
46. Berezikov E, Chung WJ, Willis J, Cuppen E, Lai EC (2007) Mammalian mirtron genes. Mol Cell 28:328–336
47. Ruby GJ, Jan CH, Bartell DP (2007) Intronic microRNA precursors that bypass Drosha processing. Nature 48:83–86

48. Okamura K, Hagen JW, Duan H, Tyler DM, Lai EC (2007) The mirtron pathway generates microRNA-class regulatory RNAs in *Drosophila*. Cell 130:89–100
49. Chung W-J, Agius P, Westholm JO, Chen M, Okamura K, Robine N, Leslie CS, Lai EC (2011) Computational and experimental identification of mirtrons in *Drosophila melanogaster* and *Caenorhabditis elegans*. Genome Res 21(2):286–300. doi: 10.1101/gr.113050.110
50. Lee Y, Kim M, Han J, Yeom K-H, Lee S, Baek SH, Kim VN (2004) MicroRNA genes are transcribed by RNA polymerase II. EMBO J 23:4051–4060
51. Denli AM, Tops BJB, Plasterk RAH, Ketting RF, Hannon GJ (2004) Processing of primary microRNAs by the microprocessor complex. Nature 432:231–235
52. Gregory RI, Yan K-P, Amuthan G, Chendrimada T, Doratotaj B, Cooch N, Shiekhattar R (2004) The microprocessor complex mediates the genesis of microRNAs. Nature 432:235–240
53. Kurihara Y, Takashi Y, Watanabe Y (2006) The interaction between DCL1 and HYL1 is important for efficient and precise processing of pri-miRNA in plant microRNA biogenesis. RNA 12:206–212
54. Yi R, Qin Y, Macara IG, Cullen BR (2003) Exportin-5 mediates the nuclear export of pre-microRNAs and short hairpin RNAs. Genes Dev 17:3011–3016
55. Lund E, Guttinger S, Calado A, Dahlberg JE, Kutay U (2004) Nuclear export of microRNA precursors. Science 303:95–98
56. Hutvagner G, Zamore PD (2002) A microRNA in a multiple-turnover RNAi enzyme complex. Science 297:2056–2060
57. Zeng Y, Cullen BR (2003) Sequence requirements for micro RNA processing and function in human cells. RNA 9:112–123
58. Zeng Y, Wagner EJ, Cullen BR (2002) Both natural and designed microRNAs can inhibit the expression of cognate mRNAs when expressed in human cells. Mol Cell 9:1327–1333
59. Cullen BR (2004) Derivation and function of small interfering RNAs and microRNAs. Virus Res 102:3–9
60. Cai X, Hagedorn CH, Cullen BR (2004) Human microRNAs are processed from capped, polyadenylated transcripts that can also function as mRNAs. RNA 10:1957–1966
61. Piriyapongsa J, Marino-Ramirez L, Jordan IK (2007) Origin and evolution of human microRNAs from transposable elements. Genetics 176:1323–1337
62. Klase Z, Kale P, Winograd R, Gupta MV, Heydarian M, Berro R, McCaffrey T, Kashanchi F (2007) HIV-1 TAR element is processed by dicer to yield a viral micro-RNA involved in chromatin remodeling of the viral LTR. BMC Mol Biol 8:63
63. Saraiya AA, Wang CC (2008) SnoRNA, a novel precursor of microRNA in giardia lamblia. PLoS Pathog 4:e1000224
64. Ender C, Krek A, Friedlander MR, Beitzinger M, Weinmann L, Chen W, Pfeffer S, Rajewsky N, Meister G (2008) A human snoRNA with microRNA-like functions. Mol Cell 32:519–528
65. Scott MS, Avolio F, Ono M, Lamond AI, Barton GJ (2009) Human miRNA precursors with box H/ACA snoRNA features. PLoS Comput Biol 5:e1000507
66. Ono M, Scott MS, Yamada K, Avolio F, Barton GJ, Lamond AI (2011) Identification of human miRNA precursors that resemble box C/d snoRNAs. Nucleic Acids Res 39(9):3879–3891
67. Zhang B, Pan X, Cobb GP, Anderson TA (2006) Plant microRNA: a small regulatory molecule with big impact. Dev Biol 289:3–16
68. Brodersen P, Voinnet O (2009) Revisiting the principles of microRNA target recognition and mode of action. Nat Rev Mol Cell Biol 10:141–148
69. Sun YH, Lu S, Shi R, Chiang VL (2011) Computational prediction of plant miRNA targets. Methods Mol Biol 744:175–186
70. Allen E, Xie Z, Gustafson AM, Sung GH, Spatafora JW, Carrington JC (2004) Evolution of microRNA genes by inverted duplication of target gene sequences in *Arabidopsis thaliana*. Nat Genet 36:1282–1290
71. Zhang W, Gao S, Zhou X, Xia J, Chellappan P, Zhou X, Zhang X, Jin H (2010) Multiple distinct small RNAs originate from the same microRNA precursors. Genome Biol 11:8
72. Shi W, Hendrix D, Levine M, Haley B (2009) A distinct class of small RNAs arises from pre-miRNA-proximal regions in a simple chordate. Nat Struct Mol Biol 16:183–189
73. Tweedie S, Ashburner M, Falls K, Leyland P, McQuilton P, Marygold S, Millburn G, Osumi-Sutherland D, Schroeder A, Seal R, Zhang H, Consortium TF (2009) Flybase: enhancing drosophila gene ontology annotations. Nucleic Acids Res 37:D555–D559
74. Stein L, Sternberg P, Durbin R, Thierry-Mieg J, Spieth J (2001) Wormbase: network access to the genome and biology of caenorhabditis elegans. Nucleic Acids Res 29(1):82–86
75. Mhuantong W, Wichadakul D (2009) MicroPC (μPC): a comprehensive resource for predicting and comparing plant microRNAs. BMC Genomics 10:366

76. Axtell MJ, Bowman JL (2008) Evolution of plant microRNAs and their targets. Trends Plant Sci 13:343–349
77. Altschul SF, Gish W, Miller W, Myers EW, Lipman DJ (1990) Basic local alignment search tool. J Mol Biol 215:403–410
78. Hertel J, de Jong D, Marz M, Rose D, Tafer H, Tanzer A, Schierwater B, Stadler PF (2009) Non-coding RNA annotation of the genome of *Trichoplax adhaerens*. Nucleic Acids Res 37:1602–1615
79. Thakur V, Wanchana S, Xu M, Bruskiewich R, Quick WP, Mosig A, Zhu XG (2011) Characterization of statistical features for plant microRNA prediction. BMC Genomics 12:108
80. Xuan P, Guo M, Liu X, Huang Y, Li W, Huang Y (2011) PlantMiRNAPred: efficient classification of real and pseudo plant pre-miRNAs. Bioinformatics 27(10):1368–1376
81. Jones-Rhoades MW (2010) Prediction of plant miRNA genes. Methods Mol Biol 592:19–30
82. Xue C, Li F, He T, Liu GP, Li Y, Zhang X (2005) Classification of real and pseudo microRNA precursors using local structure-sequence features and support vector machine. BMC Bioinformatics 6:310
83. Wu Y, Wei B, Liu H, Li T, Rayner S (2011) MiRPara: a SVM-based software tool for prediction of most probable microrna coding regions in genome scale sequences. BMC Bioinformatics 12:107
84. Kadri S, Hinman V, Benos PV (2009) HHMMiR: efficient *de novo* prediction of microRNAs using hierarchical hidden markov models. BMC Bioinformatics 10(1):S35
85. Teune JH, Steger G (2010) NOVOMIR: *de novo* prediction of MicroRNA-coding regions in a single plant-genome. J Nucleic Acids 2010:495904
86. Hofacker IL, Priwitzer B, Stadler PF (2004) Prediction of locally stable RNA secondary structures for genome-wide surveys. Bioinformatics 20:191–198
87. Giegerich R, Voss B, Rehmsmeier M (2004) Abstract shapes of RNA. Nucleic Acids Res 32:4843–4851
88. Wang X-J, Reyes JL, Chua N-H, Gaasterland T (2004) Prediction and identification of *Arabidopsis thaliana* microRNAs and their mRNA targets. Genome Biol 5:R65
89. Jones-Rhoades MW, Bartel DP (2004) Computational identification of plant microRNAs and their targets, including a stress-induced miRNA. Mol Cell 14:787–799
90. Dezulian T, Remmert M, Palatnik JF, Weigel D, Huson DH (2006) Identification of plant microRNA homologs. Bioinformatics 22:359–360
91. Wen J, Frickey T, Weiller GF (2008) Computational prediction of candidate miRNAs and their targets from *Medicago truncatula* non-protein-coding transcripts. In Silico Biol 8:291–306
92. Adai A, Johnson C, Mlotshwa S, Archer-Evans S, Manocha V, Vance V, Sundaresan V (2005) Computational prediction of miRNAs in *Arabidopsis thaliana*. Genome Res 15:78–91
93. Lindow M, Jacobsen A, Nygaard S, Mang Y, Krogh A (2007) Intragenomic matching reveals a huge potential for miRNA-mediated regulation in plants. PLoS Comput Biol 3:e238
94. Sunkar R, Girke T, Jain PK, Zhu JK (2005) Cloning and characterization of micrornas from rice. Plant Cell 17:1397–1411
95. Lu C, Jeong DH, Kulkarni K, Pillay M, Nobuta K, German R, Thatcher SR, Maher C, Zhang L, Ware D, Liu B, Cao X, Meyers BC, Green PJ (2008) Genome-wide analysis for discovery of rice microRNAs reveals natural antisense micrornas (nat-miRNAs). Proc Natl Acad Sci USA 105:4951–4956
96. Thieme CJ, Gramzow L, Lobbes D, Theißen G (2011) SplamiR—prediction of spliced miRNAs in plants. Bioinformatics 27:1215–1223
97. Pertea M, Lin X, Salzberg SL (2001) GeneSplicer: a new computational method for splice site prediction. Nucleic Acids Res 29:1185–1190
98. Sewer A, Paul N, Landgraf P, Aravin A, Pfeffer S, Brownstein MJ, Tuschl T, van Nimwegen E, Zavolan M (2005) Identification of clustered microRNAs using an *ab initio* prediction method. BMC Bioinformatics 6:267
99. Bentley D, Balasubramanian S, Swerdlow H, Smith G, Milton J, Brown C, Hall K, Evers D, Barnes C, Bignell H et al (2008) Accurate whole human genome sequencing using reversible terminator chemistry. Nature 456(7218):53–59
100. Margulies M, Egholm M, Altman W, Attiya S, Bader J, Bemben L, Berka J, Braverman M, Chen Y, Chen Z et al (2005) Genome sequencing in microfabricated high-density picolitre reactors. Nature 437(7057):376–380
101. Hoffmann S, Otto C, Kurtz S, Sharma C, Khaitovich P, Vogel J, Stadler P, Hackermüller J (2009) Fast mapping of short sequences with mismatches, insertions and deletions using index structures. PLoS Comput Biol 5(9):e1000502
102. Li H, Durbin R (2009) Fast and accurate short read alignment with Burrows–Wheeler transform. Bioinformatics 25(14):1754

103. Li R, Yu C, Li Y, Lam T, Yiu S, Kristiansen K, Wang J (2009) SOAP2: an improved ultrafast tool for short read alignment. Bioinformatics 25(15):1966
104. Knoop V (2010) When you cannot trust the DNA: RNA editing changes transcript sequences. Cell Mol Life Sci 68(4):567–586
105. Friedlaender M, Chen W, Adamidi C, Maaskola J, Einspanier R, Knespel S, Rajewsky N (2008) Discovering microRNAs from deep sequencing data using miRDeep. Nat Biotechnol 26(4):407–415
106. Hackenberg M, Sturm M, Langenberger D, Falcon-Perez J, Aransay A (2009) miRanalyzer: a microRNA detection and analysis tool for next-generation sequencing experiments. Nucleic Acids Res 37(2):W68
107. Hackenberg M, Rodríguez-Ezpeleta N, Aransay A (2011) miRanalyzer: an update on the detection and analysis of microRNAs in high-throughput sequencing experiments. Nucleic Acids Res 39(2):W132
108. Hendrix D, Levine M, Shi W (2010) miRTRAP, a computational method for the systematic identification of miRNAs from high throughput sequencing data. Genome Biol 11(4):R39
109. Shi W, Hendrix D, Levine M, Haley B (2009) A distinct class of small RNAs arises from pre-miRNA—proximal regions in a simple chordate. Nat Struct Mol Biol 16(2):183–189
110. Langenberger D, Bermudez-Santana C, Hertel J, Hoffmann S, Khaitovich P, Stadler P (2009) Evidence for human microRNA-offset RNAs in small RNA sequencing data. Bioinformatics 25(18):2298
111. Fasold M, Langenberger D, Binder H, Stadler P, Hoffmann S (2011) Dario: a ncRNA detection and analysis tool for next-generation sequencing experiments. Nucleic Acids Res 39(2):W112–W117
112. Erhard F, Zimmer R (2010) Classification of ncRNAs using position and size information in deep sequencing data. Bioinformatics 26(18):i426
113. Lu Z, Yip K, Wang G, Shou C, Hillier L, Khurana E, Agarwal A, Auerbach R, Rozowsky J, Cheng C et al (2011) Prediction and characterization of noncoding RNAs in C. elegans by integrating conservation, secondary structure, and high-throughput sequencing and array data. Genome Res 21(2):276
114. Ding J, Zhou S, Guan J (2010) MirenSVM: towards better prediction of microRNA precursors using an ensemble SVM classifier with multi-loop features. BMC Bioinformatics 11(11):S11
115. Batuwita R, Palade V (2009) Micropred: effective classification of pre-miRNAs for human miRNA gene prediction. Bioinformatics 25:989–995
116. Zhao D, Wang Y, Luo D, Shi X, Wang L, Xu D, Yu J, Liang Y (2010) PMirp: a pre-microRNA prediction method based on structure-sequence hybrid features. Artif Intell Med 49:127–132
117. Hertel J, Stadler PF (2006) Hairpins in a haystack: recognizing microRNA precursors in comparative genomics data. Bioinformatics 22:e197–e202
118. Washietl S, Hofacker IL, Stadler PF (2005) Fast and reliable prediction of noncoding RNAs. Proc Natl Acad Sci USA 102:2454–2459
119. Oulas A, Boutla A, Gkirtzou K, Reczko M, Kalantidis K, Poirazi P (2009) Prediction of novel microRNA genes in cancer-associated genomic regions—a combined computational and experimental approach. Nucleic Acids Res 37:3276–3287
120. Nawrocki EP, Kolbe DL, Eddy SR (2009) Infernal 1.0: inference of rna alignments. Bioinformatics 25(13):1713
121. Yousef M, Nebozhyn M, Shatkay H, Kanterakis S, Showe LC, Showe MK (2006) Combining multi-species genomic data for microRNA identification using a naive bayes classifier. Bioinformatics 22:1325–1334
122. Gkirtzou K, Tsamardinos I, Tsakalides P, Poirazi P (2010) Maturebayes: a probabilistic algorithm for identifying the mature miRNA within novel precursors. PLoS One 5(8):e11843
123. Grundhoff A, Sullivan CS, Ganem D (2006) A combined computational and microarray-based approach identifies novel microRNAs encoded by human gamma-herpesviruses. RNA 12:733–750
124. Alexiou P, Maragkakis M, Papadopoulos GL, Simmosis VA, Zhang L, Hatzigeorgiou AG (2010) The DIANA-mirextra web server: from gene expression data to microRNA function. PLoS One 5(2):e9171
125. Nam J-W, Shin K-R, Han J, Lee Y, Kim VN, Zhang B-T (2005) Human microRNA prediction through a probabilistic co-learning model of sequence and structure. Nucleic Acids Res 33(11):3570–3581
126. Thakur V, Wanchana S, Xu M, Bruskiewich R, Quick WP, Mosig A, Zhu X-G (2011) Characterization of statistical features for plant microRNA prediction. BMC Genomics 12:108

127. Zhu E, Zhao F, Xu G, Hou H, Zhou L, Li X, Sun Z, Wu J (2010) mirTools: microRNA profiling and discovery based on high-throughput sequencing. Nucleic acids Res 38(2):W392

128. Yang J, Shao P, Zhou H, Chen Y, Qu L (2010) DeepBase: a database for deeply annotating and mining deep sequencing data. Nucleic Acids Res 38(1):D123

Chapter 21

MicroRNA Target Finding by Comparative Genomics

Robin C. Friedman and Christopher B. Burge

Abstract

MicroRNAs (miRNAs) have been implicated in virtually every metazoan biological process, exerting a widespread impact on gene expression. MicroRNA repression is conferred by relatively short "seed match" sequences, although the degree of repression varies widely for individual target sites. The factors controlling whether, and to what extent, a target site is repressed are not fully understood. As an alternative to target prediction based on sequence alone, comparative genomics has emerged as an invaluable tool for identifying miRNA targets that are conserved by natural selection, and hence likely effective and important. Here we present a general method for quantifying conservation of miRNA seed match sites, separating it from background conservation, controlling for various biases, and predicting miRNA targets. This method is useful not only for generating predictions but also as a tool for empirically evaluating the importance of various target prediction criteria.

Key words Comparative genomics, Conservation, Natural selection, MicroRNAs, miRNAs, 3′ UTRs, Seed matches

1 Introduction

MicroRNAs (miRNAs) were first discovered using experimental genetics approaches as repressors of the expression of other RNAs in *trans* [1–4]. Since these discoveries, miRNAs have been implicated in virtually every metazoan biological process, including development, tissue definition, genetic disease, immune function, and countless others [5]. It was quickly noticed that targets of the first-discovered animal miRNAs had sequences in their 3′ UTRs with partial complementarity to their respective miRNA regulators, leading to the possibility that each miRNA could target multiple genes. Small interfering RNAs (siRNAs), first observed as intermediates in RNA interference, strongly repress RNAs with complete or near-complete complementarity by directing cleavage of the target RNA strand between the portion paired with nucleotides 10 and 11 of the siRNA. Likewise, plant miRNAs usually have enough complementarity to direct cleavage of their

Jan Gorodkin and Walter L. Ruzzo (eds.), *RNA Sequence, Structure, and Function: Computational and Bioinformatic Methods*, Methods in Molecular Biology, vol. 1097, DOI 10.1007/978-1-62703-709-9_21,

targets, making target prediction relatively simple. In contrast, animal miRNAs typically do not have enough complementarity to their targets to mediate cleavage [6, 7]. Without extensive complementarity between miRNAs and their targets, it was initially unclear what the rules of miRNA specificity might be and how to predict further targets. Lacking the strong effects of enzymatic cleavage of messages, miRNAs repress their targets by translational repression and mRNA destabilization, often totaling less than 20% repression at the protein level [8–10]. Because of the noise inherent in experimental assays for gene expression, it is difficult to evaluate such small effects for individual targets. Such experimental noise can, however, be averaged out when aggregating a group of potential targets to study principles of targeting. Because of the difficulty of experimentally verifying individual miRNA targets, many turned to computational methods to determine the rules of miRNA specificity and to predict new targets. In addition to miRNAs themselves, it was demonstrated that siRNA-mediated RNA interference (RNAi) has off-target effects through an miRNA targeting mechanism [11]. Despite the widespread use of genetic RNAi screens and the promise of therapeutic RNAi, off-target effects remain a substantial problem. Therefore, the principles of miRNA target recognition are important not only for predicting miRNA function but also for understanding siRNA off-target effects.

An early hint of target features important for specificity was found when it was noticed that the 5′ ends of some *Drosohpila* miRNAs were perfectly complementary to motifs known as the K box, Brd box, and GY box, which had previously been shown to confer post-transcriptional repression [12]. Subsequently, several groups attempted to leverage what was known about miRNA targets to computationally predict new targets [13–17]. This first generation of methods based their predictions on some combination of complementarity (often weighted towards the 5′ end of the miRNA), free energy of pairing to the entire miRNA, and evolutionary conservation of the target site. These sets of predictions had little overlap and a major weakness of these methods was soon apparent: they were based on assumptions with little or no experimental support, with some methods using the extremely small number of experimentally determined targets as training and/or test sets. For example, the free-energy-of-pairing calculation assumes two free RNA molecules in solution, but the miRNA is tightly and stably incorporated into an Argonaute protein, which constrains the possible structures that can be formed [10, 18]. Although most first-generation prediction programs were validated using a limited set of experimentally determined targets, Lewis and coworkers used conservation as an independent metric for determining the specificity of miRNA prediction and for finding features important for targeting [15]. Reasoning that bona fide miRNA target sequences that are vital for the survival of an organism would be conserved over evolutionary time, they tested potential targeting rules by comparing the conservation of miRNA complementary sequences to those complementary to shuffled miRNAs (which had been filtered to control for the number of matches in a single genome). Using the resulting "signal to noise ratio," or ratio of conserved sequences complementary to miRNAs over those complementary to shuffled controls, the authors found that matches to segments at the 5′ end of miRNAs were conserved at a much higher rate than matches to the 3′ end. They defined nucleotides 2–8 of the miRNA as the "seed" sequence. They next predicted targets by selecting sequences with conserved perfect complementarity to the seed sequence ("seed matches"), and passing a free energy criterion for predicted pairing to the rest of the miRNA.

Although the evidence for targeting by seed matches was initially based only on conservation, experimental evidence quickly confirmed their importance in target recognition [19]. Following refined analysis of preferential conservation of seed matches without requiring complementarity to the 3′ end of the miRNA [20], a striking analysis of in vivo silencing of fluorescent protein in transgenic *Drosophila melanogaster* showed that seed matches confer repression on targets without strong pairing to the 3′ end of the miRNA [21]. However, extensive pairing to the 3′ end of the miRNA could supplement a weak seed match to enhance repression or could compensate for imperfect seed matches, such as those with G:U wobble pairs. Comparing the number of conserved sequences complementary to miRNA seed regions to those complementary to shuffled miRNAs revealed that the vast majority of miRNA targets lacked substantial 3′ pairing [21, 22]. The next generation of mammalian target predictions refined the seed as nucleotides 2–7 of the miRNA and found that in mammals as well as in *Drosophila*, dropping any free energy criterion (leaving only the seed matches) provided better sensitivity without sacrificing specificity [20, 21]. This led to the following simple strategy for predicting miRNA targets: start with perfect seed matches in 3′ UTRs and filter them for perfect conservation between several species, e.g., human, mouse, rat, and dog, yielding a set of predictions that are enriched for true targets. The second generation of target predictions was based on these principles but had some variations: Lewis et al. [20] found evidence for the conservation of an adenosine opposite position 1 of the miRNA and required perfect seed matches, whereas other methods counted Watson–Crick matches at position 1 and searched for 3′ complementarity as well [21, 23, Fig. 1]. The degree of overlap for several current target prediction programs is quite high because all of them now require stringent seed pairing [10]. However, there are also

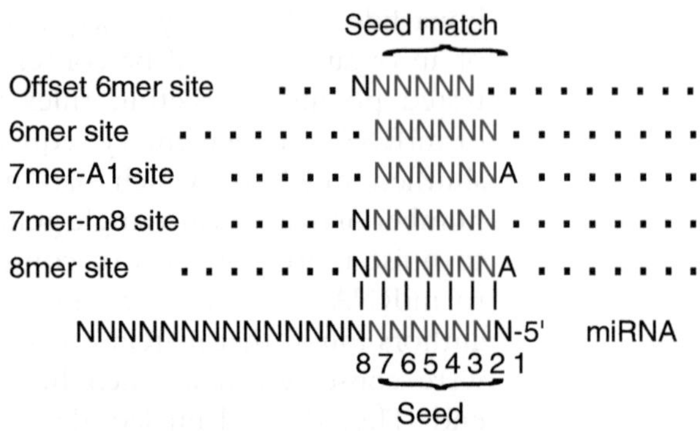

Fig. 1 Seed match types effective in mammals. Perfect Watson–Crick pairing to the seed sequence of the miRNA (nucleotides 2–7) can be supplemented by pairing to position 8 or an adenosine opposite position 1. An offset 6mer represses targets despite a lack of pairing to nucleotide 2 of the seed

minor differences due to the use of different alignments, UTR annotations, and miRNA annotations.

Several experimental approaches have provided genome-scale confirmation of the importance of miRNA seed matches. Transfections of miRNA mimics or siRNAs into human cell lines followed by microarray analysis leads to the repression of hundreds of messages containing seed matches, but more extended pairing has little additional effect [24–28]. Quantitative mass spectrometry approaches have confirmed this result at the protein level [8, 9].

Unfortunately, a quick back-of-the-envelope calculation reveals the difficulty of validating targets based simply on seed matches:

$$(20{,}000 \text{ genes})\left(\frac{1{,}000 \text{ nucleotides}}{3'\text{UTR}}\right)\left(\frac{1 \text{ 6mer match}}{4^6 \text{ nucleotides}}\right)$$
$$\approx 4{,}800 \text{ seed matches}$$

It is apparent that the average miRNA has several thousand potential targets based on seed matches. However, experimental assays based on miRNA transfection, miRNA knockouts, and smaller-scale reporter assays have suggested that seed matches are typically necessary but not sufficient to confer repression on an mRNA and have ruled out the possibility that each miRNA has several thousand targets that are substantially repressed under most conditions [8, 24, 27, 29]. Therefore there must therefore be other determinants of miRNA targeting beyond the mere presence or absence of a seed match, such as sequence context. Several of these context features have been found by the analysis of global expression datasets. The first feature is the type of the seed match itself, which has a large effect on target repression. A match to positions 2–7 of a miRNA (a 6mer seed match) typically has only a small effect on messages unless flanked by a Watson–Crick match opposite position 8 (a 7mer-m8), an adenosine opposite position 1 (a 7mer-A1), or both (an 8mer) [8, 9, 27, 28, Fig. 1]. Later, the "offset 6mer," matching nucleotides 3–8 of the miRNA but not position 2, was described [30, Fig. 1]. Another important feature is positioning of the seed match at least 15 nucleotides after the stop codon. Seed matches in 5′ UTRs and ORFs are much less effective than those in 3′ UTRs, likely because the scanning or translating ribosome interferes with RISC binding [27]. In fact, the "ribosome shadow" of 15 nucleotides past the stop codon is also subject to steric interference by the ribosome, decreasing the efficacy of miRNA target repression in this region [27]. AU-rich composition of the sequence surrounding a seed match also improves target efficacy, presumably by decreasing secondary structure and increasing accessibility of the site [27, 28, 31]. Similarly, seed matches in the middle of long UTRs are less effective than those at the ends, probably reflecting increased secondary structure [27]. Increased sequence conservation in the local vicinity of the seed match has also been associated with increased target site efficacy [28]. Multiple seed matches in the same UTR tend to multiplicatively affect downregulation, e.g., 2 sites that by themselves each downregulate expression to 80% of previous levels tend to repress expression to (80%) × (80%) = 64% when present together [27, 28]. However, seed matches located between 8 and 40 nucleotides of another seed match tend to act cooperatively, providing a potent increase in efficacy [27]. Combining these effects into a target prediction framework enables improved prediction of efficacy [8, 27].

Presumably, seed matches that are under purifying natural selection confer substantial miRNA-mediated repression, which is associated with features such as sequence context. However, seed match conservation and sequence context are complementary data sources for target prediction, because not all features conferring repression are known and because the efficacy of repression conferred by a potential miRNA target is not perfectly correlated with conservation [32]. Also, many bona fide miRNA targets presumably represent species-specific variations or adaptations that are by definition not conserved. Filtering seed matches for conservation provides a reduced set of target predictions that is enriched for targets that are subject to purifying selection, so target predictions based on context features alone may have higher sensitivity. However, repression by miRNAs in a heterologous assay does not imply endogenous targeting or function. A message with a seed match might never be co-expressed with an miRNA under physiological conditions, precluding any chance of interaction. Or, some repression of a gene might have no phenotypic impact if the relevant biological process is robust to small changes in expression (e.g., in cases of canalization). In contrast, given an appropriate null model and statistical framework, a significant signal for evolutionary conservation likely corresponds to function that is important for the fitness of the organism in some way, bypassing cases in which noisy gene expression lacks consequence and prioritizing vital interactions for further study.

2 Comparative Genomics for Conserved miRNA Targets

Here we provide a detailed description of a previously published comparative-genomics method for the prediction of miRNA targets [30]. Compared to the published description of the method, we focus more on underlying principles and motivations, technical considerations, and decision points and extensions that may be necessary when applying it in different systems.

2.1 Principles

Purifying natural selection, also called negative selection, results in conservation of functional motifs or other features. It is important to keep in mind that conserved sequences do not constitute all functional sequences because they do not include species- or lineage-specific functions, but they are greatly enriched for function. Of course, this enrichment can only be quantified by comparing to an estimate of conservation occurring by chance, often called background conservation. In fields such as population genetics, this is accomplished by estimating parameters such as mutation rate and divergence time and by using a theoretical approach, i.e., forward simulation of evolution. In the context of genomics, the prevalence of unknown parameters virtually precludes the possibility of accurate estimates of background conservation using this approach. For example, G/C content, mutation rates and patterns of nucleotide bias, gene conversion rates, and crossover rates vary strongly throughout the genome, and there are errors in multiple alignments, orthology information, and gene annotations. In contrast, an empirical approach using control sequences with similar properties to functional motifs that are similarly affected by these uncertainties has become an increasingly common tool in comparative genomics. These empirical approaches have the added benefit of being conservative with respect to sequence overlap with functional elements other than miRNA targets, which could provide spurious support for miRNA target function.

Conservation of a sequence of length k (a k-mer) can be described by two values: The level of its conservation and the level expected by chance based on that observed for control k-mers. To determine motif-level conservation, one can count the number of k-mers that are conserved at or above a particular threshold (often called the conservation signal) and compare it to the number expected by chance (called the background or noise). The signal-to-background ratio then represents the fold-enrichment of conservation over that expected by chance, and the signal minus the background (or signal above background) represents the number of *k*-mers under increased purifying selection relative to the background level. These quantities are related to the specificity and sensitivity of a conservation-based target prediction approach.

While the most obvious use of these data is to predict which miRNA targets are under purifying selection, conservation can also be used as a tool to evaluate target-prediction guidelines and algorithms themselves. For example, early target-prediction tools used the predicted free energy of interaction between the full miRNA and the target sequence, but lacked evidence that this criterion enriched for true targets. In contrast, a simpler model of consecutive base-pairing to a specific region of the 3′ end of the miRNA was found to be preferentially conserved, and therefore to enrich for true targets [27, 30]. Armed with the conservation signal and background, one can make hypotheses about specific miRNA seed match types or other target types and evaluate them using the conservation approach. MicroRNA targets should have a signal-to-background ratio and a signal above background as high as possible in 3′ UTRs, but not in control regions such as the reverse complement of 3′ UTRs. Likewise, components of the algorithm such as the method for background estimation can be compared using these criteria as well as using the variance on signal-to-background ratios. That is, few sequences should have a ratio of less than 1, and the method should yield similar results when using distinct control sets chosen based on the same criteria.

It is apparent that motif conservation can be used not just as a target prediction tool but as a powerful alternative to experiments for inference of specificity determinants that can be applied in a variety of regulatory contexts. Here, we apply this to the prediction of miRNA seed match targets, paying special attention to technical considerations and implementation.

2.2 Data Collection and Annotation

2.2.1 Selection of Species

The analysis must start with a reference species for which accurate 3′ UTR and miRNA annotations are available. The set of orthologous species considered should ideally cover a range of evolutionary distances from the reference species, but the method is valid for as few as two species. Inclusion of additional species generally adds to the sensitivity of target identification, and the method is robust to the use of multiple closely related species having redundant information, such as mouse and rat. However, species with extreme sequence divergence between their 3′ UTRs are more problematic. This is because when two species are highly divergent, occasional spurious alignments or convergent evolution can lead to overestimates of the evolutionary age of a site. For example, it is inadvisable to compare human and zebrafish 3′ UTRs, as less than 5% of 7mers are alignable between these species, with many conserved 7mers being obvious cases of mis-alignment (unpublished observation). This problem could be overcome by more sophisticated evolutionary models (*see* **Note 3.1**).

2.2.2 miRNA Families

For the purposes of this analysis, differences in miRNAs outside of the extended seed region (nucleotides 2–8 of the mature sequence) are irrelevant. Therefore, miRNAs in the reference species can first be collapsed into families sharing the same nucleotides 2–8. Because the method assumes that the seed of each miRNA family is conserved in all of the species considered, the set of miRNA families must be filtered for conservation. In practice, miRNA annotations and genomic assemblies are often incomplete, so few miRNA families will be apparently conserved between every species being considered. A reasonable alternative is to make the simplifying assumption that miRNAs that are perfectly conserved between the reference species and a highly divergent species (e.g., human and zebrafish, as annotated by miRbase) are conserved throughout the intervening phylogeny. The method is robust to minor violations of this assumption because species lacking a particular miRNA will contribute equally to the signal and background estimates. MicroRNA families should be considered conserved between two species if any miRNA having the same nucleotides 2–8 is conserved.

2.2.3 3′ UTR Annotations, Orthology, and Alignments

Comprehensive 3′ UTR annotations are available only for a limited number of organisms, so for most species orthologous regions must be approximated by regions aligning to 3′ UTRs in the reference organism. Alignments or 3′ UTRs can be quickly extracted from whole-genome multiple alignments, available for example from the UCSC genome browser or from Ensembl. Alternately, pairwise alignments of the reference species to each orthologous species can be used. Custom alignments could also be used to improve on orthology predictions (*see* **Note 3.2**).

Alternative cleavage and polyadenylation and splicing lead to a variety of 3′ UTR isoforms in metazoans, leading to tradeoffs between sensitivity and the complexity of the analysis. A simple approach involves using the longest 3′ UTR annotation for each gene and discarding the rest, but multiple isoforms can also be used so long as they are not overlapping. UTR annotations should not overlap with ORF annotations, as ORFs are subject to substantially different selective pressures. After extracting the alignments corresponding to the 3′ UTR annotations, a phylogenetic tree of the orthologous species should be constructed based on the concatenated 3′ UTR sequences, for example using the DNAML program from the PHYLIP package [33, *see* **Note 3.3**].

2.3 Conservation and Background

2.3.1 Conservation Metric

The conservation of each subsequence of a 3′ UTR can now be calculated. Here, we use a phylogenetic branch-length metric similar to that used by Stark and coworkers [34] but adapted in [30]. This simple procedure is as follows: For each site of a motif, record the set of species having that motif aligned and perfectly conserved. Next, convert this set of species to a branch length by recording the minimum set of branches required to connect the species.

An algorithm for converting species to a branch length score is as follows: (1) Root the tree at an arbitrary internal node. (2) Beginning from the reference organism, mark each ancestor until reaching the root. (3) Starting from each species in which the site is conserved, traverse each ancestor node until reaching either the root or a previously visited internal node. Upon reaching the root, continue traversing nodes towards the reference organism, stopping when reaching a previously visited internal node. Sum the branch lengths traversed during this step.

This simple conservation metric makes several assumptions, including that orthologous sequences are aligned and that no convergent mutations or back-mutations have occurred. These assumptions are probably reasonable for properly selected datasets, but more complex metrics could also be applied to compensate for violations if desired (*see* **Note 3.1**).

2.3.2 Local Conservation

Mutation rates and patterns of nucleotide substitution bias, gene conversion rates, and crossover rates vary throughout the human genome, and additionally some genes are associated with recent selective sweeps. Likewise, alignment quality and ambiguities in orthology vary for different genes and different regions of the genome. Therefore, the rate of background conservation varies widely in the human genome (and likely in all other metazoans). Each *k*-mer site must be compared to control *k*-mers in a similar conservation background, rather than an unrealistic average background over the entire genome. A simple solution is to divide 3′ UTRs into bins based on their level of background conservation. The choice for the number of bins used will depend on a trade-off between closer control of local conservation (associated with more bins) and the reduction in sampling noise resulting from having more observations, which is of particular importance for 8mers. For human 3′ UTRs, ten bins was found to be a reasonable number, whereas the smaller UTRs of *Caenorhabditis elegans* required use of just 4 bins in order to have a reasonable statistical sample for 8mers in each bin.

The conservation branch length of each nucleotide in each 3′ UTR is then scored as described in Subheading 2.3.1, treating each nucleotide as an instance of a motif of size one. Each 3′ UTR is then sorted based on the mean conservation branch length of each of its nucleotides. Then, UTRs are divided into equally sized bins based on their mean conservation. For each bin, build a new phylogenetic tree as in Subheading 2.2.3. The calculation of conservation and background for each *k*-mer as described below should then be carried out and recorded separately for each 3′ UTR bin. The results will be aggregated at the end of these calculations.

2.3.3 k-mer Conservation

The next step is to collect the distribution of conservation branch lengths for all sequences of length 6, 7, and 8. Each 6mer site in the 3′ UTRs of the reference species is scanned sequentially. For each site, the associated branch length is calculated as described in Subheading 2.3.1. For each local conservation bin, the number of occurrences of each of the 4,096 6mers at each branch length are saved into a table (*see* **Note 3.4**). The procedure is repeated for all 7mers and 8mers, recording the branch length of each occurrence. The conservation distributions for *k*-mers that are not miRNA seed matches will be useful as controls.

2.3.4 Control k-mers

To control for mutation rates, the quality of alignments and UTR annotations, control *k*-mers simulating the seed matches of mock miRNAs must be selected. Several factors influence estimates of background conservation, including GC content, dinucleotides, and matches to other short conserved motifs. Therefore, control *k*-mers should be matched to real seed matches for each of these properties. Because conservation biases due to these factors can change over different parts of the phylogeny, we will select a set of control *k*-mers for every branch length cutoff. A simple yet effective means of finding *k*-mers with similar conservation properties is to estimate their expected conservation rates based on a first-order Markov model. This estimates the expected conservation rate of a sequence using the conditional conservation rates of its constituent dinucleotides.

First, the branch length distribution is calculated for each nucleotide (A, C, G, T) and each dinucleotide (AA, AC, etc.) in the 3′ UTR alignments, as done with 6mers, 7mers, and 8mers. For every possible branch-length cutoff, nucleotide conservation rates are calculated.

Let $P(\mathbf{A}_{0.5}|\mathrm{A})$ be the probability that an adenosine is conserved at or beyond a branch-length cutoff 0.5 given that the reference organism contains an adenosine at that position, $N(\mathrm{A})$ be the number of adenosines in 3′ UTRs, and $N(\mathbf{A}_{0.5})$ be the number of 3′ UTR adenosines that are conserved at a branch-length cutoff of 0.5.

Then:

$$P(\mathbf{A}_{0.5}|\mathrm{A}) = N(\mathbf{A}_{0.5})/N(\mathrm{A})$$

For dinucleotides, conservation rates are calculated conditioned on the first nucleotide, for example:

$$P(\mathbf{AT_{0.5}}|\mathbf{A_{0.5}}\mathrm{T}) = N(\mathbf{AT_{0.5}})/N(\mathbf{A_{0.5}}\mathrm{T})$$

where $\mathbf{AT_{0.5}}$ signifies an AT dinucleotide being conserved at a branch-length of at least 0.5, and $\mathbf{A_{0.5}}\mathrm{T}$ signifies an AT dinucleotide in which the A is conserved at a branch-length at least 0.5. Finally, at each branch-length cutoff, an expected conservation of each *k*-mer is calculated using a first-order Markov model. For example, for the 4mer sequence ACGT at a branch-length cutoff of 0.5:

$$P(\mathbf{ACGT_{0.5}}|\mathrm{ACGT}) = P(\mathbf{A_{0.5}}|\mathrm{A})P(\mathbf{AC_{0.5}}|\mathbf{A_{0.5}}\mathrm{C}) \\ P(\mathbf{CG_{0.5}}|\mathbf{C_{0.5}}\mathrm{G})P(\mathbf{GT_{0.5}}|\mathbf{G_{0.5}}\mathrm{T})$$

For each miRNA seed match type of length *k* (2–7 6mer, 3–8 6mer, 7mer-A1, 7mer-m8, 8mer-A1), all sequences of length *k* are sorted by their expected conservation. The *k*-mers are then filtered so that they contain the same number of CpG sequences as the seed match, the same total of G and C nucleotides, and the same number of pumilio family motif matches (UGUA or UGUG). Finally, the 50 remaining *k*-mers with closest expected conservation are selected as controls, 25 with higher expected conservation and 25 with lower if possible. Thus, for each seed match of each type, there is a set of 50 control *k*-mers for every possible branch-length cutoff.

The estimate of the background level of conservation for each *k*-mer is calculated using this set of control *k*-mers. At every branch-length cutoff and every local conservation UTR bin the conservation rate for any *k*-mer is calculated as the fraction of sites having branch length equal to or greater than the cutoff. For example, the conservation rate of the 6mer ACGTAC at branch-length cutoff 0.5, $P(\mathbf{ACGTAC_{0.5}}|\mathrm{ACGTAC})$ is simply given by:

$$P(\mathbf{ACGTAC_{0.5}}|\mathrm{ACGTAC}) = N(\mathbf{ACGTAC_{0.5}})/N(\mathrm{ACGTAC})$$

Let C^1 to C^{50} be the 50 control 6mers for ACGTAC. Then the estimated background number of sites conserved at a branch-length cutoff of 0.5, $B(\mathbf{ACGTAC_{0.5}})$ is:

$$B(\mathbf{ACGTAC_{0.5}}) = N(\mathrm{ACGTAC})\frac{\sum_{i=1}^{50} P(\mathbf{C^i_{0.5}})}{50}$$

That is, the estimated background number of sites conserved for a *k*-mer and a particular branch-length cutoff is the total number of sites times the mean conservation rate of the control *k*-mers in the same local conservation UTR bin. For non-functional *k*-mers, the conservation rate should be close to the conservation rate of its control *k*-mers, and thus the conservation signal should roughly equal the background. These calculations should be performed separately for each local conservation bin.

2.3.5 Global Targeting Properties

For each seed match type and for each branch-length cutoff, one can sum the signal and the background for all miRNAs and for all local conservation UTR bins to yield a global signal and background. The total number of miRNA seed matches of each type conserved above background levels is given by this combined signal minus its corresponding background. The specificity can also be examined by comparing the ratio of the conservation signal to its background. The specificity should increase roughly monotonically as a function of the branch-length cutoff, whereas the highest sensitivity should be achieved at an intermediate branch-length cutoff. An estimate of the uncertainty in the background calculation can be obtained by treating the sets of control *k*-mers as independent samples and calculating an associated standard deviation. The conservation signal is typically more than two standard deviations above the background, especially in the case of 8mer and 7mer seed matches.

Evaluation of the accuracy of this estimate depends on maximizing the sensitivity while ensuring that no excess conservation is observed for negative controls. The analysis should be repeated using regions for which no miRNA targeting is expected, such as constitutive introns or the reverse-complement of the 3′ UTRs. One can also test matches to non-seed regions of the miRNA, such as the 3′ end, or shuffles (random permutations of the nucleotides) of miRNA seed matches. If there is no systematic bias in picking control *k*-mers, the signal-to-background ratios for these controls should be close to 1. One can make modifications to the algorithm, for example the number of local conservation bins used, and evaluate the results by looking for high conservation signal above background for true miRNA seed matches, but signal roughly equal to background for negative controls.

2.3.6 Nested Seed Matches

Having the conservation signal and background estimates, one can now observe the excess conservation (the imprint of natural selection) of miRNA seed matches. However, the seed match types are nested and overlapping (2–7 6mer, offset or 3–8 6mer, 7mer-m8, 7mer-A1, and 8mer), *see* Fig. 1. In order to avoid double-counting targets and to sensitively calculate the conservation signal and background for each seed match type, the nested nature must be dealt with. 8mer seed matches (with an adenosine opposite position 1 of the miRNA and a Watson–Crick match opposite position 8) are the largest motifs and thus the signal to background calculations do not require adjustment for this target type. 7mers (both m8 and A1) can occur as part of an 8mer or individually.

Therefore, at each branch-length cutoff, one must subtract the 8mer conservation signal from the 7mer conservation signal, and subtract the 8mer conservation background from the 7mer conservation background, separately for 7mer-m8 matches and 7mer-A1 matches. Thus, any conservation signal above background levels that the 8mer is responsible for is not double-counted. This scheme also maximizes sensitivity [30]. Likewise, 2–7 6mer seed matches can be a part of 7mer-A1, 7mer-m8, or 8mer matches. So the signal and background of each of these types must be subtracted from the 2–7 6mer's. The same applies to the 3–8 6mer (except that it cannot be included in a 7mer-A1 seed match). Standard deviations of the background estimate should be calculated after performing this subtraction for nested target types.

There may also be cases in which the seed match type definitions are ambiguous. For example, the 7mer-m8 match of a certain miRNA may be the same sequence as the 7mer-A1 match of a different miRNA. In these cases, it is important only to avoid double-counting. One reasonable solution is to assign conservation to 7mer-m8 rather than 7mer-A1 matches, and 2–7 6mer rather than 3–8 6mer matches. This ordering makes was selected because 7mer-m8 matches are more effective than 7mer-A1 matches [27, 28] and likewise with 2–7 6mers and 3–8 6mers [30].

2.4 Individual miRNA Targets

Up until now, we have been looking at global properties of targeting and global estimates of the number of selectively maintained miRNA seed matches. These estimates depend on the usage of global conservation rates at or above branch-length cutoffs, whereas evaluating individual targets requires a background estimate that is applicable to a single target site.

2.4.1 Probability of Conserved Targeting

This metric for evaluating the conservation of individual targets was designed to utilize all the techniques outlined above except that the background is calculated precisely for a single target site rather than in aggregate for all target sites. The difference is that rather than using a branch-length cutoff to calculate a conservation rate, a branch-length window is used instead. Thus, seed matches do not receive conservation "credit" if more well-conserved sites for the same miRNA have a high signal to background, nor are they penalized if more poorly conserved sites have a low signal to background. For a given site, the fraction of that *k*-mer's occurrences that are within the branch-length window is compared to the mean fraction in the same window for control *k*-mers, yielding a conservation signal-to-background ratio for a single site. The branch-length window should be as small as possible while maintaining a minimum of 20 occurrences (for example), in order to achieve some statistical confidence. This parameter can be tuned to fit the size of the dataset.

This signal-to-background ratio can be roughly converted to a probability that any single site will be conserved above background levels using the simple formula:

$$\hat{P}_{CT} = \frac{S/B - 1}{S/B}$$

where S/B is the conservation signal-to-background ratio and $\hat{P}_{CT}$ is the raw probability of conserved targeting for that site. To smooth out noise caused by a small number of occurrences in each branch-length window (especially relevant for 8mers), one can fit a modified sigmoid function to these raw $\hat{P}_{CT}$ scores:

$$P_{CT} = \max\left(0, \beta_0 + \frac{\beta_1}{1 + \exp(-\beta_2 x + \beta_3)}\right)$$

where x is the branch length value corresponding to each $\hat{P}_{CT}$. Thus, the final P_{CT} is a smoothed version of the $\hat{P}_{CT}$ estimates. This modified logistic function guarantees a monotonic increase from low to high P_{CT} as branch-length increases. Here, β_0 represents an offset along the y axis (P_{CT}), β_1 constrains the maximum value attainable, β_2 controls the slope of the increase in P_{CT}, and β_3 represents an offset along the x axis (branch lengths). Thus, the smoothing allows variation in the maximum and minimum P_{CT} attainable for each miRNA, as well as the position and steepness of the transition from low to high P_{CT}. In practice, we have observed that this function fits the data well.

3 Notes

3.1 Conservation Metric

There are many possible metrics for converting a list of species sharing a conserved sequence into an estimate of the level of evolutionary conservation, ranging from simple models such as binary functions to more complex ones such as Hidden Markov Model (HMM)-based classifiers. The branch length metric presented here is a compromise between computational speed and expressive power. It assumes that the sequence first appeared in the last common ancestor of the set of present-day species having the sequence. Thus, it assumes no outgroups ever had the sequence and that the sequence arose only once. Convergent evolution and back-mutation violate these assumptions, which could become a problem when comparing highly diverged species or highly conserved sequences. In such a context, any conservation metric could be substituted for the branch-length score. For example, a score from a phastCons-like hidden Markov model could be directly substituted. However, it is important to consider the conservation of the entire seed match (6mer-8mer) rather than the conservation of each nucleotide individually. This is because single mismatches and G:U wobble basepairs have a dramatic effect on miRNA efficacy, so conservation of most but not all of the seed match should not be rewarded [27, 30].

One alternative worth further discussion is a model that does not require precise alignment. Stark and coworkers considered miRNA seed matches conserved if they were aligned within a window rather than requiring perfect alignment [35]. This approach allows the possibility of recovering from misalignments due to DNA rearrangements, major insertions or deletions, or simply ambiguities. In mammals, miRNA seed matches were aligned within a small window but not precisely aligned roughly as often as control *k*-mers, meaning this would have added equally to signal and to background measurements (unpublished observation). However, the utility of this approach depends on the precise set of species, the quality of UTR definitions, and the alignment quality. In general, it is recommended to evaluate the signal-to-background ratio of non-aligned seed matches separately from those that are precisely aligned.

3.2 Orthology

The conservation metric assumes that 3′ UTRs that are aligned are orthologous. Deviations from this assumption can introduce substantial noise. This noise will apply equally to miRNA seed matches and to control *k*-mers, but is still undesirable because it can decrease the statistical confidence in measurements and distort the significance of individual conserved target sites. Roughly half of mammalian genes are one-to-one orthologs, so aligning the correct regions is relatively easy in these cases. For genes with paralogs, the decision of which regions to align is often ambiguous or arbitrary. As a result, the accuracy of whole-genome alignments is likely substantially lower for paralogous genes. If miRNA targets in paralogous gene families are of particular interest, the closest orthologs could be identified using one of a number of available methods or databases [36]. The algorithm could then be supplied with pairwise or multiple alignments of the orthologous 3′ UTR regions calculated by one of several methods [37].

3.3 PHYLIP

PHYLIP is a widely used and practical free software package for phylogenetic analysis [33]. The DNAML program estimates phylogenies from DNA sequences by maximum likelihood, and is useful in this application for building trees based specifically on 3′ UTR sequences. However, for large datasets and for batch applications, DNAML can be difficult to use. We recommend that the tree structure be fixed using a Newick-format tree (the "U" option), which drastically reduces computational complexity, takes advantage of known species relationships, and removes inconsistencies between trees for different local conservation bins. DNAML cannot handle particularly large sequence files as input. A simple workaround for large datasets is to select several random subsamples of 3′ UTR sequences and to average the branch lengths of the resulting trees. In cases when the species relationships are known, the PHAST package (http://compgen.bscb.cornell.edu/phast/) is more adept at handling large datasets.

3.4 Branch Lengths

For trees with few branches, a limited number of branch length values are possible. However, the number of possible branch length values increases exponentially as new species are added. Therefore, calculated branch lengths should be rounded (to the nearest 1/100th works well) in order to capture subtle variations while still maintaining a discrete space of limited size.

3.5 Comparison with Other Target Prediction Approaches

The target prediction algorithm described here is publicly available on the TargetScan web site (targetscan.org). This approach has been validated *in silico* by the signal-to-background calculation using rigorous control *k*-mers, but has also predicted unbiased sets of experimentally verified targets well [8, 30]. However, several other target prediction approaches are available, often based on different principles, having different features, and yielding different results [10].

Several algorithms use the same principles of stringent seed pairing but use different criteria for conservation of sites. Older methods, such as PicTar and earlier versions of TargetScan, used a binary metric for conservation, requiring a site to be conserved in a given number of species [20, 22, 38]. If there are only two species considered, then the metric presented here is binary. However, as the number of species examined increases, the branch length metric becomes more flexible and more sensitive than a binary metric. There are also several approaches that consider the sequence and structural context of the site rather than its conservation, such as the TargetScan context score [27], TargetRank [28], and PITA [31]. As described in Section 1, these approaches are complementary and are useful for different situations, such as analysis of species-specific targets.

EIMMo uses an interesting alternative conservation metric, feeding the phylogenetic distribution of aligned miRNA seed matches into a Bayesian framework for estimating the phylogenetic distribution of selection [39]. A key advantage of this method is its prediction of clade-specific targets for miRNAs lost in parts of the phylogeny. The method presented here relies instead on curation of miRNAs so that predictions are only made for miRNAs broadly conserved throughout the phylogeny considered. Disadvantages of EIMMo include its lack of controls for GC content, dinucleotide content, and for local conservation variation. It would be interesting to adapt an EIMMo-like conservation metric to the framework presented here.

Other approaches that do not require stringent seed pairing or require strong pairing to most of the miRNA sequence do not have robust empirical support for these criteria. By comparison with carefully selected controls or with genome-wide experimental data, target prediction criteria can be accepted or rejected. Luckily, with the amount of sequenced genomes and genome-wide experimental data available, the means to rigorously evaluate all target prediction criteria are currently at hand.

3.6 Areas of Future Improvement

Although this conservation method has several areas that could be improved, the general approach is flexible enough to incorporate a wide variety of modifications. Below we highlight a few areas with potential for fruitful improvement.

3.6.1 Local Conservation Models

The solution to controlling for local conservation presented here splits genes into bins based on the average branch length of their constituent nucleotides. Of course, this creates discontinuities at the edges of bins. Additionally, the average branch length is a relatively crude measure of local conservation. For example, it does not take into account which genomes are aligned to a sequence.

In the extreme, one could build a separate conservation model for each gene. However, there must be a sufficient number of occurrences of control *k*-mers within comparable regions to achieve statistical accuracy. Therefore there is a tradeoff between increased accuracy of controlling for local conservation and statistical power.

A simple modification to avoid edge effects would be to create for each gene a set of comparable background genes, i.e., each gene would be in the center of its own bin. For a slightly more sophisticated treatment of local conservation, one could filter each set of genes to ensure that they have the same pattern of missing data, i.e., they lack orthologs or 3′ UTR alignment in the same sets of species.

3.6.2 Incorporation of Context Features

Context scores predict targets by incorporating features of the sequence surrounding the seed match, such as pairing to the 3′ end of the miRNA, local A/U content, and 3′ UTR length. Here we have focused on the conservation of the seed match itself, disregarding the surrounding context. Merging these two complementary methods into a single target prediction tool would presumably yield increased accuracy. This could be accomplished by calculating the context score for each miRNA-gene pair in all orthologous species, creating a new conservation metric that used the conservation of the context score rather than the conservation of individual seed matches. Context scores could then be calculated for mock miRNAs to estimate the background distribution.

3.6.3 miRNA Targeting in the ORF

It has long been recognized that miRNA seed matches in the ORF can be conserved above background levels despite their decreased efficacy relative to 3′ UTR seed matches [20, 34, 40]. However, detecting conservation of functional RNA motifs in the ORF is challenging due to the strong background conservation at the amino acid level. This increased conservation can make evolutionary patterns such as back-mutation and convergent evolution more likely than in the 3′ UTR. Therefore, an alternate conservation metric, such as a maximum parsimony phylogenetic reconstruction, could be useful for this problem. Additionally, background conservation would have to be evaluated in the context of the specific reading frame in which a motif occurs. Thus, three separate background estimates would have to be made for the three possible frames. On the other hand, finding orthologous regions, defining region boundaries, and generating multiple alignments are all easier in the case of coding sequences compared to 3′ UTRs.

Acknowledgements

The TargetScan algorithm described was co-developed with David P. Bartel and was based on work by Kyle Kai-How Farh. The authors thank David Bartel and Vikram Agarwal for helpful discussions and the US Department of Energy Office of Science for funding the development of this algorithm.

References

1. Lee RC, Feinbaum RL, Ambros V (1993) The C. elegans heterochronic gene lin-4 encodes small RNAs with antisense complementarity to lin-14. Cell 75(5):843–854
2. Wightman B, Ha I, Ruvkun G (1993) Post-transcriptional regulation of the heterochronic gene lin-14 by lin-4 mediates temporal pattern formation in C. elegans. Cell 75(5): 855–862
3. Moss EG, Lee RC, Ambros V (1997) The cold shock domain protein LIN-28 controls developmental timing in C. elegans and is regulated by the lin-4 RNA. Cell 88(5): 637–646
4. Reinhart BJ, Slack FJ, Basson M, Pasquinelli AE, Bettinger JC, Rougvie AE, Horvitz HR, Ruvkun G (2000) The 21-nucleotide let-7 RNA regulates developmental timing in Caenorhabditis elegans. Nature 403(6772):901–906. doi:10.1038/35002607
5. Bartel DP (2004) MicroRNAs: genomics, biogenesis, mechanism, and function. Cell 116(2):281–297
6. Elbashir SM, Lendeckel W, Tuschl T (2001) RNA interference is mediated by 21- and 22-nucleotide RNAs. Genes Dev 15(2):188–200
7. Hutvágner G, Zamore PD (2002) A microRNA in a multiple-turnover RNAi enzyme complex. Science 297(5589):2056–2060. doi:10.1126/science.1073827
8. Baek D, Villén J, Shin C, Camargo FD, Gygi SP, Bartel DP (2008) The impact of microRNAs on protein output. Nature 455(7209):64–71. doi:10.1038/nature07242
9. Selbach M, Schwanhäusser B, Thierfelder N, Fang Z, Khanin R, Rajewsky, N Widespread changes in protein synthesis induced by microRNAs (2008) Nature 455(7209):58–63. doi:10.1038/nature07228
10. Bartel DP (2009) MicroRNAs: target recognition and regulatory functions. Cell 136(2):215–233. doi:10.1016/j.cell.2009.01.002

11. Doench JG, Petersen CP, Sharp PA (2003) siRNAs can function as miRNAs. Genes Dev 17(4):438–442. doi:10.1101/gad.1064703
12. Lai EC (2002) Micro RNAs are complementary to 3' UTR sequence motifs that mediate negative post-transcriptional regulation. Nat Genet 30(4):363–364. doi:10.1038/ng865
13. Stark A, Brennecke J, Russell RB, Cohen SM (2003) Identification of Drosophila microRNA targets. PLoS Biol 1(3):E60. doi:10.1371/journal.pbio.0000060
14. Enright AJ, John B, Gaul U, Tuschl T, Sander C, Marks DS (2003) MicroRNA targets in Drosophila. Genome Biol 5(1):R1. doi:10.1186/gb-2003-5-1-r1
15. Lewis BP, Shih IH, Jones-Rhoades MW, Bartel DP, Burge CB (2003) Prediction of mammalian microRNA targets. Cell 115(7):787–798
16. Rajewsky N, Socci ND (2004) Computational identification of microRNA targets. Dev Biol 267(2):529–535. doi:10.1016/j.ydbio.2003.12.003
17. John B, Enright AJ, Aravin A, Tuschl T, Sander C, Marks DS (2004) Human microRNA targets. PLoS Biol 2(11):e363. doi:10.1371/journal.pbio.0020363
18. Rajewsky N (2006) microRNA target predictions in animals. Nat Genet 38(Suppl):S8–S13. doi:10.1038/ng1798
19. Doench JG, Sharp PA (2004) Specificity of microRNA target selection in translational repression. Genes Dev 18(5):504–511. doi:10.1101/gad.1184404
20. Lewis BP, Burge CB, Bartel DP (2005) Conserved seed pairing, often flanked by adenosines, indicates that thousands of human genes are microRNA targets. Cell 120(1):15–20. doi:10.1016/j.cell.2004.12.035
21. Brennecke J, Stark A, Russell RB, Cohen SM (2005) Principles of microRNA-target recognition. PLoS Biol 3(3):e85. doi:10.1371/journal.pbio.0030085
22. Stark A, Brennecke J, Bushati N, Russell RB, Cohen SM (2005) Animal microRNAs confer robustness to gene expression and have a significant impact on 3'UTR evolution. Cell 123(6):1133–1146. doi:10.1016/j.cell.2005.11.023
23. Krek A, Grün D, Poy MN, Wolf R, Rosenberg L, EJ Epstein, MacMenamin P, da Piedade I, Gunsalus KC, Stoffel M, Rajewsky N (2005) Combinatorial microRNA target predictions. Nat Genet 37(5):495–500. doi:10.1038/ng1536
24. Lim LP, Lau NC, Garrett-Engele P, Grimson A, Schelter JM, Castle J, Bartel DP, Linsley PS, Johnson JM (2005) Microarray analysis shows that some microRNAs downregulate large numbers of target mRNAs. Nature 433(7027):769–773. doi:10.1038/nature03315
25. Birmingham A, Anderson EM, Reynolds A, Ilsley-Tyree D, Leake D, Fedorov Y, Baskerville S, Maksimova E, Robinson K, Karpilow J, Marshall WS, Khvorova A (2006) 3' UTR seed matches, but not overall identity, are associated with RNAi off-targets. Nat Method 3(3):199–204. doi:10.1038/nmeth854
26. Jackson AL, Burchard J, Schelter J, Chau BN, Cleary M, Lim L, Linsley PS (2006) Widespread siRNA "off-target" transcript silencing mediated by seed region sequence complementarity RNA. 12(7):1179–1187. doi:10.1261/rna.25706
27. Grimson A, Farh KK, Johnston WK, Garrett-Engele P, Lim LP, Bartel DP (2007) MicroRNA targeting specificity in mammals: determinants beyond seed pairing. Mol Cell 27(1):91–105. doi:10.1016/j.molcel.2007.06.017
28. Nielsen CB, Shomron N, Sandberg R, Hornstein E, Kitzman J, Burge CB (2007) Determinants of targeting by endogenous and exogenous microRNAs and siRNAs. RNA 13(11):1894–1910. doi:10.1261/rna.768207
29. Farh KK, Grimson A, Jan C, Lewis BP, Johnston WK, Lim LP, Burge CB, Bartel DP (2005) The widespread impact of mammalian microRNAs on mRNA repression and evolution. Science 310(5755):1817–1821. doi:10.1126/science.1121158
30. Friedman RC, Farh KK, Burge CB, Bartel DP (2009) Most mammalian mRNAs are conserved targets of microRNAs. Genome Res 19(1):92–105. doi:10.1101/gr.082701.108
31. Kertesz M, Iovino N, Unnerstall U, Gaul U, Segal E (2007) The role of site accessibility in microRNA target recognition. Nat Genet 39(10):1278–1284. doi:10.1038/ng2135
32. Hausser J, Landthaler M, Jaskiewicz L, Gaidatzis D, Zavolan M (2009) Relative contribution of sequence and structure features to the mRNA binding of Argonaute/EIF2C-miRNA complexes and the degradation of miRNA targets. Genome Res 19(11):2009–2020. doi:10.1101/gr.091181.109
33. Felsenstein J (1989) PHYLIP: phylogenetic inference package. Cladistics 5(2): 163–166
34. Stark A, Lin MF, Kheradpour P, Pedersen JS, Parts L, Carlson JW, Crosby MA, Rasmussen MD, Roy S, Deoras AN, Ruby JG, Brennecke J, Harvard FlyBase curators, Berkeley Drosophila Genome Project, Hodges E, Hinrichs AS, Caspi A, Paten B, Park S-W, Han MV, Maeder ML, Polansky BJ, Robson BE, Aerts S, van Helden J, Hassan B, Gilbert DG, Eastman DA, Rice M, Weir M, Hahn

MW, Park Y, Dewey CN, Pachter L, Kent WJ, Haussler D, Lai EC, Bartel DP, Hannon GJ, Kaufman TC, Eisen MB, Clark AG, Smith D, Celniker SE, Gelbart WM, Kellis M (2007a) Discovery of functional elements in 12 Drosophila genomes using evolutionary signatures. Nature 450(7167): 219–232. doi:10.1038/nature06340

35. Stark A, Kheradpour P, Parts L, Brennecke J, Emily H, Hannon GJ, Kellis M (2007b) Systematic discovery and characterization of fly microRNAs using 12 Drosophila genomes. Genome Res 17(12):1865–1879. doi:10.1101/gr.6593807

36. Altenhoff AM, Dessimoz C (2009) Phylogenetic and functional assessment of orthologs inference projects and methods. PLoS Comput Biol 5(1):e1000262. doi:10.1371/journal.pcbi.1000262

37. Chen X, Tompa M (2010) Comparative assessment of methods for aligning multiple genome sequences. Nat Biotechnol 28(6):567–572. doi:10.1038/nbt.1637

38. Lall S, Grün D, Krek A, Chen K, Wang Y-L, Dewey CN, Sood P, Colombo T, Bray N, MacMenamin P, Kao H-L, Gunsalus KC, Pachter L, Piano F, Rajewsky N (2006) A genome-wide map of conserved microRNA targets in C. elegans. Curr Biol 16(5):460–471. doi:10.1016/j.cub.2006.01.050

39. Gaidatzis D, van Nimwegen E, Hausser J, Zavolan M (2007) Inference of miRNA targets using evolutionary conservation and pathway analysis. BMC Bioinformatics 8:69. doi:10.1186/1471-2105-8-69

40. Schnall-Levin M, Zhao Y, Perrimon N, Berger B (2010) Conserved microRNA targeting in Drosophila is as widespread in coding regions as in 3'UTRs. P Natl Acad Sci USA. doi:10.1073/pnas.1006172107

Chapter 22

Bioinformatics of siRNA Design

Hakim Tafer

Abstract

RNA interference mediated by small interfering RNAs is a powerful tool for investigation of gene functions and is increasingly used as a therapeutic agent. However, not all siRNAs are equally potent, and although simple rules for the selection of good siRNAs were proposed early on, siRNAs are still plagued with widely fluctuating efficiency. Recently, new design tools incorporating both the structural features of the targeted RNAs and the sequence features of the siRNAs substantially improved the efficacy of siRNAs. In this chapter we will present a review of sequence and structure-based algorithms behind them.

Key words Accessibility, Binding sites, Computer simulation, Drug design, Gene targeting, RNA interference, RNA, small interfering/genetics, Sequence analysis, RNA structure

1 Background

RNA interference (RNAi) describes the post-transcriptional gene silencing process triggered by endogenous or exogenous double-stranded RNAs (dsRNAs). After being processed by Dicer, the dsRNAs are transferred to the RNA-Induced Silencing Complex (RISC), where one of the strands (the guide strand) is introduced while the other strand is degraded (the passenger strand). Target recognition happens through hybridization of the guide RNA with its target gene, which causes the cleavage and the subsequent degradation of the target strand.

The successful utilization of artificial dsRNAs to knockdown specific genes was first reported by Fire et al. [1]. In 2001 Elbashir et al. [2] showed that siRNA-mediated gene knockdown could also be applied in mammalian cells. Initial expectations that there were no need to search for optimal siRNA sequences [3] rapidly proved to be unfounded, as strong variations in silencing efficiency were reported for different siRNAs directed against the same target [4]. Still the potential of RNAi to transiently knockdown genes motivated the scientific community to improve the siRNA design rules (for a review *see* [5]). Elbashir et al. [6] published the

Jan Gorodkin and Walter L. Ruzzo (eds.), *RNA Sequence, Structure, and Function: Computational and Bioinformatic Methods*, Methods in Molecular Biology, vol. 1097, DOI 10.1007/978-1-62703-709-9_22,

first protocol for designing active siRNAs. They encouraged the use of 21 nucleotides long siRNAs with a G/C content of about 50% and 2 nucleotides $3'$overhangs.

In 2003, Khvorova et al. [7] and Schwarz et al. [8] proved that even though both strands of the dsRNA could serve as a guide strand [2, 6], the strand with the lower $5'$ stability was preferentially incorporated into the RISC. Subsequent studies concentrated on finding sequence patterns on the guide strand which correlated with the repression efficacy [9–14]. The majority of those studies confirmed that the relative stability of the siRNA ends was a major determinant of the functionality of siRNAs. Further improvements in the design of siRNA came from the study published by Patzel et al. [15], who showed that the siRNA efficiency directly correlate with the siRNA structuredness.

The small number of siRNAs used in those early studies led to poor agreements on the sequence patterns and to parameter overfitting [16]. The use of heterogeneous data, gathered either from previous work or from siRNA databases (e.g., siRecords [17]), did not resolve this issue, as the oligonucleotide activity is highly sensitive to biological and experimental parameters (transfection efficiency, cell type, siRNA concentration, target concentration, efficiency measure). To overcome those problems Huesken et al. [18] generated a set of 2,431 randomly selected siRNAs targeted against 34 mRNAs, which was used to train an artificial neural network for designing siRNAs. Statistical analysis of this data set confirmed some of the previously published siRNAs features (duplex asymmetry) and revealed new, highly significant sequence motives.

A long debated topic in the field of siRNA design is the influence of the target structure on the siRNA efficiency. While target site structure was recognized as an important feature in the design of antisense oligonucleotides and ribozymes [19–24], data arguing for [25–39] and against [14, 15, 40] the influence of target site accessibility on the siRNA efficiency were reported. From a thermodynamic point of view, the interaction of two RNAs can be decomposed into two stages: Binding can only occur at positions not already involved in intramolecular base pairs. Thus, base pairs within the target site have to be opened to make the site *accessible*. The energy necessary to do this is termed the disruption or breaking energy. Once the binding site is devoid of structure intermolecular helices can be formed, yielding a stabilizing interaction energy. The total binding energy is then computed as the sum of the hybridization energy and the breaking energy.

In principle such a model could directly predict the fraction of mRNAs that will be bound by siRNAs. This, however, requires knowledge of siRNA and mRNA concentrations which are in general not available. Furthermore, the model implicitly assumes that reactants are free solutes, thus neglecting possible influences of mRNA binding proteins, active translation by the ribosome, and the RISC on the siRNA binding. Still, the application of this approach on siRNA data published by Schubert et al. [34], where an siRNA was targeted to a gradually less accessible target site, showed that siRNA efficiency is directly correlated to the target site accessibility [41, 42]. Those findings were corroborated by four further studies [29, 30, 37–39] which looked specifically at the effect of local target secondary structure on RNAi efficiency based on large (100 siRNAs against 3 genes) to very large (3,084 siRNAs against 82 genes) homogeneous data sets.

The majority of the siRNA design rules mentioned above can be mapped to key events of the silencing pathways (*see* Fig. 1). The limited length of the siRNA duplex as well as the presence of 3′ end dangles allows the siRNA to evade immunorecognition [43–45]. The rules promoting the sequence/energy asymmetry [7, 8] reflect the ability of Dicer to sense the thermodynamic asymmetry between the two ends of the duplex. The negative effect of structure of the guide strand on the repression efficiency may be explained by a reduced ability of the siRNA to bind to its target and/or hindered interaction of the siRNA with RISC components [15]. Finally the influence of the target site accessibility on the siRNA efficiency derives from (1) the ability of RISC to bind to single-stranded region only and (2) the inability of RISC to unfold structured RNA [36].

In the following sections we discuss the three siRNA design tools (`OligoWalk` [37, 38], `Sirna` [29, 46], `RNAxs` [39]) which perform siRNA design aided by target accessibility criteria. Similarly we will summarize the different sequence-based approach developed to design siRNA. Finally database containing siRNA-related information will be briefly reviewed.

2 Sequence-Based siRNA Design

2.1 Rules-Based siRNA Design

The first thorough analysis of the influence of siRNA sequence characteristics on its repression efficiency was published by Elbashir in 2001 [2, 6]. Different siRNA characteristics were varied such as the length of the dsRNA, the GC content as well as the number and type of overhanging nucleotides and showed that 21 nt duplexes with 2 nt overhang and low GC content were highly efficient.

Schwarz [8] and Khvorova [7] showed that besides the length requirements, siRNAs must be asymmetric in order for the guide strand to be introduced in the RISC. Namely the siRNA strand of the dsRNA with the weakest 5′ end binding to the complement strand is preferentially included into RISC. From a sequence point of view, the siRNA to be inserted into RISC possesses a higher A/T content on its 5′ end than the passenger strand.

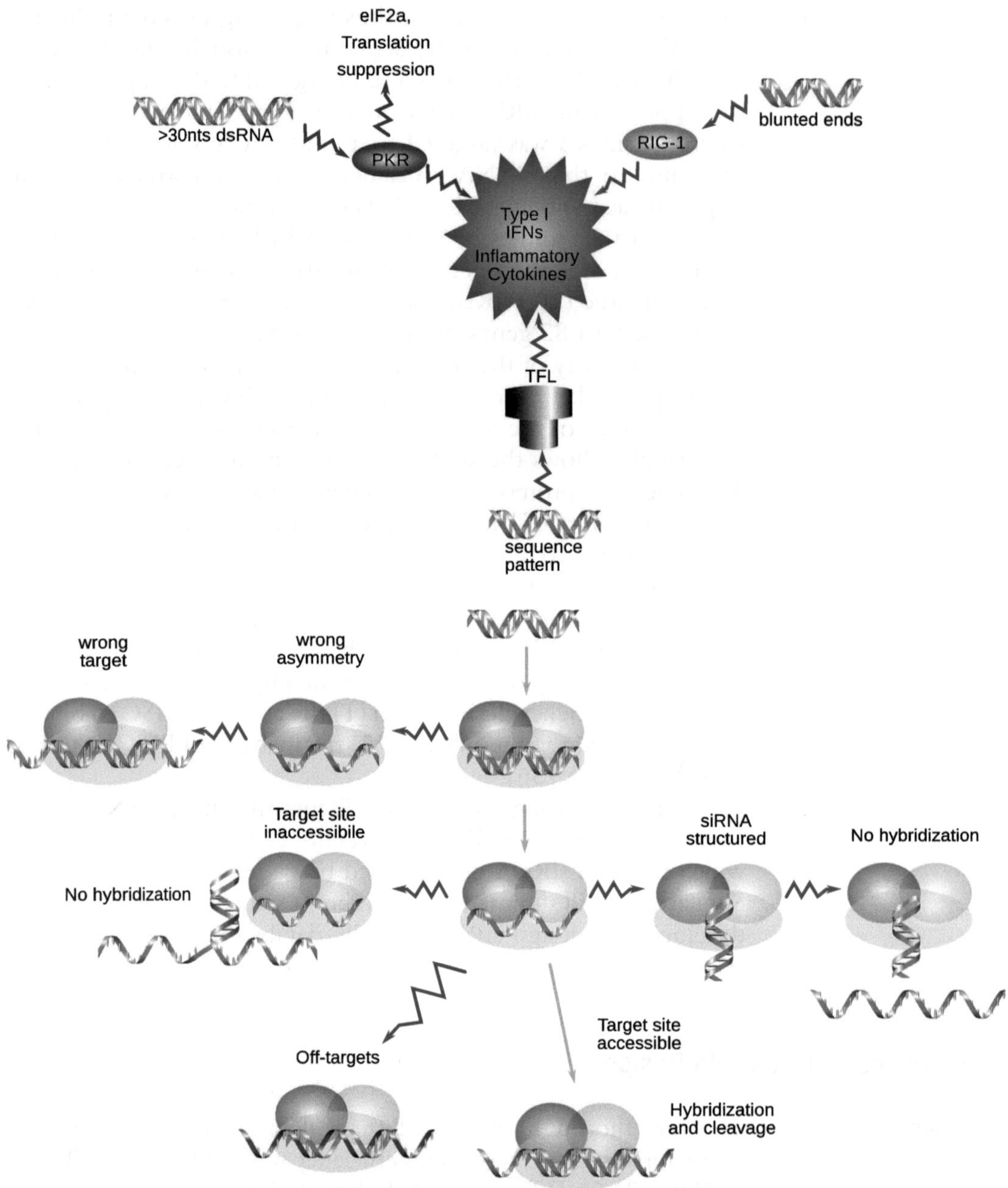

Fig. 1 Impact of siRNA characteristics along the silencing pathway. The innate immune system may be activated by dsRNAs. dsRNAs with specific sequence patterns or high "U" contents are recognized by Toll-Like Receptors (TLRs) inducing inflammatory cytokines and interferon of type I (IFN-α, IFN-β). Large dsRNAs (>30nts) are sensed by PKR (double-stranded RNA-activated protein kinase) which can induce interferon response, expression of inflammatory cytokines, and cell death. dsRNAs with 2nts overhangs escape the RIG-1 triggered cytokines and interferon response. Once into RISC, the passenger strand is separated from the guide strand. The strand with the lower 5′-end stability is incorporated into RISC, while the other strand is degraded. A wrong asymmetry results in the selection the bad siRNA strand, leading to no on-target effect. siRNAs that are highly structured are not able to hybridize to their target. Reciprocally siRNAs targeting highly structured region cannot bind to their target. Finally sequence-specific off-target effects make it more difficult to gain information from RNAi experiments

The first scoring scheme developed to design siRNAs was published by Reynolds et al. [14]. To this aim a set of 180 siRNAs targeting the mRNA of two genes was studied. Besides the confirmation of the asymmetry rule, [14] further showed that a moderate G/C content and a lack of structure in the guide siRNA favors repression efficiency. For the first time, position-specific nucleotide preference was reported. While the presence of A/U bases at the 5′ end of the antisense strand is directly correlated with the stability asymmetry of the dsRNA, the presence of U at position 10 is related to the preference of most endonucleases to cleave 3′ of U [2].

Further studies found position-specific nucleotide preferences. Amarzguioui and Prydz [9], based on a set of 46 siRNAs targeted against 4 genes, reported that the most important parameters were the sequence asymmetry, followed by duplex end asymmetry and the presence of a U at position 14 of the target site. Similarly [12] devised a scoring system based on positional features of nucleotides. Sequence asymmetry was found as being the main descriptors, followed by the presence of U at position 12 of the target site and a A at position 5. An in-depth analysis was also delivered by Ui-Tei et al. [13], who analyzed 62 siRNAs targeted against 4 genes. Here also the siRNA sequence asymmetry was stressed out, followed by the presence of at least five A/U residues in the 5′terminal and the absence of any G/C stretch longer than 9 nts.

Saetrom and Snove [47] published the first benchmark of siRNA design tools available at this time. The approach from [9] and [14] performed best, while that from [12] performed not better than a random classifier. The relative low performance of the published rules is mainly due to the low numbers of siRNAs used to find them, leading to overfitting, contradictory models penalizing or bonifying identical nucleotides at the same position in the sequence [47], and models performing poorly on unseen datasets [47]. Moreover this crude rules may hardly describe the interplay of different less significant siRNA descriptor.

2.2 Machine-Learning Aided siRNA Design

Reference 16 was the first publication to present an siRNA design tool using a machine-learning approach. Based on a dataset of 101 positive and 103 negative siRNAs, [16] applies a machine-learning approach to classify siRNAs into working and non-working based on the siRNA sequence patterns, the hybridization energy of the siRNA with its target, the asymmetry in stability between the 5′ and 3′ end of the dsRNA, and the structuredness of the siRNA strand.

A milestone publication for the topic of siRNA design was [18], where the largest gene knockdown experiments performed to date was achieved. Huesken et al. [18] generated a set of 2,431 siRNAs directed against 34 mRNAs and measured the

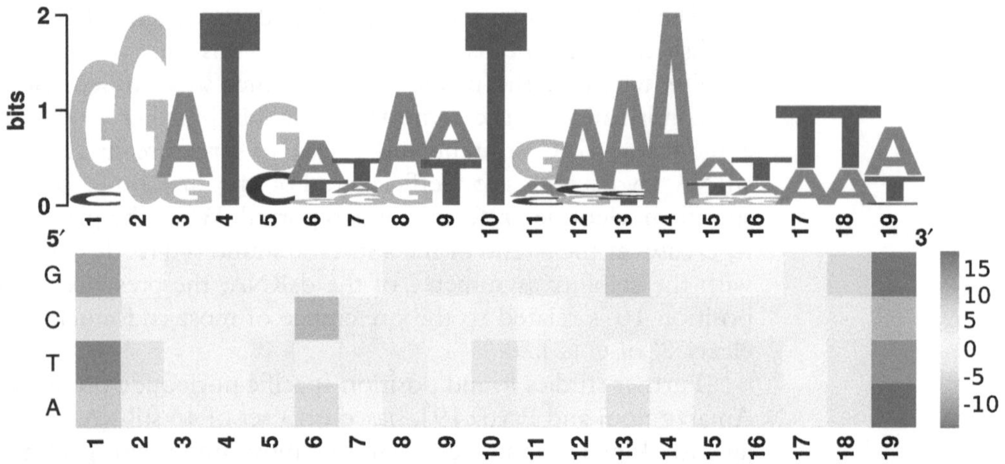

Fig. 2 *Top*: Weblogo representation of the highest scoring siRNA target sequence for the algorithm from Reynolds, Hsieh, Takasaki, Dsir, Amarzguioui, Katoh, and iScore. The sequence asymmetry is clearly seen, as well as the presence of a `T` at position 10. *Bottom*: Heatmap representation of the sum of the normalized coefficients of the models from Reynolds, Hsieh, Takasaki, Dsir, Amarzguioui, Katoh, and iScore for each nucleotide and position along the siRNA target sequence. *Dark squares* represent preferred nucleotide/position combination while *light ones* stand for unfavorable nucleotide/position. Here also the sequence asymmetry is clearly seen, with `A/T` being highly preferred at the target 3′ end, while `G/C` being overrepresented at the target 5′ end

corresponding repression efficiency under strict experimental conditions. Based on this gold-standard dataset, which is now freely available, Huesken et al. [18] first looked at overrepresented nucleotide along the siRNA sequence, recovering the asymmetry as well as the presence of `U` at position 10 of functional siRNAs. (*See* Fig. 2.) Further they trained an Artificial Neural Network on sequence characteristics of the siRNA on a test dataset containing 2,182 and applied it on a test dataset containing 249 sequences, reaching a pearson correlation coefficient of 0.66.

While the performance of the tool developed by Huesken et al. [18], *BioPredsi* performs well on the test dataset, the use of a neural network makes biological interpretation of the results difficult. Vert et al. [48] trained a linear approach on the data published by Huesken et al. [18] named `Dsir`. In this approach, each siRNA is represented in sparse and spectral form. The spectral representation counts the occurrence of mono, di, and trinucleotide inside the siRNA while the composite representation reports the presence or absence of each nucleotide at each position along the siRNA sequence. The LASSO procedure, which shrinks the coefficient of irrelevant features to zero, was used to fit the regressor data, a.k.a the representation of the siRNA, to the corresponding siRNA efficiency. While this approach perform as well as *BioPredsi*, it allows to interpret the model much more easily as the magnitude and sign of the fitted coefficients give a clue about the significance. Interestingly, Vert et al. [48] showed that the occurrence of asymmetric oligonucleotides motifs is as important as the position-specific nucleotide preferences.

In the same pursuit of interpretability, Matveeva et al. [49] presented a linear approach where in contrast to *BioPredsi* and `Dsir`, both the sequence and the stability of the siRNA duplex are taken into account. In the same publication, Matveeva et al. [49] benchmarked 10 different methods on different datasets amounting 3,336 siRNAs directed against 145 mRNAs. In this benchmark the method of [48, 49] and [18] performed best. Still, while the method [49] performed on par with `Dsir`, it uses only 22 input parameters instead of the 115 used by Vert et al. [48].

3 Accessibility-Aided siRNA Design

3.1 *siRNA*

Ding et al. [29] presented the first attempt to design siRNAs by considering target site accessibility in 2004. Their algorithm called *Sirna* selects siRNAs based on sequence and accessibility criteria. In their algorithm, accessible regions are identified with the help of *Sfold*. *Sfold* computes the accessibility along the target sequence by generating a statistically representative sample of 1,000 structures from the Boltzmann-weighted ensemble of secondary structures. The equilibrium probability of nucleotide i in the target sequence to be unbound (the accessibility of nucleotide i) is estimated by counting how often the nucleotide is unbound in the sampled structures [24]. An siRNA target site is then considered accessible if the single nucleotide accessibility averaged over the binding site is higher than 0.5. For each such site, *Sirna* selects siRNAs that further meet requirements of empirical sequence-based rules. Those are the Reynolds rules [14], a G/C content of 30–70%, the cleavage site instability rules of Khvorova et al. [7] and the asymmetry rules [7, 8]. For each of the siRNAs that passed the previous filters, a further measure of accessibility is computed. This feature, termed accessibility-weighted interaction energy, is calculated by weighting each stacked base pair in the siRNA–mRNA interaction by the probability of the dinucleotide that is involved in the stack to be unpaired. siRNAs having a weighted interaction energy ≤ -10 kcal/mol are selected.

In a further study [30], Shao et al. extended *Sirna* by using a slightly different accessibility measure. Shao defined the accessibility as the energy cost for breaking intramolecular base pairs located at the binding site. This disruption energy is the energy difference between the free energy of the original mRNA structure and the energy of the altered structure, where the siRNA target site is devoid of intramolecular base pairs. Again, this disruption energy is computed using a sample of 1,000 mRNA structures as computed by Sfold [24, 29, 50]. For each sampled structure the breaking energy is computed by removing pairs in the target site and reevaluating the energy of this modified structure. The resulting energy difference is averaged over the 1,000 structures in the sample. Note that this procedure differs somewhat from the thermodynamic model mentioned before, in that it does not allow for any refolding. Thus, it assumes that there is no time for the mRNA structure to adapt in response to siRNA binding except for opening a few base pairs within the target site.

Based on a dataset of 100 published siRNAs directed against 3 mRNAs (PTEN [28], CD54 [28], and Lamin A [51]), Shao et al. showed that their new measure of accessibility can improve the siRNA selection efficiency over a random selection process by nearly 40%, 2.5 times more than the asymmetry criterion (16%) and significantly more than the degree of improvement of the sequence-based design rules (*see* for example [14]). They advised to design siRNA against regions with disruption energies > −10 kcal/mol, with G/C content between 30% and 70% and without AAAA, UUUU, GGGG, or CCCC repeats.

3.2 OligoWalk

As in Shao et al. [30], the accessibility in OligoWalk is defined as the energy cost for removing all base pairs at the siRNA target site [37, 38]. Here, however, this cost is defined as the difference in free energy of the mRNA in the native state and for the mRNA with the hybridization site single-stranded. The latter is computed by folding the mRNA under the constraint that bases within the target site are not allowed to pair. In contrast to Shao et al. [30], this allows the mRNA structure assume different structures in the bound and unbound case. Moreover, rather than rely on a stochastic sample, free energies are computed exactly via the partition function over all possible secondary structures.

OligoWalk also uses the folding energy of the siRNA alone, as a measure of its propensity to form self-structure that may interfere with mRNA binding. In addition it uses a larger number of sequence-based rules than most other methods. In total the designs process considers 28 features [38] (5 thermodynamics features, 23 sequence features from Shabalina et al. [52]). The selection of the siRNAs is carried out by a support vector machine (SVM) trained on the Huesken dataset by considering all 28 features. On their test dataset (653 siRNAs targeting 52 mRNAs), 78% of the siRNAs predicted by OligoWalk down-regulated their target by more than 70%, an improvement of 33% over a random selection process.

3.3 RNAxs

Tafer et al. [39] analyzed the effect of the target mRNA structure on RISC function *in silico* and compared the potential of target site accessibility as an siRNA design criterion to a broad range of sequence-based design rules. Their definition of siRNA target site accessibility is the same as in `OligoWalk`, although the manner of computation is different. Accessibilities are computed using a local folding algorithm as implemented in `RNAplfold` [53, 54]. In short, the program works by sliding a window of length W over the sequence, and computing for each window the partition function under the constraint that only *local* base pairs (where the two pairing partners are separated by at most L nucleotides) are allowed (*see* Fig. 3). From this RNAplfold computes the accessibilities for all regions of length u as the probability that this stretch is unpaired in thermodynamic equilibrium. Furthermore, the accessibility for a region is averaged over all sequence windows of length W containing this region. The main advantage of this method is speed: Since it considers only structures of size L, runtime is reduced from $O(n^3)$ to $O(n \cdot L^2)$, where n is the length of the mRNA. Most importantly, while `OligoWalk` needs to perform a separate RNA folding for each potential target site, `RNAplfold` computes the accessibilities of all possible target sites in a single pass.

In order to verify if the target site accessibility, as computed by RNAplfold, can be used to discriminate between functional and non-functional siRNAs, and to determine the optimal parameters for W, L, and u, two independent siRNA data sets of measured siRNA efficacies were used [18, 39]. From both data sets, highly functional and poorly functional siRNAs were selected and the accessibility of their target sites was assessed with `RNAplfold` for different W, L, and u. Silencing efficiency correlated significantly with target site accessibility, with the most significant separation resulting from 80 nucleotides and 40 nucleotides for W and L, respectively.

When varying the length u of the unpaired region, two parameter ranges with especially good separation were observed. The first peak measures the accessibility of the 6–8 nucleotides starting at the 3′ end of the target site, and therefore corresponds to the so-called seed region. This is in agreement with previous observations that the 5′-seed region of both siRNAs and microRNAs is the major determinant for RISC-mediated target recognition [36, 55, 56]. Furthermore, a second peak was observed for u values of 12–16, reminiscent of biochemical data showing that accessibility of the first 16 nucleotides within the target site is required for highly efficient RISC-mediated cleavage [36] (*see* Fig. 3).

Accessibility was further compared to a number of additional sequence and structure features [7, 8, 14, 15]. The best two design criteria turned out to be asymmetry of the siRNA and target site accessibility. The overall best predictions resulted from the combination of the accessibility, asymmetry, self-folding (folding energy of the siRNA [15]), and free-end (folding structure of the siRNA [15]) criteria.

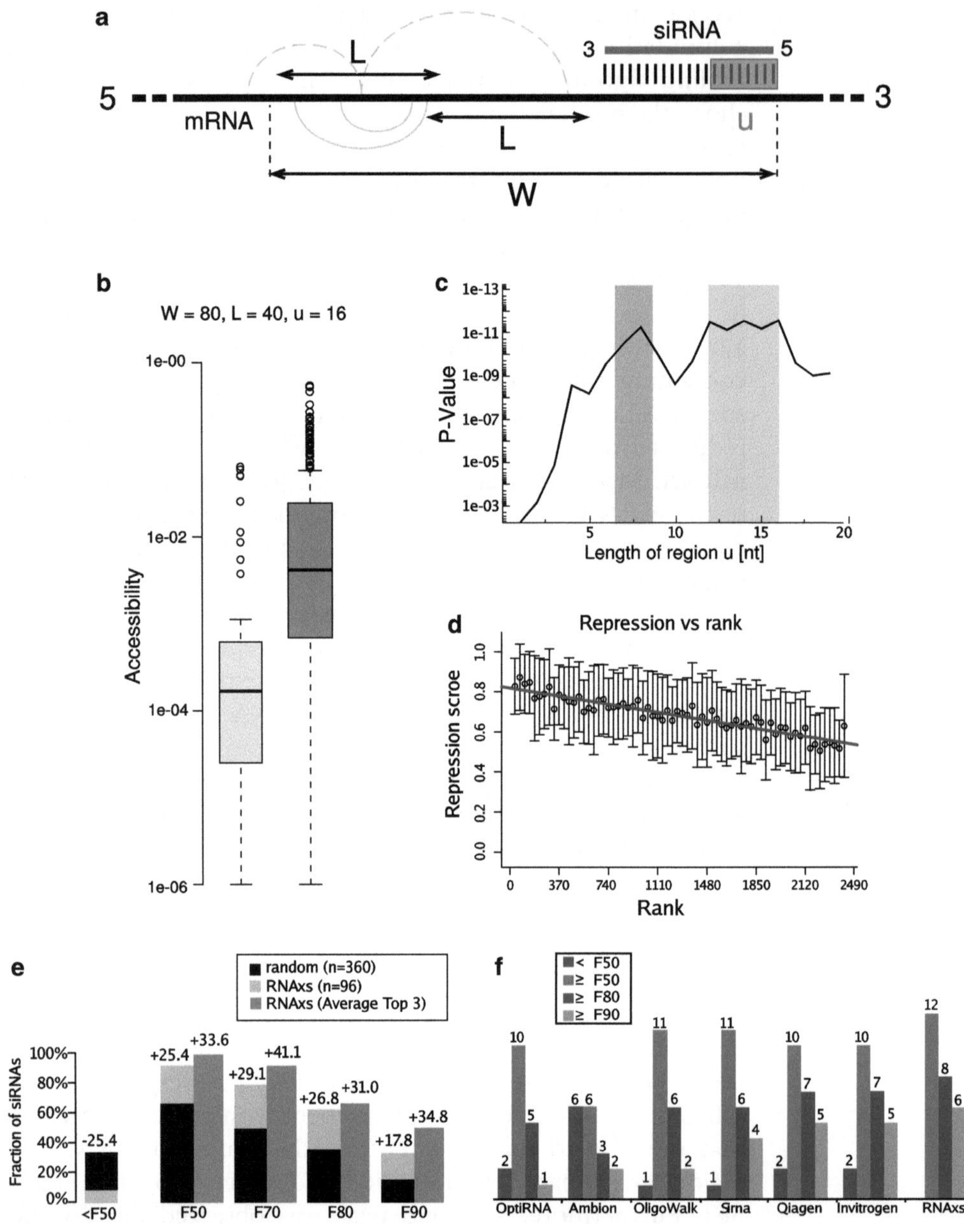

Fig. 3 (**a**) The RNA is folded locally in a sliding window approach (window size W). Within W, base pairing is restricted to a maximum distance L. u represents the stretch of consecutive nucleotides within an siRNA target site starting at its $3'$ end for which the accessibility is computed. Allowed and forbidden base pairs are shown with plain curved lines and *with dotted curved lines*, respectively. (**b**) Box-plot diagram comparing the accessibility of functional (*box left*) and non-functional (*box right*) siRNAs. The quartiles are represented by the *edges of the rectangles*; *black horizontal lines* within the boxes depict medians. The *circles* represent outliers and *dotted lines* show the standard deviation. (**c**) Accessibility distributions of functional and non-functional siRNAs are best differentiated for a length of 8 and/or 16 nucleotides (according to p-values). (**d**) Plot of the ranking of the siRNAs as defined in RNAxs and the measured repression efficiency. (**e**) Performance of RNAxs on a set of 360 siRNAs. siRNAs were grouped into five functionality classes: repression efficiency <50% (<F50), $\geq$50% ($\geq$F50), $\geq$70% ($\geq$F70), $\geq$80% ($\geq$F80) or $\geq$90% ($\geq$F90). (**f**) Performance of RNAxs compared to 6 other design tools. All tools were used with default parameters using the available web servers to design siRNAs against four genes. For each tool and each gene, the repression efficiency of the three best-ranked siRNAs was assessed

In addition to the filtering, RNAxs uses a ranking of the siRNAs according to their overall performance in all four criteria. Since different selection criteria are crucial for distinct stages in the RNAi pathway, a poor performance in one descriptor can presumably not be compensated by good values in another. Therefore a hierarchical sorting was devised that emphasizes the least favorable criterion for each siRNA, rather than constructing a combined score (*see* Fig. 3).

The algorithm was tested on an independent dataset of 360 siRNAs targeting every other position of 4 genes. On average, over 75% of the rationally selected siRNAs had a repression efficiency > 70% (+30% improvement over randomly selected siRNAs) and almost every third siRNAs reduced gene expression by more than 90% (+18%). When considering the three top ranked siRNAs for all four genes, half of them silenced the targeted gene by more than 90% (+35%) (*see* Fig. 3).

Prediction efficiency of `RNAxs` was compared to `OligoWalk` and `Sirna`, as well as three commercial siRNA selection tools (`Invitrogen`, `Ambion`, `Qiagen`) and a machine-learning method which did not use the accessibility to design siRNA (`OptiRNA`). The comparison was carried out by sorting the designed siRNAs into different functionality classes (less than 50% (<F50), more than 50% (≥F50), more than 80% (≥F80), and more than 90% (≥F90)). `RNAxs` was the only tool where all predicted siRNA had a measured repression efficiency in ≥F50. It was also the only tool to predict 50% of the siRNAs in the best functional category (16% for `OligoWalk`, 33% for `Sirna`).

4 Notes

One can only speculate whether the general level of predictive accuracy for siRNA potency is low because of the innate diversity of the siRNA silencing approach, or because some of the important explanatory features are still to be discovered. A crucial step towards answering this question is the availability of the right data, in the right format. So far, there is no consensus on how biological data is extracted or results are reported. This makes comparison of different datasets a challenging and time-consuming process. A unified framework would greatly facilitate the replication of reported results and data sharing between different groups. MIARE (Minimum Information About an RNAi Experiment, www.miare.org), a set of guidelines on the information that should be reported for every RNAi experiment, is a significant step towards this direction.

References

1. Fire A, Xu S, Montgomery MK, Kostas SA, Driver SE, Mello CC (1998) Potent and specific genetic interference by double-stranded RNA in Caenorhabditis elegans. Nature 391(6669):806–811. http://www.hubmed.org/display.cgi?uids=9486653
2. Elbashir SM, Harborth J, Lendeckel W, Yalcin A, Weber K, Tuschl T (2001) Duplexes of 21-nucleotide RNAs mediate RNA interference in cultured mammalian cells. Nature 411(6836):494–498. http://www.hubmed.org/display.cgi?uids=11373684
3. Stein CA (2001) Antisense that comes naturally. Nat Biotechnol 19(8):737–738. doi:10.1038/90783. http://dx.doi.org/10.1038/90783
4. Holen T, Amarzguioui M, Wiiger MT, Babaie E, Prydz H (2002) Positional effects of short interfering RNAs targeting the human coagulation trigger Tissue Factor. Nucleic Acids Res 30(8):1757–1766. http://www.hubmed.org/display.cgi?uids=11937629
5. Patzel V (2007) *In silico* selection of active siRNA. Drug Discov Today 12(3–4):139–148. doi:10.1016/j.drudis.2006.11.015. http://dx.doi.org/10.1016/j.drudis.2006.11.015
6. Elbashir SM, Martinez J, Patkaniowska A, Lendeckel W, Tuschl T (2001) Functional anatomy of siRNAs for mediating efficient RNAi in Drosophila melanogaster embryo lysate. EMBO J 20(23):6877–6888. http://www.hubmed.org/display.cgi?uids=11726523
7. Khvorova A, Reynolds A, Jayasena SD (2003) Functional siRNAs and miRNAs exhibit strand bias. Cell 115(2):209–216. http://www.hubmed.org/display.cgi?uids=14567918
8. Schwarz DS, Hutvágner G, Du T, Xu Z, Aronin N, Zamore PD (2003) Asymmetry in the assembly of the RNAi enzyme complex. Cell 115(2):199–208
9. Amarzguioui M, Prydz H (2004) An algorithm for selection of functional siRNA sequences. Biochem Biophys Res Commun 316(4):1050–1058. http://www.hubmed.org/display.cgi?uids=15044091
10. Hohjoh H (2004) Enhancement of RNAi activity by improved siRNA duplexes. FEBS Lett 557(1–3):193–198. http://www.hubmed.org/display.cgi?uids=14741366
11. Hsieh AC, Bo R, Manola J, Vazquez F, Bare O, Khvorova A, Scaringe S, Sellers WR (2004) A library of siRNA duplexes targeting the phosphoinositide 3-kinase pathway: determinants of gene silencing for use in cell-based screens. Nucleic Acids Res 32(3):893–901. http://www.hubmed.org/display.cgi?uids=14769947
12. Takasaki S, Kotani S, Konagaya A (2004) An effective method for selecting siRNA target sequences in mammalian cells. Cell Cycle 3(6):790–795
13. Ui-Tei K, Naito Y, Takahashi F, Haraguchi T, Ohki-Hamazaki H, Juni A, Ueda R, Saigo K (2004) Guidelines for the selection of highly effective siRNA sequences for mammalian and chick RNA interference. Nucleic Acids Res 32(3):936–948. http://www.hubmed.org/display.cgi?uids=14769950
14. Reynolds A, Leake D, Boese Q, Scaringe S, Marshall WS, Khvorova A (2004) Rational siRNA design for RNA interference. Nat Biotechnol 22(3):326–330. http://www.hubmed.org/display.cgi?uids=14758366
15. Patzel V, Rutz S, Dietrich I, Köberle C, Scheffold A, Kaufmann SH (2005) Design of siRNAs producing unstructured guide-RNAs results in improved RNA interference efficiency. Nat Biotechnol 23(11):1440–1444. doi:10.1038/nbt1151. http://dx.doi.org/10.1038/nbt1151
16. Saetrom P (2004) Predicting the efficacy of short oligonucleotides in antisense and RNAi experiments with boosted genetic programming. Bioinformatics 20(17):3055–3063. http://www.hubmed.org/display.cgi?uids=15201190
17. Ren Y, Gong W, Xu Q, Zheng X, Lin D, Wang Y, Li T (2006) siRecords: an extensive database of mammalian siRNAs with efficacy ratings. Bioinformatics 22(8):1027–1028. doi:10.1093/bioinformatics/btl026. http://dx.doi.org/10.1093/bioinformatics/btl026
18. Huesken D, Lange J, Mickanin C, Weiler J, Asselbergs F, Warner J, Meloon B, Engel S, Rosenberg A, Cohen D, Labow M, Reinhardt M, Natt F, Hall J (2005) Design of a genome-wide siRNA library using an artificial neural network. Nat Biotechnol 23(8):995–1001. http://www.hubmed.org/display.cgi?uids=16025102
19. Lima WF, Monia BP, Ecker DJ, Freier SM (1992) Implication of RNA structure on antisense oligonucleotide hybridization kinetics. Biochemistry 31(48):12055–12061
20. Vickers TA, Wyatt JR, Freier SM (2000) Effects of RNA secondary structure on cellular antisense activity. Nucleic Acids Res 28(6):1340–1347. http://www.hubmed.org/display.cgi?uids=10684928
21. Mir KU, Southern EM (1999) Determining the influence of structure on hybridization using oligonucleotide arrays. Nat Biotechnol 17(8):788–792. doi:10.1038/11732. http://dx.doi.org/10.1038/11732

22. Milner N, Mir KU, Southern EM (1997) Selecting effective antisense reagents on combinatorial oligonucleotide arrays. Nat Biotechnol 15(6):537–541. doi:10.1038/nbt0697-537. http://dx.doi.org/10.1038/nbt0697-537
23. Zhao JJ, Lemke G (1998) Rules for ribozymes. Mol Cell Neurosci 11(1–2):92–97. doi:10.1006/mcne.1998.0669. http://www.hubmed.org/display.cgi?uids=9608536
24. Ding Y, Lawrence CE (2001) Statistical prediction of single-stranded regions in RNA secondary structure and application to predicting effective antisense target sites and beyond. Nucleic Acids Res 29(5):1034–1046
25. Bohula EA, Salisbury AJ, Sohail M, Playford MP, Riedemann J, Southern EM, Macaulay VM (2003) The efficacy of small interfering RNAs targeted to the type 1 insulin-like growth factor receptor (IGF1R) is influenced by secondary structure in the IGF1R transcript. J Biol Chem 278(18):15991–15997. http://www.hubmed.org/display.cgi?uids=12604614
26. Kretschmer-Kazemi Far R, Sczakiel G (2003) The activity of siRNA in mammalian cells is related to structural target accessibility: a comparison with antisense oligonucleotides. Nucleic Acids Res 31(15):4417–4424
27. Xu Y, Zhang H-Y, Thormeyer D, Larsson O, Du Q, Elmén J, Wahlestedt C, Liang Z (2003) Effective small interfering RNAs and phosphorothioate antisense DNAs have different preferences for target sites in the luciferase mRNAs. Biochem Biophys Res Commun 306(3):712–717
28. Vickers TA, Koo S, Bennett CF, Crooke ST, Dean NM, Baker BF (2003) Efficient reduction of target RNAs by small interfering RNA and RNase H-dependent antisense agents. A comparative analysis. J Biol Chem 278(9):7108–7118. http://www.hubmed.org/display.cgi?uids=12500975
29. Ding Y, Chan CY, Lawrence CE (2004) Sfold web server for statistical folding and rational design of nucleic acids. Nucleic Acids Res 32(Web Server issue):W135–W141. doi:10.1093/nar/gkh449. http://dx.doi.org/10.1093/nar/gkh449
30. Shao Y, Chan CY, Maliyekkel A, Lawrence CE, Roninson IB, Ding Y (2007) Effect of target secondary structure on RNAi efficiency. RNA 13(10):1631–1640. doi:10.1261/rna.546207. http://dx.doi.org/10.1261/rna.546207
31. Luo KQ, Chang DC (2004) The gene-silencing efficiency of siRNA is strongly dependent on the local structure of mRNA at the targeted region. Biochem Biophys Res Commun 318(1):303–310. http://www.hubmed.org/display.cgi?uids=15110788
32. Yoshinari K, Miyagishi M, Taira K (2004) Effects on RNAi of the tight structure, sequence and position of the targeted region. Nucleic Acids Res 32(2):691–699. http://www.hubmed.org/display.cgi?uids=14762201
33. Overhoff M, Alken M, Far RK, Lemaitre M, Lebleu B, Sczakiel G, Robbins I (2005) Local RNA target structure influences siRNA efficacy: a systematic global analysis. J Mol Biol 348(4):871–881. http://www.hubmed.org/display.cgi?uids=15843019
34. Schubert S, Grünweller A, Erdmann VA, Kurreck J (2005) Local RNA target structure influences siRNA efficacy: systematic analysis of intentionally designed binding regions. J Mol Biol 348(4):883–893. http://www.hubmed.org/display.cgi?uids=15843020
35. Brown JR Sanseau P (2005) A computational view of microRNAs and their targets. Drug Discov Today 10(8):595–601. http://www.hubmed.org/display.cgi?uids=15837603
36. Ameres SL, Martinez J, Schroeder R (2007) Molecular basis for target RNA recognition and cleavage by human RISC. Cell 130(1):101–112. doi:10.1016/j.cell.2007.04.037. http://dx.doi.org/10.1016/j.cell.2007.04.037
37. Lu ZJ, Mathews DH (2008) Oligowalk: an online siRNA design tool utilizing hybridization thermodynamics. Nucleic Acids Res 36(Web Server issue):W104–W108. doi:10.1093/nar/gkn250. http://dx.doi.org/10.1093/nar/gkn250
38. Lu ZJ, Mathews DH (2008) Efficient siRNA selection using hybridization thermodynamics. Nucleic Acids Res 36(2):640–647. doi:10.1093/nar/gkm920. http://dx.doi.org/10.1093/nar/gkm920
39. Tafer H, Ameres SL, Obernosterer G, Gebeshuber CA, Schroeder R, Martinez J, Hofacker IL (2008) The impact of target site accessibility on the design of effective siRNAs. Nat Biotechnol 26(5):578–583. doi:10.1038/nbt1404. http://dx.doi.org/10.1038/nbt1404
40. Boese Q, Leake D, Reynolds A, Read S, Scaringe SA, Marshall WS, Khvorova A (2005) Mechanistic insights aid computational short interfering RNA design. Methods Enzymol 392:73–96. doi:10.1016/S0076-6879(04)92005-8. http://dx.doi.org/10.1016/S0076-6879(04)92005-8
41. Mückstein U, Tafer H, Hackermüller J, Bernhart SH, Stadler PF, Hofacker IL (2006) Thermodynamics of RNA-RNA binding. Bioinformatics 22(10):1177–1182. doi:10.1093/bioinformatics/btl024. http://www.hubmed.org/display.cgi?uids=16446276
42. Mückstein U, Tafer H, Bernhard SH, Hernandez-Rosales M, Vogel J, Stadler PF,

Hofacker IL (2008) Translational control by RNA-RNA interaction: improved computation of RNA-RNA binding thermodynamics. In: Elloumi M, Küng J, Linial M, Murphy R, Schneider K, Toma C (eds) Bioinformatics research and development. Communications in computer and information science, vol 13. Springer, Berlin, pp 114–127. doi:10.1007/978-3-540-70600-7_9

43. Hornung V, Guenthner-Biller M, Bourquin C, Ablasser A, Schlee M, Uematsu S, Noronha A, Manoharan M, Akira S, de Fougerolles A, Endres S, Hartmann G (2005) Sequence-specific potent induction of IFN-alpha by short interfering RNA in plasmacytoid dendritic cells through TLR7. Nat Med 11(3):263–270. doi:10.1038/nm1191. http://dx.doi.org/10.1038/nm1191
44. de Haro C, Méndez R, Santoyo J (1996) The eIF-2alpha kinases and the control of protein synthesis. FASEB J 10(12): 1378–1387
45. Marques JT, Williams BRG (2005) Activation of the mammalian immune system by siRNAs. Nat Biotechnol 23(11):1399–1405. doi:10.1038/nbt1161. http://dx.doi.org/10.1038/nbt1161
46. Shao XD, Wu KC, Guo XZ, Xie M-J, Zhang J, Fan D-M (2008) Expression and significance of HERG protein in gastric cancer. Cancer Biol Ther 7(1):45–50
47. Saetrom P, Snove O (2004) A comparison of siRNA efficacy predictors. Biochem Biophys Res Commun 321(1):247–253. http://www.hubmed.org/display.cgi?uids=15358242
48. Vert JP, Foveau N, Lajaunie C, Vandenbrouck Y (2006) An accurate and interpretable model for siRNA efficacy prediction. BMC Bioinformatics
49. Matveeva O, Nechipurenko Y, Rossi L, Moore B, Saetrom P, Ogurtsov AY, Atkins JF, Shabalina SA (2007) Comparison of approaches for rational sirna design leading to a new efficient and transparent method. Nucleic Acids Res
50. Ding Y, Lawrence CE (2003) A statistical sampling algorithm for RNA secondary structure prediction. Nucleic Acids Res 31(24):7280–7301
51. Harborth J, Elbashir SM, Vandenburgh K, Manninga H, Scaringe SA, Weber K, Tuschl T (2003) Sequence, chemical, and structural variation of small interfering RNAs and short hairpin RNAs and the effect on mammalian gene silencing. Antisense Nucleic Acid Drug Dev 13(2):83–105. doi:10.1089/108729003321629638. http://dx.doi.org/10.1089/108729003321629638
52. Shabalina SA, Spiridonov AN, Ogurtsov AY (2006) Computational models with thermodynamic and composition features improve siRNA design. BMC Bioinformatics 7:65. doi:10.1186/1471-2105-7-65. http://dx.doi.org/10.1186/1471-2105-7-65
53. Bernhart SH, Tafer H, Mückstein U, Flamm C, Stadler PF, Hofacker IL (2006) Partition function and base pairing probabilities of RNA heterodimers. Algorithms Mol Biol 1(1):3. doi:10.1186/1748-7188-1-3. http://www.hubmed.org/display.cgi?uids=16722605
54. Bompfünewerer AF, Backofen R, Bernhart SH, Hertel J, Hofacker IL, Stadler PF, Will S (2008) Variations on RNA folding and alignment: lessons from benasque. J Math Biol 56:119–144. doi:10.1007/s00285-007-0107-5
55. Haley B, Zamore PD (2004) Kinetic analysis of the RNAi enzyme complex. Nat Struct Mol Biol 11(7):599–606. http://www.hubmed.org/display.cgi?uids=15170178
56. Jackson AL, Bartz SR, Schelter J, Kobayashi SV, Burchard J, Mao M, Li B, Cavet G, Linsley PS (2003) Expression profiling reveals off-target gene regulation by RNAi. Nat Biotechnol 21(6):635–637. http://www.hubmed.org/display.cgi?uids=12754523

Chapter 23

RNA–Protein Interactions: An Overview

Angela Re, Tejal Joshi, Eleonora Kulberkyte, Quaid Morris, and Christopher T. Workman

Abstract

RNA binding proteins (RBPs) are key players in the regulation of gene expression. In this chapter we discuss the main protein–RNA recognition modes used by RBPs in order to regulate multiple steps of RNA processing. We discuss traditional and state-of-the-art technologies that can be used to study RNAs bound by individual RBPs, or vice versa, for both in vitro and in vivo methodologies. To help highlight the biological significance of RBP mediated regulation, online resources on experimentally verified protein–RNA interactions are briefly presented. Finally, we present the major tools to computationally infer RNA binding sites according to the modeling features and to the unsupervised or supervised frameworks that are adopted. Since some RNA binding site search algorithms are derived from DNA binding site search algorithms, we discuss the commonalities and novelties introduced to handle both sequence and structural features uniquely characterizing protein–RNA interactions.

Key words RNA binding proteins, RNA-binding domains, RNA-binding specificity, RNA-binding sites, RNA–protein interactions, Cross-Linking and ImmunoPrecipitation, RNAcompete

1 Overview

RNA binding proteins (RBPs) are involved in almost every central process in the cell and often serve essential functional roles. The interactions between protein and RNA can serve a number of purposes within their respective biological process, including the coordination and stabilization of protein complexes, processing and maturation of mRNA to the trafficking, stabilization and silencing of mature mRNA. Many of these interactions are specific and require a mechanism for recognition of the appropriate binding target. Unlike DNA binding proteins, which typically recognize features in the major groove of double-stranded B-DNA, RBPs may recognize single-stranded RNA, double-stranded RNA, three-dimensional structural features of folded RNAs or they may bind RNA non specifically. When an RBP's function depends on recognizing a specific RNA sequence, be it the primary sequence

Jan Gorodkin and Walter L. Ruzzo (eds.), *RNA Sequence, Structure, and Function: Computational and Bioinformatic Methods*, Methods in Molecular Biology, vol. 1097, DOI 10.1007/978-1-62703-709-9_23,

or secondary or tertiary structure, it may be possible to learn the determinants or principles of the binding interaction. Indeed, the development of models for RBP specificity has become an active area of bioinformatic research in part due to the relatively recent discovery of micro-RNA mediated mRNA silencing that depends on a number of RBPs. In this chapter we review and highlight the role of RBPs in a few of the biological process related to roles of mRNA. We omit the discussion or RBPs related to microRNA targeting and mRNA silencing due to its inclusion in previous chapters. Structurally characterized RNA binding domains are reviewed and considered for their suitability for RNA specificity modeling. Important new high-throughput methods to identify RNA binding sequences are introduced as they are enabling new possibilities to understand RNA binding specificity to RBPs. Finally, we review the available methods and data resources that are used to learn and model RBP–RNA interactions.

2 Functional Roles of RBPs

The following sections highlight the role of RBPs in a number of processes spanning the whole life-span of an mRNA. For instance, RBPs can function in transcription, by modulating RNA polymerases basal activities [1], by providing specificity to gene transcription regulation [2], or by terminating transcript through the recognition of the polyadenylation signal on the nascent mRNA [3]. Splicing, which is extensively regulated by RBPs, follows the initial processing of nascent transcripts and several important studies have revealed clues to the general principles of RBP-mediated RNA splicing [4]. However, in the remainder of this section we only focus on the regulation of mature RNAs, where protein–RNA interactions play a primary role. Since each section is not intended to comprehensively review RBPs in a process, we refer the reader to the appropriate reviews where possible.

2.1 mRNA Localization

The localization of mRNAs to sub-cellular sites is a spatial mechanism for regulating gene activity [5–7]. Over direct transport of protein products, mRNA trafficking can increase the efficiency and temporal resolution of protein synthesis in response to cellular cues and facilitates the formation of protein complexes due to higher local concentration of the necessary mRNAs. With the exception of nuclear export, reviewed in [8], localizing mRNAs consist of three non-mutually exclusive mechanisms [9, 10] able to sort mRNAs into their cellular sites: directed transport of mRNAs, local selective stabilization, and local trapping. Each mechanism requires distinct RBPs to recognize distinct localization signals in the mRNAs. The localization signals involved in the active and direct transport tend to appear as synergistic clustered repeats of secondary structures [11–13] although some localization signals appear to be encoded in primary sequence [14, 15]. Many localizing RBPs individually interact with low affinity and specificity with the UTRs of localized mRNAs [16], cooperative binding of multiple RBPs is believed to be important (*see*, e.g., [17]). Selective stabilization is the result of RBP mediated protection in one cellular location. The best characterized example is Hsp83 [18], which is targeted by the 3′ un-translated regions (3′ UTRs)-bound Smaug RBP for deadenylation and degradation everywhere in the Drosophila embryos except the posterior pole, where it localizes. The third mechanism uses diffusion and local entrapment; however, owing to its moderate efficiency to spatially restrict mRNAs, it generally occurs with selective stabilization, as in the case of Nanos mRNA posterior pole localization in Drosophila embryos [19].

2.2 mRNA Translation

The regulation of translation can occur on a global basis, by modifications of the translational machinery, or can specifically target selected mRNAs. The focus here is on aspects of global and mRNA-specific regulation that directly involve RNA binding proteins, and direct the reader to general reviews [20–23] on the topic. An intriguing regulatory mechanism allows mRNA-specific regulation through RBP-based modulation of the basic translational machinery [24]. For instance, mRNA-specific RBPs can inhibit the association between the ribosome 43S complex and the mRNA either by physical hindrance in a cap-dependent manner [25], or by 43S scanning arrest in a cap-independent manner, as in the case of Drosophila msl-2 mRNA by SXL [26–28]. Additionally, some RBPs enable global eIF4E structural adapters to selectively inhibit specific mRNAs, as it is the case of Smaug and Bruno RBPs that mediate the inhibitory effects of the Cup and Maskin eIF4E adaptors on nanos, oskar, and poly(A)-tailed mRNAs [29–31]. RBPs can control translation also at late initiation steps, for instance by preventing ribosomal subunits joining [32], or at post-initiation steps, as exemplified by the hnRNP E1 RBP that inhibits Dab2 and ILEI at the elongation step by 3′ UTR binding [33, 34]. Several translation-dependent quality controls involve RBPs that distinguish aberrant mRNAs from normal mRNAs and couple the translation machinery to a degradation pathway [35]. Cytoplasmic polyadenylation is a further potent system to regulate translation [36]. Several models have been proposed that invoke RBPs as place markers to recruit the catalytic complexes according to a polyadenylation dynamic combinatorial code [37, 38].

2.3 mRNA Degradation

Cells activate many different degradation mechanisms in quality surveillance, RNA maturation, and regulated mRNA turnover. The role of RBPs in quality control of nuclear RNAs is to preferentially bind aberrant RNAs in order to promote export and degradation in the cytoplasm [39], or nuclear TRAMP-dependent adenylation and 3′ to 5′ decay through the exosome [40, 41]. Cytoplasmic surveillance elicits several mechanisms such as nonsense-mediated decay (NMD) based on aberrant stop codons [42], and ribosome extension-mediated decay (REMD), when translation extends beyond the normal stop codon [43]. For instance, NMD requires RBPs in order to recognize regulatory sites in mRNA decay substrates, as typified by the exon-junction complex (EJC) [44], the poly-A binding protein 1 PABPC1 [45] and HRP1 [46]. Furthermore, RBPs can function in NMD as adaptors between associating complexes, such as Upf1 which mediates the SURF complex formation and the following association with EJC [47]. Additional RBPs like Pub1 [48], the APOBEC1–ACF editing complex [49] and several 3′ UTR helicases or chaperones [50] provide selective regulation of decay efficiency. The role of 3′ UTRs is recognized also in the REMD decay, by specifying the proper distance between termination codon and polyadenylation site [51]. Note that key factors in quality surveillance pathways are also used in conditionally regulated degradation pathways, which often depend on mRNA-specific RBPs like Staufen1 [52] and SLBP [53].

2.4 mRNA Editing

RNA editing is a post-transcriptional process that covalently alters RNA sequences, either by converting adenosines to inosines (A-to–I editing) or by converting cytidines to uridines (C-to-U editing). Adenosine-to-inosine (A-to-I) editing affects adenosines preferentially localizing to double-stranded regions of viral RNAs, cellular pre-mRNAs and non-coding RNAs [54]. A-to-I editing is catalyzed by enzymes of the adenosine deaminase acting on RNA (ADAR) family [55]. Amino(N)-terminal regions of ADARs contain dsRNA-binding motifs (dsRBMs), whereas carboxy(C)-terminal regions contain a conserved catalytic domain. ADARs target dsRNA of any sequence, but have preferences for certain neighboring nucleotides. For both ADAR1 and ADAR2, the 5′ nearest neighbor has the most influence on whether an adenosine will be edited. Although the catalytic domain largely dictates nearest neighbor preferences, for human hADAR2, the dsRBM has a role in discriminating adenosines with a 3′ G. Furthermore, bases beyond the nearest neighbor affect ADARs preferences. In addition to preferences for neighboring nucleotides, ADARs exhibit selectivity, whereby the number of adenosines edited in a dsRNA is affected by dsRNA length and whether base-pairing is interrupted by mismatches, bulges, or loops [56, 57]. A-to-I editing has been implicated in many processes, including modulation of neuronal signaling [58, 59], establishment of higher brain function [60], tuning of RNAi activity and control of microRNA biogenesis pathways [61]. C-to-U editing, converts cytidines to uridines, is catalyzed by the AID–APOBEC enzyme family. After the initial demonstration of C-to-U editing in apoB mRNA [62, 63], more comprehensive search for APOBEC1 editing targets showed that APOBEC1 editing is constrained mainly to 3′ UTRs [64]. The APOBEC1 sequence pattern supporting editing at sites in 3′ UTRs consists of a cytidine flanked on both sides by either adenosine or uridine and followed by an appropriately spaced sequence motif (WCWN$_{2-4}$WRAUYANUAU). Nonetheless, this motif consensus sequences was not targeted when present in coding sequences, with the notable exception of ApoB. APOBEC1-mediated editing in 3′ UTRs could affect post-transcriptional processes, including transcript stability, polyadenylation, subcellular localization, and translational efficiency. A-to-I and C-to-U editing show the transmission of information from DNA to RNA is a critical process. Along this line, a recent paper [65] reported an extraordinary extent of DNA-to-RNA base changes that cannot arise from classic editing, but for which the underlying mechanisms are unknown. Albeit highly controversial [66], this finding might suggest a completely different layer of gene regulation at the RNA level.

2.5 mRNA Stability

Among the strongest examples of RBPs stabilizing mRNA are AUF1, TTP, and the members of the Hu family [67], which bind to A/U-rich elements (AREs), preferentially located within mRNA 3′ UTRs. The stability of a particular mRNA results from the combinatorial effect of multiple stabilizing and destabilizing RBPs. Whether the effect of co-bound RBPs is cooperative or antagonist depends on the spatial relationship and the difference in affinity between their regulatory sites within the UTR, as well as by the relative amount of such RBPs in the cellular condition and localization wherein the binding occurs [68, 69]. In addition, RBPs and microRNAs can combine to stabilize mRNA [70]. Interestingly, RBPs and microRNAs can bidirectionally influence their stability [71, 72].

2.6 Roles of RBPs in Disease

Since RBPs are involved in most aspects of RNA metabolism, any mutation or other disruption to RBP functions can lead to a number of diseases. For example, mutation or over-expression of RBPs in cancer may result in erroneous or profuse binding to RNA at various stages of RNA metabolism, leaving a significant impact on cancer cells. During development of the nervous system, gene expression undergoes tight dynamic regulation. Deschenes-Furry and his colleagues [73] list RBPs participating in normal neuronal development and functioning.

For a more general review, Lukon et al. [74] discuss several diseases caused either by RBP loss-of-function or by detrimental RBP gain-of-function. Fragile X syndrome is associated with the expansion of CGG triplet on 5′ UTR of FMR1, resulting in the loss of function of FMR1 that is required for normal neuronal development. In the paraneoplastic neurologic syndromes (PNSs), a group of autoimmune disorders, RBP loss-of-function occurs due to autoantibodies targeting RBPs, such as Hu family proteins and Nova proteins. The neuronal specific Nova family of proteins are involved in alternative splicing of their target pre-mRNAs in regions of the central nervous system, e.g. the hindbrain and ventral spinal cord [75].

Trinucleotide repeat disorders are sometimes caused by defective RBPs. In myotonic dystrophy type 1 (DM1) insertion of multiple repeats in the 3′ UTR region of DMPK gene, whereas in myotonic dystrophy type 2, much longer repeats of tetra-nucleotide CCTG give rise to toxic mutant RNAs. Similar mechanism exists in oculopharyngeal muscular dystrophy (OPMD), an adult-onset degenerative disorder, where expansion of a GCG repeat in the exon of PABPN1 gene generates a PABPN1 mutant. The mutant gene then stimulates extension of its poly(A) tails up to the length that corresponds to that of a nascent mRNA. Transcripts with such long poly(A) tail accumulate in the nuclei of skeletal muscle filaments to cause muscular dystrophy.

The STAR family of proteins are involved in cellular differentiation and proliferation and function via RNA binding and signal transduction. Sam68 protein, a STAR protein family member, is composed of one KH domain flanked by conserved 5′ and 3′ regions. Phosphorylation of Sam68 by the breast cancer tumor tyrosine-kinase (BRK) promotes its relocation from nucleus to cytoplasm and is associated with proliferation and tumorigenicity in breast and prostate cancers [76, 77]. Other widely studied cancer-associated RBPs are ASF/SF2 and eIF4E. eIF4E is an oncogene specifically over-expressed in breast cancer which often results in poor patient survival [78]. The proto-oncogene ASF/SF2 is often over-expressed in various cancers. Over-expression of ASF/SF2 could alter splicing of crucial cell cycle regulators and tumor suppressor genes, and hence is an important target protein for cancer therapy.

Often RNA operons, the master regulators of co-expressed genes, could lose one or more target mRNA due to mutations in the USER regions that prevented its binding to RBP. Two SNPs in the 3′ UTR region of FGF20 gene are associated with Parkinson disease [79]. Similarly, SNPs in the miRNA genes or their target sites on mRNAs might result in the RBP loss-of-function.

3 Principles of Protein–RNA Binding, RNA-Binding Domain Specificities

As suggested, a code for protein–RNA recognition would be useful for developing prediction methods for RBP in vivo as well as for engineering libraries of RBPs with programmable RNA specificities in vitro. Here, we present a comparative overview

Table 1
Summary of abundant RNA-binding domains

Domain (topology)	Protein–RNA interface	Protein–RNA interactions
RRM ($\beta\alpha\beta\beta\alpha\beta$)	β-sheet;	Stacking, H, electrostatic bonding
KH-type1 ($\beta\alpha\alpha\beta\beta\alpha$)	Gly-X-X-Gly motif, flanking helices, β-strand following $\alpha 2$, variable $\beta 2$-$\beta 3$ loop form a cleft	Contacts to backbone from GXXG loop; H bonds to bases; hydrophobic interactions from nonaromatic residues confer base specificity
KH-type2 ($\alpha\beta\beta\alpha\alpha\beta$)	Gly-X-X-Gly motif, flanking helices, β-strand following $\alpha 2$, variable $\alpha 2$–$\beta 2$ loop form a cleft	Contacts to backbone from GXXG loop; H bonds to bases; hydrophobic interactions from nonaromatic residues confer base specificity
dsRBD ($\alpha\beta\beta\beta\alpha$)	Helix $\alpha 1$, helix $\alpha 2$, $\beta 1$–$\beta 2$ loop	Contacts to RNA backbone provide shape specificity
DEAD-box (S1,S2 helicase)	Motifs around 2 RecA-like domains form tight structure	Interaction to A from Q motif; contact to RNA backbone from RecA-like domains
PUF ($\alpha\alpha\alpha$)	Helix α	Stacking interactions provide binding pocket H bonds to bases by 2 amino acids provide specificity
Sm/Lsm ($\alpha\beta$)	Pore of the hexameric or heptameric ring	Stacking interactions; H bonds to bases
SAM ($\alpha\alpha\alpha\alpha\alpha\alpha$)	SAM's shallow hydrophobic core	H bonds from phosphate groups to helices side-chains; hydrogen bonds to the Watson–Crick face of G3 in the loop
ZnF-CCHH ($\alpha\beta$)	α-helix	Contacts to bases, electrostatic interactions to RNA backbone from protein side-chains
ZnF-CCCH (little 2D struct.)	Protein side-chains form binding pockets	H bonds to bases from protein backbone; H bonds to A or U from protein side chains confer specificity
ZnF-CCHC (Zinc knuckles)	Protein backbone	H bonds to bases from protein backbone

of the protein–RNA recognition modes mediated by the most abundant RNA-binding domains, summarized in Table 1.

3.1 RNA-Recognition Motif: RRM

The RRM is 90–100 amino acids in length with a compact $\beta\alpha\beta\beta\alpha\beta$ topology, which forms a four-stranded β-sheet packed against two α-helices, *see* Fig. 1. The RRM is found in 1,437 proteins and is often present in up to six copies per protein. The RRM–RNA interactions are single-stranded specific and show only low sequence specificity. The primary interaction interface is the β-sheet, from which three aromatic side-chains stack on or

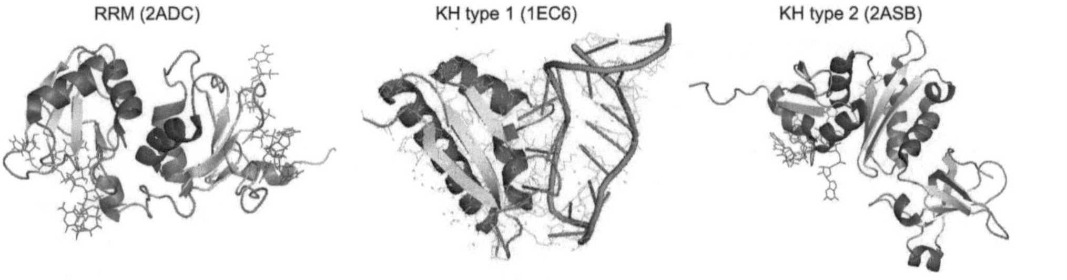

Fig. 1 Example co-crystal structures of RRM and KH domains. PDB structure 2ADC (*left*) shows the RRM of Polypyrimidine Tract Binding protein complexed with CUCUCU RNA. Structure 1EC6 (*middle*) shows Nova-2 KH3 K-homology domain complexed with a 20-mer RNA hairpin. Structure 2ASB (*right*) shows a *Mycobacterium tuberculosis* NusA fragment containing two KH type II domains (top of structure) complexed with GAACUCAAUAG

insert between the two sugar rings of two nucleotides; however, RRMs interact with RNA through a wide range of interfaces [80]. For example, the loops $\beta1$–$\beta1$, $\beta2$–$\beta3$, and $\alpha2$–$\beta4$, which vary in their size and amino acid sequences, allow the recognition of additional nucleotides. Multiple RRM copies per protein generally ensure increased affinity relative to individual domains, due to the interactions between the inter-domain linkers and the RNA and between the RRMs themselves, *see* [81].

3.2 K-Homology Domain

The K-homology (KH) domains are 70 amino acids and can fold into two topologies: $\beta\alpha\alpha\beta\beta\alpha$ (type I) and $\alpha\beta\beta\alpha\alpha\beta$ (type II). The two consecutive α-helices are linked by the GXXG motif and a variable loop, *see* Fig. 1. For both types, four single-stranded nucleotides are recognized in a cleft that is formed by the invariant Gly–X–X–Gly motif, the flanking helices, the β-strand that follows $\alpha2$ (type I) or $\alpha3$ (type II) and the variable loop between $\beta2$ and $\beta3$ (type I) or between $\alpha2$ and $\beta2$ (type II). The recognition of single-stranded RNA longer than four nucleotides can be mediated by the variable loop [82], the extension of the domain [83] or the juxtaposition of two domains [84].

3.3 Double-Stranded RNA-Binding Domain: dsRBD

The dsRBD (Fig. 2) is 70–75 amino acids in size with a $\alpha\beta\beta\beta\alpha$ topology where the two α helices are packed along one face of a three-stranded anti-parallel β-sheet. This domain occurs in up to five copies per protein and has so far been found in 645 proteins. The dsRBD sequence-independently recognizes double-stranded RNAs, by contacts to the $2'$-OH groups and the phosphate backbone from the amino-terminal part of α-helix 2 and the $\beta1$–$\beta2$ loop. The contacts with the RNA cover 15 nucleotides spanning two minor grooves separated by a major groove. The additional interaction between α-helix 1 and the RNA duplex modulates the binding specificity of dsRBDs for a variety of RNA

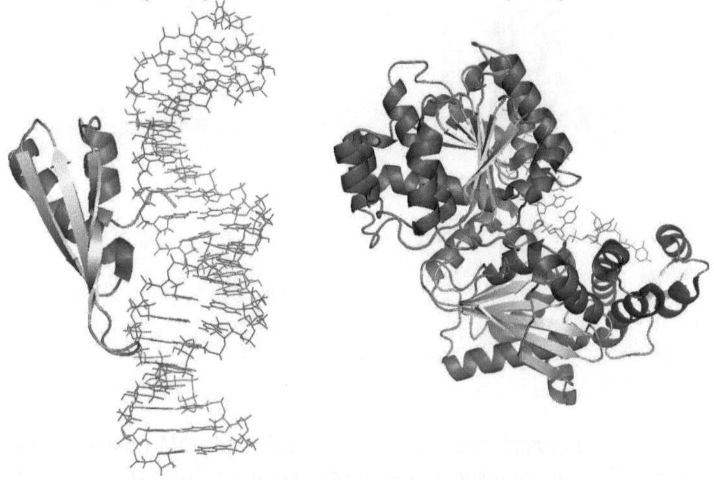

Fig. 2 Example co-crystal structures of dsRBD and DEAD domains. PDB structure 2L2K (*left*) shows the dsRBD of Adenosine Deaminase, RNA-specific, B1 protein (ADAR2) complexed with a stem-loop pre-mRNA encoding the R/G editing site of GluR-2. PDB structure 3I5X (*right*) shows the DEAD-box domain of ATP-dependent RNA helicase MSS116 complexed with an RNA oligonucleotide and the ATP analog AMP-PNP

structures, such as stem-loops, internal loops, bulges or helices. However, independence from sequence hinders the engineering of dsRBDs with distinct recognition properties.

3.4 DEAD-Box Domain

The DEAD-box domain (Fig. 2) is present in 1,154 proteins and includes a helicase core of two tandem RecA-like domains. This domain usually uses an ATP-dependent conformational change to coordinate RNA transient folding and remodeling. The interaction between at least 11 protein motifs closes the two RecA-like domains, such that they interact with each other and tightly trap the binding between an adenosine nucleotide and the single-stranded RNA. The reduction in single-stranded binding affinity upon ATP hydrolysis is important to many DEAD-box domain activities, including short helix unwinding. The DEAD-box domain does not form contacts with the nucleotides of the RNA but only with its backbone. The bound single-stranded RNA can include one or two sharp bends, which likely facilitate the internal initiation of strand separation of the nearby helix. The DEAD-box domain can be extended by additional domains that, by recognizing specific RNA structures, guide the helicase core to the unwinding site [85, 86]. For these reasons, DEAD-box containing RBPs will likely not represent good candidates for protein–RNA binding models.

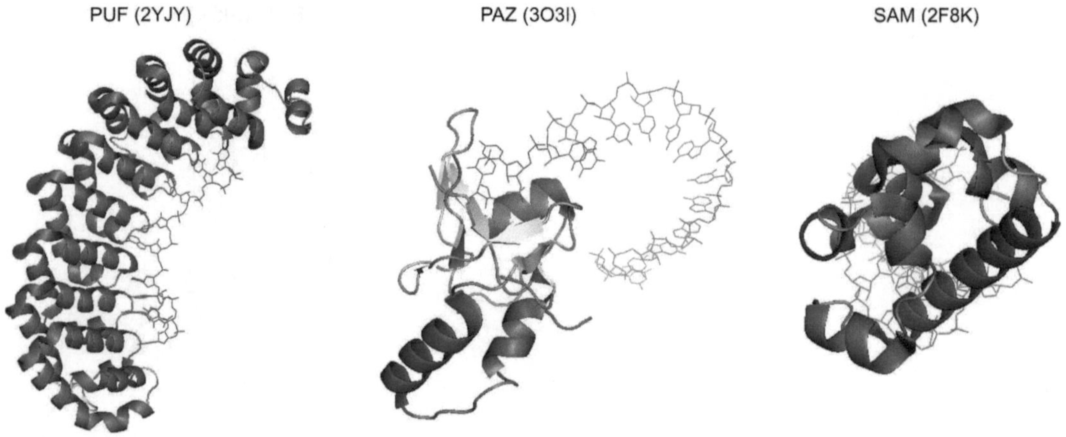

Fig. 3 Example co-crystal structures of PUF, PAZ, and SAM domains. PDB structure 2YJY (*left*) shows the PUF repeat of Pumilio Homolog 1 complexed with AUUGCAUAUA. PDB structure 3O3I (*middle*) shows the PAZ domain of Piwi-like protein 1 complexed with GCGAAUAUUCGCUU. PDB structure 2F8K (*right*) shows the SAM domain of the *S. cerevisiae* Smg homolog complexed with UAAUCUUUGACAGAUU

3.5 Pumilio Repeat Domain: PUF

The PUF domain is around 35 amino acids in size with an $\alpha\alpha\alpha$ topology. The PUF domain is much rarer in nature than the RRM (present in 111 proteins) and typically occurs in six to eight tandem repeats in a protein, packed together in a curved structure, *see* Fig. 3. The PUF repeats bind single-stranded RNAs, which lay along the inner surface with each nucleotide contacting two repeats. The RNA binding is mediated by the α-helix 2 in each repeat. For each nucleotide, the side-chain of the fourth amino acid in α-helix 2 stacks on top of the base while the side-chains of the third and seventh amino acid are hydrogen-bonded to the Watson–Crick edge. Additionally, the fourth amino acid side-chain of the following repeat is stacked underneath the base. This modularity and relatively well understood binding determinants suggest that it will be possible to develop specificity models for PUF containing proteins.

3.6 PAZ Domain

The core fold of the PAZ (Piwi/Argonaute/Zwille) domain is 110 amino acids in length and consists of a β-barrel domain juxtaposed to a small $\alpha\beta$ domain that forms a clamp-like structure in which RNA binds. The PAZ domain recognizes a 2-nt overhang at the 3' end of dsRNA, as well as the 3' end of ssRNA. In eukaryotes, the PAZ domain is only found in Argonaute and Piwi subfamilies, as well as in the RNase III family ribonuclease Dicer, all of which are essential in the RNA silencing pathway. Ago subfamily proteins constitute the core component of the RNA Induced Silencing Complex (RISC); structure-function studies have provided insights into guide-strand mediated cleavage of target RNAs within Ago complexes [87]. Piwi subfamily proteins bind to piRNAs, a class of ssRNAs of 26 to 30 nucleotides in length with 5′ phosphate and 2′-*O*-methylated 3′ ends. Piwi proteins and piRNAs are germ line specific and have a key role in protecting the genome against mobile genetic elements [88]. Nearly all residues involved in 3′ end overhang recognition in Ago1 PAZ domain are conserved in Piwi PAZ domains, particularly the Tyr and His aromatic residues that line the binding pocket. Nevertheless, the Piwi PAZ domain binding pocket accommodates the methyl-group of 2′–OCH_3 at the 3′ end of RNAs, a modification that impedes binding to the Ago1 PAZ domain. A factor possibly responsible for this main difference is the presence of a common 8 to 10 amino acids insertion between the $\beta6$ and $\beta7$ strands in the PAZ domain of all four Piwi PAZ subfamily members, that is absent in Ago1 PAZ domain and that could facilitate the generation of a more spacious binding pocket.

3.7 LSm Domain

Sm and Sm-like proteins of the RNA-binding Lsm (like Sm) domain family are found in all domains of life and are generally involved in important RNA-processing tasks [89]. Sm and Lsm proteins are characterized by the presence of the two conserved Sm motifs, Sm1 and Sm2. The Sm1 motif is clearly present in the bacterial Hfq protein, although the Sm2 motif cannot be recognized by sequence alone. The Sm fold itself consists of an N-terminal α-helix followed by a twisted five-strand β-sheet. Both Lsm proteins and Hfq exist as doughnut structures even in the absence of the target RNA, unlike the Sm complex, which assembles around single-stranded U-rich RNA regions. The Hfq binding site can accommodate either oligo(A) or oligo(U) but not oligo(C) or oligo(G). Knowledge of the binding specificities of Lsm complexes is limited. Lsm proteins can show binding preference similar to those of Hfq and the Sm complex; moreover, they can associate with unrelated sequences such as 5′-GCUGAUUA-3′ and 5′-UGUACAUAU-3′.

3.8 SAM Domain

Even if SAM domains are widely known as protein–protein interaction domains [90], their RNA binding ability has recently been proven [91, 92]. The SAM domain is present in 358 proteins. It adopts a globular fold made of six helices that are packed by a hydrophobic core such as in Vts1p, EphB2, Byr2, and p73. The SAM domain pocket is surrounded by a large electropositive patch; engineered substitutions in evolutionary conserved basic residues of this region reduce the RNA binding affinity of the domain. Albeit very similar, the SAM domain of Smaug shows a major difference residing in helix 6, which extends and contacts the PHAT domain, a protein domain required for its stabilization in Smaug. The SAM domain binds to a stem-loop termed Smg recognition element (SRE) and consisting of a base pair helix

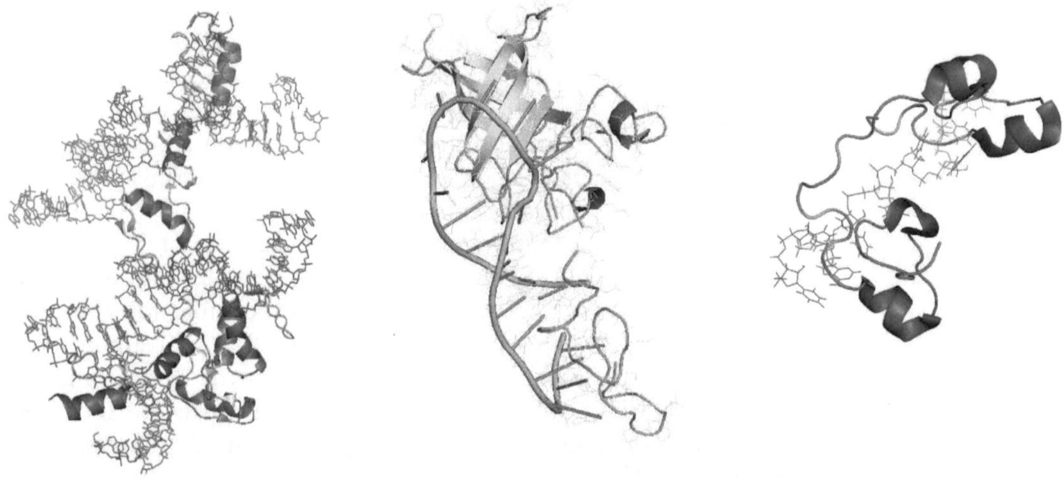

Fig. 4 Example co-crystal structures of Zinc-finger domains. PDB structure 1UN6 (*left*) shows the CCHH domain of Transcription Factor IIIA complexed with the central half of the 5S RNA comprising loop E, helix V, loop A, helix II, and part of helix IV. PDB structure 3TRZ (*middle*) shows the CCHC domain of lin-28 homolog A complexed with let-7d microRNA pre-element. PDB structure 1RGO shows the CCCH domain of Butyrate response factor 2 complexed with UUAUUUAUU

capped by a 4–7 nt loop with consensus sequence CNGGN [0-3]. Specificity of RNA binding arises from the association of a guanosine in the third position of the loop with the shallow pocket on the SAM domain and from multiple SAM domain contacts to the unique sugar-phosphate backbone structure of the loop, which is defined in part by a non-planar base pair within the loop.

3.9 Zinc-Finger Domains

The zinc-finger domains are classifiable into different types according to different combinations and topologies of zinc-binding amino acids (Cys2His2 CCHH, CCCH, or CCHC; *see* Fig. 4, Table 2). The zinc-finger domains are usually present in multiple copies per protein. The CCHH-type zinc fingers, which have so far been found in 2,026 proteins (data taken from the PROSITE database http://prosite.expasy.org/, release 20.73), have two modes of RNA binding. The first mode is based on sequence-independent interaction between the basic amino acids of α-helices and the RNA double helix backbone. In the second mode, the CCHH-type zinc fingers specifically recognize individual nucleotides that bulge out of a structurally rigid element, by side-chain contacts from the N-terminal parts of α-helices. The CCCH-type zinc fingers, which have so far been found in 486 proteins, show a third mode of RNA binding, by which a single-stranded RNA is sequence-specifically recognized. The CCHC-type zinc fingers form base-specific hydrogen bonds to guanines in single-stranded RNAs from the protein's backbone. The length of the inter-domain linker influences the spacing of

Table 2
Summary of zinc-finger domains

Domain	Copies per protein	Sequence specificity	Shape specificity	Nts per domain	No. proteins
RRM	1–6	Low	ssRNA	4,5	1,437
KH-type1	≥ 1	A, C in third position	ssRNA	4	1,296
KH-type2	≥ 1	A, C in third position	ssRNA	4	1,300
dsRBD	1–5	None	dsRNA	15	645
DEAD-BOX	1	None	dsRNA	var.	1,154
PUF	8	Gln+Asn to U, Gln+Cys to A, Glu+Ser to G	ssRNA	1	111
Sm/Lsm	1	poly-A, poly-U (Sm) poly-A, poly-U, GCUGAUUA, UGUACAUAU (Lsm)	ssRNA	var.	25
SAM	1	CNGGN[0-3]	stem-loop	4–7 nts loop	358
ZnF-CCHH	≥ 1	None	ssRNA, dsRNA	var.	2,026
ZnF-CCCH	2	poly-U, AU rich	ssRNA	4	486
ZnF-CCHC	1–2	G	ssRNA		1

the recognized guanines in ssRNAs. This implies that CCCH and some of the CCHH containing RBPs may be suited for RNA-binding models.

In summary, RNA-binding domains tend to show sequence and/or shape specificities [93]; for instance, the RRMs, dsRBDs, and the CCHH-type zinc fingers bind single-stranded RNAs, double-stranded RNAs, and RNA bulges, respectively. Nonetheless, the RNA-binding domains are capable of expanding the repertoire of bound RNAs in terms of length, sequence, and structure. This is attained in several ways: first, by amino acid changes in variable regions of the domains, e.g. in the $\beta1$–$\alpha1$ and $\beta2$–$\beta3$ loops of the RRM and in the α-helix 1 of the dsRBD; second, by combining multiple RNA-binding domains as in the zinc fingers case [94]; third, by C- and N-terminal extension of the domain, such as for the RRM of CBP20 [95].

We note in conclusion that the knowledge of the protein–RNA recognition code is extremely useful to design custom RBPs capable of targeting specific RNA sequences of interest [96]. Such RBPs can potentially be applied to modulate RNA splicing, translation and degradation, to specify alternative splicing patterns and to guide RNA intracellular localization, among others. Here we provide a brief outline of the major determinants to be accounted for in designing custom RBPs. First, protein backbone-mediated recognition likely leads to restrictions on the spectrum of sequences that could be targeted; in this regard, the PUF domain is the most promising candidate due to the extensive use of side-chain contacts. The second aspect is the interaction complexity, i.e. the number of amino acids required to recognize a nucleotide. Indeed, a specific recognition of an extended sequence might require multiple RNA-binding domains, and smaller domains capable of binding the same number of nucleotides are more tractable relative to bigger ones. The third aspect is the efficiency in recognition, defined as the number of domain variants that should be required in order to recognize any possible RNA sequence, which ranges from four, in the case of PUF, to 256, in the case of RRM.

4 Experimental Methods for RBP–RNA Interactions

This section provides the conceptual framework of assays developed to identify the RNA species bound by specific RBPs or, vice versa, subsets of RBPs that bind specific RNAs. It is organized in three parts. The first part covers in vitro methods of studying protein–RNA interactions and general principles of these experimental protocols are outlined. Additional attention will be given to recently developed techniques complementary to in vivo methods. The second part illustrates in vivo transcriptome-wide approaches, whereas the third part shortly presents a few example techniques for studying protein–RNA interactions from a structural perspective.

4.1 Identification of Protein–RNA Interactions In Vitro

Traditionally, in vitro methodologies use one of the two approaches to understand RNA–RBP interactions. First, a known RBP can be targeted to identify RNAs that may interact with it. Conventional electrophoretic mobility shift assays (EMSA) or, the so-called, supershift assays [97] are commonly used to assess RBP activity by showing that RNA migration in PAGE is retarded after incubation with protein in the presence or absence of an antibody targeting the RBP. The second approach involves identification of any RBP associated with the target RNA. This can be achieved by affinity chromatography, i.e. binding an antisense oligonucleotide to a matrix, through which a cell lysate will be passed in hopes that RBPs and its associated proteins will bind a specific target RNA. One limitation of in vitro methodologies is their ability to identify non-physiologically relevant interactions. To understand the biological importance of RNA–RBP interactions, it is necessary to measure these interactions in vivo.

4.2 Systematic Evolution of Ligands by Exponential Enrichment

Systematic Evolution of Ligands by Exponential Enrichment (SELEX) methodology [98] has refined our understanding of the basis for protein recognition in protein–RNA interactions. The in vitro selection protocol is as follows: a DNA pool is chemically synthesized with a region of random or mutagenized sequence flanked on each end by constant sequence and with a T7 RNA polymerase promoter. DNA is amplified by a few cycles of PCR and transcribed in vitro to make the RNA pool. The RNAs are partitioned based on whether they bind to the protein. The retained RNAs are eluted, reverse transcribed, amplified by PCR, transcribed, and then the entire cycle is repeated. With successive rounds of selection, the ratio of high affinity to low affinity sequences increases and the pool thus becomes dominated by the highest affinity RNA species. Sampling at an intermediate round of selection allows for the identification of sequences with a range of affinities, where the relative concentration of each sequence is proportional to its affinity, before the sequence pool becomes too biased for the highest affinity sequence.

4.3 RNAcompete

RNAcompete is a method for the rapid characterization of the binding specificity of RBPs [99]. The method relies on an RNA library designed to include all possible 8-base sequences represented at least 12 times in unstructured RNAs and all possible 6-base and 7-base loop sequences (and 60% of 8-base loops) in RNA hairpins containing unique 10-base pair stems. These sequences are synthesized on using a microarray as template to generate ssDNA which is then converted to dsDNA and amplified using PCR. Finally an in vitro transcription step is used to generate the ssRNA library from the dsDNA. Following the generation of the RNA library, a tagged RBP of interest is used to perform a single pull-down of RNA target sequences. The RBP selected RNA sequences are subsequently labeled and hybridized to a microarray of the same format as was used to generate the RNA library and subsequent computational analysis is used to assess the enrichment of selected RNAs relative to a sample from the starting library.

RNAcompete provides a systematic estimate of RBP binding affinities to short RNAs that contain a complete range of k-mers in structured and unstructured conformation. Therefore, RNAcompete can be envisaged as a valuable tool, in addition to positional weight matrices (PWMs) and consensus motifs, in order to validate and compare the in vivo methods of studying protein–RNA interactions. In summary, RNAcompete consists of three basic steps: (1) generation of an RNA pool comprising a library of RNA sequences and structures; (2) a single pull-down of the RNAs bound to a tagged RBP of interest; and (3) microarray hybridization and computational interrogation of the relative enrichment of the RNAs in the bound fraction relative to the starting pool.

4.4 Identification of Protein–RNA Interactions In Vivo

In vivo protein–RNA interaction methods allow either the characterization of the RBPs that bind to a particular RNA or the characterization of the RNAs that bind to a previously identified RBP and are complementary to each other. The following section discusses these distinct and complementary approaches.

4.5 RIP-Chip

RNP immunoprecipitation-microarray (RIP-Chip) is the first method to use immunoprecipitation in order to assay protein–RNA binding in vivo [100]. In RIP-Chip, antibodies are used to bind specific RBPs and enrich for RNA fragments that are bound to these RBPs. The associated RNA fragments can be identified by hybridization to a microarray, therefore enabling a genome-wide view of protein–RNA interactions. Some of the disadvantages of RIP-Chip are noise issues, due to the microarray technology, the possibility for additional RBPs to co-immunoprecipitate along with the RBP of interest, and the possibility that observed RBP–RNA associations do not reflect in vivo associations, resulting from re-association of RBPs and RNAs subsequent to cell lysis [101, 102]. Furthermore, such techniques do not allow the localization of RBP binding sites within the identified RNA fragments and require subsequent motif analysis to determine RNA binding preferences.

4.6 CLIP and HITS–CLIP

Ultraviolet light Cross-Linking and ImmunoPrecipitation (CLIP) enables in vivo rigorous purification of RBPs along with small fragments of RNAs, which can be amplified and sequenced, *see* Fig. 5 for method overview. UV-induced crosslinking between RBPs and RNAs is applied in vivo, before protein purification, in order to improve upon the standard immunoprecipitation approaches. For example, photo-crosslinking prevents re-association of protein–RNA complexes in vitro or other nonspecific pull-downs as can happen in co-immunoprecipitation. The covalent bonds formed by UV-crosslinking allow for more rigorous purification schemes, which result in more highly purified protein–RNA complexes and subsequent partial proteinase K digestion to better identify the actual binding sites. The reverse transcriptase (RT), used in sample preparation, can transcribe through the cross-linked sites with some frequency. Interestingly, RT errors at cross-linked sites can be exploited to precisely map the sites of protein–RNA contacts (such as by the iCLIP method). High-throughput Sequencing CLIP (HITS–CLIP) combines high-throughput sequencing with the standard CLIP procedure. Due to the quantitation by high-throughput DNA sequencing (HTS/NGS), CLIP provides greater sensitivity (number of high-confidence motif matches covered by top binding sites identified) and better spatial resolution of RBP binding sites. Currently, CLIP suffers from the challenges related to HTS techniques, such as high sequencing error rates, variable quality CLIP tag alignments, and the definition of appropriate

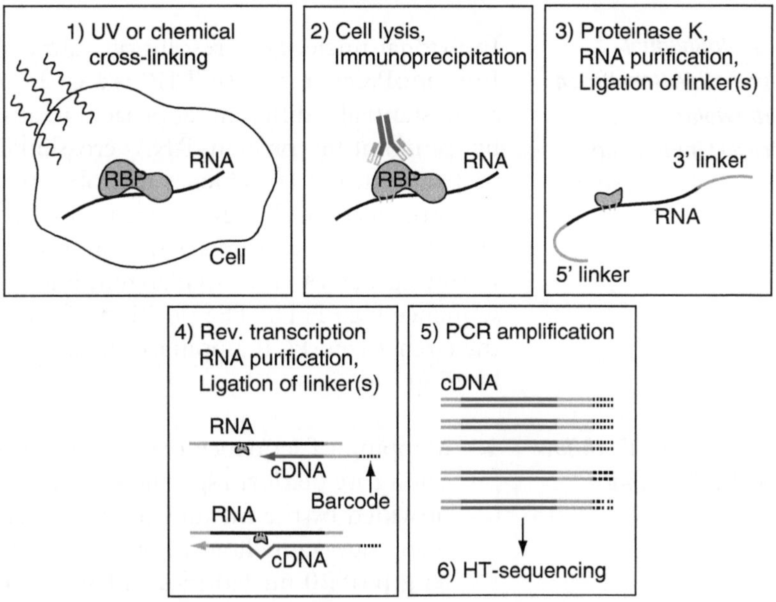

Fig. 5 Overview of CLIP-based methods. Panel 1 depicts the crosslinking stop that covalently binds RBPs to their cognate targets. Cell lysis and immunoprecipitation follow and require either an antibody specific to the RBP or the expression of an epitope tagged RBP. Panel 3 summarizes steps that digest all but the covalently bound portion of the RBP, purification of precipitated RNA and ligation of linker sequences. Panel 4 depicts the reverse transcription step used to generate a cDNA library which may make use of barcoded adaptors for multiplexed sequencing. A final PCR amplification step precedes high-throughput sequencing by any of the currently available technologies

background CLIP tag distributions for assessing the statistical significance of RBP binding sites. Differences in CLIP protocols can also bias the inferred specificity of the RBP, for example some of the RNAase used to digest unbound RNA have sequence-specificity and that this can impact the RBP binding sites represented by the CLIP tags [103]. Also, CLIP cross-linking protocol with greater sensitivity can suffer from less specificity [104].

4.7 Photo-Activatable Ribonucleoside Enhanced Cross-Linking and Immunoprecipitation

Photo-Activatable Ribonucleoside enhanced Cross-Linking and Immunoprecipitation (PAR-CLIP) is a CLIP method variant, where the introduction of photo-activated nucleosides in the media are taken up by cells and subsequently used for protein–RNA crosslinking. This modification provides advantages over standard CLIP. First, PAR-CLIP obtains 100- to 1,000-fold higher cross-linked RNA recovery, using comparable radiation intensities. The second advantage relates to UV radiation-induced T-to-C mutations characteristic of the cross-linked sites that have incorporated photo-activated nucleoside analogs. Based on this, PAR-CLIP exploits mutation analysis to improve the identification of the RBP binding site positions or footprint.

4.8 Individual-Nucleotide Resolution Ultraviolet Cross-Linking and ImmunoPrecipitation

Individual-nucleotide resolution ultraviolet Cross-Linking and ImmunoPrecipitation (iCLIP) is a CLIP method variant that takes a substantially different approach relative to all other variants, by pinpointing protein–RNA crosslinking sites during sample preparation. iCLIP achieves this objective by taking advantage of the propensity of the reverse transcription to stop before the cross-linked nucleotides due to the amino acids that remain bound. Resulting cDNAs are circularized, linearized, PCR amplified, and sequenced via HTS. The position of the adaptor sequence used in the circularized PCR amplification can be used to identify the RBP binding site.

4.9 Analytic Workflow for NGS Methods

1. Mapping of sequence reads (tags). Image processing and base calling are platform specific and mostly done using the software provided by the manufacturer. Strategy for genome alignment is an important choice. For instance, [105] uses a minimal read length of 20 nucleotides and allows for one T to C mismatch, whereas [104] allows for two T to C mutations in reads of at least 13 nucleotides.
2. Generation of clusters of mapped sequence reads. Sequence reads are clustered according to the nucleotide overlapping in their mapped genomic positions. High-confidence clusters of sequence reads can then be identified by a set of criteria such as the enrichment over the control and minimal tag density. The PAR-CLIP-based approach introduced by [105] ranks sequence read clusters according to their total number of UV radiation-induced T-to-C mutations (i.e., cross-linking positions). A more sophisticated analysis of crosslinking induced mutations is also used by HITS–CLIP in order to identify high-confidence sequence read clusters [102].
3. Optional identification of crosslinking-centered regions (CCRs) in sequence read clusters. For instance, [105] considers around 20 nucleotides to the left and right of the site of T-to-C mutation with the highest sequence reads density in a sequence read cluster. A Gaussian kernel-density-based classifier (PARalyzer) was used in each cluster identified by PARalyzer in order to more precisely delineate CCRs in a cluster. Subsequently, CCRs can be extended in order to ensure encompassing the real RBP binding site. The choice of the extension method depends on the crosslinking properties of different RBPs, as shown in [106]. Extending the region to the full underlying reads is suitable when the protein–RNA interaction site is protected from crosslinking, whereas extending by a generic window size is suitable when crosslinking occurs at the protein binding motif.

4. Motif search. Several available algorithms allow either *de novo* motif analysis, such as MEME, #ATS [107], cERMIT [108], PhyloGibbs [109] and PARalyzer [104], or analysis of a priori known motif frequency relative to the regions identified for protein–RNA interactions.

4.10 Finding the Proteins Bound to RNAs

Although the in vivo study of the protein components of protein–RNA complexes is challenging, a few approaches have been developed. The Peptide Nucleic Acid (PNA)-assisted identification of RBPs (PAIR) method was developed to address this problem by combining and optimizing standard UV-induced protein–nucleic acid crosslinking and magnetic bead-based assays [110]. The unique features of this method are the utilization of PNA oligonucleotides [111], peptide linked nucleic acid analogs capable of binding RNAs with higher specificity and selectivity than complementary DNA or RNA, along with the efficient delivery of PNA oligos to living cells. Once inside the cell, the PNAs hybridize to their cognate RNA and UV light is used to induce RBP–PNA crosslinking where targeted RNAs are bound. The RBP–PNA complexes are then isolated by magnetic beads, coupled to an antisense PNA oligo, and their proteins identified by mass spectrometry techniques. The majority of the studies on protein components of protein–RNA complexes found by protein capture methods suffer from low specificity. In contrast, the application of quantitative mass spectrometry to RNA affinity purification assays, like in [112], facilitates the detection of proteins specifically binding to the RNA of interest from background binders.

4.11 Structural Analysis of Protein–RNA Interactions

The crosslinking and mapping of protein domain (CLAMP) method allows mapping the RNA-binding domains within the RBPs that are cross-linked to specific nucleotides in the RNA. This approach is particularly useful to analyze the relative contribution of multiple RNA-binding domains to the protein–RNA interaction for proteins with multiple RBDs. CLAMP requires site-specific incorporation of a chromophore, photochemical protein–RNA crosslinking, and site-specific, chemical protein cleavage.

4.12 Structural Analysis of Protein–RNA Interactions with Mass Spectrometry

The method allows the identification of the amino acids in the protein of interest that interact with the RNA. It is based upon the differential accessibility of the primary amine-modifying reagent N-hydroxysuccinimide (NHS)-biotin to lysine residues in the free protein versus the protein–RNA complex. Subsequent MS analysis enables accurate identification of these residues. The important role of lysine-phosphate backbone contacts in formation of many nucleoprotein complexes motivates the choice of measuring lysine accessibility. Introducing sodium dodecyl sulfate polyacrylamide gel electrophoresis (SDS-PAGE) and in-gel proteolysis prior to MS is important as SDS-PAGE allows separation of individual protein subunits based on their molecular weight differences. Thereafter, contact lysines can be accurately assigned to individual components of a multi-subunit complex. Subsequent in-gel proteolysis produces short peptide fragments amenable to MS/MS analysis. The biotinylated peptide peaks can be readily identified from MS data and the modified sites accurately assigned to appropriate lysine residues by MS/MS analysis. Comparative examination reveals lysines readily modified in the free protein but protected in the context of the nucleoprotein complex.

4.13 Online Resources for Experimental Protein–RNA Interactions

The protein–RNA interaction data generated by the presented technologies have been captured in a few resources. Worthy examples are the RNA binding Protein Database at http://rbpdb.ccbr.utoronto.ca/ (RBPDB) [113] and the CLIPZ database at http://www.clipz.unibas.ch [114]. Owing to space restrictions, only some relevant features are highlighted here. RBPDB is a useful starting point to begin the study of manually collected RNA binding interactions and/or binding sites for a certain RNA binding protein. RBPDB includes interactions identified experimentally by in vitro (e.g., RNAcompete) or in vivo methods (e.g., RIP-Chip, CLIP), for all RBPs in four metazoan species (human, mouse, fly and worm). RBPDB contains functionality to scan an input RNA sequence for motifs, provided they are associated with full Positional Weight Matrices (PWMs) in RBPDB and to retrieve potential binding sites annotated by PWM scores.

In contrast to RBPDB, CLIPZ is a more specific database of RNA binding sites generated by means of HITS–CLIP and enables visualization and downstream analysis of data generated by this technique. One common downstream analysis is the prediction of RBP binding sequence motifs within RBP binding sequences (tag clusters) based on motif enrichment analysis. Putative binding site motifs are returned along with their statistical significance. Other approaches enable the investigation of spatial relationships between RBPs.

The Protein–RNA Interface Database at http://pridb.gdcb.iastate.edu/index.php (PRIDB) [115] is a database of protein–RNA interfaces extracted from complexes in the Protein Data Bank (PDB). It allows users to identify and visualize interfacing amino acids and ribonucleotides within the primary sequences of the interacting protein and RNA chains of interest. To define interfaces PRIDB uses both a distance-based rule and the ENTANGLE algorithm [116]. In addition to protein–RNA interfaces, PRIDB identifies ProSite [117] motifs in protein chains and FR3D [118] motifs in RNA chains.

The Atlas of UTR Regulatory Activity at http://aura.science.unitn.it (AURA) is a manually curated catalog of human UTRs and UTR regulatory annotations. Through its intuitive web interface, it provides full access to a wealth of information on UTRs by including phylogenetic conservation, RNA sequence and structure data, single nucleotide variation, gene expression, gene functional descriptions and by integrating non-redundant, experimentally assessed interactions of RBPs and miRs with human UTRs from literature and specialized databases [119].

5 Computational Inference of RBP Binding Sites

The identification of RNA sequence elements that act as binding sites for RBPs has been addressed by several computational approaches. Only a brief sketch of these approaches will be presented here. Detailed discussions can be found in a review of bioinformatics methods for predicting RNA-binding sites by Puton et al. [120] and in the corresponding literature. The task of inferring RNA binding sites in protein structures, albeit an active area of research [121], is not addressed here.

5.1 Binding Site Search

Position weight matrices (PWMs), containing probabilities of occurrence for each nucleic acid at each position are usually used to summarize the statistical properties of observed binding sites. PWMs are scanned over RNA sequences to predict potential RBP binding sites. Regulatory sequence analysis tools, such as RSAT at http://rsat.ulb.ac.be/rsat/, can be used to perform this search. Note that the reliability of this representation of RNA binding specificities depends on the availability of a substantial amount of experimental data.

5.2 Binding Site Models

When presenting the current techniques to model RBP binding sites, transcription factor binding site models represent important precedence for the problem of pattern discovery and prediction. Modeling binding aspects of RBPs require either novel techniques or the adaptation of existing methods. Several techniques for identifying RBP binding sites are highlighted here for their novelties and commonalities when compared to DNA binding site discovery. Similar to transcription factors, RBP binding sites are modeled by unsupervised (i.e., density estimation) and, to a lesser extent, supervised (i.e., regression) approaches. However, RBP binding sites differ from transcription factor binding sites in that RNA structure can play a role in binding, so the models can be divided into those that ignore RNA structure and those that do not. The methods that consider RNA structure can be further divided into those that model RNA structure and those that model RNA structural context.

The so-called unsupervised approaches are provided as input a set of RNA sequences enriched for binding sites of a given RBP (generated, e.g., through a SELEX procedure) as well as a background model of typical RNA sequence composition. If the influence of RNA structure is ignored, then methods for transcription factors can be applied directly to this problem with only minor modifications (i.e., replacing Us with Ts and not scanning the reverse complement sequence for binding site), *see* [122] for a comparison of these methods. For example, a popular approach is to apply MEME [123] (Multiple Expectation Maximization for Motif Elicitation) to fit a position-specific scoring matrix (PSSM) motif model by maximizing the likelihood of the observed sequence set under the PSSM (and an appropriate background model) using the expectation–maximization (EM) algorithm. This PSSM model defines a product multinomial distribution over bound k-mers that assumes that individual nucleotides are statistically independent. Note that MEME does not allow any gap in the sequence patterns, which can represent a limitation when RNA-binding domains, such as RRMs, bind to multiple, spaced and very short nucleotides in the RNA sequence (e.g., PTB [81]). Another notable example of a structure-naive method applied to RBP binding site modeling is Xie et al. [124] who assign a conservation index to all possible k-mers (RNA words of length k) to make an unbiased genome-wide search for k-mers that are surprisingly conserved in 3′ UTRs. These k-mer represent possible regulatory elements.

MEMERIS [125] is a simple modification of the MEME algorithm that considers base-pairing probability estimates assessed by RNAfold [126] when fitting PSSM motifs. These base-pairing probabilities restrict the search space for the start position of an RBP binding site. In this way, MEMERIS searches for a motif associated with a particular RNA structure context (i.e., unpaired regions); this approach differs from several methods that search for specific sequence-structure elements (e.g., stem-loops).

The motif finding algorithms that model RNA structure can be classified into three types. The first type performs RNA sequence-based alignments and uses co-variation between the aligned sequences to derive a consensus structure. The success of such methods greatly depends on the quality of the alignment, which in turn requires high sequence similarity between the input RNA sequences, which is not necessarily the case, particularly when searching within long 3′ UTRs for local patterns shared by multiple mRNAs bound by the same RBP. Alternatively, techniques like RNAProfile [127] first predict minimal free energy folds for each sequence in order to identify common folds. Here, the main problems are the limited accuracy in fold prediction, and the representation of a whole ensemble of folds via a single fold, the minimal free energy fold. The third approach uses dynamic programming to simultaneously align and fold a pair of RNA sequences and predict their common secondary structure using energy-based considerations, resulting in a structure-based alignment [128]. This pair-wise alignment is then extended to a multiple alignment using various heuristics. Whereas the RNA sequence is available, its secondary structure is most often a matter of algorithmic predictions and represents noisy input to any analysis. Probabilistic, covariance models [129], like CMfinder [130], more adequately capture the observed variation in the structure and sequence of RNA patterns. Recently RNApromo [131] has been used to model RBP sequence and structural preferences common to a set of co-regulated RNA sequences.

Supervised methods fit an RBP binding model by including it as part of a regression model trained to predict a quantitative estimate of RBP binding assigned to a set of RNA sequences. These methods have been applied more rarely to RBP binding data because of the difficulty, until recently, of acquiring the necessary input data. Early methods in this area were either structure-naive [132] or used simplistic stem-loop models [133]. More recent examples are provided by #ATS [107] and RNAcontext [134]. For example, RNAcontext learns the sequence and structure preferences of an RBP by fitting a physical model to the RBP binding affinity data provided by an in vitro assay, RNAcompete. RNAcontext is interesting for two reasons: its capability of modeling RBPs preferences to sequences in their structural contexts, and also the full exploitation of high-throughput quantitative data in order to define the model parameters. RNAcontext takes as input a set of sequences associated with their affinity estimates and works in three steps. First, it defines the probability that a word of length k is an RBP binding site as the product between two terms (whose mathematical formula is that of a logistic function); the first term describes the inferred RBP sequence preferences (as a positional weight matrix), whereas the second term describes the relative structural preferences of the RBP to different structural contexts. The second step is to estimate a sequence affinity from the affinities assigned to each word by the previous motif model; the sequence scoring function is set to the probability that at least one of its words is an RBP binding site. The third step is to learn the set of parameters that minimize the sum of the squared differences between the measured input affinities and the predicted affinities modeled as a linear function of the sequence score function. #ATS is similar to RNAcontext, except that it uses a greedy search to define a degenerate consensus sequence motif and only considers a single structural context at a time. However, #ATS is more appropriate than RNAcontext for in vivo binding assays because the sequence scoring function of the former is better suited to the longer RNA sequences associated with these assays.

6 Future Perspectives

The development of models for protein–RNA interactions will depend on experimental data from structure analysis and from the immunoprecipitation-based methods, e.g. CLIP, as highlighted here. The combination of structural data, that define the intermolecular contacts, and high-throughput sequencing methods, that define the RNA sequence specificity, may allow for the definition of predictive models for a specific RBP. Both data resources are currently in limited supply and, in most cases, do not include RBDs with variable amino acid sequence. Variation of both nucleotide and peptide sequences could be achieved through site directed mutagenesis or through studying orthologous RBPs that vary between species, to allow for mapping the relationship between amino acids and nucleotides at the binding interface in a general way.

Acknowledgements

The preparation of this chapter was partially supported by a Canadian Institute of Health Research grant to Quaid Morris (MOP-93671) and by an Italian Autonomous Province of Trento grant to Angela Re.

References

1. Dieci G, Ruotolo R, Braglia P, Carles C, Carpentieri A, Amoresano A, Ottonello S (2009) Positive modulation of RNA polymerase III transcription by ribosomal proteins. Biochem Biophys Res Commun 379(2):489–493. doi:10.1016/j.bbrc.2008.12.097
2. Wan F, Anderson DE, Barnitz RA, Snow A, Bidere N, Zheng L, Hegde V, Lam LT, Staudt LM, Levens D, Deutsch WA, Lenardo MJ (2007) Ribosomal protein S3: a KH domain subunit in NF-kappaB complexes that mediates selective gene regulation. Cell 131(5):927–939. doi:10.1016/j.cell.2007.10.009
3. Whitelaw E, Proudfoot N (1986) Alpha-thalassaemia caused by a poly(a) site mutation reveals that transcriptional termination is linked to 3' end processing in the human alpha 2 globin gene. EMBO J 5(11):2915–2922. http://www.hubmed.org/fulltext.cgi?uids=3024968
4. Lareau LF, Inada M, Green RE, Wengrod JC, Brenner SE (2007) Unproductive splicing of sr genes associated with highly conserved and ultraconserved dna elements. Nature 446(7138):926–929. doi:10.1038/nature05676
5. Holt CE, Bullock SL (2009) Subcellular mRNA localization in animal cells and why it matters. Science 326(5957):1212–1216. doi:10.1126/science.1176488
6. Bertrand E, Chartrand P, Schaefer M, Shenoy SM, Singer RH, Long RM (1998) Localization of ASH1 mRNA particles in living yeast. Mol Cell 2(4):437–445. http://www.hubmed.org/fulltext.cgi?uids=9809065
7. Lécuyer E, Yoshida H, Parthasarathy N, Alm C, Babak T, Cerovina T, Hughes TR, Tomancak P, Krause HM (2007) Global analysis of mRNA localization reveals a prominent role in organizing cellular architecture and function. Cell 131(1):174–187. doi:10.1016/j.cell.2007.08.003. http://www.hubmed.org/fulltext.cgi? uids=17923096
8. Grünwald D, Singer RH, Rout M (2011) Nuclear export dynamics of RNA-protein complexes. Nature 475(7356):333–341 doi:10.1038/nature10318

9. Wolke U, Weidinger G, Köprunner M, Raz E (2002) Multiple levels of posttranscriptional control lead to germ line-specific gene expression in the zebrafish. Curr Biol 12(4):289–294
10. Lipshitz HD, Smibert CA (2000) Mechanisms of RNA localization and translational regulation. Curr Opin Genet Dev 10(5):476–488. http://www.hubmed.org/fulltext.cgi?uids=10980424
11. Chartrand P, Meng XH, Huttelmaier S, Donato D, Singer RH (2002) Asymmetric sorting of ash1p in yeast results from inhibition of translation by localization elements in the mRNA. Mol Cell 10(6):1319–1330
12. Lewis RA, Kress TL, Cote CA, Gautreau D, Rokop ME, Mowry KL (2004) Conserved and clustered RNA recognition sequences are a critical feature of signals directing RNA localization in xenopus oocytes. Mech Dev 121(1):101–109
13. Macdonald PM, Struhl G (1988) cis-acting sequences responsible for anterior localization of bicoid mRNA in drosophila embryos. Nature 336(6199):595–598. doi:10.1038/336595a0
14. Cenik C, Chua HN, Zhang H, Tarnawsky SP, Akef A, Derti A, Tasan M, Moore MJ, Palazzo AF, Roth FP (2011) Genome analysis reveals interplay between 5'utr introns and nuclear mRNA export for secretory and mitochondrial genes. PLoS Genet 7(4). doi:10.1371/journal.pgen.1001366. http://www.hubmed.org/fulltext.cgi?uids=21533221
15. Palazzo AF, Springer M, Shibata Y, Lee CS, Dias AP, Rapoport TA (2007) The signal sequence coding region promotes nuclear export of mRNA. PLoS Biol 5(12). doi:10.1371/journal.pbio.0050322. http://www.hubmed.org/fulltext.cgi?uids=18052610
16. Arn EA, Cha BJ, Theurkauf WE, Macdonald PM (2003) Recognition of a bicoid mRNA localization signal by a protein complex containing swallow, nod, and RNA binding proteins. Dev Cell 4(1):41–51
17. Müller M, Heym RG, Mayer A, Kramer K, Schmid M, Cramer P, Urlaub H, Jansen RP, Niessing D (2011) A cytoplasmic complex mediates specific mRNA recognition and localization in yeast. PLoS Biol 9(4). doi:10.1371/journal.pbio.1000611. http://www.hubmed.org/fulltext.cgi?uids=21526221
18. Tadros W, Goldman AL, Babak T, Menzies F, Vardy L, Orr-Weaver T, Hughes TR, Westwood JT, Smibert CA, Lipshitz HD (2007) Smaug is a major regulator of maternal mRNA destabilization in drosophila and its translation is activated by the PAN GU kinase. Dev Cell 12(1):143–55. doi:10.1016/j.devcel.2006.10.005
19. Forrest KM, Gavis ER (2003) Live imaging of endogenous RNA reveals a diffusion and entrapment mechanism for nanos mRNA localization in drosophila. Curr Biol 13(14):1159–1168
20. Gebauer F, Hentze MW (2004) Molecular mechanisms of translational control. Nat Rev Mol Cell Biol 5(10):827–835. doi:10.1038/nrm1488
21. Jackson RJ, Hellen CUT, Pestova TV (2010) The mechanism of eukaryotic translation initiation and principles of its regulation. Nat Rev Mol Cell Biol 11(2):113–127. doi:10.1038/nrm2838
22. Loh PG, Song H (2010) Structural and mechanistic insights into translation termination. Curr Opin Struct Biol 20(1):98–103. doi:10.1016/j.sbi.2009.12.005
23. Gingold H, Pilpel Y (2011) Determinants of translation efficiency and accuracy. Mol Syst Biol 7:481. doi:10.1038/msb.2011.14
24. Dever TE (2002) Gene-specific regulation by general translation factors. Cell 108(4):545–556
25. Muckenthaler M, Gray NK, Hentze MW (1998) Irp-1 binding to ferritin mRNA prevents the recruitment of the small ribosomal subunit by the cap-binding complex eif4f. Mol Cell 2(3):383–388
26. Gebauer F, Grskovic M, Hentze MW (2003) Drosophila sex-lethal inhibits the stable association of the 40S ribosomal subunit with msl-2 mRNA. Mol Cell 11(5):1397–1404
27. Grskovic M, Hentze MW, Gebauer F (2003) A co-repressor assembly nucleated by sex-lethal in the 3'utr mediates translational control of drosophila msl-2 mRNA. EMBO J 22(20):5571–5581. doi:10.1093/emboj/cdg539
28. Beckmann K, Grskovic M, Gebauer F, Hentze MW (2005) A dual inhibitory mechanism restricts msl-2 mRNA translation for dosage compensation in drosophila. Cell 122(4):529–540. doi:10.1016/j.cell.2005.06.011
29. Nelson MR, Leidal AM, Smibert CA (2004) Drosophila cup is an eif4e-binding protein that functions in smaug-mediated translational repression. EMBO J 23(1):150–159. doi:10.1038/sj.emboj.7600026
30. Nakamura A, Sato K, Hanyu-Nakamura K (2004) Drosophila cup is an eif4e binding protein that associates with bruno and regulates oskar mRNA translation in oogenesis. Dev Cell 6(1):69–78
31. Stebbins-Boaz B, Cao Q, de Moor CH, Mendez R, Richter JD (1999) Maskin is a

cpeb-associated factor that transiently interacts with elf-4e. Mol Cell 4(6):1017–1027
32. Ostareck DH, Ostareck-Lederer A, Shatsky IN, Hentze MW (2001) Lipoxygenase mRNA silencing in erythroid differentiation: The 3'utr regulatory complex controls 60S ribosomal subunit joining. Cell 104(2):281–290
33. Chaudhury A, Hussey GS, Ray PS, Jin G, Fox PL, Howe PH (2010) TGF-beta-mediated phosphorylation of hnRNP E1 induces EMT via transcript-selective translational induction of Dab2 and ILEI. Nat Cell Biol 12(3):286–293. doi:10.1038/ncb2029
34. Hussey GS, Chaudhury A, Dawson AE, Lindner DJ, Knudsen CR, Wilce MCJ, Merrick WC, Howe PH (2011) Identification of an mRNP complex regulating tumorigenesis at the translational elongation step. Mol Cell 41(4): 419–431. doi:10.1016/j.molcel.2011.02.003
35. Doma MK, Parker R (2007) RNA quality control in eukaryotes. Cell 131(4):660–668. doi:10.1016/j.cell.2007.10.041
36. Villalba A, Coll O, Gebauer F (2011) Cytoplasmic polyadenylation and translational control. Curr Opin Genet Dev 21(4):452–457. doi:10.1016/j.gde.2011. 04.006
37. Piqué M, López JM, Foissac S, Guigó R, Méndez R (2008) A combinatorial code for cpe-mediated translational control. Cell 132(3):434–448. doi:10.1016/j.cell.2007.12.038
38. Kim KW, Nykamp K, Suh N, Bachorik JL, Wang L, Kimble J (2009) Antagonism between gld-2 binding partners controls gamete sex. Dev Cell 16(5):723–733. doi:10.1016/j.devcel.2009.04.002
39. Reed R, Hurt E (2002) A conserved mRNA export machinery coupled to pre-mRNA splicing. Cell 108(4):523–531
40. Houseley J, LaCava J, Tollervey D (2006) RNA-quality control by the exosome. Nat Rev Mol Cell Biol 7(7):529–539 doi:10.1038/nrm1964.
41. LaCava J, Houseley J, Saveanu C, Petfalski E, Thompson E, Jacquier A, Tollervey D (2005) RNA degradation by the exosome is promoted by a nuclear polyadenylation complex. Cell 121(5):713–724. doi:10.1016/j.cell.2005.04.029
42. Isken O, Maquat LE (2008) The multiple lives of nmd factors: balancing roles in gene and genome regulation. Nat Rev Genet 9(9):699–712. doi:10.1038/nrg2402
43. Kong J, Liebhaber SA (2007) A cell type-restricted mRNA surveillance pathway triggered by ribosome extension into the 3' untranslated region. Nat Struct Mol Biol 14(7):670–676. doi:10.1038/nsmb1256
44. Hir HL, Gatfield D, Izaurralde E, Moore MJ (2001) The exon-exon junction complex provides a binding platform for factors involved in mRNA export and nonsense-mediated mRNA decay. EMBO J 20(17):4987–4997. doi:10.1093/emboj/20.17.4987
45. Behm-Ansmant I, Gatfield D, Rehwinkel J, Hilgers V, Izaurralde E (2007) A conserved role for cytoplasmic poly(a)-binding protein 1 (pabpc1) in nonsense-mediated mRNA decay. EMBO J 26(6):1591–601. doi:10.1038/sj.emboj.7601588
46. González CI, Ruiz-Echevarría MJ, Vasudevan S, Henry MF, Peltz SW (2000) The yeast hnRNP-like protein Hrp1/Nab4 marks a transcript for nonsense-mediated mRNA decay. Mol Cell 5(3):489–499
47. Hwang J, Sato H, Tang Y, Matsuda D, Maquat LE (2010) Upf1 association with the cap-binding protein, cbp80, promotes nonsense-mediated mRNA decay at two distinct steps. Mol Cell 39(3):396–409. doi:10.1016/j.molcel.2010.07.004
48. Ruiz-Echevarría MJ, Peltz SW (2000) The RNA binding protein pub1 modulates the stability of transcripts containing upstream open reading frames. Cell 101(7):741–751
49. Chester A, Somasekaram A, Tzimina M, Jarmuz A, Gisbourne J, O'Keefe R, Scott J, Navaratnam N (2003) The apolipoprotein b mRNA editing complex performs a multifunctional cycle and suppresses nonsense-mediated decay. EMBO J 22(15)3971–3982. doi:10.1093/emboj/cdg369
50. Stalder L, Mühlemann O (2008) The meaning of nonsense. Trends Cell Biol 18(7):315–321. doi:10.1016/j.tcb.2008.04.005
51. Inada T, Aiba H (2005) Translation of aberrant mRNAs lacking a termination codon or with a shortened 3'-UTR is repressed after initiation in yeast. EMBO J 24(8):1584–1595. doi:10.1038/sj.emboj.7600636
52. Kim YK, Furic L, Desgroseillers L, Maquat LE (2005) Mammalian Staufen1 recruits Upf1 to specific mRNA 3'UTRs so as to elicit mRNA decay. Cell 120(2):195–208. doi:10.1016/j.cell.2004.11.050
53. Kaygun H, Marzluff WF (2005) Translation termination is involved in histone mRNA degradation when DNA replication is inhibited. Mol Cell Biol 25(16):6879–6888. doi:10.1128/MCB.25.16.6879-6888.2005
54. Wulff BE, Sakurai M, Nishikura K (2011) Elucidating the inosinome: global approaches to adenosine-to-inosine RNA editing. Nat Rev Genet 12(2):81–85. doi:10.1038/nrg2915. http://www.hubmed.org/fulltext.cgi?uids=21173775

55. Bass BL (2002) RNA editing by adenosine deaminases that act on RNA. Annu Rev Biochem 71:817–846. doi:10.1146/annurev.biochem.71.110601.135501. http://www.hubmed.org/fulltext.cgi?uids=12045112
56. Eggington JM, Greene T, Bass BL (2011) Predicting sites of adar editing in double-stranded RNA. Nat Commun 2:319–319. doi:10.1038/ncomms1324. http://www.hubmed.org/fulltext.cgi?uids=21587236
57. Hundley HA, Bass BL (2010) Adar editing in double-stranded utrs and other noncoding RNA sequences. Trends Biochem Sci 35(7):377–383. doi:10.1016/j.tibs.2010.02.008. http://www.hubmed.org/fulltext.cgi?uids=20382028
58. Jepson JE, Reenan RA (2008) RNA editing in regulating gene expression in the brain. Biochim Biophys Acta 1779(8):59–470. doi:10. 1016/j.bbagrm.2007.11.009. http:// www.hubmed.org/fulltext.cgi?uids=18086576
59. Hoopengardner B, Bhalla T, Staber C, Reenan R (2003) Nervous system targets of RNA editing identified by comparative genomics. Science 301(5634):832–836. doi:10.1126/science.1086763. http://www.hubmed.org/fulltext.cgi? uids=12907802
60. Paz-Yaacov N, Levanon EY, Nevo E, Kinar Y, Harmelin A, Jacob-Hirsch J, Amariglio N, Eisenberg E, Rechavi G (2010) Adenosine-to-inosine RNA editing shapes transcriptome diversity in primates. Proc Natl Acad Sci USA 107(27):12174–12179. doi:10.1073/pnas.1006183107. http://www.hubmed.org/fulltext.cgi?uids=20566853
61. Nishikura K (2010) Functions and regulation of RNA editing by adar deaminases. Annu Rev Biochem 79:321–349. doi:10.1146/annurev-biochem-060208-105251. http://www.hubmed.org/fulltext.cgi?uids=20192758
62. Chen SH, Habib G, Yang CY, Gu ZW, Lee BR, Weng SA, Silberman SR, Cai SJ, Deslypere JP, Rosseneu M (1987) Apolipoprotein b-48 is the product of a messenger RNA with an organ-specific in-frame stop codon. Science 238(4825):363–366. http://www.hubmed.org/fulltext.cgi?uids=3659919
63. Powell LM, Wallis SC, Pease RJ, Edwards YH, Knott TJ, Scott J (1987) A novel form of tissue-specific RNA processing produces apolipoprotein-b48 in intestine. Cell 50(6):831–840. http://www.hubmed.org/fulltext.cgi?uids=3621347
64. Rosenberg BR, Hamilton CE, Mwangi MM, Dewell S, Papavasiliou FN (2011) Transcriptome-wide sequencing reveals numerous apobec1 mrna-editing targets in transcript 3' UTRs. Nat Struct Mol Biol 18(2):230–236. doi:10.1038/nsmb.1975. http://www.hubmed.org/fulltext.cgi?uids=21258325
65. Li M, Wang IX, Li Y, Bruzel A, Richards AL, Toung JM, Cheung VG (2011) Widespread RNA and DNA sequence differences in the human transcriptome. Science 333(6038):53–58. doi:10.1126/science.1207018. http://www.hubmed.org/fulltext.cgi?uids=21596952
66. Schrider DR, Gout JF, Hahn MW (2011) Very few RNA and DNA sequence differences in the human transcriptome. PLoS One 6(10). doi:10.1371/journal.pone.0025842. http://www.hubmed.org/fulltext.cgi?uids=22022455
67. Barreau C, Paillard L, Osborne HB (2005) Au-rich elements and associated factors: are there unifying principles? Nucleic Acids Res 33(22):7138–7150. doi:10.1093/nar/gki1012
68. Lal A, Mazan-Mamczarz K, Kawai T, Yang X, Martindale JL, Gorospe M (2004) Concurrent versus individual binding of HuR and AUF1 to common labile target mRNAs. EMBO J 23(15):3092–3102. doi:10.1038/sj.emboj.7600305
69. Sobue S, Murakami M, Banno Y, Ito H, Kimura A, Gao S, Furuhata A, Takagi A, Kojima T, Suzuki M, Nozawa Y, Murate T (2008) v-Src oncogene product increases sphingosine kinase 1 expression through mRNA stabilization: alteration of AU-rich element-binding proteins. Oncogene 27(46):6023–6033. doi:10.1038/onc.2008.198
70. Kedde M, Strasser MJ, Boldajipour B, Oude Vrielink JA, Slanchev K, le Sage C, Nagel R, Voorhoeve PM, van Duijse J, Ørom UA, Lund AH, Perrakis A, Raz E, Agami R (2007) RNA-binding protein Dnd1 inhibits microRNA access to target mRNA. Cell 131(7):1273–1286. doi:10.1016/j.cell.2007.11.034
71. Burns DM, D'Ambrogio A, Nottrott S, Richter JD (2011) Cpeb and two poly(a) polymerases control mir-122 stability and p53 mRNA translation. Nature 473(7345):105–108. doi:10.1038/nature09908
72. Abdelmohsen K, Hutchison ER, Lee EK, Kuwano Y, Kim MM, Masuda K, Srikantan S, Subaran SS, Marasa BS, Mattson MP, Gorospe M (2010) miR-375 inhibits differentiation of neurites by lowering HuD levels. Mol Cell Biol 30(17):4197–4210. doi:10.1128/MCB.00316-10

73. Deschenes-Furry J, Perrone-Bizzozero N, Jasmin BJ (2006) The RNA-binding protein HuD: a regulator of neuronal differentiation, maintenance and plasticity. Bioessays 28:822–833
74. Lukong K, Chang K, Khandjian E, Richard S (2008) RNA-binding proteins in human genetic disease. Trends Genet 24(8):416–425. ISSN 01689525. doi:10.1016/j.tig.2008.05.004. http://linkinghub.elsevier.com/retrieve/pii/S016895250800173X
75. Yang YY, Yin GL, Darnell RB (1998) The neuronal RNA-binding protein Nova-2 is implicated as the autoantigen targeted in POMA patients with dementia. Proc Natl Acad Sci USA 95:13254–13259
76. Song L, Wang L, Li Y, Xiong H, Wu J, Li J, Li M (2010) Sam68 up-regulation correlates with, and its down-regulation inhibits, proliferation and tumourigenicity of breast cancer cells. J Pathol 222:227–237
77. Busa R, Paronetto MP, Farini D, Pierantozzi E, Botti F, Angelini DF, Attisani F, Vespasiani G, Sette C (2007) The RNA-binding protein sam68 contributes to proliferation and survival of human prostate cancer cells. Oncogene 26(30):4372–4382. ISSN 0950-9232. http://dx.doi.org/10.1038/sj.onc.1210224
78. Wang R, Geng J, Wang JH, Chu XY, Geng HC, Chen LB (2009) Overexpression of eukaryotic initiation factor 4E (eIF4E) and its clinical significance in lung adenocarcinoma. Lung Cancer 66:237–244
79. Keene JD (2007) RNA regulons: coordination of post-transcriptional events. Nat Rev Genet 8(7):533–543. ISSN 1471-0056. doi:10.1038/nrg2111. http://www.nature.com/doifinder/10.1038/nrg2111
80. Cléry A, Blatter M, Allain FH (2008) RNA recognition motifs: boring? not quite. Curr Opin Struct Biol 18(3):290–298. doi:10.1016/j.sbi.2008.04.002. http://www.hubmed.org/fulltext.cgi?uids=18515081
81. Oberstrass FC, Auweter SD, Erat M, Hargous Y, Henning A, Wenter P, Reymond L, Amir-Ahmady B, Pitsch S, Black DL, Allain FH (2005) Structure of PTB bound to RNA: specific binding and implications for splicing regulation. Science 309(5743):2054–2057. doi:10.1126/science.1114066. http://www.hubmed.org/fulltext.cgi?uids=16179478
82. Lewis HA, Musunuru K, Jensen KB, Edo C, Chen H, Darnell RB, Burley SK (2000) Sequence-specific RNA binding by a Nova KH domain: implications for paraneoplastic disease and the fragile X syndrome. Cell 100(3):323–332
83. Liu Z, Luyten I, Bottomley MJ, Messias AC, Houngninou-Molango S, Sprangers R, Zanier K, Krämer A, Sattler M (2001) Structural basis for recognition of the intron branch site RNA by splicing factor 1. Science 294(5544):1098–1102. doi:10.1126/science.1064719
84. Beuth B, Pennell S, Arnvig KB, Martin SR, Taylor IA (2005) Structure of a mycobacterium tuberculosis nusa-rna complex. EMBO J 24(20):3576–3587. doi:10.1038/sj.emboj.7600829
85. Diges CM, Uhlenbeck OC (2001) Escherichia coli DbpA is an RNA helicase that requires hairpin 92 of 23S RNA. EMBO J 20(19):5503–5512. doi:10.1093/emboj/20.19.5503
86. Wang S, Hu Y, Overgaard MT, Karginov FV, Uhlenbeck OC, McKay DB (2006) The domain of the Bacillus subtilis DEAD-box helicase YxiN that is responsible for specific binding of 23S rRNA has an RNA recognition motif fold. RNA 12(6):959–967. doi:10.1261/rna.5906
87. Wang Y, Juranek S, Li H, Sheng G, Wardle GS, Tuschl T, Patel DJ (2009) Nucleation, propagation and cleavage of target rnas in ago silencing complexes. Nature 461(7265):754–761. doi:10.1038/nature08434
88. Siomi MC, Sato K, Pezic D, Aravin AA (2011) Piwi-interacting small RNAs: the vanguard of genome defence. Nat Rev Mol Cell Biol 12(4):246–258. doi:10.1038/nrm3089
89. Wilusz CJ, Wilusz J (2005) Eukaryotic Lsm proteins: lessons from bacteria. Nat Struct Mol Biol 12(12):1031–1036. doi:10.1038/nsmb1037. http://www.hubmed.org/fulltext.cgi?uids=16327775
90. Qiao F, Bowie JU (2005) The many faces of SAM. Sci STKE 2005(286). doi:10.1126/stke.2862005re7. http://www.hubmed.org/fulltext.cgi?uids=15928333
91. Aviv T, Lin Z, Ben-Ari G, Smibert CA, Sicheri F (2006) Sequence-specific recognition of RNA hairpins by the SAM domain of Vts1p. Nat Struct Mol Biol 13(2):168–176. doi:10.1038/nsmb1053. http://www.hubmed.org/fulltext.cgi?uids=16429151
92. Oberstrass FC, Lee A, Stefl R, Janis M, Chanfreau G, Allain FH (2006) Shape-specific recognition in the structure of the Vts1p SAM domain with RNA. Nat Struct Mol Biol 13(2):160–167. doi:10.1038/nsmb1038. http://www.hubmed.org/fulltext.cgi?uids=16429156
93. Stefl R, Skrisovska L, Allain FH-T (2005) RNA sequence- and shape-dependent recognition by proteins in

the ribonucleoprotein particle. EMBO Rep 6(1):33–38. doi:10.1038/sj.embor.7400325
94. Lunde BM, Moore C, Varani G (2007) RNA-binding proteins: modular design for efficient function. Nat Rev Mol Cell Biol 8(6):479–490. doi:10.1038/nrm2178
95. Mazza C, Segref A, Mattaj IW, Cusack S (2002) Large-scale induced fit recognition of an m(7)GpppG cap analogue by the human nuclear cap-binding complex. EMBO J 21(20):5548–5557
96. Mackay JP, Font J, Segal DJ (2011) The prospects for designer single-stranded rna-binding proteins. Nat Struct Mol Biol 18(3):256–261. doi:10.1038/nsmb.2005
97. Gagnon KT, Maxwell ES (2011) Electrophoretic mobility shift assay for characterizing rna-protein interaction. Method Mol Biol 703:275–291. doi:10.1007/978-1-59745-248-9-19
98. Tuerk C, Gold L (1990) Systematic evolution of ligands by exponential enrichment: RNA ligands to bacteriophage t4 DNA polymerase. Science 249(4968):505–510. http://www.hubmed.org/fulltext.cgi?uids=2200121
99. Ray D, Kazan H, Chan ET, Peña Castillo L, Chaudhry S, Talukder S, Blencowe BJ, Morris Q, Hughes TR (2009) Rapid and systematic analysis of the rna recognition specificities of RNA-binding proteins. Nat Biotechnol 27(7):667–670. doi:10.1038/nbt.1550
100. Keene JD, Komisarow JM, Friedersdorf MB (2006) Rip-chip: the isolation and identification of mRNAs, microRNAs and protein components of ribonucleoprotein complexes from cell extracts. Nat Protoc 1(1):302–307. doi:10.1038/nprot.2006.47
101. Mili S, Steitz JA (2004) Evidence for reassociation of rna-binding proteins after cell lysis: implications for the interpretation of immunoprecipitation analyses. RNA 10(11):1692–1694. doi:10.1261/rna.7151404
102. Zhang C, Darnell RB (2011) Mapping in vivo protein-RNA interactions at single-nucleotide resolution from hits-clip data. Nat Biotechnol 29(7):607–614. doi:10.1038/nbt.1873. http://www.hubmed.org/fulltext.cgi?uids=21633356
103. Kishore S, Jaskiewicz L, Burger L, Hausser J, Khorshid M, Zavolan M (2011) A quantitative analysis of CLIP methods for identifying binding sites of RNA–binding proteins. Nat Method 8(7):559–564. doi:10.1038/nmeth.1608. http://www.hubmed.org/fulltext.cgi?uids=21572407
104. Corcoran DL, Georgiev S, Mukherjee N, Gottwein E, Skalsky RL, Keene JD, Ohler U (2011) PARalyzer: definition of RNA binding sites from PAR–CLIP short-read sequence data. Genome Biol 12(8). doi:10.1186/gb-2011-12-8-r79. http://www.hubmed.org/fulltext.cgi?uids=21851591
105. Hafner M, Landthaler M, Burger L, Khorshid M, Hausser J, Berninger P, Rothballer A, Ascano M Jr, Jungkamp A-C, Munschauer M, Ulrich A, Wardle GS, Dewell S, Zavolan M, Tuschl T (2010) Transcriptome-wide identification of RNA-binding protein and microRNA target sites by par-clip. Cell 141(1):129–141. doi:10.1016/j.cell.2010.03.009
106. Silness J, Berge M (1990) Changes over time in the clientele and restoration pattern in a dental school prosthodontic department. Int Dent J 40(2):109–116
107. Li X, Quon G, Lipshitz HD, Morris Q (2010) Predicting in vivo binding sites of RNA-binding proteins using mRNA secondary structure. RNA 16(6):1096–1107. doi:10.1261/rna.2017210. http://www.hubmed.org/fulltext.cgi?uids=20418358
108. Georgiev S, Boyle AP, Jayasurya K, Ding X, Mukherjee S, Ohler U (2010) Evidence-ranked motif identification. Genome Biol 11(2):R19. doi:10.1186/gb-2010-11-2-r19
109. Siddharthan R, Siggia ED, van Nimwegen E (2005) Phylogibbs: a gibbs sampling motif finder that incorporates phylogeny. PLoS Comput Biol 1(7):e67. doi:10.1371/journal.pcbi.0010067
110. Zielinski J, Kilk K, Peritz T, Kannanayakal T, Miyashiro KY, Eiríksdóttir E, Jochems J, Langel U, Eberwine J (2006) In vivo identification of ribonucleoprotein-RNA interactions. Proc Natl Acad Sci USA 103(5):1557–1562. doi:10.1073/pnas.0510611103. http://www.hubmed.org/fulltext.cgi?uids=16432185
111. Nielsen PE, Egholm M, Berg RH, Buchardt O (1991) Sequence-selective recognition of DNA by strand displacement with a thymine-substituted polyamide. Science 254(5037):1497–1500. http://www.hubmed.org/fulltext.cgi?uids=1962210
112. Butter F, Scheibe M, Mörl M, Mann M (2009) Unbiased RNA-protein interaction screen by quantitative proteomics. Proc Natl Acad Sci USA 106(26):10626–10631. doi:10.1073/pnas.0812099106. http://www.hubmed.org/fulltext.cgi?uids=19541640
113. Cook KB, Kazan H, Zuberi K, Morris Q, Hughes TR (2011) RBPDB: a database of RNA-binding specificities. Nucleic Acids Res 39(Database issue):301–308. doi:10.1093/nar/gkq1069. http://www.hubmed.org/fulltext.cgi?uids=21036867
114. Khorshid M, Rodak C, Zavolan M (2011) Clipz: a database and analysis environment

for experimentally determined binding sites of RNA-binding proteins. Nucleic Acids Res 39(Database issue):245–252. doi:10.1093/nar/gkq940. http://www.hubmed.org/fulltext.cgi?uids=21087992
115. Lewis BA, Walia RR, Terribilini M, Ferguson J, Zheng C, Honavar V, Dobbs D (2011) PRIDB: a protein-RNA interface database. Nucleic Acids Res 39(Database issue):277–282. doi:10.1093/nar/gkq1108. http://www.hubmed.org/fulltext.cgi?uids=21071426
116. Allers J, Shamoo Y (2001) Structure-based analysis of protein-RNA interactions using the program entangle. J Mol Biol 311(1):75–86. doi:10.1006/jmbi.2001.4857
117. Sigrist CJ, Cerutti L, de Castro E, Langendijk-Genevaux PS, Bulliard V, Bairoch A, Hulo N (2010) Prosite, a protein domain database for functional characterization and annotation. Nucleic Acids Res 38(Database issue):D161–D166. doi:10.1093/nar/gkp885
118. Sarver M, Zirbel CL, Stombaugh J, Mokdad A, Leontis NB (2008) Fr3d: finding local and composite recurrent structural motifs in RNA 3D structures. J Math Biol 56(1–2):215–252. doi:10.1007/s00285-007-0110-x
119. Paradis E, Claude J, Strimmer K (2004) Ape: analyses of phylogenetics and evolution in R language. Bioinformatics 20(2):289–290. http://www.hubmed.org/fulltext.cgi?uids=14734327
120. Puton T, Kozlowski L, Tuszynska I, Rother K, Bujnicki JM (2011) Computational methods for prediction of protein-RNA interactions. J Struct Biol. doi:10.1016/j.jsb.2011.10.001
121. Shulman-Peleg A, Shatsky M, Nussinov R, Wolfson HJ (2008) Prediction of interacting single-stranded RNA bases by protein-binding patterns. J Mol Biol 379(2):299–316. doi:10.1016/j.jmb.2008.03.043
122. Tompa M, Li N, Bailey TL, Church GM, De Moor B, Eskin E, Favorov AV, Frith MC, Fu Y, Kent WJ, Makeev VJ, Mironov AA, Noble WS, Pavesi G, Pesole G, Régnier M, Simonis N, Sinha S, Thijs G, van Helden J, Vandenbogaert M, Weng Z, Workman C, Ye C, Zhu Z (2005) Assessing computational tools for the discovery of transcription factor binding sites. Nat Biotechnol 23(1):137–144. doi:10.1038/nbt1053. http://www.hubmed.org/display.cgi?uids=15637633
123. Bailey TL, Elkan C (1994) Fitting a mixture model by expectation maximization to discover motifs in biopolymers. Proc Int Conf Intell Syst Mol Biol 2:28–36. http://www.hubmed.org/display.cgi?uids=7584402
124. Xie X, Lu J, Kulbokas EJ, Golub TR, Mootha V, Lindblad-Toh K, Lander ES, Kellis M (2005) Systematic discovery of regulatory motifs in human promoters and 3' UTRs by comparison of several mammals. Nature 434(7031):338–345. doi:10.1038/nature03441
125. Hiller M, Pudimat R, Busch A, Backofen R (2006) Using RNA secondary structures to guide sequence motif finding towards single-stranded regions. Nucleic Acids Res 34(17). doi:10.1093/nar/gkl544. http://www.hubmed.org/fulltext.cgi?uids=16987907
126. Bompfünewerer AF, Backofen R, Bernhart SH, Hertel J, Hofacker IL, Stadler PF, Will S (2008) Variations on RNA folding and alignment: lessons from benasque. J Math Biol 56(1–2):129–144. doi:10.1007/s00285-007-0107-5. http://www.hubmed.org/fulltext.cgi?uids=17611759
127. Pavesi G, Mauri G, Stefani M, Pesole G (2004) Rnaprofile: an algorithm for finding conserved secondary structure motifs in unaligned rna sequences. Nucleic Acids Res 32(10):3258–3269. doi:10.1093/nar/gkh650
128. Sankoff D (1985) Simultaneous solution of the RNA folding, alignment, and proto-sequence problems. SIAM J Appl Math 45:810–825
129. Eddy SR, Durbin R (1994) RNA sequence analysis using covariance models. Nucleic Acids Res 22(11):2079–2088. http://www.hubmed.org/fulltext.cgi?uids=8029015
130. Yao Z, Weinberg Z, Ruzzo WL (2006) Cmfinder–a covariance model based RNA motif finding algorithm. Bioinformatics 22(4):445–452. doi:10.1093/bioinformatics/btk008
131. Michal Rabani, Michael Kertesz, Eran Segal (2008) Computational prediction of RNA structural motifs involved in posttranscriptional regulatory processes. Proc Natl Acad Sci USA 105(39):14885–14890. doi:10.1073/pnas.0803169105
132. Foat BC, Houshmandi SS, Olivas WM, Bussemaker HJ (2005) Profiling condition-specific, genome-wide regulation of mRNA stability in yeast. Proc Natl Acad Sci USA 102(49):17675–17680. doi:10.1073/pnas.0503803102. http://www.hubmed.org/fulltext.cgi?uids=16317069

133. Foat BC, Stormo GD (2009) Discovering structural cis-regulatory elements by modeling the behaviors of mRNAs. Mol Syst Biol 5:268–268. doi:10.1038/msb.2009.24. http://www.hubmed.org/fulltext.cgi?uids=19401680

134. Kazan H, Ray D, Chan ET, Hughes TR, Morris Q (2010) Rnacontext: a new method for learning the sequence and structure binding preferences of RNA-binding proteins. PLoS Comput Biol 6:e1000832. doi:10.1371/journal.pcbi.1000832

Index

Jan Gorodkin and Walter L. Ruzzo (eds.), *RNA Sequence, Structure, and Function: Computational and Bioinformatic Methods*, Methods in Molecular Biology, vol. 1097, DOI 10.1007/978-1-62703-709-9,

D

N

O

Q

R

T

U

V

W

X

Y

Z

MIX
Papier aus verantwortungsvollen Quellen
Paper from responsible sources
FSC® C105338

If you have any concerns about our products,
you can contact us on
ProductSafety@springernature.com

In case Publisher is established outside the EU,
the EU authorized representative is:
Springer Nature Customer Service Center GmbH
Europaplatz 3, 69115 Heidelberg, Germany

Printed by Libri Plureos GmbH
in Hamburg, Germany